Weissermel, Arpe
Industrial Organic Chemistry

A Wiley company

Klaus Weissermel
Hans-Jürgen Arpe

Industrial
Organic Chemistry

Translated by
Charlet R. Lindley

Third Completely
Revised Edition

VCH
A Wiley company

Prof. Dr. Klaus Weissermel
Hoechst AG
Postfach 80 03 20
D-65926 Frankfurt
Federal Republic of Germany.

Prof. Dr. Hans-Jürgen Arpe
Dachsgraben 1
D-67824 Feilbingert
Federal Republic of Germany

This book was carefully produced. Nevertheless, authors and publisher do not warrant the information contained therein to be free of errors. Readers are advised to keep in mind that statements, data, illustrations, procedural details or other items may inadvertently be inaccurate.

1st edition 1978
2nd edition 1993
3rd edition 1997

Published jointly by
VCH Verlagsgesellschaft mbH, Weinheim (Federal Republic of Germany)
VCH Publishers, Inc., New York, NY (USA)

Editorial Director: Karin Sora
Production Manager: Dipl.-Ing. (FH) Hans Jörg Maier

British Library Cataloguing-in-Publication Data: A catalogue record for this book is available from the British Library.

Deutsche Bibliothek Cataloguing-in-Publication Data:
Weissermel, Klaus:
Industrial organic chemistry / Klaus Weissermel ; Hans-Jürgen Arpe. Transl. by Charlet R. Lindley. – 3., completely rev. ed. – Weinheim : VCH, 1997
  Dt. Ausg. u. d. T.: Weissermel, Klaus: Industrielle organische Chemie
  ISBN 3-527-28838-4 Gb.

Composition: Filmsatz Unger & Sommer GmbH, D-69469 Weinheim
Printing: betz-druck gmbh, D-64291 Darmstadt
Bookbindung: Wilhelm Osswald & Co., D-67433 Neustadt
Printed in the Federal Republic of Germany.

Preface to the Third Edition

In the few years that have passed since the publication of the 2nd English edition, it has become clear that interest in Industrial Inorganic Chemistry has continued to grow, making a new English edition necessary.

In the meantime, further translations have been published or are in preparation, and new editions have appeared.

The availability of large amounts of new information and up-to-date numerical data has prompted us to modernize and expand the book, at the same time increasing its scientific value. Apart from the scientific literature, a major help in our endeavors was the support of colleagues from Hoechst AG and numerous other chemical companies. Once again we thank VCH Publishers for the excellent cooperation.

February 1997

K. Weissermel
H.-J. Arpe

Preface to the Second Edition

The translation of "Industrial Organic Chemistry" into seven languages has proved the worldwide interest in this book. The positive feedback from readers with regard to the informational content and the didactic outline, together with the outstanding success of the similar work "Industrial Inorganic Chemistry" have encouraged us to produce this new revised edition.

The text has been greatly extended. Developmental possibilities appearing in the 1st Edition have now been revised and uptated to the current situation. The increasingly international outlook of the 1st Edition has been further extended to cover areas of worldwide interest. Appropriate alterations in nomenclature and style have also been implemented.

A special thank you is extended to the Market Research Department of Hoechst AG for their help in the collection of numerical data. It is also a pleasure to express our gratitude to VCH Verlagsgesellschaft for their kind cooperation and for the successful organization and presentation of the books.

February 1993
K. Weissermel
H.-J. Arpe

Preface to the First Edition

Industrial organic chemistry is exhaustively treated in a whole series of encyclopedias and standard works as well as, to an increasing extent, in monographs. However, it is not always simple to rapidly grasp the present status of knowledge from these sources.

There was thus a growing demand for a text describing in a concise manner the most important precursors and intermediates of industrial organic chemistry. The authors have endeavored to review the material and to present it in a form, indicative of their daily confrontation with problems arising from research and development, which can be readily understood by the reader. In pursuing this aim they could rely, apart from their industrial knowledge, on teaching experience derived from university lectures, and on stimulating discussions with many colleagues.

This book addresses itself to a wide range of readers: the chemistry student should be able to appreciate from it the chemisty of important precursors and intermediates as well as to follow the development of manufacturing processes which he might one day help to improve. The university or college lecturer can glean information about applied organic syntheses and the constant change of manufacturing processes and feedstocks along with the resulting research objectives. Chemists and their colleagues from other disciplines in the chemical industry — such as engineers, marketing specialists, lawyers and industrial economists — will be presented with a treatise dealing with the complex technological, scentific and economic interrelationships and their potential developments.

This book is arranged into 14 chapters in which precursors and intermediates are combined according to their tightest possible correlation to a particular group. A certain amount of arbitrariness was, of course, unavoidable. The introductory chapter reviews the present and future energy and feedstock supply.

As a rule, the manufacturing processes are treated after general description of the historical development and significance of a product, emphasis being placed on the conventional processes and the applications of the product and its simportant derivatives. The sections relating to heavy industrial organic products are frequently followed by a prognosis concerning potential developments. Deficiencies of existing technological or chemical processes, as well as possible future improvements or changes to other more economic or more readily available feedstocks are briefly discussed.

The authors endeavored to provide a high degree of quality and quantity of information. Three types of information are at the reader's disposal:

1. The main text.
2. The synopsis of the main text in the margin.
3. Flow diagrams illustrating the interrealationship of the products in each chapter.

These three types of presentation were derived from the widespread habit of many readers of underlining or making brief notes when studying a text. The reader has been relieved of this work by the marginal notes which briefly present all essential points of the main text in a logical sequence thereby enabling him to be rapidly informed without having to study the main text.

The formula or process scheme (flow diagram) pertaining to each chapter can be folded out whilst reading a section in order that its overall relevance can be readily appreciated. There are no diagrams of individual processes in the main text as this would result in frequent repetition arising from recurring process steps. Instead, the reader is informed about the significant features of a process.

The index, containing numerous key words, enables the reader to rapidly find the required information.

A first version of this book was originally published in the German language in 1976. Many colleagues inside and outside Hoechst AG gave us their support by carefully reading parts of the manuscript and providing valuable suggestions thereby ensuring the validity of the numerous technological and chemical facts. In particular, we would like to express our thanks to Dr. H. Friz, Dr. W. Reif (BASF); Dr. R. Streck, Dr. H. Weber (Hüls AG); Dr. W. Jordan (Phenolchemie); Dr. B. Cornils, Dr. J. Falbe, Dr. W. Payer (Ruhrchemie AG); Dr. K. H. Berg, Dr. I. F. Hudson (Shell); Dr. G. König, Dr. R. Kühn, Dr. H. Tetteroo (UK-Wesseling).

We are also indebted to many colleagues and fellow employees of Hoechst AG who assisted by reading individual chapters, expanding the numerical data, preparing the formula diagrams and typing the manuscript. In particular we would like to thank Dr. U. Dettmeier, M. Keller, Dr. E. I. Leupold, Dr. H. Meidert, and Prof. R. Steiner who all carefully read and corrected or expanded large sections of the manuscript. However, decisive for the choice of material was the access to the experience and the world-wide information sources of Hoechst AG.

Furthermore, the patience and consideration of our immediate families and close friends made an important contribution during the long months when the manuscript was written and revised.

In less than a year after the first appearance of 'Industrielle Organische Chemie' the second edition has now been published. The positive response enjoyed by the book places both an obligation on us as well as being an incentive to produce the second edition in not only a revised, but also an expanded form. This second edition of the German language version has also been the basis of the present English edition in which the numerical data were updated and, where possible, enriched by data from several leading industrial nations in order to stress the international scope.

Additional products were included along with their manufacturing processes. New facts were often supplemented with mechanistic details to facilitate the reader's comprehension of basic industrial processes.

The book was translated by Dr. A. Mullen (Ruhrchemie AG) to whom we are particularly grateful for assuming this arduous task which he accomplished by keeping as closely as possible to the original text whilst also managing to evolve his own style. We would like to thank the Board of Ruhrchemie AG for supporting this venture by placing the company's facilities at Dr. Mullen's disposal.

We are also indebted to Dr. T. F. Leahy, a colleague from the American Hoechst Corporation, who played an essential part by meticulously reading the manuscript.

Verlag Chemie must also be thanked − in particular Dr. H. F. Ebel − for its support and for ensuring that the English edition should have the best possible presentation.

Hoechst, in January 1978   K. Weissermel
                                              H.-J. Arpe

Contents

1. Various Aspects of the Energy and Raw Material Supply

The availability and price structure of energy and raw materials have always determined the technological base and thus the expansion and development of industrial chemistry. However, the oil crisis was necessary before the general public once again became aware of this relationship and its importance for the world economy.

Coal, natural gas, and oil, formed with the help of solar energy during the course of millions of years, presently cover not only the energy, but also to a large extent chemical feedstock requirements.

There is no comparable branch of industry in which there is such a complete interplay between energy and raw materials as in the chemical industry. Every variation in supply has a double impact on the chemical industry as it is one of the greatest consumers of energy. In addition to this, the non-recoverable fossil products, which are employed as raw materials, are converted into a spectrum of synthetic substances which we meet in everyday life.

The constantly increasing demand for raw materials and the limited reserves point out the importance of safeguarding future energy and raw material supplies.

All short- and medium-term efforts will have to concentrate on the basic problem as to how the flexibility of the raw material supply for the chemical industry on the one hand, and the energy sector on the other hand, can be increased with the available resources. In the long term, this double function of the fossil fuels will be terminated in order to maintain this attractive source of supply for the chemical industry for as long as possible.

In order to better evaluate the present situation and understand the future consumption of primary energy sources and raw materials, both aspects will be reviewed together with the individual energy sources.

fossil fuels
natural gas, petroleum, coal
have two functions:

1. energy source
2. raw material for chemical products

long range aims for securing industrial raw material and energy supply:

1. extending the period of use of the fossil raw materials
2. replacing the fossil raw materials in the energy sector

1.1. Present and Predictable Energy Requirements

primary energy consumption (in 10^{12} kwhr)

	1964	1974	1984	1989	1994
World	41.3	67.5	82.6	95.2	93.6
USA	12.5	15.4	19.5	23.6	24.0
W. Europe	7.9	10.7	11.6	13.0	13.2

During the last twenty-five years, the world energy demand has more than doubled and in 1995 it reached 94.4×10^{12} kwhr, corresponding to the energy from 8.12×10^9 tonnes of oil (1 tonne oil $= 11\,620$ kwhr $= 10 \times 10^6$ kcal $= 41.8 \times 10^6$ kJ). The average annual increase before 1974 was about 5%, which decreased through the end of the 1980s, as the numbers in the adjacent table illustrate. In the 1990s, primary energy consumption has hardly changed due to the drop in energy demand caused by the economic recession following the radical changes in the former East Bloc.

However, according to the latest prediction of the International Energy Agency (IEA), global population will grow from the current 5.6 to 7×10^9 people by the year 2010, causing the world energy demand to increase to 130×10^{12} kwhr.

In 1989, the consumption of primary energy in the OECD (Organization for Economic Cooperation and Development) countries was distributed as follows:

31% for transport
34% for industrial use
35% for domestic and agricultural use, and other sectors

energy consumption of the chemical industry:

6% of total consumption, *i. e.*, second greatest industrial consumer

The chemical industry accounts for 6% of the total energy consumption and thereby assumes second place in the energy consumption scale after the iron processing industry.

changes in primary energy distribution worldwide (in %):

	1964	1974	1984	1995
oil	41	48	42	38
coal	37	28	27	23
natural gas	15	18	19	19
nuclear energy	6	6	7	6
water power/ others	1	3	5	14

(others include, e. g., biomass)

Between 1950 and 1995, the worldwide pattern of primary energy consumption changed drastically. Coal's share decreased from ca. 60% in 1950 to the values shown in the accompanying table. In China and some of the former Eastern Bloc countries, 40% of the energy used still comes from coal. Oil's share amounted to just 25% of world energy consumption in 1950, and reached a maximum of nearly 50% in the early 1970s. Today it has stabilized at ca. 38%, and is expected to decrease slightly to 37% by 2000.

reasons for preferred use of oil and natural gas as energy source:

1. economic recovery
2. versatile applicability
3. low transportation and distribution costs

restructuring of energy consumption not possible in the short term

oil remains main energy source for the near future

The reasons for this energy source structure lie with the ready economic recovery of oil and natural gas and their versatile applicability as well as lower transportation and distribution costs.

In the following decades, the forecast calls for a slight decrease in the relative amounts of energy from oil and natural gas, but a small increase for coal and nuclear energy. An eventual transition to carbon-free and inexhaustible energy sources is desirable, but this development will be influenced by many factors.

In any event, oil and natural gas will remain the main energy sources in predictions for decades, as technological reorientation will take a long time due to the complexity of the problem.

1.2. Availability of Individual Sources

1.2.1. Oil

New data show that the proven and probable, *i.e.*, supplementary, recoverable world oil reserves are higher than the roughly 520×10^9 tonnes, or 6040×10^{12} kwhr, estimated in recent years. Of the proven reserves (1996), 66% are found in the Middle East, 13% in South America, 3% in North America, 2% in Western Europe and the remainder in other regions. With about 26% of the proven oil reserves, Saudi Arabia has the greatest share, leading Iraq, Kuwait and other countries principally in the Near East. In 1996, the OPEC countries accounted for ca. 77 wt% of worldwide oil production. Countries with the largest production in 1994 were Saudi Arabia and the USA.

oil reserves (in 10^{12} kwhr):

	1986	1989	1995
proven	1110	1480	1580
total	4900	1620	2470

A further crude oil supply which amounts to ten times the above-mentioned petroleum reserves is found in oil shale, tar sand, and oil sand. This source, presumed to be the same order of magnitude as mineral oil only a few years ago, far surpasses it.

There is a great incentive for the exploitation of oil shale and oil sand. To this end, extraction and pyrolysis processes have been developed which, under favorable local conditions, are already economically feasible. Large commercial plants are being run in Canada, with a significant annual increase (for example, production in 1994 was 17% greater than in 1993), and the CIS. Although numerous pilot plants have been shut down, for instance in the USA, new ones are planned in places such as Australia. In China, oil is extracted from kerogen-containing rock strata. An additional plant with a capacity of 0.12×10^6 tonnes per year was in the last phase of construction in 1994.

reserves of "synthetic" oil from oil shale and oil sands (in 10^{12} kwhr):

	1989	1992
proven	1 550	1 550
total	13 840	12 360

kerogen is a waxy, polymeric substance found in mineral rock, which is converted to "synthetic" oil on heating to $>500\,°C$ or hydrogenation

oil consumption (in 10^9 t of oil):

	1988	1990	1994
World	3.02	3.10	3.18
USA	0.78	0.78	0.81
W. Europe	0.59	0.60	0.57
CIS	0.45	0.41	n.a.
Japan	0.22	0.25	n.a.

n.a. = not available

At current rates of consumption, proven crude oil reserves will last an estimated 43 years (1996). If the additional supply from oil shale/oil sands is included, the supply will last for more than 100 years.

aids to oil recovery:

recovery phase	recovery agent	oil recovered (in %)
primary	well head pressure	10–20
secondary	water/gas flooding	→30
tertiary	chemical flooding (polymers, tensides)	→50

However, the following factors will probably help ensure an oil supply well beyond that point: better utilization of known deposits which at present are exploited only to about 30% with conventional technology, intensified exploration activity, recovery of difficult-to-obtain reserves, the opening up of oil fields under the seabed as well as a restructuring of energy and raw material consumption.

1.2.2. Natural Gas

natural gas reserves (in 10^{12} kwhr):

	1985	1989	1992	1995
proven	944	1190	1250	1380
total	2260	3660	3440	3390

($1\,m^3$ natural gas $= 9.23$ kwhr)

The proben and probable world natural gas reserves are somewhat larger than the oil reserves, and are currently estimated at 368×10^{12} m^3, or 3390×10^{12} kwhr. Proven reserves amount to 1380×10^{12} kwhr.

In 1995, 39% of these reserves were located in the CIS, 14% in Iran, 5% in Qatar, 4% in each of Abu Dhabi and Saudi Arabia, and 3% in the USA. The remaining 31% is distributed among all other natural gas-producing countries.

at the present rate of consumption the proven natural gas reserves will be exhausted in ca. 55 years

Based on the natural gas output for 1995 (25.2×10^{12} kwhr), the proven worldwide reserves should last for almost 55 years.

In 1995, North America and Eastern Europe were the largest producers, supplying 32 and 29%, respectively, of the natural gas worldwide.

rapid development in natural gas consumption possible by transport over long distances by means of:

1. pipelines
2. specially designed ships
3. transformation into methanol

Natural gas consumption has steadily increased during the last two decades. Up until now, natural gas could only be used where the corresponding industrial infrastructure was available or where the distance to the consumer could be bridged by means of pipelines. In the meantime, gas transportation over great distances from the source of supply to the most important consumption areas can be overcome by liquefaction of natural gas (LNG = **l**iquefied **n**atural **g**as) and transportation in specially built ships as is done for example in Japan, which supplies itself almost entirely by importing LNG. In the future, natural gas could possibly be transported by first converting it into methanol — via synthesis gas — necessitating, of course, additional expenditure.

substitution of the natural gas by synthetic natural gas (SNG) not before 2000 (*cf.* Section 2.1.2)

The dependence on imports, as with oil, in countries with little or no natural gas reserves is therefore resolvable. However, this situation will only fundamentally change when synthesis gas technology — based on brown (lignite) and hard coal — is established and developed. This will probably take place on a larger scale only in the distant future.

1.2.3. Coal

As far as the reserves are concerned, coal is not only the most widely spread but also the most important source of fossil energy. However, it must be kept in mind that the estimates of coal deposits are based on geological studies and do not take the mining problems into account. The proven and probable world hard coal reserves are estimated to be 61920×10^{12} kwhr. The proven reserves amount to 4610×10^{12} kwhr. Of this amount, ca. 35 % is found in the USA, 6 % in the CIS, 13 % in the Peoples' Republic of China, 13 % in Western Europe, and 11 % in Africa. In 1995, 3.4×10^6 tonnes of hard coal were produced worldwide, with 56 % coming out of the USA and China.

In 1989, the world reserves of brown coal were estimated at 6800×10^{12} kwhr, of which 860×10^{12} kwhr are proven reserves. By 1992, these proven reserves had increased by ca. 30 %.

With the huge coal deposits available, the world's energy requirements could be met for a long time to come. According to studies at several institutes, this could be for several thousand years at the current rate of growth.

1.2.4. Nuclear Fuels

Nuclear energy is − as a result of its stage of development − the only realistic solution to the energy supply problem of the next decades. Its economic viability has been proven.

The nuclear fuels offer an alternative to fossil fuels in important areas, particularly in the generation of electricity. In 1995, 17 % of the electricity worldwide was produced in 437 nuclear reactors, and an additional 59 reactors are under construction. Most nuclear power plants are in the USA, followed by France and Japan. The uranium and thorium deposits are immense and are widely distributed throughout the world. In 1995, the world production of uranium was 33 000 tonnes. Canada supplied the largest portion with 9900 tonnes, followed by Australia, Niger, the USA and the CIS.

In the low and medium price range there are ca. 4.0×10^6 tonnes of uranium reserves, of which 2.2×10^6 tonnes are proven; the corresponding thorium reserves amount to around 2.2×10^6 tonnes.

hard coal reserves (in 10^{12} kwhr):

	1985	1989	1992	1995
proven	5600	4090	5860	4610
total	54500	58600	67800	61920

"hard coal" also includes tar coal and anthracite

brown coal reserves (in 10^{12} kwhr):

	1985	1989	1992	1994
proven	1360	860	1110	1110
total	5700	6800	n.a.	n.a.

n.a. = not available

energy sources for electricity (in %):

	USA		Western Europe		World	
	1975	1987	1974	1995	1975	1995
natural gas/oil	76	13	36	21	35	25
coal		53	34	29	37	38
nuclear energy	9	17	6	35	5	17
hydroelectric/others	15	17	24	15	23	20

reserves of nuclear fuels (in 10^6 tonnes):

	uranium	thorium
proven	2.2	n.a.
total	4.0	2.2

n.a. = not available

energy content of uranium reserves
(in 10^{12} kwhr):

690	with conventional reactor technology
80 000	by full utilization via breeders

function of fast breeders
(neutron capture):

$^{238}U \rightarrow {}^{239}Pu$
$^{232}Th \rightarrow {}^{233}U$

Employing uranium in light-water reactors of conventional design in which essentially only ^{235}U is used (up to 0.7% in natural uranium) and where about 1000 MWd/kg ^{235}U are attained means that 4×10^6 tonnes uranium correspond to ca. 690×10^{12} kwhr. If this uranium were to be fully exploited using fast breeder reactors, then this value could be very considerably increased, namely to ca. 80000×10^{12} kwhr. An additional 44000×10^{12} kwhr could be obtained if the aforementioned thorium reserves were to be employed in breeder reactors. The significance of the fast breeder reactors can be readily appreciated from these figures. They operate by synthesizing the fissionable ^{239}Pu from the nonfissionable nuclide ^{238}U (main constituent of natural uranium, abundance 99.3%) by means of neutron capture. ^{238}U is not fissionable using thermal neutrons. In the same way fissionable ^{233}U can be synthesized from ^{232}Th.

The increasing energy demand can be met for at least the next 50 years using present reactor technology.

reactor generations:

light-water reactors
high temperature reactors
breeder reactors

advantage of high temperature reactors:

high temperature range (900 – 1000 °C)
process heat useful for strongly endothermic
chemical reactions

The dominant reactor type today, and probably for the next 20 years, is the light-water reactor (boiling or pressurized water reactor) which operates at temperatures up to about 300 °C. High temperature reactors with cooling medium (helium) temperature up to nearly 1000 °C are already on the threshold of large scale development. They have the advantage that they not only supply electricity but also process heat at higher temperatures (*cf.* Sections 2.1.1 and 2.2.2). Breeder reactors will probably become commercially available in greater numbers as generating plants near the end of the 1990s at the earliest, since several technological problems still confront their development.

In 1995, Japan and France were the only countries that were still using and developing breeder reactors.

essential prerequisites for the use of nuclear
energy:

1. reliable supply of nuclear energy
2. technically safe nuclear power stations
3. safe disposal of fission products and
   recycling of nuclear fuels (reprocessing)

It is important to note that the supply situation of countries highly dependent on energy importation can be markedly improved by storing nuclear fuels due to their high energy content. The prerequisite for the successful employment of nuclear energy is not only that safe and reliable nuclear power stations are erected, but also that the whole fuel cycle is completely closed. This begins with the supply of natural uranium, the siting of suitable enrichment units, and finishes with the waste disposal of radioactive fission products and the recycling of unused and newly bred nuclear fuels.

Waste management and environmental protection will determine the rate at which the nuclear energy program can be realized.

1.3. Prospects for the Future Energy Supply

As seen in the foregoing sections, oil, natural gas, and coal will remain the most important primary energy sources for the long term. While there is currently little restriction on the availability of energy sources, in light of the importance of oil and natural gas as raw materials for the chemical industry, their use for energy should be decreased as soon as possible.

with the prevailing energy structure, oil and natural gas will be the first energy sources to be exhausted

competition between their energy and chemical utilization compels structural change in the energy pallette

The exploitation of oil shales and oil sands will not significantly affect the situation in the long term. The substitution of oil and natural gas by other energy sources is the most prudent solution to this dilemma. By these means, the valuable fossil materials will be retained as far as possible for processing by the chemical industry.

In the medium term, the utilization of nuclear energy can decisively contribute to a relief of the fossil energy consumption. Solar energy offers an almost inexhaustible energy reserve and will only be referred to here with respect to its industrial potential. The energy which the sun annually supplies to the earth corresponds to thirty times the world's coal reserves. Based on a simple calculation, the world's present primary energy consumption could be covered by 0.005% of the energy supplied by the sun. Consequently, the development of solar energy technology including solar collectors and solar cell systems remains an important objective. At the same time, however, the energy storage and transportation problems must be solved.

possible relief for fossil fuels by generation of energy from:

1. nuclear energy (medium term)
2. solar energy (long term)
3. geothermal energy (partial)
4. nuclear fusion energy (long term)

The large scale utilization of the so-called unlimited energies — solar energy, geothermal energy, and nuclear fusion — will become important only in the distant future. Until that time, we will be dependent on an optimal use of fossil raw materials, in particular oil. In the near future, nuclear energy and coal will play a dominant role in our energy supply, in order to stretch our oil reserves as far as possible. Nuclear energy will take over the generation of electricity while coal will be increasingly used as a substitute for petroleum products.

possible substitution of oil for energy generation by means of:

1. coal
2. nuclear energy
3. combination of coal and nuclear energy
4. hydrogen

Before the energy supply becomes independent of fossil sources — undoubtedly not until the next century — there will possibly be an intermediate period in which a combination of nuclear energy and coal will be used. This combination could utilize nuclear process heat for coal gasification leading to the greater employment of synthesis gas products (*cf.* Section 2.1.1).

Along with the manufacture of synthesis gas via coal gasification, nuclear energy can possibly also be used for the manufacture of hydrogen from water via high temperature steam electrolysis or chemical cyclic processes. The same is true of water electrolysis using solar energy, which is being studied widely in several countries. This could result in a wide use of hydrogen as an energy source (hydrogen technology) and in a replacement of hydrogen manufacture from fossil materials (*cf.* Section 2.2.2).

long-term aim:

energy supply solely from renewable sources; raw material supply from fossil sources

This phase will lead to the situation in which energy will be won solely from renewable sources and oil and coal will be employed only as raw materials.

1.4. Present and Anticipated Raw Material Situation

characteristic changes in the raw material base of the chemical industry:

feedstocks until 1950:

1. coal gasification products (coking products, synthesis gas)
2. acetylene from calcium carbide

The present raw material situation of the chemical industry is characterized by a successful and virtually complete changeover from coal to petroleum technology.

The restructuring also applies to the conversion from the acetylene to the olefin base (*cf.* Sections 3.1 and 4.1).

1.4.1. Petrochemical Primary Products

feedstocks after 1950:

1. products from petroleum processing
2. natural gas
3. coal gasification products as well as acetylene from carbide and light hydrocarbons

expansion of organic primary chemicals was only possible due to conversion from coal to oil

return to coal for organic primary chemicals is not feasible in short and medium term

primary chemicals are petrochemical basis products for further reactions; *e. g.*, ethylene, propene, butadiene, BTX aromatics

primary chemicals production (10^6 tonnes)

	1989	1991	1993
USA	37.1	39.5	41.7
W. Europe	35.4	38.3	39.4
Japan	15.9	19.2	18.4

feedstocks for olefins and aromatics:

Japan/WE: naphtha (crude gasoline)

USA: liquid gas ($C_2 - C_4$)
       and, increasingly, naphtha

feedstocks for synthesis gas ($CO + H_2$):

methane and higher oil fractions

The manufacture of carbon monoxide and hydrogen via gasification processes together with the manufacture of carbide (for welding and some special organic intermediates), benzene, and certain polynuclear aromatics are the only remaining processes of those employed in the 1950s for the preparation of basic organic chemicals from coal. However, these account for only a minor part of the primary petrochemical products; currently ca. 95% are based on oil or natural gas. Furthermore, there is no doubt that the expansion in production of feedstocks for the manufacture of organic secondary products was only possible as a result of the changeover to oil. This rapid expansion would not have been possible with coal due to inherent mining constraints. It can thus be appreciated that only a partial substitution of oil by coal, resulting in limited broadening of the raw material base, will be possible in the future. The dependence of the chemical industry on oil will therefore be maintained.

In Japan and Western Europe, naphtha (or crude gasoline) is by far the most important feedstock available to the chemical industry from the oil refineries. A decreasing availability of natural gas has also led to the increasing use of naphtha in the USA. Olefins such as ethylene, propene, butenes, and butadiene as well as the aromatics benzene, toluene, and xylene can be obtained by cracking naphtha. Of less importance are heavy fuel oil and refinery gas which are employed together with natural gas for the manufacture of synthesis gas. The latter forms the basis for the manufacture of ammonia, methanol, acetic acid, and 'oxo' products. The process technology largely determines the content and yield of the individual cuts.

This technology has been increasingly developed since the oil crisis, so that today a complex refinery structure offers large quantities of valuable products. Thus heavy fuel oil is partially converted to lower boiling products through thermal cracking processes such as visbreaking and coking processes. Furthermore, the residue from the atmospheric distillation can, following vacuum distillation, be converted by catalytic or hydrocracking. This increases the yield of lighter products considerably, although it also increases the energy needed for processing.

trend in demand for lighter mineral oil products necessitates more complex oil processing, *e. g.*, from residual oils

restructuring of refineries by additional conversion plants such as:

1. thermocrackers
2. catalytic crackers
3. hydrocrackers

The spectra of refinery products in the USA, Western Europe, and Japan are distinctly different due to the different market pressures, yet they all show the same trend toward a higher demand for lighter mineral oil fractions:

markets 1973/93 for mineral oil products show characteristic drop in demand:

1. total of 16–31%
2. heavy fuel oil of 36–54%

Table 1-1. Distribution of refinery products (in wt %).

	USA			Western Europe			Japan		
	1973	1983	1993	1973	1985	1993	1973	1983	1993
Refinery & liquid gas	9	10	8	4	4	3	6	11	3
Motor gasoline, naphtha	44	49	47	24	26	29	21	24	20
Jet fuel	6	7	9	4	5	7	8	11	15
Light fuel oil, diesel oil	19	20	20	32	38	37	12	17	32
Heavy fuel oil	16	9	8	33	22	21	50	33	23
Bitumen, oil coke	6	5	8	3	5	3	3	4	7
Total refinery products (in 10^6 tonnes)	825	730	690	730	527	577	260	220	179

The aforementioned development toward lower boiling products from mineral oil was influenced by the fuel sector as well as by the chemical industry. Even though in principle all refinery products are usable for the manufacture of primary chemicals such as olefins and the BTX (benzene–toluene–xylene) aromatics, there is still a considerable difference in yield. Lowering the boiling point of the feedstock of a cracking process increases not only the yield of $C_2 - C_4$ olefins, but also alters the olefin mixture; in particular, it enhances the formation of the main product ethylene, by far the most important of the chemical building blocks (*cf.* adjacent table).

olefin yields from moderate severity cracking (in wt%)

	ethane	naphtha	oil
ethylene	82	30	20
propene	2	17	14
C_4-olefins	3	11	9

remainder: fuel gas, gasoline from cracking, oil residue

saving oil as an energy source is possible in several ways:

1. increased efficiency during conversion into energy
2. gradual substitution by coal or nuclear energy
3. gradual substitution as motor fuel by, *e. g.*, methanol, ethanol

future supplies of primary chemicals increasing due to countries with inexpensive raw material base, *e. g.*, oil producing countries

typical production, *e. g.*, in Saudi Arabia (starting in 1984)
ethylene
ethanol
ethylene glycols
dichloroethane
vinyl chloride
styrene
starting in 1993, *e. g.*, MTBE
$(0.86 \times 10^6$ tonnes per year)

Independent of the higher supply of refinery fractions preferred by the chemical industry through expanded processing technology, by and large the vital task of reducing and uncoupling the dual role of oil as a supplier of both energy and raw materials remains.

A first step toward saving oil could be to increase the efficiency of its conversion into electricity, heat, and motive power.

In the industrial sector, currently only 55% of the energy is actually used. Domestic and small consumers, who represent not only the largest but also the expanding consumption areas, use only 45%, while transport uses only 17%. The remainder is lost through conversion, transport, and waste heat.

The gradual replacement of oil in energy generation by coal and nuclear energy could have an even greater effect (*cf.* Section 1.3). This includes the partial or complete replacement of gasoline by methanol (*cf.* Section 2.3.1.2) or by ethanol, perhaps from a biological source (*cf.* Section 8.1.1).

Over and above this, there are other aspects of the future of the primary raw chemical supply for the chemical industry. First among these is the geographic transfer of petrochemical production to the oil producing countries. Saudi Arabia has emerged in the last few years as a large-scale producer of primary chemicals and the most important olefins, in order to (among other things) make use of the petroleum gas previously burned off. A number of nonindustrialized and newly industrialized nations have followed this example, so that in the future they will be able to supply not only their domestic requirements, but also the established production centers in the USA, Western Europe, and Japan.

Thus it can be expected that the capacity for production of primary chemicals in these newly industrialized countries will increase continuously. This is a challenge to the industrialized countries to increase their proportion of higher valued products.

In 1995, the world production capacity for the total area of petrochemical products was about 200×10^6 tonnes per year. Of this, about 29% was in Western Europe, 23% in the USA, 17% in Southeast Asia, 10% in Japan, and 21% in the remaining areas.

1.4.2. Coal Conversion Products

The chemical industry uses appreciable amounts of coal only as a raw material for recovery of benzene, naphthalene, and other condensed aromatics.

Measured against the world demand, coal furnishes up to 11% of the requirements for benzene, and more than 95% of the requirements for polynuclear and heteroaromatics.

In addition, coal is the source for smaller amounts of acetylene and carbon monoxide, and is the raw material for technical carbon, *i.e.*, carbon black and graphite.

The changing situation on the oil market brings up the question to what extent precursors and secondary products from petrochemical sources can be substituted by possible coal conversion products. In general, the organic primary chemicals produced from oil could be manufactured once again from coal using conventional technology. However, the prerequisite is an extremely low coal price compared to oil or natural gas. In Europe, and even in the USA with its relatively low coal costs, it is currently not economical to manufacture gasoline from coal.

Viewed on the longer term, however, coal is the only plausible alternative to petroleum for the raw material base. To fit the current petrochemical production structure and to enhance profitability, earlier proven technologies must be improved to increase the yield of higher valued products.

Basically, the following methods are available for the manufacture of chemical precursors from coal:

1. Gasification of brown or hard coal to synthesis gas and its conversion into basic chemicals (*cf.* Section 2.1.1)

2. Hydrogenation or hydrogenative extraction of hard coal

3. Low temperature carbonization of brown or hard coal

4. Reaction of coal with calcium carbonate to form calcium carbide, followed by its conversion to acetylene

The state of the art and possible future developments will be dealt with in detail in the following sections.

In the future, incentive for the gasification of coal, which requires a considerable amount of heat, could result from the availability of nuclear process heat.

The application of nuclear process heat in the chemical industry is aimed at directly utilizing the energy released from the nuclear

coal as raw material:

currently up to 11% worldwide of the benzene-aromatics, but ca. 95% of the condensed aromatics, are based on coal gasification

substitution of oil by coal assumes further development of coal gasification and conversion processes

extremely low coal costs required

coal however remains sole alternative to oil

coal chemistry processes:

1. gasification
2. hydrogenation (hydrogenative extraction)
3. low temperature carbonization
4. manufacture of acetylene (carbide)

new process technologies coupling coal gasification with process heat under development

reactors for chemical reactions, and not by supplying it indirectly via electricity. This harnessing of nuclear process heat for chemical reactions is only possible under certain conditions. With the light-water reactors, temperatures up to about 300 °C are available, and application is essentially limited to the generation of process steam.

The development of high temperature reactors in which temperatures of $800 - 1000$ °C are attained presents a different situation. It appears feasible that the primary nuclear process heat can be used directly for the steam- or hydrogasification of coal, methane cracking, or even for hydrogen generation from water in chemical cyclic processes. The first-mentioned processes have the distinct advantage that coal and natural gas are employed solely as raw material and not simultaneously as the energy source. By this means up to 40% more gasification products can be obtained.

nuclear coal gasification results in up to 40% more gasification products

In the long term the advent of nuclear coal gasification can make a decisive contribution to guaranteeing the energy supply. In these terms, the consumption of the chemical industry is minimal; however − in light of their processing possibilities − chemistry is compelled to take a deeper look at coal gasification products.

exploitation of nuclear coal gasification by chemical industry only sensible in combination with power industry

technical breakthrough not expected before 2000, due to necessary development and testing periods

From the standpoint of the chemical industry, the dovetailing of energy and raw material needs offers the opportunity to develop high temperature reactors attractive to both sectors. Since the development of the high temperature reactors is not yet complete this stage will not be reached for 10 to 20 years. Furthermore, the coupling of the chemical section to the reactor will also involve considerable developmental work (*cf.* Section 2.1.1.1).

At the same time, this example illustrates the fact that the new technologies available at the turn of the century will be those which are currently being developed. This aspect must be taken into account in all plans relating to long-term energy and raw material supply.

2. Basic Products of Industrial Syntheses

2.1. Synthesis Gas

Nowadays the term synthesis gas or syn gas is mainly used for gas mixtures consisting of CO and H_2 in various proportions which are suitable for the synthesis of particular chemical products. At the same time, this term is also used to denote the $N_2 + 3 H_2$ mixture in the ammonia synthesis.

On account of their origin or application, several CO/H_2 combinations are denoted as water gas, crack gas, or methanol synthesis gas, and new terms such as oxo gas have evolved.

2.1.1. Generation of Synthesis Gas

The processes for the manufacture of synthesis gas were originally based on the gasification of coke from hard coal and low temperature coke from brown coal by means of air and steam. After World War II, the easy-to-handle and valuable liquid and gaseous fossil fuels — oil and natural gas — were also employed as feedstocks. Their value lay in their high hydrogen content (*cf.* Section 2.2.2); the approximate H : C ratio is 1 : 1 for coal, 2 : 1 for oil, 2.4 : 1 for petroleum ether and a maximum of 4 : 1 for methane-rich natural gas.

Recently, the traditional coal gasification processes have regained significance in a modern technological form. The capacity of the synthesis gas plants based on coal, only 3% in 1976, had already risen to about 12% by the end of 1982 and is now at approximately 16%. Somewhat more than half of this capacity is attributable to the Fischer–Tropsch factory in South Africa (Sasol).

Alternate feedstocks for the manufacture of synthesis gas, including peat, wood, and other biomass such as urban or agricultural waste, are currently being examined.

Many proposals for chemical recycling processes are also based on synthesis gas recovery from used plastics by addition of acid and water.

nowadays synthesis gas denotes mainly CO/H_2 mixtures in various proportions

alternative names for CO/H_2 mixtures:

1. according to origin:
   'water gas' (CO + H_2) from steam and coal
   'crack gas' (CO + $3H_2$) from steam reforming of CH_4

2. according to application:
   'methanol synthesis gas' (CO + $2H_2$) for the manufacture of CH_3OH
   'oxo gas' (CO + H_2) for hydroformylation

raw materials for synthesis gas generation:

brown coal
hard coal

natural gas, petroleum gas
mineral oil fractions

natural gas and light oil fractions are best suited for synthesis gas due to high H_2 content

renaissance of coal gasification already underway in favorable locations following the oil crisis

chemical recycling methods to convert used plastics to liquid or gaseous raw materials such as synthesis gas

2.1.1.1. Synthesis Gas *via* Coal Gasification

coal gasification can be regarded physically as gas/solid reaction and chemically as partial oxidation of C or as reduction of H_2O with C

total process is much more complex and only describable using numerous parallel and secondary reactions

In the gasification of coal with steam and O_2, that is, for the conversion of the organic constituents into gaseous products, there are several partly interdependent reactions of importance. The exothermic partial combustion of carbon and the endothermic water gas formation represent the actual gasification reactions:

partial combustion

$$2\,C + O_2 \rightleftarrows 2\,CO \quad \left(\Delta H = -\begin{smallmatrix}60 \text{ kcal}\\246 \text{ kJ}\end{smallmatrix}/\text{mol}\right) \tag{1}$$

heterogeneous water gas reaction

$$C + H_2O \rightleftarrows CO + H_2 \left(\Delta H = +\begin{smallmatrix}28 \text{ kcal}\\119 \text{ kJ}\end{smallmatrix}/\text{mol}\right) \tag{2}$$

In addition to the above, other reactions take place, the most important of which are shown below:

Boudouard reaction

$$C + CO_2 \rightleftarrows 2\,CO \quad \left(\Delta H = +\begin{smallmatrix}38 \text{ kcal}\\162 \text{ kJ}\end{smallmatrix}/\text{mol}\right) \tag{3}$$

homogenous water gas reaction
(water gas shift)

$$CO + H_2O \rightleftarrows CO_2 + H_2 \left(\Delta H = -\begin{smallmatrix}10 \text{ kcal}\\42 \text{ kJ}\end{smallmatrix}/\text{mol}\right) \tag{4}$$

hydrogenative gasification

$$C + 2\,H_2 \rightleftarrows CH_4 \left(\Delta H = -\begin{smallmatrix}21 \text{ kcal}\\87 \text{ kJ}\end{smallmatrix}/\text{mol}\right) \tag{5}$$

methanation

$$CO + 3\,H_2 \rightleftarrows CH_4 + H_2O \left(\Delta H = -\begin{smallmatrix}49 \text{ kcal}\\206 \text{ kJ}\end{smallmatrix}/\text{mol}\right) \tag{6}$$

for C gasification a strong heat supply at a high temperature level is essential, as

1. heterogeneous water gas reaction is strongly endothermic and involves high energy of activation
2. the reaction velocity must be adequately high for commercial processes

General characteristics of the coal gasification processes are the high energy consumption for the conductance of the endothermic partial reactions and the high temperature necessary (at least $900 - 1000\,°C$) to achieve an adequate reaction velocity. The heat supply results either from the reaction between the gasification agent and the coal, *i. e.*, autothermal, or from an external source, *i. e.*, allothermal.

The various gasification processes can be characterized on the one hand by the type of coal used, such as hard or brown coal, and its physical and chemical properties. On the other hand, the processes differ in the technology involved as for example in the heat supply [allothermal (external heating) or autothermal (self heating)] and in the type of reactor (fixed-bed, fluidized-bed, entrained-bed). Furthermore, the actual gasification reaction and the gas composition are determined by the gasification agent (H_2O, O_2 or air, CO_2, H_2), the process conditions (pressure, temperature, coal conversion), and reaction system (parallel or counter flow).

important factors in the industrial gasification of hard or brown coal:

1. physical and chemical properties of the coal
2. allothermal or autothermal heat supply
3. type of reactor
4. gasification agent
5. process conditions

The Winkler gasification, Koppers–Totzek gasification, and the Lurgi pressure gasification are established industrial processes.

conventional industrial gasification processes:

Winkler
Koppers–Totzek
Lurgi

In addition, second-generation gasification processes such as the Rheinbraun hydrogenative gasification and the Bergbau-Forschung steam gasification in Germany, the Kellogg coal gasification (molten Na_2CO_3) and the Exxon alkali carbonate catalyzed coal gasification in the USA, and the Sumitomo (recently in cooperation with Klöckner–Humboldt–Deutz) coal gasification (molten iron) in Japan have reached a state of development where pilot and demonstration plants have been in operation for several years.

more recent pilot plant tested gasification processes:

Rheinbraun (H_2)
Bergbau-Forschung (steam)
Kellogg (molten Na_2CO_3)
Sumitomo/Klöckner-Humboldt-Deutz (molten iron)

Several multistage processes developed in, e.g., England (Westinghouse), the USA (Synthane, Bi-Gas, Hy-Gas, U-Gas, Hydrane), and Japan are designed primarily for the production of synthetic natural gas (SNG = substitute natural gas).

multistep SNG processes:

US Bureau of Mines (Synthane)
Bituminous Coal Res. (Bi-Gas)
Institute of Gas Technology (Hy-Gas, U-Gas)

The *Winkler process* employs fine grain, nonbaking coals which are gasified at atmospheric pressure in a fluidized-bed (Winkler generator) with O_2 or air and steam. The temperature depends on the reactivity of the coal and is between 800 and 1100 °C (generally 950 °C). Brown coal is especially suitable as feed. The H_2:CO ratio of the product gas in roughly 1.4:1. This type of gasification was developed in Germany by the Leunawerke in 1931. Today this process is in operation in numerous plants throughout the world.

Winkler gasification:

fluidized-bed generator (pressure-free) with O_2 + H_2O used commercially in numerous plants

Newer process developments, particularly the gasification under higher pressure (10 – 25 bar) at 1100 °C, have resulted in better economics. The reaction speed and the space–time yield are increased, while the formation of byproducts (and thus the expense of gas purification) is decreased. An experimental plant of this type has been in operation by Rheinbraun since 1978, and a large-scale plant with the capacity to process 2.4×10^6 tonnes per year of brown coal was brought on line in 1985 with a first

modification of Winkler gasification:

HTW-process (high temperature Winkler) under higher temperatures/pressures, e.g., Rheinbraun-Uhde coal dust particles up to 6 mm 1100 °C, up to 25 bar, fluidized bed

Koppers–Totzek gasification:

flue dust with O_2 + H_2O
pressure-free, 1400–2000 °C

first commercial plant 1952. By 1984, 19 plants in 17 countries

modification of Koppers–Totzek gasification:

Shell process, Krupp Koppers PRENFLO process at higher pressure, *e. g.,* higher throughput

Texaco process as developed by Ruhrchemie/Ruhrkohle:

C/H_2O suspension, 1200–1600 °C, 20–80 bar in entrained-bed reactor

First plants in FRG, China, Japan, USA

Lurgi pressure gasification:

(20–30 bar) in moving fixed-bed with 2 characteristic temperatures:

1. 600–750 °C predegassing
2. ca. 1000–1200 °C (depending on O_2/H_2O) for main gasification

advantage of process:

raw gas under pressure ideal for further processing to synthesis gas or SNG

raw gas composition (in vol%) with open-burning coal feed (Ruhr):
    9–11 CH_4
    15–18 CO
    30–32 CO_2
    38–40 H_2

in 1979, already 16 Lurgi gasification plants in operation

currently 90% of all gasified coal treated in Lurgi process

run of 0.6×10^6 tonnes per year. A combination with a steam turbine power plant began operation in 1989.

In the *Koppers–Totzek process,* flue dust (powdered coal or petroleum coke) is gasified at atmospheric pressure with a parallel flow of O_2 and H_2O at 1400 to 2000 °C. The reaction takes place accompanied by flame formation. This high gasification temperature eliminates the formation of condensable hydrocarbons and thus the resulting synthesis gas has an 85–90% content of CO and H_2. Brown coal is also suitable as a feedstock. The first commercial plant was constructed in Finland in 1952. Since then, this process has been in operation in several countries.

A further development of the Koppers–Totzek process was made by Shell and also by Krupp Koppers (PRENFLO process = **Pr**essurized **En**trained **Flo**w gasification). Here the gasification is still carried out at temperatures of 2000 °C, but at higher pressures of up to 40 bar.

A similar principle for flue dust gasification is employed in the Texaco process that has been used commercially by Ruhrchemie/Ruhrkohle AG since 1978. The coal is fed to the reactor as an aqueous suspension (up to 71% coal) produced by wet milling. With the high temperatures (1200–1600 °C) and pressures (20–80 bar), high C-conversions of up to 98% and high gas purity can be attained. Many plants using this process have been built or are planned.

The origin of the *Lurgi pressure gasification* goes back to 1930 and, as a result of continuous development, this process is the most sophisticated. The Lurgi gasification operates according to the principle of a fixed bed moved by a rotating blade where lumpy hard coal or brown coal briquets are continously introduced. Initially, degassing takes place at 20–30 bar and 600–750 °C. Coal with a caking tendency forms solid cakes which are broken up by the blades. O_2 and H_2O are fed in from the base and blown towards the coal, and synthesis gas is generated under pressure at about 1000 °C. This gas is ideally suited for further processing to SNG, for example, as it has a relatively high methane content. However, the other substances present (benzene, phenols and tar) necessitate byproduct treatment.

There are several large scale Lurgi plants in operation throughout the world. One location is Sasolburg/South Africa where synthesis gas is used to manufacture hydrocarbons by the Fischer–Tropsch process. The African Explosives & Chem. Ind. (South

Africa) has also been employing synthesis gas for the manufacture of methanol since 1976. In this case, the ICI process is used and the plant has an annual capacity of 33000 tonnes. Further methanol plants based on synthesis gas from coal are planned in other countries, *e. g.,* in the USA and Western Europe (*cf.* Section 7.4.2).

Further development of the Lurgi pressure gasification process has been carried out by various firms with the object of increasing the efficiency of the reactors. The newest generation of Lurgi processors (Mark-V gasifiers) have a diameter of almost 5 m and produce ca. 100000 m³/h.

In all gasification processes dealt with up to now, part of the coal (30−40%) is combusted to provide the necessary process heat. For this reason other more economical sources of heat are now being studied so that the coal load can be reduced.

The application of process heat from gas-cooled high temperature nuclear reactors for the gasification of brown coal is being studied in Germany by Rheinbraun in cooperation with Bergbau-Forschung and the nuclear research plant in Jülich. The helium emerging from the pebble-bed reactor at a temperature of 950 °C supplies the necessary heat for the gasification process. With a brown coal feed, the minimum temperature necessary in the gasification generator is regarded to be 800 °C.

This advantageous conservation of the fossil raw material coal can only be obtained by the expensive commercial coupling of two technologies, and thus a "third generation" gasification process will not be established quickly.

further development of Lurgi gasification aims at higher reactor efficiency, *e. g.,* by increase in diameter from present 3.70 to 5.00 m and increased pressures of 50−100 bar, or decreasing the O_2/H_2O ratio to 1:1 at higher temperatures and yielding liquid slag

conventional gasification processes consume about 1/3 of coal for the generation of:

1. steam as gasification agent
2. heat for the gasification process

therefore developments to substitute combustion heat from fossil sources by process heat from nuclear reactors

promising concept:
HTR = **h**igh **t**emperature **r**eactors

example:

pebble-bed reactor currently at pilot plant stage:

heat transfer agent − He
gas outlet temperature − approx. 950 °C

2.1.1.2. Synthesis Gas *via* Cracking of Natural Gas and Oil

The production of synthesis gas from natural gas and oil in the presence of steam is analogous to coal gasification, since there is a coupling of endothermic and exothermic gasification reactions:

$$-CH_2- + 0.5O_2 \rightarrow CO + H_2 \quad \left(\Delta H = -\frac{22 \text{ kcal}}{92 \text{ kJ}}/\text{mol}\right) \quad (7)$$

$$-CH_2- + H_2O \rightarrow CO + 2H_2 \quad \left(\Delta H = +\frac{36 \text{ kcal}}{151 \text{ kJ}}/\text{mol}\right) \quad (8)$$

synthesis gas manufacture from natural gas or crude oil according to two principles:

1. allothermal catalytic cracking with H_2O (steam cracking or reforming)

2. autothermal catalyst-free cracking with $H_2O + O_2$ (+CO_2)

to 1:

ICI process most well known steam reforming based on Schiller process of IG Farben

feedstock 'naphtha' also known as 'chemical petrol'

three process steps:

1.1. hydrodesulfurization using $CoO-MoO_3/Al_2O_3$ or $NiO-MoO_3/Al_2O_3$ at 350–450 °C until S content < 1 ppm

olefins are simultaneously hydrogenated

1.2. catalytic reforming in primary reformer with $Ni-K_2O/Al_2O_3$ at 700–830 °C and 15–40 bar

1.3. autothermal reforming of residual CH_4 in the secondary reformer *i.e.*, another partial combustion of gas due to high heat requirement

conductance of processes (1.3):

lined chamber reactor with heat resistant Ni catalyst (>1200 °C) CH_4 content lowered to 0.2–0.3 vol%

sensible heat recovered as steam

The simultaneous attainment of the Boudouard water gas and methane-formation equilibria corresponds in principle to the coal gasification reaction.

Both natural gas and crude oil fractions can be converted into synthesis gas using two basically different methods:

1. With the allothermal steam reforming method, catalytic cracking takes place in the presence of water vapor. The necessary heat is supplied from external sources.

2. With the autothermal cracking process, heat for the thermal cracking is supplied by partial combustion of the feed, again with H_2O and possibly recycled CO_2 to attain a desired CO/H_2 ratio.

Process Principle 1:

Today, the most well known large-scale steam reforming process is ICI's which was first operated in 1962. Hydrocarbon feeds with boiling points up to ca. 200 °C (naphtha) can be employed in this process.

The ICI process consists of three steps. Since the $Ni-K_2O/Al_2O_3$ reforming catalyst is very sensitive to sulfur, the naphtha feed must be freed from sulfur in the first step. To this end it is treated with H_2 at 350–450 °C using a $CoO-MoO_3/Al_2O_3$ catalyst. The resulting H_2S is adsorbed on ZnO. Simultaneously, any olefins present are hydrogenated. In the second step, the catalytic reforming takes place in catalyst-filled tubes at 700–830 °C and 15–40 bar. The reforming tubes are heated by burning natural gas or ash-free distillates.

At a constant temperature, an increase in pressure causes the proportion of methane – an undesirable component in synthesis gas – remaining in the product gas to increase. However, due to construction material constraints, temperatures higher than ca. 830 °C cannot be reached in externally heated reforming tubes. For this reason, the product gas is fed into a lined chamber reactor filled with a high-temperature-resistant Ni catalyst. A portion of the gas is combusted with added air or oxygen whereby the gas mixture reaches a temperature of over 1200 °C. Methane is reacted with steam at this temperature until only an insignificant amount remains (0.2–0.3 vol%). This is the third step of the process.

The tube furnace is called the 'primary reformer' and the lined chamber reactor the 'secondary reformer'. The sensible heat from the resulting synthesis gas is used for steam generation.

The advantage of the ICI process is that there is no soot formation, even with liquid crude oil fractions as feed. This makes catalyst regeneration unnecessary.

advantages of ICI process:

no soot and thus little loss in catalyst activity

Process Principle 2:

Synthesis gas manufacture by partial oxidation of crude oil fractions was developed by BASF, Texaco and Hydrocarbon Research. A modified version was also developed by Shell. All hydrocarbons from methane to crude oil residues (heavy fuel oil) can be used as feedstock.

The preheated feeds are reacted with H_2O and less than the stoichiometric amounts of O_2 in the combustion sector of the reactor at $30-80$ bar and $1200-1500\,°C$. No catalyst is used. The heat generated is used to steam reform the oil. Soot formed from a small portion of the oil is removed from the synthesis gas by washing with H_2O or oil and is made into pellets. In 1986, the Shell gasification process was in operation in 140 syn gas plants.

to 2:

well-known autothermal processes:

BASF/Lurgi (Gassynthan)
Texaco
Hydrocarbon Research
Shell (gasification process)

process operation (Shell):

catalyst-free, $1200-1500\,°C$, $30-80$ bar
resulting soot converted into fuel oil pellets

advantage:

various crude oil fractions possible as feedstock

2.1.2. Synthesis Gas Purification and Use

Synthesis gas from the gasification of fossil fuels is contaminated by several gaseous compounds which would affect its further use in different ways. Sulfur, present as H_2S or COS, is a poison for many catalysts that partly or completely inhibits their activity. CO_2 can either directly take part in the chemical reaction or it can interfere by contributing to the formation of excess inert gas.

A large number of different processes are available to purify the synthesis gas by removing H_2S, COS and CO_2. The Rectisol process of Lurgi and Linde for example is widely used and involves pressurized washing with methanol. Another example is the Selexol process (Allied; now UCC) which exploits the pressure-dependent solubility of the acidic gases in the dimethyl ethers of poly(ethylene glycol) (*cf.* Section 7.2.4). The Shell Sulfinol process employs mixtures of sulfolan/diisopropylamine/water, while the Lurgi Purisol process uses N-methylpyrrolidone. Also employed in other processes are diethanolamine, diglycolamine, propylene carbonate or alkali salts of amino acids such as N-methylaminopropionic acid (Alkazid process).

Pressurized washes with K_2CO_3 solutions (Benfield, Catacarb) as well as adsorption on molecular sieves (UCC) are frequently used.

synthesis gas aftertreatment:

removal of H_2S, COS, CO_2

purification processes for synthesis gas:

pressurized washing with:

1. CH_3OH (Rectisol process)
2. poly(ethylene glycol) dimethyl ether (Selexol process)
3. sulfolan/diisopropanolamine/H_2O (Sulfinol process)
4. NMP (Purisol process)
5. numerous other organic and inorganic solvent systems
6. molecular sieves

Benfield process (developed in 1950 by Benson and Field):

$K_2CO_3 + CO_2 + H_2O \rightleftharpoons KHCO_3$
20 bar, $105\,°C \rightleftharpoons 1$ bar, $105\,°C$

Claus process:

$$H_2S + 1.5\,O_2 \rightarrow SO_2 + H_2O$$

$$SO_2 + 2\,H_2S \xrightarrow{\text{(cat)}} 3\,S + H_2O$$

adjustment of required CO/H_2 ratio in synthesis gas possible:

1. during gasification by altering amount of H_2O and O_2
2. after gasification by CO conversion $CO + H_2O \rightarrow CO_2 + H_2$ and removal of CO_2

to 2:

conversion catalysts:

Fe–Cr-oxide multistep at 350–400 °C
CO up to 3 vol%
CuO–ZnO, single-step at 190–260 °C,
CO up to 0.1 vol%

synthesis gas applications:

1.   chemical feedstock for syntheses
1.1. CH_3OH
1.2. aldehydes, alcohols from olefins
1.3. hydrocarbons *via* Fischer–Tropsch

Fischer-Tropsch technology

start   1954 Sasol I
         1980 Sasol II
         1983 Sasol III
         1993 Shell Malaysia
         1993 Sasol suspension reactor

The regeneration of the absorption/adsorption systems is accomplished in different ways, mainly by physical processes such as degassing at high temperatures or low pressures. The H_2S is generally converted to elemental sulfur in the Claus oven. Here some of the H_2S is totally oxidized to SO_2, which is reduced to sulfur with additional H_2S in a following step. This second step requires a catalyst, which is frequently based on Al_2O_3.

The resulting pure synthesis gas must have a particular CO/H_2 ratio for the conversion which follows; e. g., methanol formation, or reaction with olefins to produce aldehydes/alcohols in 'oxo' reactions. This ratio may be defined by the stoichiometry or by other considerations. It can be controlled in several gasification processes by adjusting the proportion of hydrocarbon to H_2O and O_2. If the CO content is too high then the required CO/H_2 ratio can be obtained by a partial catalytic conversion analogous to equation 4 using shift catalysts — consisting of Fe–Cr-oxide mixtures — which are employed at 350–400 °C. In this way, the CO content can be reduced to about 3–4 vol%. An increased CO conversion is necessary if synthesis gas is to be used for the manufacture of pure hydrogen (*cf.* Section 2.2.2). In this case, more effective low temperature catalysts (*e. g.,* Girdler's G-66 based on Cu–Zn-oxide) is employed. Their operating temperature lies between 190 and 260 °C. In the water gas equilibrium only 0.1 vol% CO is present at this temperature.

In addition to the very important applications of synthesis gas as feedstock for the manufacture of methanol (*cf.* Section 2.3.1) or for aldehydes/alcohols from olefins *via* hydroformylation (*cf.* Section 6.1), it is also used by Sasol in South Africa for the manufacture of hydrocarbons *via* the Fischer-Tropsch process. The hydrocarbons manufactured there are based on synthesis gas from coal (Lurgi gasification process) supplied from their own highly mechanized mines. Two different Fischer–Tropsch syntheses are operated. With the Arge process (**Ar**beits**ge**meinschaft (joint venture) Ruhrchemie-Lurgi), higher boiling hydrocarbons such as diesel oil and wax are produced in a gas-phase reaction at 210–250 °C over a fixed bed of precipitated iron catalyst. The Synthol process (a further development of the original Kellogg process) yields mainly lower boiling products such as gasoline, acetone and alcohols using a circulating fluidized bed (flue dust with circulation of the iron catalyst) at 300–340 °C and 23 bar. The expansion of the original Sasol I plant with Sasol II made a total annual production of 2.5×10^6 tonnes of liquid products in 1980 possible. Sasol III, a duplicate of Sasol II, began production in 1983, increasing the total capacity to 4.5×10^6 tonnes per year.

Until recently, Sasol used a suspension reactor in which an active iron catalyst was suspended in heavy hydrocarbons with turbulent mixing. This gives a better conversion and selectivity at higher temperatures; the reaction product contains fewer alcohols, but more higher olefins.

The first Fischer-Tropsch plant outside of Africa was started up by Shell in Malaysia in 1993. It is based on natural gas, and has a production capacity for mid-distillation-range hydrocarbons of 0.5×10^6 tonnes per year.

Even though these aforementioned applications of synthesis gas are still the most important, other uses of synthesis gas, of the component CO, or of secondary products like methanol or formaldehyde have received increasing attention, and replacement processes based on coal are already in industrial use.

Examples include modifications of the Fischer-Tropsch synthesis for production of C_2-C_4 olefins, olefin manufacture from methanol (*cf.* Section 2.3.1.2), the homologation of methanol (*cf.* Section 2.3.1.2), and the conversion of synthesis gas to ethylene glycol (*cf.* Section 7.2.1.1) or to other oxygen-containing C_2 products (*cf.* Section 7.4.1.4).

However, the use of synthesis gas as a source for carbon monoxide and hydrogen (*cf.* Sections 2.2.1 and 2.2.2) and, after methanation (*cf.* eq 6), as an energy source (synthetic natural gas – SNG) remains unchanged.

In the nuclear research plant (KFA) at Jülich, Germany, a concept for a potential future energy transport system was proposed based on the exothermic CO/H_2 conversion to CH_4. In the so-called ADAM-EVA circulation process, methane is steam reformed (endothermic) into a CO/H_2 mixture using helium-transported heat from a nuclear reactor (EVA), the gas mixture is supplied to the consumer by pipeline and there methanated (exothermic; ADAM). The methane formed is fed back to the EVA reformer. In 1979, an ADAM-EVA pilot plant was brought on line in KFA-Jülich; in 1981, it was expanded to a capacity of 10 MW.

Analogous to crude oil, CO/H_2 mixtures could function as feedstocks for the chemical industry and as an energy source for household and industrial consumers.

Synthesis gas is being used increasingly as a reduction gas in the manufacture of pig iron.

Reactor versions in Sasol plants:

1. tubular fixed-bed reactor
2. circulating fluidized-bed reactor
3. suspension reactor

1.4. olefin-selective
 Fischer–Tropsch synthesis

2. raw material for CO and H_2 recovery
3. raw material for CH_4 manufacture, as SNG for public energy supply
4. possible energy carrier

 'ADAM-EVA' project of Rheinbraun/ KFA Jülich

ADAM (**A**nlage mit **D**rei **A**diabaten **M**ethanisierungsreaktoren – Unit with three adiabatic methanation reactors)
EVA (**E**inzelrohr-**V**ersuchs-**A**nlage – Single tube experimental unit)

principle:

methanation reaction is reversible on supplying energy, *i.e.*, instead of electricity, CO/H_2 is transported to consumer and CH_4 is returned for reforming

5. reduction gas in pig iron manufacture

2.2. Production of the Pure Synthesis Gas Components

CO and H_2 as mixture and also the pure components important large scale industrial chemical precursors

Carbon monoxide and hydrogen, both as synthesis gas and individually, are important precursors in industrial chemistry. They are the smallest reactive units for synthesizing organic chemicals and play a decisive role in the manufacture of several large-scale organic chemicals. Furthermore, hydrogen in particular could become an important energy source in meeting the demand for heat, electricity and motor fuel for the transportation sector.

H_2 in future perhaps also source of energy for:

1. heating
2. electricity
3. motor fuel

2.2.1. Carbon Monoxide

CO produced from:

1. coal-coking gases
2. hydrocarbon crack gases from natural gas up through higher oil fractions

The raw materials for CO are the gas mixtures (synthesis gas) which result from the carbonization of hard coal, the low temperature carbonization of brown coal or the steam reforming of hydrocarbons.

CO separation *via* two processes:

The CO can be separated from the above gas mixtures using essentially one of two processes:

1. physically, by partial condensation and distillation
2. chemically, *via* Cu(I)–CO complexes

1. Low temperature separation
2. Absorption in aqueous copper salt solutions

to 1:

example – Linde process:

raw gas preliminary purification in two steps:

1.1. CO_2 with $H_2NCH_2CH_2OH$ (reversible carbonate formation *via* ΔT)
1.2. H_2O and residual CO_2 on molecular sieves (reversible *via* ΔT)

The low temperature separation, *e. g.,* according to the Linde or Air Liquide process, requires that several process steps involving gas treatment occur before the pure $H_2/CO/CH_4$ mixture is finally separated.

The raw gas, *e. g.,* from the steam reforming of natural gas, is freed from CO_2 by scrubbing with ethanolamine solution until the CO_2 concentration reaches ca. 50 ppm. The remaining CO_2 and H_2O are removed by molecular sieve adsorbents. Both products would cause blockages due to ice formation.

Moreover, the gas mixture should be free from N_2 as, due to similar vapor pressures, a separation would be very involved and expensive. The N_2 separation from the natural gas thus occurs before the steam reforming.

separation of gas mixture ($H_2/CO/CH_4$) in two steps:

1. partial condensation of CH_4 and CO
2. fractional distillation of CH_4 and CO with CO overhead

The actual low temperature separation takes place essentially in two steps. Initially, CH_4 and CO are removed by condensing from the gas mixture after cooling to ca. $-180\,°C$ at 40 bar. The CO and CH_4 are depressurized to 2.5 bar during the next step in the CO/CH_4 separation column. The CO is removed overhead, the CH_4 content being less than 0.1% by volume. The process is characterized by a very effective recycling of all gases in order to exploit the refrigeration energy.

The CO absorption in CuCl solution, acidified with hydrochloric acid, or alkaline ammonium Cu(I) carbonate or formate solution is conducted at a pressure of up to ca. 300 bar. The desorption takes place at reduced pressure and ca. $40-50\,°C$. There are essential differences in the concentrations of the Cu salt solutions depending on whether CO is to be recovered from gas mixtures or whether gases are to be freed from small amounts of CO.

to 2:

absorption by Cu(I) salt based on pressure-dependent reversible complex formation: $[Cu(CO)]^{\oplus}$

e. g., Uhde process with Cu salt/NH_3–H_2O

A more modern Tenneco Chemicals process called 'Cosorb' employs a solution of CuCl and $AlCl_3$ in toluene for the selective absorption of CO from synthesis gas. The Cu(I)–CO complex is formed at ca. $25\,°C$ and up to 20 bar. The CO is released at $100-110\,°C$ and $1-4$ bar. Water, olefins, and acetylene affect the absorption solution and must therefore be removed before the separation. Many large-scale plants are — following the startup of a prototype in 1976 — in operation worldwide.

in the Tenneco 'Cosorb' process, the temperature dependence of the CO complex formation with $Cu[AlCl_4]$ in toluene is employed

gas pretreatment (removal of H_2O and C_2H_2) essential:

H_2O hydrolyzes $AlCl_3$
C_2H_2 forms acetylide

A newer technology applicable to separation processes uses semipermeable membranes to enrich the CO in a gas mixture.

applications of CO:

1. in combination with H_2
1.1. synthesis gas chemistry (CH_3OH and Fischer–Tropsch hydrocarbons)
1.2. hydroformylation ('oxo' aldehydes, 'oxo' alcohols)

The use of carbon monoxide in a mixture with hydrogen (*cf.* Sections 2.1.2 and 6.1) is more industrially significant than the reactions with pure CO. Examples of the latter include the carbonylation of methanol to acetic acid (*cf.* Section 7.4.1.3) and of methylacetate to acetic anhydride (*cf.* Section 7.4.2).

Another type of carbonylation requires, in addition to CO, a nucleophilic partner such as H_2O or an alcohol. These reactions are employed industrially to produce acrylic acid or its ester from acetylene (*cf.* Section 11.1.7.1) and propionic acid from ethylene (*cf.* Section 6.2).

2. in combination with nucleophilic partner (H_2O, ROH)

2.1. Reppe carbonylation (acrylic acid, propionic acid and its esters)

2.2. Koch reaction (branched carboxylic acids)

A special case of hydrocarbonylation is the Koch synthesis for the manufacture of branched carboxylic acids from olefins, CO and H_2O (*cf.* Section 6.3).

Furthermore, CO is reacted with Cl_2 to produce phosgene, which is important for the synthesis of isocyanates (*cf.* Section 13.3.3).

3. directly in reactions
3.1. phosgene formation with Cl_2 (isocyanates, carbonates)
3.2. carbonyl formation with metals (catalysts)

Carbonylation reactions sometimes require metal carbonyl catalysts such as Fe, Co, Ni or Rh carbonyls. They are either separately synthesized with CO or form from catalyst components and CO *in situ*.

2.2.2. Hydrogen

industrial H_2 sources
fossil fuels
H_2O

large scale H_2 production:

petro-, coal-based- and electrochemistry

world H_2 production (in %):

	1974	1984	1988
cracking of crude oil	48	} 77	} 80
cracking of natural gas	30		
coal gasification	16	18	16
electrolysis	3	4	
miscellaneous			} 4
processes	3	1	

total (in 10^6 tonnes)   24.3 ca.45   ca.45

H_2O reduction with fossil fuels combines both H_2 sources

Hydrogen is present in fossil fuels and water in sufficient amounts to fulfill its role as a reaction component in industrial organic syntheses. It can be produced from these sources on a large scale by three different methods:

1. petrochemical processes
2. coal-based chemical processes
3. electrochemical processes (electrolysis)

The percentage of the ca. 45×10^6 tonnes of hydrogen manufactured worldwide in 1988 that derived from petrochemicals is essentially the same as in 1974 (*cf.* adjacent table). The percentage from the gasification of coal and coke (primarily from coke-oven gas) rose slightly during this period, while the part from electrolytic processes (mainly chlor-alkali electrolysis) remained practically constant.

Petrochemical Processes:

The principle of hydrogen generation from the reaction of water with fossil fuels was presented previously under the manufacture of synthesis gas. Synthesis gas can be obtained from lighter hydrocarbons though allothermal catalytic reforming, and from heavier oil fractions by autothermal partial oxidation. The H_2 share is particularly high when a methane-rich natural gas is available for synthesis gas manufacture:

$$CH_4 + H_2O \rightleftharpoons 3H_2 + CO \quad \left(\Delta H = +\frac{49 \text{ kcal}}{205 \text{ kJ}} / \text{mol} \right) \qquad (9)$$

H_2O reduction with CH_4 supplies
1/3 H_2 from H_2O
2/3 H_2 from CH_4

H_2 manufacture *via* refinery conversion process for light oil distillates

use of H_2 from refinery gas for refinery processes such as:

hydrofining
hydrotreating
hydrocracking

The steam reforming of hydrocarbons is in principle a reduction of water with the carbon of the organic starting material. In the case of methane, 1/3 of the hydrogen is supplied by water. This share increases with higher hydrocarbons.

A second possibility of producing H_2 *via* a chemical reaction is offered by processing light crude oil distillates in refineries where H_2 is released during aromatization and cyclization reactions. The refinery gas which results is an important H_2 source for internal use in the refinery (hydrofining, hydrotreating). The wide-scale application of hydrocrackers in the USA (*cf.* Section 1.4.1) will mean that more H_2 will have to be manufactured in the refineries by steam reforming of hydrocarbons to meet the increased demand for H_2.

A third method for the manufacture of H_2 is the electric arc cracking of light hydrocarbons, used primarily for the production of acetylene (*cf.* Section 4.2.2.), but which yields a raw gas whose major component is H_2.

Several other purification steps must follow the desulfurization of the CO/H_2 mixture (*cf.* Section 2.1.2) if pure H_2 (ca. 99.999 vol%) is to be isolated from synthesis gas.

Initially, CO conversion with steam takes place to form CO_2 and H_2 (*cf.* eq 4). The CO_2 is removed by washing under pressure. After that, a finishing purification follows during which traces of CO are removed by methanation, *i. e.*, by reacting with H_2 to CH_4 and H_2O (*cf.* eq 6). This step is conducted at 300–400°C in the presence of a supported Ni catalyst.

The isolation of H_2 from refinery gas can be accomplished like the recovery of CO through low temperature fractional condensation; or through adsorption of all impurities on zeolite or carbon-containing molecular sieves and **p**ressure **s**wing **a**dsorption (PSA) processes from, *e. g.*, Bergbau-Forschung, Linde, or UCC. This also applies to isolation of H_2 from coke-oven gas.

A newer process developed by Monsanto for the purification or isolation of H_2 from gas mixtures is based on separation through semipermeable membranes. The gas mixture flows around an array of hollow fibers, and only hydrogen and helium can diffuse into the fibers through their semipermeable polysulfone-coated walls. Separation from CH_4, CO, O_2, and N_2 can be accomplished in this way. Monsanto has installed these H_2 separation units (Prism separators) on an industrial scale worldwide for the regulation of the H_2/CO ratio for oxo synthesis and for the recovery of H_2 from flue gases in hydrogenation and NH_3 plants.

Ube (Japan) has developed a similar H_2 separation technology using aromatic polyimide membranes. This technology is already being used commercially.

Electrochemical Processes:

Hydrogen is also manufactured industrially by electrolysis either directly as in the case of H_2O, HF, and 22–25% hydrochloric acid or chlor-alkali electrolysis (diaphragm-cell process); or indirectly in a secondary chemical reaction, as in chlor-alkali electrolysis (mercury process *via* sodium amalgam). With the exception of the H_2O electrolysis, hydrogen is merely a welcome byproduct which helps to make the process − involving high electrical costs − more economical. However, the hydrogen

---

manufacture of H_2 in C_2H_2 electric arc cracking

H_2 separation from synthesis gas in multistep process:

1. H_2S, COS removal
2. CO conversion
3. CO_2 removal
4. finish purification by methanation, *i. e.*, CO traces removed *via* Ni-catalyzed reaction:
   $CO + H_2 \rightarrow CH_4 + H_2O$

H_2 separation from refinery gas *via* fractional low temperature condensation or adsorption on molecular sieve

selective diffusion of H_2 through membranes (Monsanto, polysulfone; Ube, polyarimide) in the form of hollow fibers

H_2 manufacture by electrolytic processes:

$2 H_2O \rightarrow 2 H_2 + O_2$
$2 HF \rightarrow H_2 + F_2$
$2 HCl/H_2O \rightarrow H_2 + Cl_2$
$2 NaCl/H_2O \rightarrow H_2 + Cl_2 (+ 2 NaOH)$

produced by electrolysis only accounts for a small percentage of the total H_2 production.

H_2 separation by electrolytic processes:

mechanical by separation of anode and cathode chambers

In contrast to the steam reforming of hydrocarbons, the hydrogen from the electrolysis is very pure (>99 vol%) thus eliminating any costly purification steps.

H_2 manufacture using novel technologies

In view of the possible future importance of hydrogen in transporting and storing energy, in the generation of heat and electricity, as a motor fuel and as a chemical raw material, research on novel technologies and the use of nuclear process heat to improve H_2 production by both chemical and electrochemical processes is being done in many places all over the world.

Several of the most promising proposals for possible future developments are described briefly below.

1. H_2 from H_2O cleavage with nuclear process heat in thermochemical cyclic processes, *e. g.*, Fe/Cl cycle, which avoid direct thermolysis requiring temperatures over 2500 °C

Instead of cracking water by reduction with hydrocarbons or by electrolysis, it is possible in principle to conduct thermal cracking with nuclear process heat. However, this cracking is not possible in one process step because water has such a high thermal stability (enthalpy of formation of H_2O from $H_2 + 0.5\,O_2 = -68$ kcal/mol $= -285$ kJ/mol), and the temperatures available from the high-temperature nuclear reactors (He outlet temperature 905 °C) are too low. A series of reactions in the form of a thermochemical cycle is necessary.

general principle of a thermochemical cyclic process:

$H_2O + X \rightarrow XO + H_2$
$XO \qquad \rightarrow X + 0.5\,O_2$

$H_2O \qquad \rightarrow H_2 + 0.5\,O_2$

An example taken from a whole series of suggested cyclic processes – whose economic and technical feasibility has not yet been proven – is the Fe/Cl cycle (Mark 9) below:

$$6\,FeCl_2 + 8\,H_2O \xrightarrow{\;650\,°C\;} 2\,Fe_3O_4 + 12\,HCl + 2\,H_2$$
$$2\,Fe_3O_4 + 3\,Cl_2 + 12\,HCl \xrightarrow{\;200\,°C\;} 6\,FeCl_3 + 6\,H_2O + O_2$$
$$6\,FeCl_3 \xrightarrow{\;420\,°C\;} 6\,FeCl_2 + 3\,Cl_2$$

$$2\,H_2O \longrightarrow 2\,H_2 + O_2 \qquad (10)$$

The heat necessary for the individual reaction steps can be obtained from nuclear process heat of less than 1000 °C. A fundamental problem of all previously introduced cyclic processes is the large quantities to be treated, which can amount to 200 to 3000 times the amount of H_2 produced.

2. H_2 from H_2O *via* pressure or steam electrolysis with higher efficiency than conventional electrolytic processes through development of electrodes, electrolytes, and cell arrangement

Other examples of future developments relating to electrolytic processes are the pressure electrolysis at >30 bar and 90 °C as well as steam electrolysis at $500-900$ °C with solid electrolytes (*e. g.*, from ZrO_2 doped with Y oxides) or with improved cell geometry. These processes are distinguished in particular by the higher efficiencies obtained, *i. e.*, by their more effective exploitation of electrical energy compared to conventional electrolysis. Further examples with respect to the use of H_2 as an energy carrier within the framework of a future hydrogen technology are summarized in Section 1.3.

The most important applications of hydrogen in the chemical industry worldwide and in several countries are listed in the following table. Clearly, most is used for ammonia synthesis and in refinery processes. Only 20–30% is used in true organic synthesis or secondary processes, *i. e.*, hydrogenation of organic intermediates:

greatest H_2 consumers:

NH_3 synthesis
refinery reprocessing and conversion processes

Table 2-1. Use of hydrogen (in %).

	Worldwide			USA			Japan		WE	
	1974	1987	1993	1981	1988	1993	1980	1993	1987	1993
Ammonia synthesis	54	57	67	53	49	47	34	27	48	53
Mineral oil processing (hydrotreating, -fining, -cracking)	22	29	20	30	37	37	40	72	28	20
Methanol synthesis	6	7	9	7	7	9	10		8	7
Oxo processes and hydrogenation (*e. g.*, benzene to cyclohexane, nitrobenzene to aniline, hydrogenation of fats, etc.)	7	7	4	6	3	7	12	1	8	20
Miscellaneous	11			4	4		4		8	
Total use of H_2 (in 10^6 tonnes)	24.2	20.0	29.4	5.7	5.5	5.8	2.1	1.2	4.9	4.2

2.3. C_1-Units

2.3.1. Methanol

Methanol is one of the most important industrial synthetic raw materials. Worldwide, about 90% is used in the chemical industry and the remaining 10% is used for energy. In 1996 the world capacity for methanol synthesis was about 29.1×10^6 tonnes per year; capacities in the USA, Western Europe, and Japan were 9.0, 2.8, and 0.20×10^6 tonnes, respectively. An increasing demand for methanol, which will rise 6% per year through 2000, has led to a planned expansion in production

CH_3OH production (in 10^6 tonnes):

	1990	1992	1994
USA	3.62	4.01	5.52
W. Europe	1.97	n.a.	2.55
Japan	0.08	0.02	0.04

n.a. = not available

capacity. For example, Saudi Arabia will have a capacity of over 3.6×10^6 tonnes per year by 1998. Production figures for several countries are given in the adjacent table.

2.3.1.1. Manufacture of Methanol

CH_3OH manufacture:

from highly compressed synthesis gas (CO/H_2) with many variations of process conditions and catalyst composition

modified oxide system based on $ZnO-Cr_2O_3$ as first-generation, $CuO-ZnO/Al_2O_3$ as second-generation catalysts

Around 1913, A. Mittasch at BASF noticed the presence of oxygen-containing compounds during experiments on the NH_3 synthesis. Systematic research and development in Germany led in 1923 to the first large scale methanol manufacture based on synthesis gas:

$$CO + 2\,H_2 \rightleftharpoons CH_3OH \quad \left(\Delta H = -\frac{22\ \text{kcal}}{92\ \text{kJ}}/\text{mol} \right) \tag{11}$$

adjustment of the H_2/CO stoichiometry in synthesis gas possible by addition of CO_2

If the synthesis gas is manufactured from methane-rich natural gas, then its composition ($CO + 3\,H_2$) does not correspond to stoichiometric requirements. In these cases, CO_2 is added to the feed as it consumes more H_2 (*cf.* eq 12) than CO (*cf.* eq 11).

$$CO_2 + 3\,H_2 \rightleftharpoons CH_3OH + H_2O \quad \left(\Delta H = -\frac{12\ \text{kcal}}{50\ \text{kJ}}/\text{mol} \right) \tag{12}$$

process modifications for CH_3OH manufacture:

1. BASF high pressure process (approx. 340 bar, 320–380°C) with $ZnO-Cr_2O_3$ catalysts

process characteristics:

low synthesis gas conversion (12–15% per pass) demands recycling process with CH_3OH removal by condensation

The BASF process is conducted at 320–380 °C and ca. 340 bar. The $ZnO-Cr_2O_3$ catalyst has a maximum activity when the Zn/Cr ratio is about 70:30. Cold gas is injected at many places in the catalyst bed to avoid high temperatures that would shift the equilibrium away from methanol. Being very resistant to the usual catalyst poisons at low concentrations, the oxide mixture has a service life of several years. Byproducts such as dimethyl ether, methyl formate, and the higher alcohols are separated by distillation with a low- and high-boiling column. In order to suppress side reactions, a short residence time (1–2 s), which completely prevents equilibration from taking place, is employed. Conversions of only 12–15% are usual for a single pass through reactor.

The industrial process has been made highly efficient. The use of high pressure centrifugal compressors − normally employed in NH_3 plants − has made a particularly strong contribution.

2. UK Wesseling high pressure process (300 bar, 350 °C) with $ZnO-Cr_2O_3$ catalysts

UK Wesseling has developed a process operating at a low CO partial pressure (13 bar in gas recycle). The reaction conditions (300 bar and 350 °C) are similar to those of the BASF process.

The $ZnO-Cr_2O_3$ catalyst is arranged in stages in the reactor. Carbon steel can serve as construction material as no $Fe(CO)_5$ is formed at the low CO partial pressure used. Methanol is obtained in high purity with only a small amount of byproducts. Using this high pressure modification, more than a million tonnes of methanol had been produced worldwide by the end of 1970. Recently, the conventional processes have been complemented by others operating at low and medium pressures. This transition to lower operating pressure was made possible by the introduction of more active Cu-based catalysts. These are, however, extremely sensitive to sulfur and require that the total sulfur content in synthesis gas be less than 1 ppm.

process characteristics:

similar to the BASF high pressure process but lower CO partial pressure (up to 20 bar) avoids formation of $Fe(CO)_5$; *i.e.,* no catalyst for methanation reaction

The ICI low pressure process, first operated in a pilot plant in 1966, plays a dominant role. Today, about 65% of the world methanol production is based on the ICI process, which is characterized by lower investment and process costs. In 1972, the prototype of a large scale plant (310000 tonnes per year) went on stream in Billingham, United Kingdom. Modern plants have an annual production capacity of about one million tonnes. The Cu–Zn–Al-oxide-based catalyst requires a synthesis gas particularly free of sulfur and chlorine. A new generation of catalysts with a 50% longer lifespan was recently introduced.

3. ICI low pressure process (50−100 bar, 240−260 °C)

process characteristics:

Cu–Zn–Al-oxide catalyst requires S- and Cl-free synthesis gas and then supplies very pure CH_3OH

The usual process conditions in the converter are 50–100 bar and 240-260 °C. Methanol can be obtained with a purity up to 99.99 wt%. The reactor is extremely simple; it contains only one catalyst charge which can be quickly exchanged. As in the high pressure process, cold gas is introduced at many places to absorb the reaction heat.

A similar low pressure process with a tubular bundle reactor was developed by Lurgi. The temperature is controlled by flowing boiling water around the entire length of the tubes. It employs a modified $CuO-ZnO$ catalyst at 50−80 bar and 250−260 °C.

4. Lurgi low pressure process

process characteristics:

CuO–ZnO catalyst (requires S-free CO/H_2 mixture) arranged in a tube reactor, *i.e.*, optimal temperature regulation by water vapor cooling; steam self-sufficient

In 1973, a 200000 tonne-per-year plant, in combination with a 415000 tonne-per-year NH_3 plant, went into operation employing this principle. In the period following, Lurgi reactors were used more and more, and in 1984 their market share had already reached 50% of all methanol plants.

Recently, there has been a trend toward a medium pressure process. A number of companies are employing Cu as well as Zn–Cr oxide-based catalysts as shown in the table below:

recent developments in CH_3OH manufacture from synthesis gas: medium pressure with Cu and Zn–Cr-oxide catalysts with advantages including higher rate of reaction, *i.e.,* higher space–time yields

Table 2-2. Methanol manufacture by the medium pressure process.

Company	Catalyst	Temperature (°C)	Pressure (bar)
Haldor Topsoe	$CuO-ZnO-Cr_2O_3$	230–260	100–150
Vulcan	$ZnO-Cr_2O_3$	270–330	150–250
Pritchard	CuO	unknown	100–250
Catalyst and Chemical Inc.	$CuO-ZnO-Al_2O_3$	240–250	100–250
BASF	$CuO-ZnO-Al_2O_3$	200–350	50–250
Mitsubishi Gas Chemical	CuO + promotor	200–280	50–150

For the sake of completeness, it must be mentioned that since 1943 Celanese (now Hoechst Celanese) has produced not only acetic acid, formaldehyde and acetaldehyde, but also methanol and numerous other components from the oxidation of a propane-butane mixture. The reaction products must undergo an involved distillative separation.

2.3.1.2. Applications and Potential Applications of Methanol

conventionally favored uses for CH_3OH:

1. formaldehyde
2. dimethyl terephthalate (DMT)

The methanol consumption worldwide, in the USA, Western Europe, and Japan can be summarized as follows:

Table 2-3. Uses of methanol (in %).

Product	World 1988	World 1994	USA 1982	USA 1988	USA 1994	Western Europe 1982	Western Europe 1988	Western Europe 1994	Japan 1982	Japan 1988	Japan 1994
Formaldehyde	39	38	30	27	24	50	44	44	47	43	36
Acetic acid	6	7	12	14	10	5	7	7	10	8	10
Methyl halides	7	7	9	6	5	6	7	6	3	5	7
Methyl *tert*-butyl ether	12	22	8	24	38	5	10	24	–	5	4
Dimethyl terephthalate	3	3	4	4	2	4	3	2	1	1	1
Methylamines	4	4	4	3	3	4	4	4	2	4	4
Methyl methacrylate	3	3	4	4	3	3	4	4	6	8	8
Solvents	9	4	10	7	7	6	1	1	6	4	3
Others	17	12	19	11	8	17	15	8	25	22	27
Total use (in 10^6 tonnes)	17.3	21.5	3.2	5.0	7.2	3.3	4.5	5.7	1.1	1.6	1.9

meanwhile, DMT position in many countries replaced by:

MTBE, AcOH, or methyl halides

Methanol uses lumped under "Others" include, for example, its use as a fuel (2% in Western Europe and 3% in the USA in 1990), and all methyl esters and ethers (*e.g.,* methyl acetate, glycol methyl ethers, etc.).

This already changing classic range of methanol applications will expand considerably during the coming decade. The raw material and energy situation with regard to the pricing policy of oil-based products, the expected per capita increase in energy consumption despite limited natural gas and oil deposits, and the undiminished population explosion with the simultaneously decreasing reserves of animal protein will compel development of new strategies.

In these considerations methanol occupies a key position. It can simultaneously function as a motor fuel or component in gasoline, an energy source, a raw material for synthesis as well as a basis for protein. Methanol must therefore be available in sufficient amounts.

potential future or developing uses for CH_3OH:

1. motor fuel or gasoline component
2. energy source
3. feedstock for synthesis
4. C-source for petro-protein

The initial plans envisage plants with an annual production of 10×10^6 tonnes (megamethanol plants) near cheap sources of raw material as, for example, in the oil fields of the Middle East where the petroleum gas is presently being burnt. These plans have not yet, however, been realized to the extent expected since the assumed price increase for petroleum did not occur.

medium-term, CH_3OH supply *via* economically operating mammoth plants (up to 10^7 tonnes per year), *e. g.,* to utilize gas at the oil fields

The first phase of expanded methanol use will take place in the motor fuel sector. Only then will its use in other areas, for example as a feedstock for synthesis, increase.

Methanol can be used either directly or after suitable conversion as a fuel or a component of gasoline.

In the medium term, the mixing of methanol (up to 15 vol%, M15) with conventional motor fuels appears promising from economic and ecological aspects.

1. CH_3OH as future motor fuel or gasoline additive
1.1. CH_3OH in M15 or M100

Even on using pure methanol (M100) as motor gasoline very few carburetor modifications are required. However, the mixture must be preheated and a larger tank is necessary. These alterations are a result of methanol's increased heat of vaporization – three times greater than gasoline's – and the approximately 50% lower energy content.

requirements:

certain technical structural modifications necessary due to higher heat of vaporization and lower energy content

advantages:

high octane number and 'clean' combustion

Moreover, there are several advantages. Among these are an improvement in the knocking behavior because of the high octane number (>110), and a decrease in pollutants in the exhaust gas through almost complete removal of the low formaldehyde content by the catalytic exhaust unit.

disadvantages:

with pure gasoline/CH_3OH mixtures, small amounts to water lead to phase separation

In the addition of methanol to gasoline, one must be careful that a second, aqueous phase does not form in the tank. This can be avoided by the addition of other components like alcohols, in particular *sec-* and *tert-*butanol, as aids to solution.

Thus it would be desirable to use crude methanol for higher alcohols, or to modify its manufacture from synthesis gas in such a way that longer-chain alcohols were produced along with the methanol.

1.2. CH_3OH as starting product for higher alcohols

1.2.1. by Co/Rh-catalyzed homologation:

$CH_3OH + nCO + 2nH_2 \rightarrow$
$CH_3(CH_2)_nOH + nH_2O$

1.2.2. by modified process ($CO/H_2 = 1$) and catalyst (composition and structure) parameters other than those customary for methanol synthesis

The first type of reaction is called homologation. Reaction with synthesis gas in the presence of Co- or Rh-containing catalysts leads to the formation of ethanol and higher alcohols.

The second variant is being developed by many firms, for example Haldor Topsoe and Lurgi (Octamix process). It is already possible to produce up to 17% of higher alcohols and other oxygen-containing products by altering the conditions in conventional methanol manufacturing, in particular, by lowering the CO/H_2 ratio to less than that required for methanol.

With a modified Cu catalyst, and in particular with the Octamix process at 300 °C, 50–100 bar and a CO/H_2 ratio approaching 1, it is possible to obtain a product (fuel methanol) that is about 50 wt% methanol and 45 wt% higher alcohols.

Neither of these processes has been practiced commercially.

1.3. CH_3OH as starting product for gasoline

Mobil MTG process, *e. g.*, methanol conversion to hydrocarbons on ZSM-5 zeolite with 0.6 nm pore diameter giving shape selectivity for HCs up to C_{10}

durene (1,2,4,5-tetramethylbenzene), undesirable in gasoline (mp. 79 °C), must be minimized

Methanol can also be dehydrated over a special zeolite catalyst (ZSM-5) in the MTG process (**m**ethanol **t**o **g**asoline) from Mobil R&D. This process produces a hydrocarbon mixture, most of which lies in the boiling range of gasoline.

In the first step, methanol is catalytically transformed into an equilibrium mixture with dimethyl ether and water. In the second step this is reacted over ZSM-5 zeolite catalyst at about 400 °C and 15 bar with further formation of water to yield a mixture of alkanes, aromatics, alkenes and cycloalkanes. In this way, 40 wt% N- and S-free gasoline with an octane number of 95 can be produced.

The catalyst must be regenerated after about three weeks by burning off the accumulated coke with air diluted with N_2.

industrial implementation of MTG process in fixed bed and in fluidized bed

One of the first commercial units using the Mobil process went into operation in New Zealand in 1986. This plant has an annual capacity of ca. 600 000 tonnes of gasoline, and uses methanol derived from natural gas.

fluidized-bed MTG process with advantages like:
lower pressure
reaction heat at higher temperature
continual ZSM-5 regeneration

A fluidized-bed version of the MTG process has been developed by Mobil/Union Kraftstoff/Uhde since 1983 in a 13 tonne-per-day unit. The methanol is transformed at 410–420 °C and 2–4 bar over a finely divided ZSM catalyst which flows upwards in the reactor. Part of the catalyst is diverted to a regenerator, where it is oxidatively regenerated.

1.4. CH_3OH transformed to $CH_3OC(CH_3)_3$ (MTBE) or CH_3O-*tert*-amyl (TAME) to raise octane number

A further possibility for using methanol in motor fuel results from the low-lead or lead-free gasoline. The necessary increase in the octane rating can be reached by adding, for example, methyl *tert*-butyl ether (MTBE) or methyl *tert*-amyl ether. These can be obtained by the acid-catalyzed addition of methanol to isobutene and isopentene, respectively (*cf.* Section 3.3.2).

Furthermore, MTBE and TAME have the advantage of being sparingly soluble in water so that, in contrast to methanol, there is no danger of a second water-rich phase forming in the gasoline tank. Recent tests have shown that MTBE brings about not only an improvement in the octane number but also a reduction in emission of CO and hydrocarbons.

There are two alternative forms in which methanol can be used as an energy carrier outside the motor fuel sector: it can easily be converted into SNG and fed into the natural gas distribution pipelines, or it can be used directly as "methyl fuel" in power stations to generate electricity.

2. CH_3OH as energy carrier by endothermic decomposition to SNG
$$4\ CH_3OH \rightarrow 3\ CH_4 + CO_2 + 2\ H_2O$$
or directly as "methyl fuel"

The development by Mobil of the zeolite-catalyzed transformation of methanol also shows the transition in its use from a fuel and energy carrier to a synthetic raw material.

3. CH_3OH as raw material for synthesis through zeolite-catalyzed reactions

By increasing the residence time and raising the temperature and pressure, the fraction of aromatics in the gasoline produced by the MTG process can be increased from the usual ca. 30 wt%. This provides an interesting route to aromatic hydrocarbons.

3.1. modified process conditions in MTG technology lead to aromatics

In contrast to the MTG process, if methanol or dimethyl ether is converted on a ZSM catalyst that has been doped with a metal such as Mn, Sb, Mg, or Ba, then olefins − predominantly ethylene and propene − are formed with a selectivity of up to 80%.

3.2. different metal doping leads to lower olefins

Processes of this type have been developed by, e.g., BASF, Hoechst, and Mobil.

Due to a number of advantages, methanol − along with higher alkanes, higher alcohols and methane − is a potential source of carbon for protein production (SCP = single cell protein).

4. CH_3OH as future C source for SCP

alternative basis for SCP:

gas oil
alkanes
ethanol
methane

Microorganisms, in particular yeasts and bacteria, can synthesize proteins from the above C sources in the presence of aqueous nutrient salt solutions containing the essential inorganic sulfur, phosphorus, and nitrogen compounds. The first industrial plants were built by BP in Scotland (4000 tonnes per year; basis, alkanes) and in France (16000 tonnes per year; basis, gas oil) for the production of the so-called alkane yeast (Toprina®). The French plant was converted from gas oil to an alkane basis (30000 tonnes per year), and then shut down in 1975. The unit in Scotland discontinued operation in 1978.

SCP manufacturing process based on alkanes:

BP, Dainippon Inc, Kanegafuchi

principles of the fermentation processes for SCP:

Alkanes are also carbon sources for SCP plants in operation in, for example, Rumania and the CIS. In 1990 the CIS produced ca. 1.3×10^6 tonnes of SCP.

1. cell build-up to proteins from organic C source and inorganic S, P, N salts (by-products: carbohydrates, fats, nucleic acids, vitamins)

   energy gained by degradation of substrate to CO_2

2. technological:

   multiphase system from
   aqueous solution
   organic basis
   air
   cellular substance

While gaining energy, microorganisms (*e. g.,* Candida yeast) degrade paraffins step by step to CO_2 and at the same time produce protein-rich cellular substances. Approximately one tonne of yeast with an amino acid pattern similar to fish meal results from one tonne alkane feed. Industrial SCP manufacture takes place in continuously operated fermentation apparatus under aseptic conditions with good mixing and heat removal. The cellular substances are continuously separated by centrifuging.

The microbiological degradation, starting with methane, passes through the following stages:

$$CH_4 \rightarrow CH_3OH \rightarrow HCHO \rightarrow HCOOH \rightarrow CO_2 \qquad (13)$$

advantages of CH_3OH:

to 1:

simple degradation without biological barrier $CH_4 \rightarrow CH_3OH$, lower O_2 requirement

Biologically, the most difficult step is from methane to methanol. For this reason, and because of the lower O_2 requirement, it is advantageous to start with methanol. Furthermore, methanol forms a homogeneous solution with the nutrient salt solution. Compared to the paraffins, the energy costs for stirring and aeration are less.

to 2:

due to H_2O solubility only three phases, better distribution and separation

Moreover, methanol can be economically manufactured in sufficient purity and easily separated from the bulk of the product after the fermentation process. According to results from ICI for the fermentation of methanol with a bacterial culture (pseudomonas), the dried cells consist of up to 81% protein. The protein contains a range of amino acids, in particular aspartic and glutamic acid, as well as leucine and alanine.

uses of single cell protein:

after being supplemented with missing essential amino acids, as an animal fodder additive; in the long term also for human nutrition

The optimal biological balance must be achieved by the addition of those amino acids which are insufficiently represented in the single cell proteins. Together with fish meal and soybeans, the single cell proteins will gain importance in increasing the protein content of fodder concentrate. At present, this is the only way in which SCP is being introduced into the food chain. In the long term, SCP will also have to be used for human nutrition as the present method of producing protein *via* animals leads to considerable loss of protein. For this purpose it is necessary to lower the nucleic acid content in the biomass, which can amount to 5−8% with yeasts and 10−22% with bacteria. A nucleic acid content of 1% in foods is regarded as the upper limit. This can only be achieved through special processing techniques, *e. g.,* extraction.

Besides ICI, who started operation of a commercial plant to produce 70 000 tonnes of SCP from 100 000 tonnes of methanol annually, but stopped production in 1987 for economic reasons, other companies such as Hoechst-Uhde, Phillips Petroleum, Shell, Mitsubishi Gas Chemical and Dor Chemicals in Israel either operate or are planning pilot plants with methanol as substrate. A newer goal is the production of flavor- or aroma-enhancing SCP types such as yeast-based Provesteen, which is produced by Phillips Petroleum in a plant with a production capacity of 1400 tonnes per year.

CH_3OH protein processes, e. g.,
ICI (Pruteen)
Hoechst-Uhde (Probion)
Phillips Petroleum (Provesteen)

2.3.2. Formaldehyde

At room temperature, formaldehyde is a colorless gas which readily polymerizes in the presence of minute amounts of impurities. Therefore, there are three commercial forms usually available in place of the monomer:

commercial forms of HCHO:

1. The aqueous $35-55\%$ solution in which over 99% of the formaldehyde is present as the hydrate or as a mixture of oxymethylene glycol oligomers.

1. hydrate, $HCHO \cdot H_2O$
$HO-CH_2-OH$
$H(OCH_2)_n OH$
$n < 10$

2. The cyclic trimeric form called trioxane obtained from the acid-catalyzed reaction of formaldehyde.

2. trioxane

3. The polymer form of formaldehyde called paraformaldehyde which results from the evaporation of aqueous formaldehyde solution and which can be reversibly converted to the monomer by the action of heat or acid.

3. paraformaldehyde
$H(OCH_2)_n OH$
$n > 10$

Methanol was first commercially dehydrogenated to formaldehyde in 1888, and has evolved into the dominant feedstock worldwide.

HCHO feedstocks:

today: CH_3OH
earlier: also C_3/C_4 alkanes,
          CH_3OCH_3
future: perhaps CH_4

Formaldehyde has also been manufactured from the free radical oxidation of propane and butane; for example, this technology accounted for ca. 20% of the formaldehyde produced in the USA in 1978. This unselective process is no longer practiced.

In Japan, dimethyl ether was also oxidized to formaldehyde for several years.

Despite extensive research, the partial oxidation of methane has not been successful. The slow rate of reaction of methane below 600 °C and the high rate of decomposition of formaldehyde above 600 °C require process conditions with extremely short residence times that are an obstacle to commercial application.

selective oxidation of CH_4 to HCHO only possible at temperatures which trigger HCHO decomposition

HCHO production (in 10^6 tonnes):

	1989	1991	1992
W. Europe	1.30	1.37	1.40
USA	0.78	0.81	0.86
Japan	0.36	0.39	0.38

The production figures for formaldehyde (as the pure chemical) in several industrialized countries are listed in the adjacent table. The worldwide production capacity for formaldehyde was about 4.9×10^6 tonnes in 1995 including 1.3, 2.2, and 0.51×10^6 tonnes per year in the USA, Western Europe, and Japan, respectively. Single plants have a capacity of up to 200 000 tonnes per year. Borden is the largest producer of formaldehyde worldwide.

2.3.2.1. Formaldehyde from Methanol

two principles for HCHO manufacture from CH_3OH

Formaldehyde can be manufactured from methanol *via* two different reaction routes:

1. dehydrogenation or oxydehydrogenation over Ag or Cu

1. Dehydrogenation or oxidative dehydrogenation in the presence of Ag or Cu catalysts.

2. oxidation over $MoO_3 + Fe_2O_3$

2. Oxidation in the presence of Fe-containing MoO_3 catalysts.

characteristic differences because of explosion range for CH_3OH/air between 6.7 – 36.5 vol% CH_3OH

The oxidative dehydrogenation process with Ag or Cu metal is operated above the explosion limit with a small quantity of air.

to 1:

small amount of O_2, *i.e.*, CH_3OH content > 36.5 vol%

In the oxidation process, a small amount of methanol is reacted below the explosion limit with a large excess of air. The thermal energy balance of the reaction is essential for process design in this manufacturing method.

to 2:

O_2 excess, *i.e.*, CH_3OH content < 6.7 vol%

To 1:

1. HCHO manufacture by CH_3OH dehydrogenation or oxydehydrogenation

Ag catalysts are preferred for the dehydrogenation and oxidative dehydrogenation of methanol. In the BASF, Bayer, Borden, Celanese, Degussa, Du Pont, ICI, Mitsubishi Gas, Mitsui Toatsu, and Monsanto processes the catalyst (silver crystals or gauze) is arranged as a few centimeter thick layer in the reactor. In the CdF Chimie process, silver is supported on carborundum.

catalyst mainly employed in three forms:

1. 0.5 – 3 mm crystals
2. gauze
3. impregnated SiC

In the initial step methanol is dehydrogenated:

mechanism of oxidative dehydrogenation:

primary dehydrogenation on metallic Ag or Cu

$$CH_3OH \rightleftharpoons HCHO + H_2 \left(\Delta H = +\frac{20 \text{ kcal}}{84 \text{ kJ}}/\text{mol} \right) \qquad (14)$$

secondary H_2 combustion with O_2

Hydrogen can be combusted exothermically on addition of air $\left(\Delta H = -\frac{58 \text{ kcal}}{243 \text{ kJ}}/\text{mol} \right)$, resulting in the following formal equation for the oxidative dehydrogenation:

$$CH_3OH + 0.5\,O_2 \rightarrow HCHO + H_2O$$

$$\left(\Delta H = -\begin{smallmatrix}38\ \text{kcal}\\159\ \text{kJ}\end{smallmatrix}\Big/\text{mol}\right) \quad (15)$$

A less-than-stoichiometric amount of air is employed in the industrial process. The air is fed so that the reaction temperature remains constant ($\pm 5\,°C$) in the range $600-720\,°C$. At temperatures of about $600-650\,°C$, conversion of methanol is incomplete, and a methanol recycle is necessary. At higher temperatures of $680-720\,°C$ and with the addition of H_2O, there is almost complete conversion of the methanol in a single pass. The water has another favorable effect on the life of the Ag catalyst in that the steam delays the deactivation caused by sintering of the thin layer of red-hot silver crystals. The catalyst has a lifetime of $2-4$ months and the spent catalyst can be easily regenerated electrolytically without loss of Ag. The catalyst is sensitive to traces of other metals as well as halogens and sulfur.

The hot gases from the reaction are very quickly cooled to ca. $150\,°C$ and washed in a counterflow with H_2O in several absorption stages. The solution is stabilized towards polymerization with a residual amount of methanol ($1-2$ wt%). A distillation can follow to produce concentrated formaldehyde solutions (37 to 42 wt%). The yield of formaldehyde exceeds 92% with a selectivity* of over 98%.

The byproducts are CO and CO_2. There is virtually no formic acid present.

To 2:

In the oxidation process, the formaldehyde formation occurs practically as a pure methanol oxidation (eq 15). A mixture of $18-19$ wt% Fe_2O_3 and $81-82$ wt% MoO_3 is employed as catalyst. Under carefully controlled conditions, it is converted into the catalytically active iron(III)-molybdate.

Excess MoO_3 is frequently added in order to make up for losses resulting from formation of molybdenum blue. This compound goes to the cooler end of the catalyst bed where it lowers both the catalytic activity and selectivity. Cr and Co oxide can be used as promoters.

process characteristics:

control of oxidation with O_2 feed (adiabatic method) without external heat supply or removal at $600-720\,°C$, *i.e.,* red-hot Ag

H_2O addition has several effects:

1. increases CH_3OH conversion
2. interferes with Ag recrystallization
3. lowers the C deposition on Ag surfaces

therefore, increase in lifetime of catalyst

Ag reactivation by electrolytic regeneration:

Ag dissolved at anode and deposited at cathode

thermal decomposition of HCHO to CO + H_2 is suppressed through:

1. short residence time in thin catalyst layer
2. rapid cooling ($0.1-0.3$ s) to $150\,°C$

2. HCHO manufacture by CH_3OH oxidation:

catalyzed by metal oxides based on $Fe_2O_3-MoO_3$, possibly with Cr_2O_3 or CoO as promoters

effective catalyst:

$Fe_2(MoO_4)_3$ sensitive to heat due to Mo_2O_5 discharge. Excess MoO_3 replenishes losses and avoids drop in activity

---

*) *cf.* "Definitions of Conversion, Selectivity and Yield", Section 15.2.

process characteristics of the oxidative HCHO manufacture:

advantages:

1. low temperature reduces problems with materials
2. higher HCHO concentration attainable directly without subsequent distillation

disadvantages:

great excess of air means higher investment and energy costs compared to Ag process, waste gas with HCHO noncombustible and requires special HCHO removal treatment

In the industrial process, methanol vapor together with a large excess of air is passed over the catalyst in a tubular reactor at $350-450\,°C$. The heat of reaction is removed by a liquid which surrounds the tubes. After cooling to $100\,°C$ the gases from the reaction are scrubbed with H_2O in a bubble column. By adjusting the amount of water, the concentration of the formaldehyde solution can be varied between 37 and 50 wt%, or, with a new development from Nippon Kasei, up to a maximum of 55 wt%.

An aqueous urea solution can also be used in the column to absorb the formaldehyde and produce urea-formaldehyde precondensates, which can be converted to thermosetting resins.

The methanol conversion is roughly $95-99\%$ and the formaldehyde selectivity reaches $91-94\%$.

The byproducts are CO, CO_2 and formic acid. The formic acid can be removed in a coupled ion exchanger.

lifetime determined by mechanical stability of the catalyst pellets

The lifetime of the catalyst is roughly two years.

The Perstorp-Reichhold ('Formox'), Hiag-Lurgi, Montecatini, SBA, IFP-CdF Haldor Topsoe, Nippon Kasei, and Lummus processes were developed in accordance with this principle.

2.3.2.2. Uses and Potential Uses of Formaldehyde

HCHO applications:

1. formaldehyde polycondensation products such as phenolic resins (novolacs, resols, Bakelite), amino resins, and others

Apart from the direct applications of aqueous formaldehyde solutions (Formalin®, Formol®), *e. g.*, as disinfectant and preservative, and as an auxiliary agent in the textile, leather, fur, paper, and wood industries, most formaldehyde is used for resin manufacture with urea, phenols, and melamine. Polycondensation products currently consume, depending on the country, between 40 and 65% of formaldehyde produced (*cf.* Table 2−4):

Table 2−4. Use of formaldehyde (in %).

	World		USA		Western Europe		Japan	
	1984	1989	1984	1995	1984	1995	1984	1995
Urea resins	32	33	27	32	45	50	37	27
Phenolic resins	11	11	23	24	11	10	8	6
Melamine resins	4	4	4	4	6	6	6	6
Pentaerythritol	6	5	7	6	6	6	8	6
1,4-Butanediol	2	2	11	12	6	7	—	—
Methylenediisocyanate	2	3	4	6	3	6	2	5
Others*	43	42	24	16	23	15	39	50
Total use (in 10^6 tonnes)	3.3	3.9	0.71	1.0	1.1	1.6	0.32	0.51

* *e. g.,* trioxane/polyformaldehyde, hexamethylenetetramine, etc.

Water-free pure formaldehyde or its trimer (trioxane) can be used to manufacture high molecular thermoplastics (polyoxymethylene).

Furthermore, crossed aldol condensations with formaldehyde open a synthetic route to polyhydric alcohols such as pentaerythritol, trimethylolpropane and neopentyl glycol (*cf.* Section 8.3).

Moreover, formaldehyde is employed in the Reppe manufacture of butynediol (*cf.* Section 4.3), of isoprene *via* 4,4-dimethyl-1,3-dioxane (*cf.* Section 5.2.2), and of β-propiolactone (*cf.* Section 11.1.7.1).

2. trioxane for polyoxymethylene, used as engineering resin

3. aldol reactions:
   pentaerythritol, trimethylolpropane, neopentyl glycol

4. butynediol manufacture:
   $HOCH_2C{\equiv}CCH_2OH$

5. miscellaneous, including

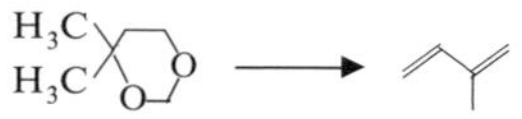

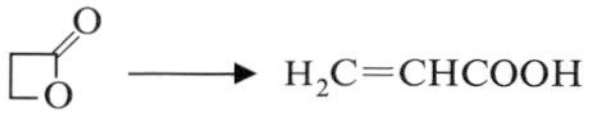

$$\text{(β-propiolactone)} \longrightarrow H_2C{=}CHCOOH$$

6. older, modified uses for formaldehyde can be extended

In the USA formaldehyde has also been employed in the manufacture of ethylene glycol. In this process, developed by Du Pont, glycolic acid is initially prepared from formaldehyde by hydrative carbonylation in the presence of sulfuric acid. After esterification, the product is hydrogenated to ethylene glycol:

three-step Du Pont glycol process:

1. hydrative carbonylation 200–250 °C, 300–700 bar, H_2SO_4
2. esterification with CH_3OH
3. hydrogenation

$$HCHO + CO + H_2O \xrightarrow{\text{[cat.]}} HOCH_2COOH$$
$$\xrightarrow[\text{2. }+H_2]{\text{1. }+CH_3OH} HOCH_2CH_2OH \qquad (16)$$

This Du Pont process for glycol manufacture was shut down in the 1960s. Only the production of glycolic acid (first stage) was continued to a limited extent (capacity ca. 60000 tonnes per year).

With the increased interest in extension of C_1 chemistry, new processes for the manufacture of glycolic acid and glycolaldehyde from formaldehyde have been developed by many companies.

new glycolic acid/glycolaldehyde production using HCHO as secondary product from CO/H_2 as a result of worldwide interest in C_1 chemistry

Catalysts such as HF, HF/BF_3, and acid exchangers (Chevron, Mitsubishi) have been suggested for the carbonylation of formaldehyde in the presence of water. These catalysts are already effective at low CO pressures of under 100 bar.

1. hydrative carbonylation of HCHO to $HOCH_2COOH$ in the presence of HF, HF/BF_3, or strongly acidic (perfluorinated) ion exchange resins

Another catalytic method has been worked out by Exxon. A 70% yield of glycolic acid can be obtained from formaldehyde and CO/H_2O at 150°C on a Nafion membrane, a comparatively thermally stable, strongly acidic ion exchanger made of a perfluorosulfonic acid resin.

2. hydroformylation of HCHO to HOCH$_2$CHO in homogeneous Co, Rh, or Ru phosphine ligand systems; *e. g.*, HRh(CO)$_2$[P(C$_6$H$_5$)$_3$]$_2$

Typical hydroformylation systems based on Co, Rh, or Ru with, e. g., phosphine ligands have been used by various firms (Ajinomoto, Mitsubishi, Chevron, National Distillers, Monsanto) for the synthesis of glycolaldehyde from formaldehyde/CO, H$_2$ at pressures of 50 – 350 bar in a homogeneous phase, i. e., generally in organic solvents.

None of these methods has been applied industrially, although production of ethylene glycol by hydrogenation of glycol aldehyde, glycolic acid, or glycolic acid ester stands a better chance of being realized than does direct manufacture from CO/H$_2$ (*cf.* Section 7.2.1.1).

Glycolic acid is used as a cleaning agent for boiler plants and pipelines, for the chelation of Ca and Fe ions in boiler feed water, for textile, leather, and fur processing, as well as a paint solvent (after esterification).

2.3.3. Formic Acid

industrial importance of formic acid:

applications arise from structure-based properties, *i.e.*, strong acid and reducing "hydroxyaldehyde"

Formic acid, the simplest carboxylic acid, is present – at times in remarkable amounts – not only in the animal and plant worlds but also as a constituent of the inanimate world. Its industrial significance is based on its properties as a carboxylic acid and its reducing properties as a hydroxyaldehyde.

HCOOH production (in 1000 tonnes):

	1991	1993	1995
W. Europe	186	194	205
USA	12	15	19
Japan	11	12	17

The world capacity for formic acid production in 1995 was about 390000 tonnes per year, of which in Western Europe, USA, and Japan were 218000, 20000, and 12000 tonnes per year, respectively. The largest producers worldwide are BASF, BP, and Kemira.

Production figures for these countries are given in the adjacent table.

HCOOH synthetic possibilities:

1. direct synthesis
2. undesired oxidative degradations

The numerous manufacturing processes for HCOOH can be divided into those in which formic acid is the main product of the process and those where it is obtained as a byproduct. At present, formic acid is primarily manufactured by direct synthesis.

HCOOH manufacturing process:

1. from CO + H$_2$O
2. from CO + ROH
   with subsequent hydrolysis

When manufacturing formic acid directly, one starts with CO, which is either hydrolyzed to HCOOH or reacted with alcohols to form formic acid esters:

$$CO + \begin{matrix} H_2O \\ ROH \end{matrix} \quad \overset{[cat.]}{\rightleftharpoons} \quad \begin{matrix} HCOOH \\ HCOOR \end{matrix} \tag{17}$$

In hydrolysis or alcoholysis, the above equilibrium is displaced towards the formate side by bases such as NaOH or Ca(OH)$_2$. Consequently, the process can be operated at 8–30 bar CO pressure and 115–150 °C. Synthesis gas can be employed instead of pure CO. The free HCOOH is obtained from its salts by acidification followed by distillation or extraction with, for example, diisopropyl ether. This process has been used since 1989 in a 40000 tonne-per-year plant in the CIS.

principle of process:

base-catalyzed reaction (under pressure) of CO (formally the anhydride of HCOOH) with H$_2$O to acid or with ROH to ester

The reaction of CO with alcohols, preferably methanol, is the first step in the most well-known production route for formic acid. This, in contrast to the carbonylation of methanol to acetic acid, can be formally regarded as the insertion of CO in the O–H bond of methanol, and takes place with catalytic amounts of sodium methylate at about 70 °C and 20–200 bar. With an excess of methanol, CO conversions of up to 95% with nearly 100% selectivity to methyl formate can be obtained.

characteristics of HCOOCH$_3$ hydrolysis: HCOOH catalyzed (*i.e.*, autocatalyzed) establishment of equilibrium, *e.g.*, in a single phase with HCOOCH$_3$/H$_2$O ratio of 2:1 to 4:1, or H$_2$O excess and rapid removal of CH$_3$OH by distillation (*i.e.*, minimum CH$_3$OH/HCOOH contact time)

The best known methyl formate processes practiced commercially have been developed by BASF, Halcon-SD, and Leonard. These are very similar in the carbonylation step, but differ from each other in the formic acid autocatalyzed hydrolysis at 80–140 °C and 3–18 bar. The largest differences are to be found in the operational procedures used to minimize the re-esterification of the product formic acid with the recycle methanol.

In order to prevent this re-esterification, a detour is generally made by synthesizing formamide by reaction of methyl formate with NH$_3$ at 80–100 °C and 4–6 bar. Formamide is then hydrolyzed.

variation of the direct HCOOCH$_3$ hydrolysis by ester ammonolysis to formamide followed by amide hydrolysis

The hydrolysis of formamide takes place continuously above 80 °C with 70% H$_2$SO$_4$ to form HCOOH and (NH$_4$)$_2$SO$_4$:

characteristics of the formamide hydrolysis:

stoichiometric amount of H$_2$SO$_4$ consumed forming (NH$_4$)$_2$SO$_4$

$$CH_3OH + CO \rightarrow HCOOCH_3 \xrightarrow{\ NH_3\ } HCONH_2 + CH_3OH$$
$$\xrightarrow{\ H_2SO_4\ } HCOOH + (NH_4)_2SO_4 \tag{18}$$

The separation of the reaction product takes place in a drum-type furnace. The purification of the stripped acid is carried out using a stainless steel or polypropene column with a silver or graphite condenser.

A process of this type was developed by BASF and operated commercially until 1982, when it was replaced by direct hydrolysis.

newer HCOOCH$_3$ production by CH$_3$OH dehydrogenation

A newer production route for methyl formate has been developed by Mitsubishi. In this method, methanol is dehydrogenated to methyl formate in the gas phase with a Cu catalyst under conditions that have not yet been disclosed:

$$2 \; CH_3OH \; \underset{}{\overset{[cat.]}{\rightleftharpoons}} \; HCOOCH_3 + 2 \; H_2 \qquad\qquad (19)$$

HCOOH separation as byproduct in nonselective processes

Formic acid is often formed, in uneconomically low concentrations, as a byproduct of undesired oxidative degradations.

For example, the separation of the limited quantities of HCOOH which result from the oxidation of acetaldehyde to acetic acid would require corrosion-resistant titanium columns.

oxidation of light naphtha to acetic acid (BP process) supplies HCOOH in economically recoverable concentrations

If, however, formic acid is present at ca. 18 wt% in the oxidized medium in addition to the main product acetic acid, as is the case in the oxidation of light naphtha or butane (*cf.* Section 7.4.1.2), then the distillative separation is worthwhile since the isolated formic acid contributes to the economics of the process.

HCOOH also coproduct in Cannizzaro reactions of HCHO (*e. g.*, CH$_3$CHO + HCHO to pentaerythritol)

Furthermore, formic acid can occur as a coproduct in the Cannizzaro reactions of formaldehyde (*cf.* Section 8.3.1).

HCOOH applications:

1. as free acid for silage manufacture, preservation, pH adjustment in tanks and pickling baths for various industries

The main use of HCOOH is in aiding lactic acid fermentation in the manufacture of green fodder silage for animal husbandry.

Today, it is only used in limited amounts for the conservation of food.

2. in the form of its salts in textile and leather processing. Na formate for synthesis of oxalic acid:

$$2 \; HCOONa \; \underset{melt}{\overset{NaOH}{\longrightarrow}} \; \begin{array}{l} COONa \\ | \\ COONa \end{array} + H_2$$

3. as alkylester for introduction of formyl group, *e. g.*, in industrial manufacture of vitamin B$_1$

4. as HCOOCH$_3$, prospective synthetic unit in C$_1$ chemistry for:

HOCH$_2$COOCH$_3$
HOCH$_2$CH$_2$OH
CH$_3$CH$_2$COOH
CH$_3$CHO
CH$_3$COOH
CH$_3$OCOOCH$_3$

Formic acid is also used for the acidification of dye and pickling baths as well as for purposes of disinfection. Its salts, such as Al and Na formate, are mainly employed in the leather and textile industries as auxiliary agents. Sodium formate is an intermediate in the manufacture of oxalic acid. Formic acid esters are employed in numerous organic syntheses. In addition, methyl formate is used as a solvent and as an insecticide.

In the future, methyl formate manufactured from synthesis gas could find additional applications within the framework of C$_1$ chemistry.

For example, the reaction of methyl formate with formaldehyde at 70–200 °C and atmospheric pressure in the presence of Brönstedt or Lewis acids yields methyl glycolate, which can be converted to ethylene glycol by, *e. g.*, Cu-catalyzed hydrogenation:

$$HCOOCH_3 + HCHO \xrightarrow{\text{[cat.]}} HOCH_2COOCH_3$$

$$\xrightarrow[+2\,H_2]{\text{[cat.]}} HOCH_2CH_2OH + CH_3OH \qquad (20)$$

Another example is the manufacture of methyl propionate from methyl formate and ethylene in polar solvents at $190-200\,°C$ in the presence of Ru or Ir complexes:

$$HCOOCH_3 + H_2C{=}CH_2 \xrightarrow{\text{[cat.]}} CH_3CH_2COOCH_3 \qquad (21)$$

Other methyl formate reactions of potential industrial interest are Rh- or Ir-catalyzed hydroisomerization to acetaldehyde, isomerization to acetic acid, and the Se-catalyzed oxidative reaction with methanol to dimethyl carbonate.

Formamide and its N-methyl derivatives are industrially versatile and – because of their polarity – important selective solvents and extracting agents (*cf.* Sections 12.2.2.2, 4.2.2, and 5.1.2). They are also useful as aprotic solvents for chemical reactions and as intermediates in various syntheses. Dimethylformamide is one of the few solvents suitable for the preparation of polyacrylonitrile solutions for synthetic fiber manufacture. The world production of dimethylformamide is estimated to be roughly 220000 tonnes per year. BASF is the largest producer. N-Methyl- and N,N-dimethylformamide are obtained – analogous to the ammonolysis of methyl formate (*cf.* eq 18) – from the reaction with methyl and dimethylamine, respectively:

5. amide and N-CH₃ derivatives as:

   polar solvents
   selective extracting agents
   aprotic reaction media
   intermediate products (*e. g.*, for Vilsmeier synthesis)

manufacture of amide and methylamides of HCOOH according to two processes:

1. ammonolysis or aminolysis of HCOOCH₃

$$HCOOCH_3 + \begin{matrix} NH_2CH_3 & HCONHCH_3 \\ \rightarrow & \\ NH(CH_3)_2 & HCON(CH_3)_2 \end{matrix} + CH_3OH \qquad (22)$$

Formamide, N-methyl- and N,N-dimethylformamide can also be synthesized directly from CO and NH₃, methyl- or dimethylamine in methanolic solution at $20-100$ bar and $80-100\,°C$ in the presence of alcoholates, as for example:

2. reaction of CO with NH₃, NH₂CH₃, NH(CH₃)₂

$$CO + NH(CH_3)_2 \xrightarrow{\text{[NaOCH}_3\text{]}} HCON(CH_3)_2 \qquad (23)$$

The yield of N,N-dimethylformamide, for instance, can be up to 95%. This route is used industrially in many plants, for example, in plants using the Leonard process.

2.3.4. Hydrocyanic Acid

Hydrocyanic acid (hydrogen cyanide) is an important synthetic unit in organic chemistry and is thus justifiably placed in the series of the C_1 basic products. The two processes below are suitable for the direct manufacture of hydrocyanic acid:

HCN manufacture according to two process principles:

1. dehydration of formamide
2. ammoxidation or dehydrogenation of various C compounds

1. Dehydration of formamide
2. Oxidative or dehydrogenative reaction of NH_3 with hydrocarbons, preferably with methane.

HCN recovery as byproduct in ammoxidation of propene

Hydrocyanic acid is also obtained to a large extent as a byproduct in the manufacture of acrylonitrile *via* the ammoxidation of propene (*cf.* Section 11.3.2).

The ratio of synthetic to byproduct hydrocyanic acid differs greatly from country to country; however, there is an overall tendency toward making the hydrocyanic acid supply more independent of the market for acrylonitrile by specific manufacture of hydrocyanic acid.

HCN byproduct capacity (in %):

	1995
USA	23
W. Europe	19
Japan	62

In 1995, the production capacity for synthetic and byproduct hydrocyanic acid in the USA, in Western Europe and in Japan was about 800000, 510000, and 90000 tonnes per year, respectively. The fraction of this which was byproduct is indicated in the adjacent table.

To 1:

$HCONH_2$ dehydration catalyzed with Fe, Al phosphates containing promoters, at reduced pressure and raised temperature; once common, now insignificant

The dehydration of formamide is conducted in iron catalyst tubes at $380-430\,°C$ under reduced pressure using modern vacuum processes. The tubes are filled with Fe or Al phosphate catalyst, which contains Mg, Ca, Z or Mn as promoters:

$$HCONH_2 \rightarrow HCN + H_2O \left(\Delta H = + \begin{array}{l} 18 \text{ kcal} \\ 75 \text{ kJ} \end{array} \Big/ \text{mol} \right) \tag{24}$$

The reaction gas, with its high HCN content of $60-70$ vol%, is suitable for direct liquefaction. The selectivity to HCN is $92-95\%$. Formamide processes were developed by BASF, Degussa and Knapsack, but today only BASF still operates a (21000 tonnes per year, 1991) unit.

To 2:

ammoxidation or ammono-dehydrogenation:

1. alkane + NH_3 + O_2 (−H_2O)
2. CH_4 + NH_3 (−H_2)

Synthesis components other than ammonia include methane (Andrussow and Degussa processes) and higher alkanes (Shawinigan, now Gulf Oil, process). The Andrussow technology — originally developed by BASF — is currently preferred for the manufacture of HCN. In principle, it is an ammoxidation of methane:

$$CH_4 + NH_3 + 1.5\,O_2 \xrightarrow{\text{[cat.]}} HCN + 3\,H_2O$$

$$\left(\Delta H = -\frac{113\ \text{kcal}}{473\ \text{kJ}}\Big/\text{mol}\right) \qquad (25)$$

The catalyst is usually platinum, either as a gauze or on a support, with additives such as rhodium. The reaction takes place at atmospheric pressure and 1000–1200 °C with a very short residence time. The reaction gas is rapidly quenched in order to avoid decomposition of HCN. After an acid wash, pure HCN is obtained by distillation from the diluted aqueous solution. Selectivity to HCN reaches about 88% (CH_4) and 90% (NH_3).

Numerous variations of the Andrussow process have been developed including those by American Cyanamid, Du Pont, Goodrich, ICI, Mitsubishi, Monsanto, Montecatini, Nippon Soda, and Rohm & Haas.

Methanol and formaldehyde have also been investigated for the manufacture of HCN by, *e. g.,* Sumitomo. The ammoxidation of methanol with Mo–Bi–P–oxide catalyst at 460 °C has a selectivity to HCN of 84% (CH_3OH).

In the Degussa BMA (**B**lausäure-**M**ethan-**A**mmoniak, *or* hydrocyanic acid–methane–ammonia) process, CH_4 and NH_3 are reacted without the addition of air or O_2:

$$CH_4 + NH_3 \xrightarrow{\text{[cat.]}} HCN + 3\,H_2$$

$$\left(\Delta H = +\frac{60\ \text{kcal}}{251\ \text{kJ}}\Big/\text{mol}\right) \qquad (26)$$

Sintered corundum tubes with a layer of Pt, Ru, or Al serve as catalyst. HCN selectivities of 90–91% (CH_4) and 83–84% (NH_3) are reached at 1200–1300 °C. The reaction of NH_3 is almost complete, and the small amount remaining is removed from the HCN/H_2 mixture with H_2SO_4. HCN is separated by absorption in water. Production plants are being operated by Degussa (Germany, Belgium, USA) and Lonza (Switzerland).

In the Shawinigan process, hydrocarbons from CH_4 to light petrol, *e. g.,* propane, are reacted with NH_3 at 1300–1600 °C in a fluidized bed of fine coke:

Andrussow process:

CH_4 ammoxidation with Pt–Rh gauze

process characteristics:

low thermal stress as short residence time avoids total oxidation of HCN, though low HCN concentration (6–7 vol%) due to inert gas

Degussa process:

CH_4 ammono-dehydrogenation on Pt, Ru, or Al coated α-Al_2O_3 tubes (corundum)

process characteristics:

O_2-free production leads to CO_2-free HCN (easier separation), H_2 is valuable byproduct, however more complicated reactor

Shawinigan process:

alkane ammono-dehydrogenation in coke fluidized bed

$$C_3H_8 + 3\,NH_3 \rightarrow 3\,HCN + 7\,H_2 \tag{27}$$

process characteristics:

high energy consumption because of electric heating, limits economic use

The fluidized bed is electrically heated by immersed graphite electrodes. Because of the high energy consumption, this process only operates economically in locations with a cheap source of electrical energy. With propane, a HCN selectivity of about 87% (C_3H_8) is attained.

This process was operated by Gulf Oil in Canada for several years, and there is still a commercial plant in South Africa.

ammoxidation process
(*e. g.*, Sohio process):

HCN merely undesired byproduct from the manufacture of acrylonitrile

Hydrocyanic acid is also produced as a byproduct in the ammoxidation of propene to acrylonitrile, as for example in the extensively applied Sohio process.

characteristics of HCN isolation as by-product:

inflexible connection to major product with respect to alterations in selectivity and production

Depending on the type of process, about 10–24 wt% of HCN is obtained relative to acrylonitrile. Improved catalysts (*e. g.*, Sohio catalyst 41) lead, however, to a noticeable decrease in HCN formation. Although hydrocyanic acid is then available as an inexpensive byproduct, the supply is too inflexible when coupled with the production of, and therefore the fluctuations in demand for, acrylonitrile.

uses of HCN:

1. HCN for C chain extension, *e. g.*,

$$(CH_3)_2CO \rightarrow (CH_3)_2C\begin{array}{c}OH\\CN\end{array}$$

$$\rightarrow H_2C{=}C(CH_3)CN \rightarrow -COOH$$
$$H_2C{=}CH{-}CH{=}CH_2 \rightarrow NC(CH_2)_4CN$$

2. HCN for intermediate products, *e. g.*, for cyanogen chloride and its cyclic trimer cyanuric chloride (trichloro-*s*-triazine or 2,4,6-trichloro-1,3,5-triazine)

Hydrocyanic acid is largely used for the manufacture of methacrylonitrile and methacrylic esters *via* the cyanohydrin of acetone (*cf.* Section 11.1.4.2). Furthermore, the formation of adiponitrile *via* the hydrocyanation of butadiene is increasing in importance (*cf.* Section 10.2.1.1). Hydrocyanic acid can also be employed in the manufacture of methionine (*cf.* Section 11.1.6).

Cyanogen chloride is an important secondary product of hydrocyanic acid. Industrially, it is manufactured by the reaction of chlorine with hydrocyanic acid in aqueous solution at $20-40\,°C$. Cyanogen chloride is then separated as a gas from the resulting aqueous hydrochloric acid:

$$HCN + Cl_2 \rightarrow ClCN + HCl \left(\Delta H = -\begin{array}{c}21\ kcal\\89\ kJ\end{array}\Big/ mol\right) \tag{28}$$

cyanuric chloride manufacture:

catalytic exothermic gas-phase trimerization of ClCN

By far the most important industrial application of cyanogen chloride is the manufacture of its cyclic trimer, cyanuric chloride. The most frequently used synthetic route is the gas-phase trimerization (above $300\,°C$) of dried cyanogen chloride on pure activated charcoal, or occasionally on activated charcoal with metal salts as promoters, in a fixed or fluidized bed:

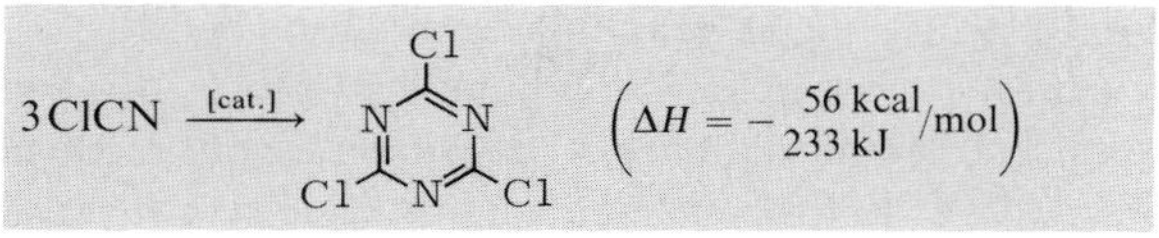

$$3\,ClCN \xrightarrow{\text{[cat.]}} \quad \left(\Delta H = -\frac{56\,\text{kcal}}{233\,\text{kJ}}/\text{mol}\right) \qquad (29)$$

Cyanuric chloride is obtained as a melt or dissolved in a solvent with a selectivity of 95% (ClCN). With a worldwide production of more than 100 000 tonnes per year, it is one of the quantitatively most significant, inexpensive, and versatile heterocycles.

Cyanuric chloride's dramatic growth is due to its application as precursor for 1,3,5-triazine herbicides, which consumes about 80% of the worldwide production. 2,4-Bis(ethylamino)-6-chloro-1,3,5-triazine (Simazin®) is a typical herbicide of this class. Triazine fungicides are considerably less important. More than 10% of the cyanuric chloride output goes toward the production of optical brighteners. In addition, cyanuric chloride is used to introduce a reactive group to chromophoric components, enabling dyes to be chemically bound to fibers ('substitution' dyes) *via* the reactivity of the chlorine atoms in the cyanuric part of the dye.

Cyanuric acid amide (melamine) can be synthesized in principle by reacting cyanuric chloride with ammonia.

For a long time, multistep industrial processes used only dicyandiamide (from calcium cyanamide) for melamine manufacture. Today, it is mainly manufactured from urea in a single-step process either in a melt at 90–150 bar and 380–450 °C, or at 1–10 bar and 350–400 °C over a modified aluminium oxide or aluminosilicate in a fixed-bed or fluidized-bed reactor:

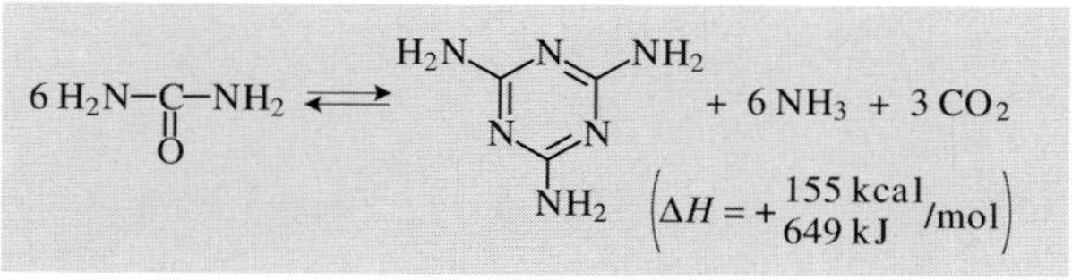

$$6\,H_2N-\overset{\text{O}}{\underset{\|}{C}}-NH_2 \rightleftharpoons \quad + 6\,NH_3 + 3\,CO_2 \qquad (30)$$

$$\left(\Delta H = +\frac{155\,\text{kcal}}{649\,\text{kJ}}/\text{mol}\right)$$

uses of cyanuric chloride:

triazine herbicides, *e. g.*, Simazin

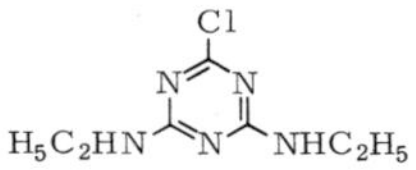

fungicides, reactive dyes, optical brighteners (melamine)

industrial melamine manufacture:

1. trimerization of dicyandiamide

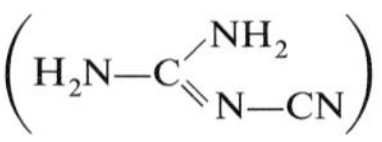

$$\left(H_2N-C\overset{NH_2}{\underset{N-CN}{\diagdown}}\right)$$

decreasing in importance
2. cyclization of urea with loss of CO_2/NH_3 increasing in importance worldwide

NH_3 and CO_2 which keep the catalyst in a fluidized state are subsequently reconverted to urea.

The melamine yield is more than 95% (urea). The worldwide capacity for melamine in 1994 was about 610 000 tonnes per year, with 240 000, 110 000 and 110 000 tonnes per year in Western Europe, Japan, and the USA, respectively. Production in these countries is reported in the adjacent table. The largest producer worldwide is DSM with a capacity of over 100 000 tonnes per year. Other low-pressure processes have been

Melamine production (in 1000 tonnes):

	1990	1992	1994
W. Europe	186	176	194
Japan	106	94	103
USA	86	98	111

developed by BASF and Chemie Linz. Montedison and Nissan, among others, have developed high-pressure processes. The major application of melamine is in polycondensation reactions with formaldehyde to form melamine resins, which can be used as thermosetting resins, glues, and adhesives. The areas of use are divided as follows:

Table 2-5. Use of melamine (in %).

Use	USA 1990	USA 1994	W. Europe 1988	W. Europe 1994	Japan 1988	Japan 1994
Molding compounds	24	32	54	53	56	59
Surface treatments	36	38	9	10	13	15
Textile/paper auxiliary	22	10	9	2	4	3
Others	18	20	28	35	27	23
Total (in 1000 tonnes)	86	69	143	220	62	58

3. HCN for alkali cyanides and cyano complexes

Hydrocyanic acid is also used for the manufacture of alkali cyanides, *e. g.*, for cyanide leaching, and for cyano complexes.

4. HCN for oxamide with application as slow-release fertilizer (high N content with low H_2O solubility)

A new and interesting application of HCN has resulted from the synthesis of 'oxamide', the diamide of oxalic acid. In a process developed by Hoechst, HCN can be catalytically oxidatively dimerized with simultaneous hydration in a one-step reaction with O_2 and $Cu(NO_3)_2$ in aqueous organic solution. Yields are very high:

$$2\,HCN + 0.5\,O_2 + H_2O \xrightarrow{[Cu^{2\oplus}]} H_2N-\underset{\underset{O}{\|}}{C}-\underset{\underset{O}{\|}}{C}-NH_2 \tag{31}$$

$$\left(\Delta H = -\frac{105\,\text{kcal}}{440\,\text{kJ}}\Big/\text{mol}\right)$$

The first industrial use of this process, a 10000 tonne-per-year plant in Italy, has been in operation since 1990.

alternate oxamide technology (Ube) is ammonolysis of dialkyl oxalate

An alternate process for the manufacture of oxamide has been developed by Ube. In this process, methyl or *n*-butyl oxalates from the oxidative carbonylation of CH_3OH or n-C_4H_9OH (*cf.* Section 7.2.1.1) are converted to oxamide and the corresponding alcohol by ammonolysis. Ube began commercial operation of this process in 1981.

Because of its low water solubility, oxamide is employed as a slow-release fertilizer.

2.3.5. Methylamines

The methylamines assume about fifth place, in quantitative terms, amongst the secondary products of methanol. Therefore, the preferred commercial manufacturing process — the stepwise methylation of NH_3 with CH_3OH — is well characterized.

importance of methylamines:

secondary products of CH_3OH, after HCHO, AcOH, methyl halides, MTBE; currently about 3–5% of CH_3OH usage

The traditional manufacturers of methylamines are BASF, ICI, Montedison, and recently UCB in Europe, whereas Du Pont and Rohm & Haas are the main suppliers in the USA. The present production capacity for methylamines is estimated to be about 600 000 tonnes per year worldwide, with 170 000, 160 000, and 45 000 tonnes per year in the USA, Western Europe, and Japan, respectively.

In commercial production, methanol and NH_3 are reacted together at 350–500 °C and 15–30 bar in the presence of aluminum oxide, silicate or phosphate:

manufacture of methylamines:

stepwise NH_3 methylation with CH_3OH

$$CH_3OH + NH_3 \xrightarrow{\text{[cat.]}} CH_3NH_2 + H_2O$$

$$\left(\Delta H = -\frac{5 \text{ kcal}}{21 \text{ kJ}}/\text{mol} \right) \qquad (32)$$

As pressure has only an insignificant effect on the course of reaction, it is usually conducted at about 20 bar in accordance with industrial requirements.

The alkylation does not stop at the monomethylamine stage; because of the thermodynamic equilibria, all three methylamines are obtained together. The mono- and dialkylation are favored by excess NH_3 and by the addition of H_2O together with the recycling of the trimethylamine. At 500 °C with a NH_3/methanol ratio of 2.4:1, for example, 54% mono-, 26% di-, and 20% trimethylamine are obtained. Side reactions include cleavage to produce CO, CO_2, CH_4, H_2, and N_2. The total selectivity to methylamines reaches about 94%.

methylation with CH_3OH using dehydration catalyst such as Al_2O_3
$Al_2O_3 \cdot SiO_2$
$AlPO_4$

mono- and dimethylation are favored by:

1. $NH_3 : CH_3OH > 1$
2. addition of H_2O
3. recycling of $(CH_3)_3N$
   (partial cleavage with H_2O)

Due to azeotrope formation, the reaction products are separated in a combination of pressure and extractive distillations.

isolation of the methylamines:

combination of pressure and extractive distillations

An important commercial process for methylamines is the Leonard process, with a worldwide capacity of more than 270 000 tonnes per year (1993).

In a process carried out by Nitto Chemical, the equilibrium of formation can be shifted by use of an acid zeolite catalyst with a particular pore structure, allowing up to 86 mol% dimethyl-

because of their structural (shape) selectivity, zeolites allow for specific production of dimethylamine

applications of methylamines:

$(CH_3)_2NH$ for $HCON(CH_3)_2$

CH_3NH_2 for $O=C\overset{\displaystyle NHCH_3}{\underset{\displaystyle NHCH_3}{}}$

$(CH_3)_3\overset{\oplus}{N}$ for $[HOCH_2CH_2\overset{\oplus}{N}(CH_3)_3]Cl^{\ominus}$
possibly for

$(CH_3)_2NCH_3 + CO \rightarrow (CH_3)_2NCCH_3$
$\qquad\qquad\qquad\qquad\qquad\overset{\displaystyle \|}{O}$

industrially important halogens and halogen combinations in methane derivatives:

1. Cl
2. F, Cl
3. F, Br
4. F, Cl, Br

chloromethane capacity, 1994 (in 1000 tonnes)

	CH_3Cl	CH_2Cl_2	$CHCl_3$	CCl_4
World	862	738	626	710
USA	327	277	232	70
W. Europe	307	340	288	321
Japan	73	70	78	69

chloromethane production, 1993 (in 1000 tonnes)

	CH_3Cl	CH_2Cl_2	$CHCl_3$	CCl_4
USA	295	237	247	182
W. Europe	274	161	226	140
Japan	106	86	53	40

amine as well as 7 mol% each of mono- and trimethylamine to be generated. A commercial plant has been in operation in Japan since 1984.

The three methylamines are important intermediates for the manufacture of solvents, insecticides, herbicides, pharmaceuticals, and detergents. The demand for the individual methylamines has developed independent of one another. Quantitatively, dimethylamine is the most important due to its use in the manufacture of N,N-dimethylformamide (*cf.* Section 2.3.3) and N,N-dimethylacetamide, which find wide application as solvents. Methylamine ranks second in terms of demand. It is mainly used in the further reaction to dimethyl urea and N-methylpyrrolidone (*cf.* Section 4.3), as well as for methyltaurine which is employed in CO_2 washes or as a raw material for detergents.

Trimethylamine plays only a minor role; it is, for example, used in the manufacture of choline chloride. This lack of need for trimethylamine is illustrated by the attempts to convert it into N,N-dimethylacetamide *via* carbonylation. There has been no industrial application so far.

2.3.6. Halogen Derivatives of Methane

Chlorine and fluorine are the most important industrial halogens for a partial or complete substitution of the hydrogen atoms in methane. Fluorine-containing methane derivatives usually also contain chlorine. Bromine occurs in only a few fluorine or fluorine- and chlorine-substituted methane derivatives which are of commercial interest.

2.3.6.1. Chloromethanes

The relative importance of the four chlorinated methanes can be readily appreciated from the manufacturing capacities worldwide, in the USA, Western Europe, and Japan. Data for 1994 are given in the adjacent table. Production figures for the USA, Western Europe, and Japan in 1993 are also given.

Chloromethanes are manufactured by two different routes:

1. All four chlorinated derivatives are manufactured together *via* thermal chlorination or catalytic oxychlorination of methane.

2. Special processes and other raw materials are used for specific manufacture of CCl_4, the most commercially important product, and CH_3Cl, used as an intermediate for further chlorination or in other reactions.

manufacture of the chloromethanes according to two variations:

1. CH_4 chlorination or oxychlorination leads to mixture of all chloroderivatives

2. alternative manufacturing routes for CCl_4 and CH_3Cl using different precursors

To 1:

The first industrial gas-phase chlorination of methane was performed by Hoechst in 1923. Today, their manufacturing capacity for chlorinated C_1 compounds is about 180000 tonnes per year. The strongly exothermic free radical reaction is conducted without external heat and usually in the absence of a catalyst (*i.e.*, without addition of radical forming substances) at 400–450 °C and slightly raised pressure. The chlorination is thermally initiated *via* the homolysis of chlorine molecules; it can, however, also be initiated photochemically.

principles of the CH_4 chlorination:

catalyst-free radical gas-phase chlorination in which product composition is determined by process conditions and CH_4/Cl_2 ratio

initiation of reaction *via* Cl_2 homolysis to 2 $Cl\cdot$, either thermally, photochemically, or with initiator

If methyl chloride is to be preferentially produced then a large excess of methane (about tenfold) must be used in order to obtain a satisfactory yield, as methyl chloride is more rapidly chlorinated than methane. On the other hand, when an equimolar Cl_2/CH_4 ratio is employed, all possible chlorinated methanes are formed together in the mole percents given:

$$CH_4 + Cl_2 \xrightarrow[-HCl]{440°C} \underset{37}{CH_3Cl} + \underset{41}{CH_2Cl_2} + \underset{19}{CHCl_3} + \underset{3}{CCl_4} \qquad (33)$$

Desired higher degrees of chlorination can be obtained by recycling the lower chlorinated products. During the treatment of the reaction mixture in the majority of the industrial processes, the resulting HCl is first scrubbed with water or with azeotropic hydrochloric acid. The chlorinated products are then condensed using a system of low temperature condensers, separated from CH_4, and isolated in pure form by distilling under pressure.

isolation of the chloromethanes:

1. HCl scrubbing with H_2O
2. condensation of the chloromethanes and CH_4 separation
3. chloromethanes purified by distillation

The byproducts are hexachloroethane and small quantities of trichloroethylene.

The selectivity to chlorinated C_1 products is more than 97%.

process problems in the CH_4 chlorination:

1. critical temperature level, *i.e.*, limited range between initiation temperature and decomposition temperature
2. high molar enthalpy per chlorination step

	kcal/mol	kJ/mol
CH_3Cl	24.7	103.5
CH_2Cl_2	24.5	102.5
$CHCl_3$	23.7	99.2
CCl_4	22.5	94.2

3. corrosive hydrochloric acid demands expensive construction materials

principle of the CH_4 oxychlorination:

catalytic reaction of CH_4 with HCl and O_2 in molten $CuCl_2/KCl$

first industrial operation:

Lummus 'Transcat' process in Japan (indirect oxychlorination)

process characteristics: two-step operation

1. CH_4 chlorination with the melt (reaction) as chlorine source
2. oxychlorination of the melt (regeneration) with catalytic effect of its components

alternative CCl_4 manufacture with different precursors, processes, and selectivities:

Asahi Glass, Dow, Hüls, Montecatini, and Scientific Design have all developed various modifications for industrial operation. These differ in the technological solutions to problems characteristic of the strongly exothermic chlorination of CH_4, which is first initiated at 250–270 °C and can proceed explosively in the industrially important temperature range of 350–550 °C.

Solutions include reactor construction with backmixing characteristics and heat removal (loop-type bubble column, Hoechst; fluidized-bed reactor, Asahi Glass; tubular reactor, C. F. Braun), high CH_4/Cl_2 ratio or addition of an inert gas (N_2, Montecatini), reaction temperature (thermal initiation of radical chains with most manufacturers, photochemical initiation *via* UV irradiation at Dow), and the workup of the reaction products (1. HCl removal, then 2. pressurized distillation, Hoechst and Hüls; or 1. CH_3Cl/CCl_4 extraction of the reaction gases, then 2. HCl removal by scrubbing, Dow).

The oxychlorination of methane is a second route to the manufacture of a mixture of all the chlorinated methane products. Direct oxychlorination has not yet been used commercially; however, a process which can be viewed as an indirect oxychlorination was developed by Lummus and put on stream in a 30000 tonne-per-year unit by Shinetsu in Japan in 1975. A modified version is also suitable for the manufacture of vinyl chloride (*cf.* Section 9.1.1.3).

The process operates with a melt consisting of $CuCl_2$ and KCl, which acts simultaneously as catalyst and as chlorine source. The melt first chlorinates methane to the four chloromethanes and is subsequently fed into an oxidation reactor, where it is rechlorinated in an oxychlorination – also known as oxyhydrochlorination – reaction with hydrogen chloride or hydrochloric acid and air. More detailed process conditions are not known to date. This process enables the utilization of the waste product (hydrochloric acid) in accordance with the following equation:

$$CH_4 + HCl + O_2 \rightarrow CH_3Cl + \text{higher chloromethanes} + H_2O \quad (34)$$

To 2:

There are four methods available for a direct synthesis of carbon tetrachloride. They can be readily characterized by the very different precursors required:

1. Carbon disulfide
2. Propane–propene mixtures
3. Organic residues containing chlorine
4. Elemental carbon, *e. g.*, low temperature coke

1. CS_2 route
2. C_3 chlorinolysis
3. chlorinolysis of residues containing Cl
4. elemental synthesis

To 2.1:

In several countries such as the USA, Italy, United Kingdom, and Mexico, carbon disulfide is chlorinated to CCl_4 in the liquid phase at 30 °C and atmospheric pressure in the presence of metallic iron, $FeCl_3$, or without catalyst. CS_2 was the only carbon source used for CCl_4 until the 1950s, when chlorination of methane and chlorinolysis of hydrocarbons were introduced as new sources of CCl_4. Today, about 25% of the production worldwide and 30% in the USA (1990) is still based on CS_2. When stoichiometric amounts of chlorine are used, the by-product is sulfur, which can be recycled and used for the manufacture of CS_2. Sulfur monochloride, which is obtained with excess chlorine, is also of interest industrially; it can also be reacted with CS_2 to give CCl_4 and sulfur:

principles of the CS_2 route:

Fe catalyst, exchange of S for Cl with equilibrium displacement due to crystallization of sulfur

$$CS_2 + \begin{matrix} 2\,Cl_2 \\ 3\,Cl_2 \end{matrix} \rightleftarrows CCl_4 + \begin{matrix} 2\,S \\ S_2Cl_2 \end{matrix} \qquad (35)$$

The CCl_4 selectivities are 90% (CS_2) and 80% (Cl_2).

To 2.2:

Propane–propene mixtures can be converted into the C_1 and C_2 fragments carbon tetrachloride and perchloroethylene *via* cracking coupled with chlorination (chlorinolysis) at 600 – 700 °C and 2 – 5 bar (*cf.* Section 9.1.4):

principles of C_3 chlorinolysis:

gas-phase cracking of C_3 coupled with chlorination forming perchlorinated C_1 + C_2 fragments (low-pressure chlorinolysis)

$$C_3H_6 + Cl_2 \rightarrow CCl_4 + Cl_2C{=}CCl_2 + HCl \qquad (36)$$

The ratio of CCl_4 to perchloroethylene can be varied between 65 : 35 and 35 : 65 depending on reaction conditions and the ratio of the starting materials. The selectivities for both products are about 90% (C_3H_6 and Cl_2). Industrially operated processes were developed by Progil-Electrochimie and Scientific Design.

Many plants are in operation in Western Europe.

process characteristics:

pressure- and temperature-dependent shift of the equilibrium

$2\,CCl_4 \rightleftarrows Cl_2C{=}CCl_2 + 2\,Cl_2$

determines product ratio

principles of residue chlorinolysis:

gas-phase cracking of chlorinated hydrocarbons with simultaneous chlorination, forming mixtures of CCl_4, $Cl_2C=CCl_2$ and $Cl_2C=CHCl$, or CCl_4 only with two variants:

1. catalytic oxychlorination with HCl/O_2 as chlorine source
2. catalyst-free chlorination directly with Cl_2 (high pressure chlorinolysis)

To 2.3:

The most economically interesting feedstocks for the manufacture of carbon tetrachloride *via* chlorinolysis are chlorine-containing organic residues. Particularly suitable residues (due to their high chlorine content) result, for example, from the chlorination of methane, manufacture of vinyl chloride, allyl chloride, and chlorobenzene, and from propylene oxide *via* chlorohydrin.

The chlorine required for the chlorinolysis can be introduced into the process either as HCl/air, *e. g.*, as in the PPG oxychlorination process, or preferably directly as elemental chlorine. Numerous companies (Diamond Shamrock, Stauffer Chemical, Hoechst, *etc.*) have developed processes of the latter type.

Important process variables such as pressure, temperature, residence time, and the Cl_2/hydrocarbon ratio determine the selectivity of the chlorinolysis, *i. e.*, whether carbon tetrachloride is formed exclusively or, instead, mixtures of CCl_4, $Cl_2C=CCl_2$, and $Cl_2C=CHCl$ are produced.

process example of a residue chlorinolysis:

Hoechst CCl_4 process

process characteristics:

high temperature   (600 °C)
high pressure   (up to 200 bar)
short residence time (ca. 1 min)
no byproducts

Almost 100% selectivity to CCl_4 is obtained with the chlorinolysis process at $120-200$ bar and 600 °C developed by Hoechst. Short residence times and high pressure prevent the formation of an equilibrium between CCl_4 and perchloroethylene which would otherwise occur. With aromatic residues as feedstock, hexachlorobenzene is formed as an isolable intermediate, and with aliphatic residues the corresponding intermediate is hexachloroethane. Hoechst brought one plant on stream in West Germany in early 1976, and another in the CIS in 1984.

To 2.4:

principle of the synthesis from elements:

chlorination of coke at high temperatures

Coal, with its low H content, would be an interesting feedstock for the manufacture of carbon tetrachloride as very little chlorine would be lost by the simultaneous formation of HCl. However, the principal disadvantage lies with the low reactivity of the coal which necessitates temperatures of about 800 °C. Processes for the direct chlorination of carbon have been frequently described, but have not been developed beyond the pilot plant stage.

process characteristics:

low HCl formation but also low reactivity

alternative CH_3Cl manufacture *via* selective uncatalyzed or catalyzed esterification of CH_3OH with HCl in liquid or gas phase (hydrochlorination)

Besides the chlorination of methane, methyl chloride can also be prepared by esterification of methanol with hydrogen chloride. This reaction is performed either in the liquid phase at $100-150$ °C with a slight excess pressure, either uncatalytically or in the presence of *e. g.*, $ZnCl_2$ or $FeCl_3$, or, preferentially, in the gas phase at 300–380 °C and 3–6 bar, with a catalyst like $ZnCl_2$, $CuCl_2$, or H_3PO_4 on a support such as SiO_2, or with Al_2O_3, in a fixed or fluidized bed:

$$CH_3OH + HCl \rightleftharpoons CH_3Cl + H_2O \quad \left(\Delta H = - \frac{8 \text{ kcal}}{33 \text{ kJ}} \big/ \text{mol} \right) \quad (37)$$

This reaction has a high selectivity, almost 98 % relative to CH_3OH. Only a small amount of dimethyl ether is formed as byproduct.

Today, with low-cost methanol and the growing surplus of HCl from numerous chlorination processes, hydrochlorination of methanol has become the most important commercial route to methyl chloride.

Methyl chloride can also be used instead of methane for the synthesis of higher chloromethanes. The thermal gas-phase reaction is preferred, although since 1979 a catalytic liquid-phase reaction at about 100 °C developed by Tokuyama Soda Co. has also been employed. The resulting hydrochloric acid can be employed in a co-process for the esterification of methanol.

CH_3Cl chlorination to higher chloro-methanes makes coupled processes use the HCl formed in the chlorination for CH_3OH esterification

All chlorinated methanes are widely used as solvents. Apart from their good dissolving power, they possess the additional advantage of being nonflammable (except CH_3Cl). However, due to their toxicity, special precautions must be taken when using them. Moreover, their significance as intermediates or reaction components is steadily increasing in various fields. For example, methyl chloride is used for the methylation and etherification of phenols, alcohols and cellulose, and it continues to be used in the manufacture of methylchlorosilanes (silicones), tetramethyllead (TML), and quarternary ammonium salts. CCl_4 and $CHCl_3$ are mainly used as feedstocks for chlorofluoromethanes, but their use for this purpose will decrease considerably due to the worldwide limitation, or total cessation, of the production of chlorofluorocarbons.

uses of chlorinated methanes:

1. solvents
2. intermediate products, *e. g.*,

2.1. CH_3Cl for methylations
   + Si (Rochow process) →
   $(CH_3)_2SiCl_2$ (silicones)

   + Pb → $(CH_3)_4Pb$ (next to $(C_2H_5)_4Pb$ important antiknocking agent)

   + Na cellulose → methylcellulose

2.2. CCl_4 and $CHCl_3$ for synthesis of hydro-carbons containing F and Cl

2.3.6.2. Chlorofluoromethanes

In the industrial processes for the manufacture of chlorofluoro-methanes, suitable chloromethanes are generally used as starting materials. The chlorine is then replaced stepwise by fluorine using HF. Thus, with increasingly vigorous reaction conditions (amount of HF, temperature, pressure, amount and type of catalyst) the series $CFCl_3$, CF_2Cl_2, and CF_3Cl can be prepared from CCl_4. Similarly $CHFCl_2$, CHF_2Cl, and CHF_3 can be synthesized from $CHCl_3$. The chlorofluorohydrocarbons are designated industrially using a numerical code which consists of

principles of the manufacture of hydro-carbons containing F and Cl:

stepwise substitution of Cl for F in chloro-hydrocarbons

$$-\overset{|}{\underset{|}{C}}-Cl + HF \rightarrow -\overset{|}{\underset{|}{C}}-F + HCl$$

Cl/F exchange leads from CCl_4 and $CHCl_3$ to two industrially important series of Cl, F-methanes

two numbers for the chlorofluoromethanes, and three numbers for the higher alkanes. The last digit is the number of fluorine atoms, the next-to-the-last digit is one more than the number of hydrogen atoms, and the first digit is one less than the number of carbon atoms in the molecule; any remaining atoms are chlorine.

The highest volume compounds CF_2Cl_2 and $CFCl_3$ are designated 12 and 11, respectively. As methane derivatives, they are denoted by only two numbers.

manufacturing process for Cl, F-methanes, from Cl methanes + HF, with two modifications:

1. heterogeneously-catalyzed gas-phase reaction with Al–Cr fluorides

2. homogeneously-catalyzed liquid-phase reaction with Sb fluorides

The manufacture of the chlorofluoromethanes takes place in a catalytic reaction, e. g., CCl_4 with HF in the gas phase at 150 °C on fixed-bed catalysts consisting of Al fluoride, Cr fluoride or Cr oxyfluorides.

Halogen exchange can also take place in the liquid phase in stainless steel-lined autoclaves. Pressure of $2-5$ bar is applied and the reaction is catalyzed with Sb fluorides at ca. 100 °C.

$$CCl_4 + HF \xrightarrow{\text{[cat.]}} CF_2Cl_2 + CFCl_3 + HCl \qquad (38)$$

important variable:

ratio CCl_4/HF determines the ratio of the reaction products

The gaseous reaction products are then freed from the bulk of the hydrogen chloride and washed with alkali solution until acid-free. After drying, the gas is liquified by compression and fractionally distilled at a pressure of $6-8$ bar. The reaction is virtually quantitative.

The whole process, including the carbon tetrachloride manufacture *via*, for example, the chlorination of methane or chlorinolysis, is regarded as being two-step.

single-step process possible (Montedison) *via* combination of:

1. chlorination of CH_4 to CCl_4 by substitution and simultaneous
2. halogen exchange forming Cl, F-methanes

A single-step process for the manufacture of CF_2Cl_2 and $CFCl_3$ was developed by Montedison and first operated (1969) in Italy. The plant has been expanded since then to 13 000 tonnes per year. In this process, the chlorination and fluorination of methane occur simultaneously at 370–470 °C and 4–6 bar on a fluidized-bed catalyst:

$$CH_4 + Cl_2 + HF \xrightarrow{\text{[cat.]}} CF_2Cl_2 + CFCl_3 + HCl \qquad (39)$$

The selectivity to the combined fluorine compounds is 99% (CH_4), 97% (Cl_2), and 94% (HF).

After separating the HCl and recovering the unreacted HF, followed by washing with water and drying, the products are separated in two stages by distillation. Byproducts such as CCl_4 and perchloroethylene are fed back into the reaction.

With increasing fluorine content, the fluoro- and chlorofluoro-derivatives of methane and ethane possess high thermal and chemical stability. Furthermore, they are nonflammable and nontoxic. These important properties have contributed to their widespread use as propellants for aerosols, as spray and foam agents, as refrigerants in refrigerating plants, as well as solvents for chemical dry cleaning (*cf.* Table 2–6):

treatment of fluorinated mixtures:

1. separation of HCl
2. recovery of HF
3. washing and drying
4. distillation

applications of Cl, F-alkanes:

1. propellants and refrigerants
2. solvents (chemical dry cleaning)
3. intermediate products, *e. g.*, for tetrafluoroethylene perfluoropropene Br, F-alkanes

Table 2–6. Industrial applications of important Cl-, F-alkanes.

Compound	Code CFC	Applications			
		Refri-gerant	Pro-pellant	Dry Cleaning solvent	Intermediate product for
$CFCl_3$	11	+	+	+	
CF_2Cl_2	12	+	+		
CF_3Cl	13	+			
$CHFCl_2$	21	+			
CHF_2Cl	22	+	+		$F_2C=CF_2$
CHF_3	23				CF_3Br
$CF_2Cl—CFCl_2$	113	+		+	$F_2C=CFCl$
$CF_2Cl—CF_2Cl$	114	+	+		

The chlorofluoroalkanes are known in Germany by the trade names Frigen® (Hoechst) and Kaltron® (Kalichemie). In the USA, the trade names are Freon® (Du Pont), Genetron® (Allied), Isotron® (Pensalt Chem. Equipment) and Ucon® (UCC).

Production figures for the most important fluorocarbons 11 and 12 are presented for the USA in the accompanying table. Due to self-restrictions and bans on the use of fluorocarbons as aerosol propellants because of their possible influence on the earth's atmospheric ozone layer, production in many countries has decreased significantly.

CHF_2Cl is becoming increasingly important because it is employed in the manufacture of tetrafluoroethylene $F_2C=CF_2$ (*cf.* Section 9.1.5) and perfluoropropene $F_2C=CF—CF_3$. Tetrafluoroethylene is mainly used as a monomer for the manufacture of perfluorinated polymers.

CFC 22 production (in 1000 tonnes):

	1991	1993	1995
USA	143	132	119

quantitatively most important products with example of trade names

CF_2Cl_2 Freon® 12
$CFCl_3$ Freon® 11

the neutral designations R 12, R 11 (R = refrigerant) or CFC 12, CFC 11 (CFC = chlorofluorocarbon) also possible instead of trade names.

CFC 11 and 12 production (in 1000 tonnes):

	1990	1992	1994
USA	156	118	66

uses of Br, Cl, F-alkanes:

fire extinguishing agents (Halons)
inhalation narcotic (Halothan, Fluothane®)

Mixed halogenated methanes find further applications in fire extinguishing systems, especially when high quality goods are to be safeguarded. Bromotrifluoromethane (CF_3Br = Halon 1301) and 1,2-dibromo-1,1,2,2-tetrafluoroethane (CF_2Br—CF_2Br = Halon 2402) are highly effective. The bromofluoromethanes and -ethanes are commercially known as Halons. Halon 1301 is produced by Asahi Glass, Du Pont and Pechiney Ugine Kuhlmann. Since the Halons also show to certain extent an even higher potential for ozone damage than the chlorofluorocarbons, reduction, and a eventual cessation, of production and use is anticipated.

2-Bromo-2-chloro-1,1,1-trifluoroethane has been employed worldwide since 1956 as an anesthetic (*e. g.*, Halothan 'Hoechst', Fluothane® ICI).

The worldwide consumption of fully halogenated fluorocarbons in 1988 was approximately 1.1×10^6 tonnes, of which the proportion of Halons was about 180 000 tonnes.

high-temperature cleavage of CFCs (secondary recycling) in H_2/O_2 mixture leads to other useful products.

For the disposal of unwanted CFCs, Hoechst has developed a thermal cleavage in an oxyhydrogen flame at 2100 °C in an acid-resistant graphite reactor. Cracking of the CFCs leads to a mixture of HCl, HF, CO_2, H_2O and Cl_2 with more than 99.99 % conversion. After quick cooling of the products to 40 °C, they are separated and made available for other reactions. Hoechst plans to build an 8000 tonne-per-year plant

3. Olefins

3.1. Historical Development of Olefin Chemistry

The development of olefin chemistry after World War II is closely related to the vigorous growth in petrochemistry.

Tar and acetylene chemistry had dominated the scene up to this period – particularly in Germany – and had generated a wide range of solvents, raw materials for paints, elastomers, thermoplastics, thermosetting plastics, and synthetic fibers. Expansion to mass production was, however, only first possible with the advancement of olefin chemistry and the development of industrial methods for olefin processing and polymerization.

In the USA, the expansion of olefin chemistry was supported by rapid developments in motorization. This caused such a steadily increasing demand for gasoline in the 1930s that the refineries were required to produce additional motor fuel by thermal cracking of the higher boiling crude oil fractions. Olefins were present as byproducts and were initially only employed in the manufacture of alkylate and polymer gasoline to improve the gasoline quality.

After 1948, there was strong economic growth in Western Europe, and consequently the feedstocks for olefin manufacture were available due to the increasing refinery capacity.

At the same time, manufacturing processes for obtaining monomers from olefins were being developed, and methods of polymerization were being improved. Without these advances the rapid growth in olefin chemistry would have been inconceivable.

two basic factors in the growth of olefin chemistry:

1. the olefin supply from cracking processes increases with rising demand for gasoline and was used initially only for improvement in its quality (alkylate and polymer gasoline); subsequently the expansion of petrochemistry, based on crude oil, led to development of cracking processes for olefin manufacture

2. chemical research concentrates on conversion of inexpensive olefins and discovers new types of catalysts for monomer manufacture and polymerization

3.2. Olefins *via* Cracking of Hydrocarbons

Due to their high reactivity, olefins, if present at all, only occur in very limited amounts in natural gas and crude oil. They must be manufactured in cleavage or cracking processes.

olefins – virtually absent in fossil fuels – mainly result from cracking processes

oil conversion processes can be classified as follows:

1. catalytic cracking
2. hydrocracking
3. thermal cracking

process characteristics of catalytic cracking:

acidic cracking catalysts lead, *via* carbonium ion reactions, to formation of saturated branched cyclic and aromatic hydrocarbons

traditional cracking catalysts:

amorphous Al-silicates, formerly 10–15, later 25, wt% Al_2O_3

modern cracking catalysts:

5–40 wt% crystalline Al-silicates (zeolites) with ordered three-dimensional network of AlO_4 and SiO_4 tetrahedra, exchanged with rare earth cations for stabilization, in a mixture with amorphous silicates

process characteristics of hydrocracking:

bifunctional catalysts (hydrogenation–dehydrogenation and acidic function) favor formation of saturated branched hydrocarbons *via* carbonium ion reactions (dehydrogenation, isomerization, hydrogenation)

Desulfurization and denitrification (*i.e.*, refining) occur simultaneously during hydrocracking

Refinery technology employs essentially three different approaches in converting the range of products naturally occurring in crude oil to those meeting market requirements. These are the catalytic, the hydrocatalytic and the thermal cracking processes.

In catalytic cracking, higher boiling distillation fractions are converted into saturated branched paraffins, naphthenes (cycloparaffins), and aromatics. As the proportion of olefins is small, catalytic cracking is primarily used for producing motor fuels. Various technologies are used, but generally a fluidized bed (FCC = **f**luidized **c**atalytic **c**racking) or a reactor with rising catalyst (riser cracking) is used. The usual process conditions are 450–500 °C and a slight excess pressure. Formerly, aluminum silicates with activators such as Cr_2O_3 (TCC catalyst = **T**hermofor **c**atalytic **c**racking process of Mobile Oil) or MnO (Houdry catalyst) were used. Mixtures of crystalline aluminum silicates (in the form of zeolites) with amorphous, synthetic, or naturally occurring aluminum silicates are now employed, either in the acidic form or their cations exchanged with rare earths; those exchanged with rare earths have a higher thermal stability. The zeolite catalysts increase the gasoline yield through shape selectivity (product selectivity determined by the geometry of the pore system) and lowered coke deposition. As in all catalytic cracking processes, the catalyst must be reactivated by burning off the deposited coke layer; cracking catalysts therefore often contain small amounts of platinum metal to promote C/CO oxidation to CO_2.

Catalytic cracking is important worldwide and has become the largest industrial user of zeolites. Leading processes have been developed by such companies as Exxon, Gulf, Kellogg, Stone and Webster, Texaco, and UOP.

With hydrocracking (catalytic cracking in the presence of hydrogen), residues, as well as higher boiling distillation fractions, can be converted into lower boiling products by various processes. The product does not contain olefins, and its composition can be determined over a wide range by the choice of feed, type of catalyst, and process conditions. With an LPG feed (**l**iquefied **p**etroleum **g**as), the process can be optimized for production of isobutane, gasoline, naphtha, or fuel oil. Bifunctional catalyst systems, consisting of metallic hydrogenation-dehydrogenation (*e.g.*, Co-Mo or Pd, Pt) and acidic (*e.g.*, $Al_2O_3 \cdot SiO_2$, preferably as zeolite) cracking components, are employed in the presence of hydrogen. Relatively high investment costs are required for the process which operates at 270–450 °C and 80–200 bar. Additionally, between 300–500 m^3 of hydrogen must be supplied per tonne of oil feed. This must be manufactured separately as it is not available from the refinery.

Thermal cracking plays an important role in olefin manufacture. This process which involves a radical cleavage of hydrocarbons takes place under pressure and starts at about $400-500\,°C$.

process characteristics of thermal cracking:

uncatalyzed radical cracking reactions lead to a high proportion of olefins

The basic mechanism of a cracking reaction can be envisaged as follows, using *n*-octane as an example. Thermal cracking is initiated by homolysis of a C—C bond to form two free radicals, *e. g.*,

$$CH_3(CH_2)_6CH_3 \rightarrow CH_3CH_2CH_2CH_2CH_2 \cdot + \cdot CH_2CH_2CH_3 \qquad (1)$$

Each alkyl radical can abstract a hydrogen atom from an *n*-octane molecule to produce an octyl radical and a shorter alkane, *e. g.*,

$$CH_3CH_2CH_2CH_2CH_2 \cdot + CH_3(CH_2)_6CH_3 \rightarrow \qquad (2)$$
$$CH_3(CH_2)_3CH_3 + CH_3(CH_2)_5\dot{C}HCH_3$$

Abstraction of secondary hydrogen is favored over primary hydrogen due to the lower C—H bond energy, with an equal probability for removal of each secondary hydrogen. Any of these radicals can also undergo β-cleavage to form ethylene or propene and a shorter alkyl radical:

$$CH_3CH_2CH_2-CH_2CH_2 \cdot \rightarrow CH_3CH_2CH_2 \cdot + H_2C=CH_2 \qquad (3)$$
$$CH_3(CH_2)_3CH_2-CH_2\dot{C}HCH_3 \rightarrow CH_3(CH_2)_3CH_2 \cdot +$$
$$H_2C=CHCH_3$$

The cracking reactions thus involve changes in the H_2 content as well as in the carbon skeleton. Dehydrogenation and H_2 transfer from H_2-rich hydrocarbon fractions (low boiling) belong to the former, and chain cracking of H_2-deficient higher molecular fractions to the latter category.

cracking processes can be classified as follows:

1. primary reactions such as dehydrogenation, H_2 transfer, C-chain homolysis, isomerization, cyclization

2. secondary reactions such as olefin polymerization alkylation, condensation of aromatics

Primary reactions involving the carbon skeleton include not only chain shortening, but also isomerization and cyclization. Secondary reactions include olefin polymerization, alkylation, and condensation of aromatics to form polynuclear aromatic compounds.

the free energy of formation (ΔG_f) provides a measure of the relative thermal stability of all hydrocarbons. Positive ΔG_f means:

at thermodynamic equilibrium, decomposition into C + H_2 is favored (*e. g.*, for ethane $\Delta G_f > 0$ for $T > 230\,°C$)

in practice, shortened reaction periods allow intermediate equilibria to appear

olefin yields depend on three variables and their interrelationships:

1. temperature
2. residence time
3. partial pressures

1. effect of temperature:

1.1. higher temperatures favor C_2/C_3 olefins over higher olefins

1.2. higher temperatures increase cracking rate and therefore require shorter residence times or lower partial pressures

2. effect of residence time:

'long-period' cracking leads to secondary reactions

'short-period' cracking increases olefin share

3. effect of partial pressure:

decrease in hydrocarbon partial pressure increases cracking (Le Chatelier-Braun) as 2 – 3 moles of crack products result from 1 mole

H_2O is the preferred foreign gas (steam cracking)

advantages:
easily condensable and thus separable, C deposition is lowered

disadvantage:
increased consumption of energy for heating and cooling

Thermodynamically, all saturated and unsaturated hydrocarbons are unstable with respect to their elements at the applied cracking temperatures. That is, if pyrolysis were allowed to go to thermodynamic equilibrium, all hydrocarbons would completely decompose into graphite and molecular hydrogen.

Accordingly, in a commercial cracking process large amounts of energy must be transferred at high temperatures within a time period sufficient to allow cracking to occur but insufficient for decomposition into the elements.

Hydrocarbon cracking is optimized by regulating three kinetic parameters:

1. Cracking temperature
2. Residence time
3. Partial pressures of the hydrocarbons

To 1:

Temperature affects the cracked gas composition. At about 400 °C the carbon chains are preferentially cracked in the center of the molecule. With increasing temperature, the cracking shifts towards the end of the chain, leading to formation of more low molecular weight olefins.

Also, the cracking rate increases with temperature, as higher radical concentrations are generated.

To 2:

The residence time affects the ratio of primary to secondary products for a constant cracking temperature.

With a short residence time, primary reactions resulting in olefin formation dominate. A longer residence time allows the increase of secondary reactions such as oligomerization and coke deposition.

To 3:

In the desired cracking reactions, there is an increase in the number of moles, and so the partial pressure of the hydrocarbons has a powerful effect. A high partial pressure favors polymerization and condensation reactions and a low partial pressure improves the olefin yield. In order to lower the partial pressure of the hydrocarbons, a foreign gas – usually steam – is mixed with the hydrocarbon fraction for pyrolysis (steam cracking or reforming). As the steam content is increased, the yield of olefin rises, while carbon deposition diminishes.

In conclusion, the manufacture of low molecular weight olefins is favored by high temperature, short residence time, and low partial pressures.

Essentially two processes, differing in the severity of conditions employed, have been developed for steam cracking:

1. Low-severity cracking below 800 °C with 1 second residence time.
2. High-severity cracking approaching 900 °C with roughly 0.5 second residence time.

The $C_2/C_3/C_4$ olefin distribution can be controlled with these variables.

production of lower olefins favored by:

1. high cracking temperature
2. short residence time
3. steam cracking

two basic cracking modifications:

1. 'low severity'
   < 800 °C, 1 s residence time

2. 'high severity'
   up to 900 °C, 0.5 s residence time

3.3. Special Manufacturing Processes for Olefins

3.3.1. Ethylene, Propene

In terms of quantity, ethylene and propene are among the most important basic organic chemicals. Ethylene is the feedstock for roughly 30% of all petrochemicals. Production figures for ethylene and propene in several countries are summarized in the adjacent tables.

In 1995, the worldwide ethylene production capacity was about 79×10^6 tonnes per year, including 22.9, 19.7, and 7.2×10^6 tonnes per year in the USA, Western Europe, and Japan, respectively; in 1994, propene production capacity worldwide was about 43.1×10^6 tonnes per year, including 12.7, 13.5, and 4.9×10^6 tonnes per year in the USA, Western Europe, and Japan.

Ethylene was originally manufactured by partial hydrogenation of acetylene, dehydration of ethanol, or separation from coke-oven gas. These processes are insignificant for countries with a developed petrochemical industry. However, in developing countries such as in South America, Asia, and Africa, dehydration of ethanol produced by fermentation still provides an important supplement to ethylene derived from petroleum. Propene only gained importance in the chemical industry after its manufacture from crude oil fractions or natural gas was possible.

Today, both olefins are obtained predominantly from the thermal cracking of saturated hydrocarbons. Ethylene–propene cracking plants have recently reached capacities of about 750 000 tonnes per year ethylene and over 450 000 tonnes per year propene.

ethylene production (10^6 tonnes):

	1990	1992	1995
USA	17.1	18.3	21.3
W. Europe	15.3	15.5	18.4
Japan	5.8	6.1	6.9

propene production (10^6 tonnes):

	1990	1992	1995
USA	9.5	10.3	11.7
W. Europe	9.1	9.7	12.1
Japan	4.2	4.5	5.0

traditional C_2H_4 manufacture:

1. partial C_2H_2 hydrogenation
2. dehydration of C_2H_5OH
3. low temperature decomposition of coke-oven gas

traditional C_3H_6 manufacture not of industrial significance

modern C_2H_4/C_3H_6 manufacture *via* thermal cracking of natural gas, refinery gas, or crude oil fractions

Table 3-1. Feedstocks for ethylene (wt% C_2H_4 production).

	USA		W. Europe		Japan		World	
	1979	1991	1981	1991	1981	1991	1981	1994
Refinery gas	1	3	} 4	2	} 10	} 2	} 31	} 42
LPG[1], NGL[2]	65	73		14				
Naphtha	14	18	80	72	90	98	58	46
Gas oil	20	6	16	12	0	0	11	12

[1] LPG = **l**iquified **p**etroleum **g**as (propane, butane)
[2] NGL = **n**atural **g**as **l**iquid (ethane, LPG, light naphtha)

feedstock for C_2H_4/C_3H_6 varies from country to country

USA: LPG
      NGL

characterized by maximum C_2H_4 yield of 81 wt% (from C_2H_6) and low C_3H_6 production

W. Europe and Japan:

naphtha (bp 80–200 °C)

characterized by maximum C_2H_4 yield around 30 wt% and C_2H_4 : C_3H_6 weight ratio of about 2.1 : 1

supplementing olefin requirements by alkane dehydrogenation according to

Houdry (Catofin)
UOP (Oleflex)
Phillips (STAR)
Snamprogetti (FBD-4)
Linde

broadening of raw material base by employing higher boiling fractions as feed (gas oil, bp. >200 °C)

characterized by lower absolute amount of C_2H_4 + C_3H_6, with a maximum C_2H_4 yield of ~25 wt%, but a higher percentage propene (C_2H_4 : C_3H_6 ≈ 1.7 : 1) and many by-products.

extension of feedstock base *via* direct employment of deasphalted oil using new cracking technology, *e.g.*, ACR technology with the features:

In the USA, the dominant feedstocks for the manufacture of ethylene and propylene are LPG and NGL gas mixtures rich in ethane, propane and butane, with a contribution from light naphtha resulting from the refining of petroleum. Because these feedstocks have a higher hydrogen content than naphtha, more ethylene is produced.

Western Europe historically lacked ethane-rich natural gas, so here, as in other countries such as Japan, naphtha was used as the primary feedstock for olefins.

Increasing use of both domestic and imported natural gas has lowered the naphtha fraction in Western Europe to ~72% at present.

The different requirements for particular olefins has led in some countries to supplementary selective production processes. For example, there are many plants operating in several countries to produce C_3/C_4 monoolefins by the Houdry-Catofin process (commercialized by Lummus Crest) or by the Oleflex process from UOP. Other propane-dehydrogenation processes have been developed by Phillips, Snamprogetti and Linde. With only small changes in the operating conditions, butane (*cf.* Section 3.3.2) or higher alkanes can be dehydrogenated in place of propane.

To expand the raw material base, processes were developed worldwide in which higher boiling crude oil fractions, such as gas oils with boiling points above 200 °C, can be employed in steam cracking (*cf.* Section 1.4.1). New cracking technologies have been developed to make cracking more economical. A process developed by UCC and Chiyoda, based on Kureha Chem. Ind. technology, is a characteristic example.

In this **A**dvanced **C**racking **R**eactor (ACR) technology, various feedstocks, including those containing sulfur, are cracked in a ceramic-lined reactor in the presence of combustion gases from process fuel and steam at temperatures up to 2000 °C and 3.5 bar.

Substantially more ethylene is obtained (*e. g.*, 38 wt% with a naphtha feed) than with conventional cracking technology.

The procedure for naphtha cracking, the most important process for manufacture of ethylene worldwide, can be subdivided into the following steps:

1. Cracking of naphtha in tube furnaces
2. Quenching
3. Cracked gas compression and purification
4. Drying, cooling and low temperature distillation

To 1:

Naphtha is vaporized with superheated steam, and fed into the tubes of the cracking furnace. The chromium nickel tubes, 50 – 200 m long and 80 – 120 mm wide, are arranged vertically in modern high-severity cracking furnaces. They are heated directly to about 1050 °C at the hottest point by combustion of gases or oils.

To 2:

The cracked products exit the furnace at about 850°C, and must be rapidly quenched to around 300°C to avoid secondary reactions. Initial quenching with steam generation in transfer-line heat exchangers is followed by cooling with an oil spray.

To 3:

In this stage, process water and pyrolysis gasoline are separated. The gaseous components are condensed in the raw gas compressor and washed with a caustic solution (*e.g.*, 5 – 15% caustic soda) to remove H_2S and CO_2.

To 4:

Before the actual workup, careful drying must take place so that the subsequent low temperature distillation will not be interrupted by ice formation. The dry raw gas is then cooled in several stages and fractionally distilled in a series of columns.

After separation, ethylene still contains acetylene and ethane. As acetylene interferes with the polymerization of ethylene, it must be either selectively hydrogenated with a catalyst or removed by extractive distillation, *e.g.*, with dimethylformamide or N-methyl-pyrrolidone. Subsequent ethylene–ethane separation demands a particularly effective column due to the close proximity of their boiling points. Ethylene can be obtained with a purity of 99.95% (polymerization grade).

Sour crude oil can also be cracked with added steam at temperatures up to 2000 °C in H_2/CH_4 incineration zones; high C_2H_4 yield

characteristics of naphtha cracking (steam cracking):

homogeneous endothermic gas-phase reaction in which the dissociation energy for the C—C cleavage is supplied by high temperature level

1. steam cracking in vertical Cr-Ni tubes which are directly heated with gas or oil to about 1050 °C

2. rapid cooling to prevent further reaction takes place in two stages:
2.1. transfer-line heat exchangers coupled with steam generation
2.2. spraying with quenching oil

3. pre-separation into gas and condensate as well as removal of H_2S and CO_2 from raw gas (NaOH, ethanolamine)

4. drying of the gas mixture (diethylene glycol or Al_2O_3 or SiO_2) and separation by low temperature distillation

polymerization-grade olefins are obtained by two alternative processes:

1. selective hydrogenation

$$HC\equiv CH \rightarrow H_2C=CH_2$$
$$CH_3C\equiv CH \rightarrow CH_3CH=CH_2$$
$$H_2C=C=CH_2 \rightarrow CH_3CH=CH_2$$

2. extractive distillation with *e.g.*, DMF or NMP as extraction solvent

attainable purities:

C_2H_4: 99.95%
C_3H_6: 99.9%

i.e., polymerization grade

naphtha steam cracking supplies, along with C_2/C_3 olefins, raw materials for C_4/C_5 olefins and for aromatics

Similarly, after separating the C_3 fraction, the propyne and allene present must be converted (by selective hydrogenation) into propene or propane before the propane–propene separation. Propene can be recovered in about 99.9% purity.

C_4 and C_5 fractions and pyrolysis gasoline are also obtained from naphtha cracking. The former serve as feedstocks for higher olefins (*cf.* Section 3.3.2) and the latter for aromatics (*cf.* Section 12.2).

Table 3–2 illustrates a typical product distribution obtained from the steam cracking of naphtha under high-severity conditions when ethane and propane are recycled.

Table 3–2. Product distribution from high-severity naphtha cracking.

Product	wt%
Residual gas (CH_4, H_2)	16.0
Ethylene	35.0
Propene	15.0
C_4 fraction	8.5
C_5 fraction and higher boiling fractions (pyrolysis gasoline, residual oil)	25.5

olefin ratio adjustable:

1. in cracking process
1.1. *via* feedstock
1.2. *via* cracking severity
2. in a subsequent 'Triolefin process'

The ratio of the industrially most important C_2, C_3, and C_4 olefins produced in naphtha cracking can be modified not only by the cracking severity (low-severity conditions increase share of higher olefins), but also by coupling a 'Triolefin process' to the cracking unit (*cf.* Section 3.4).

olefins used for polyolefins (wt%):

	PE			PP		
	1985	1992	1994	1985	1992	1994
USA	50	54	53	35	41	40
Western Europe	55	58	57	34	47	48
Japan	48	47	48	44	51	50

Ethylene is primarily used in all industrialized countries for polymerization. The fraction of propene used for polypropylene has been considerably smaller historically, but has increased in the last few years. The adjacent table gives an overview of the proportion of olefin used for polyofin (PE = polyethylene, PP = polypropylene) in several countries. A more detailed breakdown of the consumption figures is given in Chapter 7 (ethylene) and Chapter 11 (propene).

3.3.2. Butenes

C_4 olefin formation:

byproducts formed in refinery process and in hydrocarbon cracking

coproduct in C_2H_4 oligomerization

The bulk of the butenes (*n*-butenes, isobutene) are produced as inevitable byproducts in the refining of motor fuel and from the various cracking processes of butane, naphtha, or gas oil. The butene production in the USA, Western Europe, and Japan in 1989 was about 8.0, 5.7 and 1.9×10^6 tonnes, respectively.

A smaller amount of 1-butene is produced as a coproduct in the oligomerization of ethylene to *a*-olefins (*cf.* Section 3.3.3.1).

Older, direct manufacturing processes, *e. g.,* from butanols or acetylene, have become insignificant.

The Phillips triolefin process, in which butenes and ethylene are produced by the disproportionation of propene, was discontinued after nine years of operation by its sole industrial user, Shawinigan in Canada, for economic reasons.

direct C_4 olefin manufacture:

industrial propene disproportionation ($2 C_3 \rightleftharpoons C_2 + C_4$), used only briefly

An expected increase in the use of butenes to produce high-octane components for gasoline led to the development of new direct production technologies. In one of these, the Alphabutol process of IFP, ethylene is dimerized selectively to 1-butene using a homogeneous titanium catalyst in the liquid phase. A co-catalyst inhibits the formation of Ti(III), preventing Ziegler polymerization from occurring. A 95% selectivity to 1-butene is achieved at a 50% C_2H_4 conversion. Several Alphabutol plants are already in operation, and many others are being planned.

future synthetic routes to C_4 olefins by homogeneously-catalyzed C_2H_4 dimerization; *e.g.*, processes of IFP, Phillips Petroleum

With other technologies, *e.g.*, Phillips Petroleum, ethylene can be homogeneously dimerized to 2-butene.

Cracking processes for the production of gasolines and olefins will remain the most important source of butenes. Different processes predominate in different countries, depending on gasoline/olefin requirements. For example, 75% of the butene produced in the USA in 1989 came from the refinery processes of catalytic cracking and reforming, while only 6% derived from steam cracking. In Western Europe and Japan, steam cracking accounted for 30% and 53%, respectively, of the 1989 butene production.

major C_4 olefin sources:

1. Fluid Catalytic Cracking (FCC) of gas oil
2. steam cracking of C_2H_6 ... gas oil

The absolute amounts and compositions of the C_4 fraction obtained from the cracking of hydrocarbons is substantially affected by three factors:

1. Type of cracking process
2. Severity of cracking conditions
3. Feedstock

As shown in Table 3 – 3, a significantly higher fraction of butenes is obtained from steam cracking of naphtha than from catalytic cracking of gas oil. Therefore naphtha steam cracking is the more interesting technology for production of unsaturated C_4 compounds.

most valuable C_4 olefin source:

steam cracking of naphtha

decrease of total C_4 and C_4 olefins, but increase of butadiene under high severity cracking conditions (consequence of desired higher C_2H_4 yield)

As the cracking severity increases, both the total yield of the C_4 fraction and the proportion of butenes decrease, while the proportion of butadiene increases due to its higher stability:

Table 3–3. Composition of C_4 fractions from steam cracking of naphtha and catalytic cracking of gas oil (in wt%).

Cracked products	Steam cracking		Catalytic cracking (FCC) zeolite catalyst
	Low severity	High severity	
1,3-Butadiene	26	47	0.5
Isobutene	32	22	15
1-Butene	20	14	12
trans-2-Butene	7	6	12
cis-2-Butene	7	5	11
Butane	4	3	13
Isobutane	2	1	37
Vinylacetylene Ethylacetylene 1,2-Butadiene	} 2	} 2	} –

butadiene is stable even under high-severity conditions (thermodynamically favored because of the conjugation energy of $\begin{smallmatrix} 3.5 \text{ kcal} \\ 14.7 \text{ kJ} \end{smallmatrix} \Big/ \text{mol}$)

principle of separation of C_4 fraction:

similar boiling points demand more selective physical and chemical separation methods than distillation

separation procedures for the C_4 fraction:

1st stage of C_4 separation:

separation of $H_2C{=}CH{-}CH{=}CH_2$ *via* extraction or extractive distillation with selective solvents
remaining C_4 fraction = raffinate

today frequently denoted Raffinate I to distinguish it from Raffinate II, which is the C_4 fraction after separation of isobutene in the 2nd stage

In steam cracking, the yield of the C_4 fraction parallels the boiling range of the feedstock, beginning with $2-4$ wt% from ethane/propane, reaching a maximum of $10-12$ wt% from naphtha, and decreasing to $8-10$ wt% from gas oil.

The C_4 fraction cannot be separated into its components economically by simple distillation due to the close proximity of their boiling points. Therefore, more effective and selective physical and chemical separation procedures must be employed. The processing of a C_4 fraction begins with the separation of the butadiene, as described in Section 5.1.2.

After extraction of the major portion and removal of the residual butadiene, for example by selective hydrogenation (*e.g.*, 'Bayer cold hydrogenation' or IFP selective hydrogenation), a mixture is obtained consisting essentially of isobutene, the *n*-butenes and butane. This is termed the C_4 raffinate and has the composition shown below:

Table 3–4. Typical composition of a C_4 raffinate.

Components	vol%
Isobutene	44–49
1-Butene	24–28
2-Butene (*cis* and *trans*)	19–21
n-Butane	6– 8
Isobutane	2– 3

Isobutene is preferably isolated next in the separation of the C_4 raffinate, since it differs from the remaining C_4 components in its branching and higher reactivity.

All current industrial processes rely on the further chemical reaction of isobutene, and shape-selective isolation is not practiced.

The molecular sieve separation process exploits the methyl branching which makes isobutene too bulky to be adsorbed in the very uniform 3–10 Å pores of the molecular sieve ('Olefin-Siv' process of UCC). Only *n*-butenes and butane are adsorbed, and then desorbed using a higher boiling hydrocarbon. In this way, isobutene with 99% purity can be obtained from the C_4 raffinate.

Isobutene is the most reactive compound in the C_4 raffinate and this property can be utilized for a chemical separation. In practice, four processes have been successfully commercialized. Of these, only the addition of water or alcohols (earlier methanol, but now also isobutanol) can be used as a reversible process.

1. Hydration of isobutene in the presence of dilute mineral acid or an acid ion exchanger to form *tert*-butanol, possibly followed by cleavage to regenerate isobutene and water:

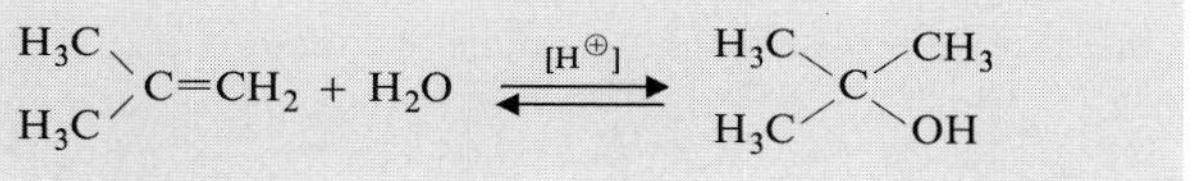

(4)

2. Addition of methanol to isobutene over an acid ion exchange resin to form methyl *tert*-butyl ether (MTBE), and possible regeneration of isobutene and CH_3OH:

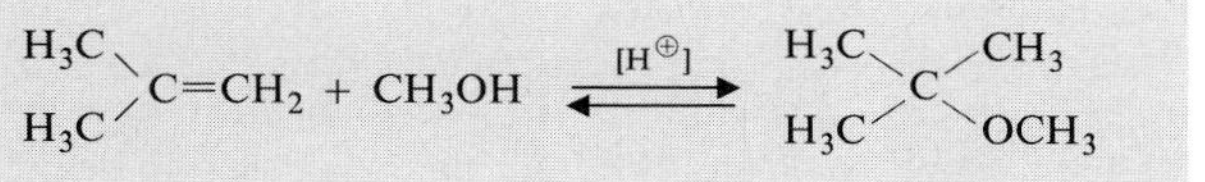

(5)

3. Oligomerization of isobutene with acidic catalysts to form 'diisobutene', a mixture of the double bond isomers of 2,4,4-trimethylpentene, preferentially:

(6)

Margin notes:

2nd stage of C_4 separation:

removal of $H_2C{=}C\overset{\nearrow CH_3}{\searrow_{CH_3}}$

principles of separation process:

methyl branching allows physical separation *via* selective molecular sieve adsorption

higher rate of reaction of isobutene allows chemical separation *via* selective reactions

example of molecular sieve adsorption:

UCC 'Olefin-Siv process'
isobutene passes through the molecular sieve
n-butenes are adsorbed

four methods for selective reaction of isobutene:

1. reversible proton-catalyzed hydration with *tert*-butanol as an intermediate; also addition of ethanol or isobutanol with possible reverse reaction

2. reversible proton-catalyzed addition of CH_3OH to form MTBE

3. irreversible proton-catalyzed oligomerization with preferential formation of di- and triisobutene

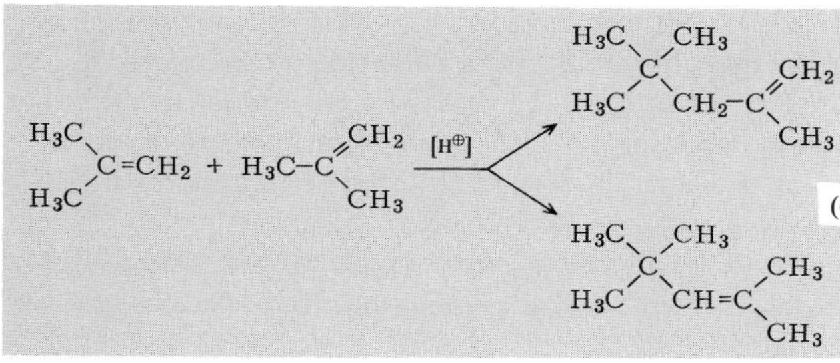

Further isobutene can add to the 'diisobutene', resulting in 'triisobutene' and higher oligomers.

4. Polymerization of isobutene in the presence of a Lewis acid catalyst to form polyisobutene:

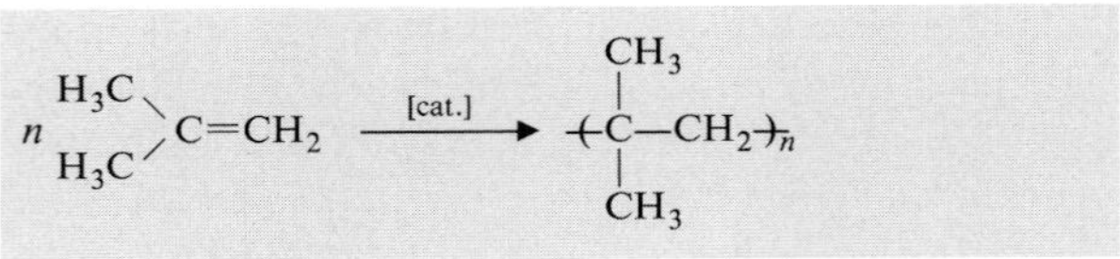

$$n \; \begin{matrix} H_3C \\ \\ H_3C \end{matrix} \!\! \diagdown \!\!\!\! \diagup \!\! C\!=\!CH_2 \quad \xrightarrow{\text{[cat.]}} \quad \begin{matrix} CH_3 \\ | \\ \!\!+\!\!C\!-\!CH_2\!\!\!\rightarrow_{\!\!n} \\ | \\ CH_3 \end{matrix} \tag{7}$$

The C_4 raffinate is the feedstock for all four processes.

Isobutene can be removed by reaction and then regenerated as pure isobutene (Routes 1 and 2). On the other hand, it is irreversibly oligomerized in Route 3, and irreversibly polymerized in Route 4.

To 1:

In the commercial hydration of isobutene (*cf.* Section 8.1.3), $50-60\%$ H_2SO_4 is generally used. Isobutene is removed from the C_4 raffinate as *tert*-butanol in a countercurrent extraction at $10-20°C$. After dilution with H_2O the *tert*-butanol is vacuum distilled from the acidic solution and used as an intermediate (*cf.* Section 8.1.3) or cleaved to regenerate isobutene.

A process using a cation exchange resin (analogous to the manufacture of MTBE) has recently been developed by Hüls, and is already in commercial use.

Nippon Oil conducts the hydration of isobutene with aqueous HCl in the presence of a metal salt. During the extraction of the C_4 raffinate, *tert*-butanol and *tert*-butyl chloride are formed. Both can be cleaved to regenerate isobutene.

One commercial process for the dehydration of *tert*-butanol, which is, *e. g.*, also obtained as a cooxidant in the Oxirane process for the manufacture of propylene oxide (*cf.* Section 11.1.1.2), was developed at Arco. In this process, the reaction is carried out in the gas phase at $260-370°C$ and about 14 bar in the presence of a modified Al_2O_3 catalyst (*e. g.*, with impregnation of SiO_2) with a high surface area. The conversion of *tert*-butanol is about 98%, with a high selectivity to isobutene. Other processes take place in the liquid phase at $150°C$ in the presence of a heterogeneous catalyst.

To 2:

The manufacture of methyl *tert*-butyl ether takes place in the liquid phase at 30–100°C and slight excess pressure on an acid ion exchange resin. Either two separate reactors or a two-stage shaft reactor are used to obtain nearly complete conversion (>99%) of the isobutene. Because of the pressure-dependent azeotrope formed from methanol and MTBE, preparation of pure MTBE requires a multistep pressure distillation. Alternatively, pure MTBE can be obtained by adsorption of methanol on a scavenger in a process recently developed by Erdölchemie and Bayer. Methanol can be separated from organic solvents (*e.g.*, MTBE) by pervaporation, a new separation technique. Pervaporation uses a membrane that works like a sieve, holding back particles larger than a particular size. Pervaporation is already widely used commercially for the removal of water from organic solvents.

industrial operation of isobutene etherification:

heterogeneous catalysis (cation exchange resin in H-form), selective reaction of isobutene from the C_4 Raffinate I in the liquid phase with process-specific workup of the pressure-dependent MTBE/CH_3OH azeotrope

separation of methanol from MTBE also possible through the filtering effect of membrane pores, *i.e.,* pervaporation

In the reaction of isobutene with methanol described above, all other components of the C_4 fraction (Raffinate II) remain unchanged, except a small portion of the diolefins and alkynes which polymerize and shorten the lifetime of the ion exchanger. Erdölchemie has recently developed a bifunctional catalyst containing Pd which, in the presence of small amounts of hydrogen, catalyzes the hydration of diolefins and acetylenes. The etherification of isobutene is not affected, except that the lifetime of the catalyst is increased.

Ethyl *tert*-butyl ether (ETBE) is produced commercially from isobutene and ethanol in a similar process.

C_2H_5OH addition to isobutene yields ETBE

The catalytic addition of isobutanol to isobutene is used by BASF. The ether is then separated from the C_4 fraction by distillation, and then catalytically cleaved back into isobutanol and isobutene. The isobutanol is recycled for formation of more ether, while isobutene is used for polymerization.

$(CH_3)_2CHCH_2OH$ addition to isobutene for isolation from C_4 fraction, then cleavage (reactive separation)

The reaction of isopentene from the C_5 fraction with methanol to produce *tert*-amyl methyl ether (TAME) can be performed over an acid ion exchange resin or the new bifunctional catalyst from Erdölchemie, analogous to the production of MTBE. The first plant began production in the United Kingdom in 1987. Additional TAME plants have either been put into operation or are scheduled.

addition of CH_3OH to $(CH_3)_2C=CHCH_3$, analogous to MTBE manufacture, also possible

MTBE production (10^6 tonnes):

	1986	1988	1991
USA	1.53	2.12	4.35
W. Europe	0.70	1.23	2.42
Japan	n.a.	n.a.	0.08

n.a. = not available

After the first commercial manufacture of MTBE in Western Europe by Snamprogetti/Anic in 1973 and Hüls in 1976, global capacity had already reached 22.2×10^6 tonnes per year in 1996 with 11.0, 3.4, and 0.30×10^6 tonnes per year in the USA,

MTBE produced worldwide in 25 countries, expanding to 33 countries in 1995

Western Europe, and Japan, respectively, making MTBE the ether with the highest production rate worldwide.

MTBE cleavage rarely performed industrially; instead MTBE used as a component of gasoline because of its high octane number

Although it is possible to regenerate isobutene over acidic oxides in the gas phase at 140–200 °C (still practiced by Exxon and Sumitomo), most MTBE is added to gasoline to increase the octane number (cf. Section 2.3.1.2).

MTBE manufacture from Raffinate I became an alternative processing method

Even as the processing of the C_4 fraction was increasingly shifted to removal of isobutene as MTBE, growing demand for MTBE necessitated process development to supply larger amounts of isobutene.

supplementary isobutene manufacture for MTBE:

1. by isomerization of n-butenes in Raffinate II

The C_4 fraction remaining after MTBE production (Raffinate II) contains n-butenes that can be isomerized to give additional isobutene. An Al_2O_3 catalyst whose surface has been modified with SiO_2 is used in a process from Snamprogetti. At 450–490 °C, n-butenes isomerize with a conversion of ca. 35% and selectivity to isobutene of ca. 81%. An analogous process was also developed by, e. g., Kellogg.

2. by combined isomerization and dehydrogenation of n-butane

Additional isobutene can also be obtained from n-butane, wich is first isomerized to isobutane in, e. g., the Butamer process (UOP), the ABB Lummus Crest process, or the Butane Isom process (BP), and then dehydrogenated, e. g., with the Catofin process (Houdry, ABB Lummus Crest) or other processes from UOP, Phillips Petroleum, Snamprogetti and others (cf. Section 3.3.1).

To 3:

The oligomerization of isobutene is acid-catalyzed and takes place at temperatures around 100 °C.

industrial operation of isobutene oligomerization:

H_2SO_4 or ion-exchange catalyst (Bayer process)

characteristics of Bayer process:

high isobutene conversion to dimers and trimers, side reaction – double bond isomerization of the n-butenes

$$H_2C=CHCH_2CH_3 \rightarrow CH_3CH=CHCH_3$$

The process developed by Bayer uses an acidic ion-exchanger as catalyst, at 100 °C and about 20 bar, suspended in the liquid phase. Isobutene is dimerized and trimerized in a strongly exothermic reaction. Conversion is 99% with a dimer : trimer ratio of 3 : 1. The catalyst is centrifuged off and the mixture of n-butenes, C_8 and C_{12} olefins is separated by distillation. The isobutene content of n-butene is thereby lowered to about 0.7 wt%.

The advantages of this process lie in its simple technology. However, the simultaneous isomerization of the double bond – i. e., formation of 2-butene from 1-butene – can be disadvantageous. This isomerization occurs only to a limited extent in the hydration process.

BF_3-catalyzed polymerization of isobutene (BASF process)

In a BASF process, the isobutene recovered from the C_4 fraction by formation of an ether with isobutanol is polymerized with a special BF_3 catalyst system to give polyisobutene with a molecular weight between 1000 and 2500. This polyisobutene is used for the manufacture of additives to fuels and lubricants. A plant with a production capacity of 60000 tonnes per year (1995) is in operation in Belgium.

To 4:

The polymerization of isobutene, *e. g.,* by the Cosden process, is conducted in the liquid phase with $AlCl_3$ as catalyst. Polyisobutene with molecular weight between 300 and 2700 is obtained. Only a small amount of *n*-butenes copolymerize.

industrial operation of isobutene polymerization:

liquid-phase reaction in presence of $AlCl_3$

After removing isobutene, the remaining fraction contains, in addition to the *n*-butenes, only butanes. A further separation is not generally undertaken since the saturated hydrocarbons remain unchanged during some of the further reactions of the butenes, *e. g.,* the hydration to butanols, and can be removed as inert substances. In principle, a separation of the 2-butenes (*cis* and *trans*) and 1-butene is possible by distillation. The butanes can be separated from the *n*-butenes by extractive distillation.

3rd stage of C_4 separation:

3.1. separation of *n*-butenes,butanes unnecessary due to selective *n*-butene reactions (butanes as unerts)

3.2. separation of *n*-butenes by distillation, of butanes and *n*-butenes by extractive distillation

New developments have led to another possibility for separating the C_4 fraction. Here the 1-butene in the butadiene-free Raffinate I, consisting mainly of butenes, is catalytically isomerized to 2-butene. A modified Pd catalyst in the presence of H_2 is used, either in the gas phase (UOP process) or in the liquid phase (IFP process). The boiling points of isobutene and 2-butene are sufficiently different (*cf.* Section 5.1.2) to allow their separation by fractionation. The first firms to treat the C_4 fraction in this way were Phillips Petroleum and Petro-Tex. They use the 2-butene for production of alkylate gasoline with a higher octane number than that produced from isobutene.

alternative C_4 raffinate separation:

isomerization of 1-butene to 2-butene followed by fractionation of isobutene and 2-butene

All butenes are valuable starting materials for industrial syntheses.

Most butene − *i. e.,* currently more than half of the worldwide production − is used to manufacture high antiknock polymer and alkylate gasoline. Isobutene is essentially dimerized to 'diisobutene', and then either hydrogenated or used to alkylate isobutane. Isobutene, 'diisobutene', and *n*-butenes are also suitable alkylation agents for aromatics.

uses of butenes:

1. for alkylation uses of isobutane or aromatics (polymer/alkylate gasoline)

Butenes are also feedstocks for polymers and copolymers; for example, isobutene for polyisobutene, isobutene together with isoprene for butyl rubber (*cf.* Section 5.1.4), 1-butene for isotactic polybutene as well as for ethylene-1-butene copolymers.

2. for homo- and copolymers (isobutene, 1-butene)

3. for the synthesis of chemical intermediates:
3.1. hydration to alcohols (all butenes)
3.2. hydroformylation to C_5 aldehydes
and alcohols (all butenes)
3.3. oxidation to maleic anhydride
(*n*-butene) and to methacrylic acid
(isobutene)
3.4. CH_3OH addition to give methyl-*tert*-
butyl ether (isobutene)
3.5. ammoxidation to methacrylonitrile
(isobutene)
3.6. dehydrogenation to butadiene
(*n*-butenes)
3.7. Prins reaction and thermolysis to
isoprene (isobutene)

The most important secondary reactions of the butenes in the manufacture of chemical intermediates are hydration to alcohols (*n*-butenes → *sec*-butanol, isobutene → *tert*-butanol, *cf.* Section 8.1.3), hydroformylation to C_5 aldehydes and alcohols (*cf.* Section 6.1), oxidation of *n*-butene to maleic anhydride (*cf.* Section 13.2.3.2), and the previously described reaction of methanol with isobutene to give methyl-*tert*-butyl ether. The oxidative degradation of *n*-butenes to acetic acid (*cf.* Section 7.4.1.2), the ammoxidation of isobutene to form methacrylonitrile (*cf.* Section 11.3.2), and the oxidation of isobutene to methacrylic acid (*cf.* Section 11.1.4.2) are of less importance. Dehydrogenation of the *n*-butenes to butadiene is described in Section 5.1.3 and the conversion of isobutene into isoprene is dealt with in Section 5.2.2.

3.3.3. Higher Olefins

higher olefins of industrial significance:

C range between 5 to ca. 18

characteristics of separation of higher
olefins:

1. C_5 olefins:

from naphtha C_5 cracking fraction,
instead of complete separation (*cf.* butene,
Section 3.3.2) only isolation of important
components such as:

isoprene
cyclopentadiene
and increasing *tert*-amyl methyl ether
(TAME)

Since olefins with more than four carbon atoms have a rapidly increasing number of isomers, separation of mixtures such as those formed in cracking into individual components is no longer feasible. Moreover, only a few components have any industrial significance. Thus, out of the C_5 fraction from naphtha cracking, consisting mostly of *n*-pentane, isopentanes, *n*-pentenes, isopentenes, isoprene, cyclopentene, cyclopentadiene, and pentadiene, only isoprene (*cf.* Section 5.2.1) and cyclopentadiene (*cf.* Section 5.4) are separated on an industrial scale. In the future, isopentenes will increasingly be etherified with methanol to *tert*-amyl methyl ether, a high-octane component of gasoline.

2. olefins $> C_5$ (unbranched):

no separation of individual components
but homologous fractions:
$C_6 - C_9$
$C_{10} - C_{13}$
$C_{14} - C_{18}$

processes for unbranched olefins limit
number of double bond isomers

Higher olefins are important industrially up to about C_{18}. In the manufacturing of unbranched olefins, only homologous mixtures of olefins are obtained; for example, $C_6 - C_9$, $C_{10} - C_{13}$, and $C_{14} - C_{18}$ olefin fractions. The position of the double bond in the unbranched olefin is largely determined by the type of manufacturing process.

3. olefins $> C_5$ (branched):

separation of individual components
from limited number of isomers possible
due to direct synthesis

On the other hand, branched olefins can be manufactured directly, *i.e.*, with a limited number of isomers.

division of higher olefins into two main
groups:

1. unbranched (straight chain, linear)
2. branched

higher unbranched olefins more important
industrially than branched

Higher olefins can thus be divided into two main groups, the branched and the unbranched. Interest in the latter (the straight chain or linear higher olefins) has been increasing, since their linearity gives rise to products with special advantages like biodegradability.

3.3.3.1. Unbranched Higher Olefins

The linear higher olefins can be further subdivided by the position of the double bond into terminal (with the double bond at the end) and internal olefins.

The linear *a*-olefins (LAO) are the most important group industrially. World manufacturing capacity for LAOs was about 2.15×10^6 tonnes per year in 1994, with the USA, Western Europe, and Japan accounting for 1.36, 0.48, and 0.10×10^6 tonnes per year, respectively.

Linear olefins are manufactured using two distinct chemistries:

1. ethylene oligomerization, either in the Ziegler process to form the so-called Ziegler olefins, or using a newer process with organometallic mixed catalysts.

2. dehydrogenation of *n*-paraffins by various methods.

Olefin oligomerization supplies mainly α-olefins, while *n*-paraffin dehydrogenation leads predominantly to α-olefins only in cracking processes.

To 1:

The Ziegler process for the manufacture of α-olefins (Alfen process) consists of a controlled ethylene growth reaction carried out in the presence of triethylaluminum and takes place in two steps:

1. The building of the carbon chain takes place at relatively low temperatures between $90-120\,°C$ and about 100 bar. Ethylene from triethylaluminum is inserted into an alkyl chain bonded to aluminum. A mixture of higher trialkyl-aluminums is formed in accordance with the following simplified scheme:

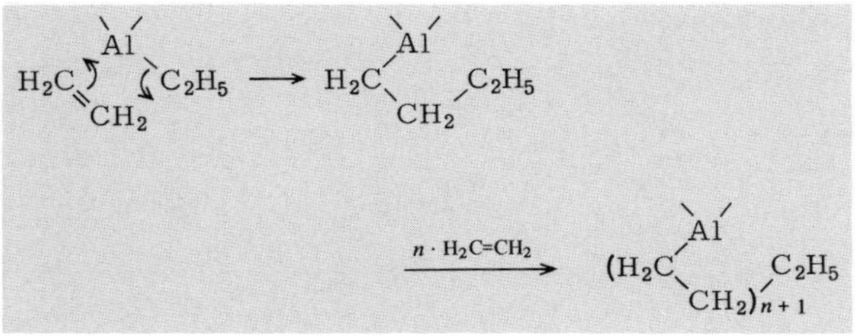

$$(8)$$

2. A mixture of ethylene oligomers is released and the catalyst is regenerated in a displacement reaction at more elevated temperatures ($200-300\,°C$) and lower pressure (50 bar):

Margin notes:

unbranched olefins further subdivided according to position of double bond:

1. terminal or α (double bond at end of chain)
2. internal

manufacturing processes for linear olefins:

1. $H_2C=CH_2$ oligomerization with two process variants for α-olefins

2. dehydrogenation of *n*-paraffins

reaction principles of synthesis:

$Al(C_2H_5)_3$-catalyzed C_2H_4 growth reaction, which is in temperature- and pressure-dependent equilibrium with displacement reaction leading to linear α-olefins

with C_2H_4, displacement reaction occurs when alkyl group contains $10-20$ C atoms; with propene and butene, already after dimerization

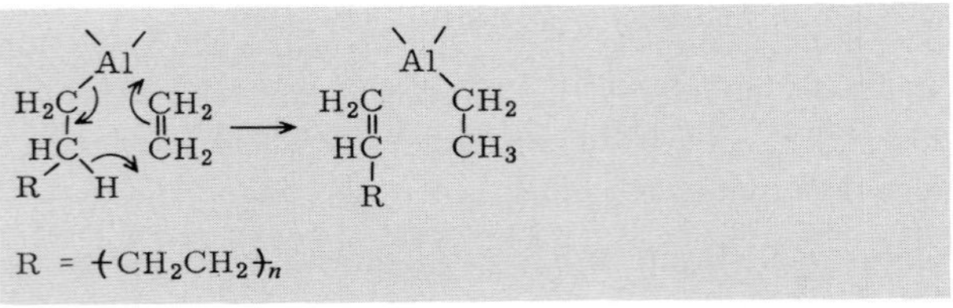

$$R = \{CH_2CH_2\}_n$$

(9)

characteristics of Ziegler olefins:

unbranched olefins with terminal double bond and even C number

common basis for Alfen and Alfol process:

mixtures of higher trialkylaluminum compounds as intermediates

variants of industrial α-olefin manufacture:

1. two-step process with separation of olefin synthesis and displacement

2. single-step process with olefin synthesis and simultaneous displacement

characteristics of single-step process:

1. Al-triethyl as catalyst
2. catalyst hydrolysis product remains in α-olefins
3. broad chain-length distribution with the olefins

The alkyl groups are cleaved as straight chain α-olefins with an even number of carbon atoms. These olefins are obtained in high purity as no isomerization can occur. The first part of the Ziegler olefin synthesis is thus analogous to the Alfol synthesis (cf. Section 8.2.2). Both processes have the same unisolated intermediate, the trialkylaluminum compound $Al[(CH_2CH_2)_n-C_2H_5]_3$. The triethylaluminum, employed in stoichiometrical amounts, must be recovered in an additional process step before being reintroduced.

The industrial manufacture of α-olefins can take place in two steps as already described, i.e., with separate synthesis and subsequent (short period) high-temperature displacement. Furthermore, the synthesis can be coupled with a transalkylation step to produce a higher share of $C_{12}-C_{18}$ olefins, important in alcohol manufacture (cf. Section 8.2.2).

The a-olefin manufacture can also be conducted as a single-step process. The high-temperature ethylene oligomerization is carried out at 200 °C and 250 bar with catalytic amounts of triethylaluminum (ca. 0.5 wt%). The ethylene is restricted in order to obtain a high proportion of unbranched α-olefins. The catalyst is destroyed by alkaline hydrolysis after the reaction and remains in the olefin product. A further characteristic of the single-step process is the broader carbon number distribution (between C_4 and C_{30}) of the resulting olefin mixture compared to the two-step route, as shown in the following table:

Table 3−5. α-Olefin distribution in ethylene oligomerization (in wt%).

α-Olefin	High-temperature process	Low-temperature process
C_4	5	5
C_6-C_{10}	48	50
C_{12}, C_{14}	20	30
C_{16}, C_{18}	13	12
$C_{20}, >C_{20}$	14	3

The α-olefin mixtures are separated by distillation into the required olefin fractions in both processes.

The two-step process is operated by the Ethyl Corporation in the USA (cap. 470000 tonnes per year, 1994), and the single-step process by Chevron in the USA (cap. 295000 tonnes per year, 1994), by Idemitsu in Japan (cap. 50000 tonnes per year, 1994) and by Mitsubishi Chemical in Japan (cap. 50000 tonnes per year, 1994).

industrial use:

1. two-step process:
   Ethyl Corporation
2. single-step process:
   Chevron (Gulf development)
   Idemitsu
   Mitsubishi Chemical (formerly Mizushima Petrochemical)

Other manufacturing processes for α-olefins use ethylene as the feedstock, but differ in the catalyst used for the oligomerization. Exxon has developed a process where ethylene is oligomerized to very pure linear α-olefins with a soluble alkylaluminum chloride/titanium tetrachloride catalyst at temperatures between -70 to $+70\,°C$ in organic solvents. The molecular weights of the olefins increase with higher reaction temperature and decreasing polarity of the solvent. The α-olefins formed are in the range $C_4 - C_{1000}$.

modifications of ethylene oligomerization based on other catalyst systems:

Ti mixed catalysts (Exxon, Mitsui)

Zr mixed catalyst (Idemitsu)

Ni phosphine complexes (Shell)

Variations of a-olefin manufacture have also been developed by Mitsui Petrochemical (mixed catalyst containing titanium), by Idemitsu (mixed catalyst containing zirconium and organic ligands), and by Shell (nickel phosphine complex catalyst). The Shell SHOP process (**S**hell **H**igher **O**lefin **P**rocess) was first used industrially in the USA in 1977, and has since been employed in other plants (*cf.* Section 3.4).

Shell SHOP process as a combination of C_2H_4 oligomerization and olefin metathesis for synthesis of α-olefins

To 2:

n-Paraffins mixed with branched paraffins, present in the petroleum or diesel-oil fractions of crude oils with a paraffin base, are available in sufficient quantities for dehydrogenation. The branched paraffins are usually undesirable components (high pour point) and can be removed from the mixture by freezing. To do this, the oil fraction is diluted and then cooled. Suitable solvents are mixtures of methyl ethyl ketone and benzene (or toluene), or methylene chloride and 1,2-dichloroethane, or liquid propane under pressure. A mixture of branched, unbranched and cyclic paraffins crystallizes out and is filtered. The straight chain component can be separated by adsorption processes.

feedstock for *n*-paraffins to be dehydrogenated:

paraffin mixtures in higher boiling fractions from paraffin-based crude oil

separation of paraffin mixtures:

dilution with solvents followed by cooling at atmospheric or higher pressure

isolation of linear paraffins from mixtures containing branched and cyclic components by two processes:

1. zeolite adsorption
2. urea extractive crystallization

Two processes have proven industrially useful:

to 1:

1. Adsorption on molecular sieves (5 Å zeolites). This can be done either from the gas phase using a carrier gas such as N_2 or H_2, as in processes from BP, Exxon (Ensorb), Leuna (Parex),

alternating adsorption from the gas or liquid phase and desorption by reduced pressure or solvent extraction utilizing the stereoselectivity of the zeolites

Mobil, Texas Oil Company (TSF), UCC (IsoSiv); or from the liquid phase, *e. g.*, UOP (Molex).

The C_5 to C_{24} range of the *n*-paraffins is preferentially adsorbed in the very uniform 5 Å pores of the zeolites, while the bulky cycloparaffins and isoparaffins are unable to penetrate the pores (shape selectivity).

After saturation of the molecular sieves, the *n*-paraffins are desorbed either by reducing the pressure or by extraction with low-boiling hydrocarbons such as *n*-pentane at elevated temperatures and slight pressure. The Leuna extraction process with a NH_3/H_2O mixture leads to a comparatively high rate of desorption.

to 2:

formation of crystalline inclusion compounds (clathrates) with excess urea (*e. g.*, for $C_{24}H_{50}$, ca. 18 mol urea), filtering off and decomposing with steam

2. The urea extractive crystallization process. This is mainly used for the separation of higher *n*-paraffins from C_{15} to C_{30}.

Two well-known processes are the Japanese Nurex process (used until 1979) in which the crystallization takes place by admixing solid urea with a paraffin solution, and the German Edeleanu process which uses a saturated aqueous urea solution.

Pure *n*-paraffins can be recovered and isolated from the separated crystalline urea inclusion compounds by heating to 75 °C. The product comprises 98% linear paraffins.

n-paraffin production (in 1000 tonnes)

	1985	1988	1992
W. Europe	620	650	560
USA	309	406	352
Japan	136	136	157

In 1992, the world production capacity of paraffin was about 1.6×10^6 tonnes per year, with 0.93, 0.39, and 0.16×10^6 tonnes per year in Western Europe, the USA, and Japan, respectively. Production figures for these countries are given in the adjacent table.

three dehydrogenation processes for *n*-paraffins to linear olefins differ in location of double bond:

Three processes have proven commercially useful for converting *n*-paraffins into olefins:

1. Thermal cracking (also steam cracking)
2. Catalytic dehydrogenation
3. 'Chemical dehydrogenation' – chlorination followed by dehydrochlorination

Only thermal cracking produces primarily *a*-olefins; the other two processes yield olefins with internal double bonds.

To 1:

1. thermal dehydrogenation under cracking conditions leads to *a*-olefins of varying chain length, therefore distillative separation into C_6-C_9-, $C_{10}-C_{13}$-, $C_{14}-C_{18}$- fractions

Higher olefins can also be obtained from steam cracking, analogous to the manufacture of lower olefins. Since $C_{20}-C_{30}$ paraffin fractions with a high wax content are frequently employed as the feed, this process is also known as wax cracking. The paraffin cracking is carried out at 500–600°C slightly above atmospheric pressure and a relatively long residence time of 7–15 seconds, and usually in the presence of steam. The conversion rate is adjusted to 25% so that the linearity is retained as far as possible and the double bond is preferentially formed in the terminal, *i. e.*, *α*-position. As thermal cracking of the C−C bond can occur at any part of the molecule, a mixture of olefin homologues results, consisting of 90 to 95% *α*-olefins. The remainder is branched olefins, diolefins, and naphthenes. The medium and higher olefins are isolated from the light gaseous products and then distilled into industrially useful fractions as, for example, C_6-C_9, $C_{10}-C_{13}$, and $C_{14}-C_{18}$. Shell and Chevron, which operate wax cracking processes for *α*-olefin production, have capacities of 310000 and 45000 tonnes per year (1985), respectively. Due to the limited flexibility in providing specifically desired olefins, Shell has since shut down all their plants.

industrial applications of wax cracking:

formerly Shell in Western Europe
Chevron in USA

To 2:

UOP developed the catalytic paraffin dehydrogenation or Pacol-Olex process whereby *n*-paraffins (C_6-C_{19}) are dehydrogenated to olefins in the presence of H_2. The incomplete paraffin conversion (ca. 10%) is carried out in the gas phase on a fixed-bed catalyst (Pt/Al_2O_3 + promoter) at $400-600\,°C$ and 3 bar. The product consists of about 96 wt% linear monoolefins with an internal, statistically distributed double bond. Separation of the unreacted paraffins takes place by means of reversible adsorption of the olefins on solid adsorbents, *e.g.*, molecular sieves.

2. catalytic dehydrogenation leads to internal olefins with UOP 'Pacol Olex' process, combining **p**araffin **c**atalytic **o**lefin **m**anufacture and **o**lefin e**x**traction

In 1994, 25 plants were using this technology and additional plants were under construction. These olefins cannot compete with *α*-olefins from the Ziegler synthesis or the ethylene oligomerization as far as their linearity is concerned. They are, however, markedly cheaper.

This UOP process has recently been extended to the dehydrogenation of C_3-C_5 alkanes (*cf.* Section 3.3.1).

analogous use of C_3-C_5 paraffins in the Oleflex process

To 3:

This route involves a thermally initiated radical chain monochlorination of the *n*-paraffins in the liquid phase at $90-120\,°C$, followed by catalytic dehydrochlorination at $250-350\,°C$ over Al silicate or metallic packing in steel columns:

3. chlorinating dehydrogenation to form internal olefins in a two-step process:

free-radical chlorination in the liquid phase with paraffin conversion up to 40%:

dehydrochlorination in gas phase on $Al_2O_3 \cdot SiO_2$ with metal salt promoters or on metallic tower packing

$$R^1CH_2CH_2R^2 \xrightarrow[-HCl]{+Cl_2} \begin{cases} R^1-CH-CH_2R^2 \\ \qquad | \\ \qquad Cl \\ \\ R^1CH_2-CH-R^2 \\ \qquad\qquad | \\ \qquad\qquad Cl \end{cases} \qquad (10)$$

$$\xrightarrow{-HCl} R^1-CH=CH-R^2$$

$$R^1, R^2 = \text{alkyl}$$

process characteristics:

internally substituted monochloroparaffins favored over terminally substituted so that virtually only internal olefins result from dehydrochlorination

The position of the double bond depends on the position of the chlorine atom in the paraffin. Since the CH_2 groups are more reactive than the CH_3 groups, the double bond is virtually always in the middle. In addition, chlorination beyond mono-substitution must be limited by means of a low chlorination temperature and low conversions (40% at the most). This technology is practiced by Hüls with a capacity of 80000 tonnes per year, feeding the n-paraffins $C_{10}-C_{13}$. Until 1982, the only olefin plant using this process was operated by Shell in the USA.

multiple chlorination leads to mixtures of polychloroalkanes which become increasingly inert and wax-like as the chlorine content increases

In addition to the monochloroparaffins and their use in n-olefin manufacture, polychloroparaffins with varying chlorine content (15–70 wt%) are of industrial significance. To manufacture them, n-paraffins in the range $C_{10}-C_{30}$ are chlorinated in a bubble column reactor at 60–120 °C and slight pressure with 100% chlorine conversion. These chloroalkanes are used as plasticizers for PVC and raw materials for paint, as mineral oil additives and as impregnation agents for flame-resistant and water-repellant textile finishes. The use of unbranched paraffins for production of chloroalkanes in Western Europe, the USA, and Japan is given in the adjacent table. Other uses for n-paraffins include the manufacture of fatty acids and fatty alcohols (cf. Section 8.2).

Use of linear alkanes for chloroalkanes (in 1000 tonnes):

	1988	1990	1992
W. Europe	70	68	68
USA	9	10	12
Japan	10	10	11

uses of linear olefins:

C_6-C_{10} olefins – via hydroformylation and hydrogenation – form 'oxo' alcohols, useful as plasticizers or for plasticizer manufacture

internal olefins and α-olefins produce similar aldehyde mixtures after hydroformylation due to simultaneous double bond isomerization (preferential reaction at end of chain)

Typical uses of the linear olefins depend on their chain length. The C_6-C_{10} olefins are converted into C_7-C_{11} alcohols by hydroformylation and hydrogenation (cf. Section 6.1). These alcohols are used as solvents or, after esterification with phthalic anhydride, as plasticizers. Olefins with an internal double bond are just as suitable as the a-olefins, since isomerization of the double bond to the terminal position takes place during hydroformylation. As a result, α-olefins and linear olefins possessing an internal double bond give virtually the same aldehyde product composition.

C_6-C_{10} olefins are also converted to a variety of **a**-olefin **oligo**mers (AOO), depending on the type of catalyst and the experimental conditions, which are particularly well suited as high quality lubricating oils.

C_6-C_{10} olefins for oligomers

The C_{10} to roughly C_{13} fractions are employed in the alkylation of benzene (*cf.* Section 13.1.4). A widely used UOP process is based on homogeneous catalysis with HF, which in newer plants has been improved by the use of an acidic heterogeneous catalyst. In the presence of a Lewis or proton (Brönstedt) acid catalyst, benzene alkylation and partial isomerization of the olefin occur simultaneously, resulting in a statistical distribution of the benzene ring along the paraffin chain. **L**inear **a**lkyl**b**enzenes (LAB) from *a*-olefins are thus similar to those obtained from internal linear olefins.

$C_{10}-C_{13}$ olefins for alkylation of benzene (LAB manufacture)

process characteristics of benzene alkylation:

simultaneous double bond isomerization of olefin leads to statistical distribution of phenyl group along linear C-chain

$$H_3C \diagup (CH_2)_n \diagdown CH \diagup (CH_2)_m \diagdown CH_3$$

Another manufacturing route for linear alkylbenzenes involves the direct alkylation of benzene with chloroparaffins, without initial dehydrochlorination to olefins (*cf.* Section 13.1.4). The consumption of *n*-paraffins for linear alkyl benzenes using both technologies is given in the adjacent table for Western Europe, the USA and Japan.

alternate benzene alkylation with monochloroparaffins

Use of linear alkanes for alkylbenzenes (in 10^6 tonnes):

	1988	1990	1992
W. Europe	0.31	0.33	0.30
USA	0.26	0.28	0.16
Japan	0.11	0.14	0.11

After sulfonation these *sec*-alkylbenzenes yield raw materials for biodegradable, anion-active detergents.

Furthermore, *a*-olefins can be directly reacted with SO_3 to alkene sulfonates (or *a*-**o**lefin sulfonates, AOS). The sultones, which are formed simultaneously, must be hydrolyzed to hydroxyalkylsulfonates to improve their water solubility:

$C_{14}-C_{18}$ olefins for SO_3 reaction to alkene sulfonates (AOS)

$$RCH_2CH{=}CH_2 + SO_3$$

$$\longrightarrow RCH_2CH{=}CHSO_3H$$

$$\longrightarrow RCH \diagup{CH_2} \diagdown CH_2 \xrightarrow{+H_2O} RCHCH_2CH_2SO_3H$$
$$\quad \diagdown O{-}S \diagup \quad\quad\quad\quad\quad |$$
$$\quad\quad\quad O_2 \quad\quad\quad\quad\quad\quad\quad OH$$

(11)

Alkene sulfonates are being produced on a larger scale only by the Lion Fat & Oil Co. (Japan). The alkene sulfonates are slightly less sensitive to hardening than the alkyl sulfonates, but there is no difference in their biodegradability and washing properties.

applications of alkyl benzenesulfonates and alkenesulfonates:

biodegradable detergent bases

However, alkyl sulfonates (or alkane sulfonates) have achieved considerably greater importance as active detergent substances (WAS = **w**ash **a**ctive **a**lkylsulfonates) due to their inexpensive manufacture from *n*-paraffins ($C_{12}-C_{18}$). Sulfochlorination

alternate raw materials for detergents:

alkyl sulfonates (alkane sulfonates) manufacture of alkyl sulfonates:
1. sulfochlorination
2. sulfoxidation

and sulfoxidation are generally employed for their large scale manufacture.

process principles of sulfochlorination:

UV initiated, radical reaction of $C_{12}-C_{18}$ paraffins with SO_2/Cl_2 to *sec*-alkyl sulfochlorides

In the continuously-operated sulfochlorination process, the *n*-paraffin mixtures are converted into *sec*-alkyl sulfochlorides with SO_2/Cl_2 mixtures at $20-35\,°C$ and mercury vapor lamps, and are then hydrolyzed with caustic soda to the corresponding alkyl sulfonates:

$$\begin{array}{l} \underset{R^1}{\overset{R}{\diagdown}}CH_2 + SO_2 + Cl_2 \xrightarrow[-HCl]{h\nu} \underset{R^1}{\overset{R}{\diagdown}}CHSO_2Cl \\[2mm] \xrightarrow[-NaCl]{+2\,NaOH} \underset{R^1}{\overset{R}{\diagdown}}CHSO_3Na \end{array} \tag{12}$$

Paraffin conversion is limited to about 30% in order to avoid the undesired formation of disulfochlorides. After the hydrolysis step and subsequent purification, unreacted paraffins are recycled to sulfochlorination.

process principles of sulfoxidation:

UV initiated, radical reaction of $C_{12}-C_{18}$ paraffins with SO_2/O_2 to *sec*-alkyl sulfonic acids

Sulfoxidation of *n*-paraffin mixtures is accomplished with UV light and SO_2/O_2 mixtures at $25-30\,°C$:

$$\underset{R^1}{\overset{R}{\diagdown}}CH_2 + SO_2 + 0.5\,O_2 \xrightarrow{h\nu} \underset{R^1}{\overset{R}{\diagdown}}CHSO_3H \tag{13}$$

commercial application of sulfoxidation:

Hoechst light–water process in Germany, France, Netherlands

Hüls plant in Germany

The Hoechst light–water process and a similar process from Hüls with a different conversion technique are used commercially. Here water, continuously introduced into the reaction (bubble) column, acts with SO_2 to convert the intermediate alkyl sulfonyl hydroperoxide to alkyl sulfonic acid and sulfuric acid:

$$\underset{R^1}{\overset{R}{\diagdown}}CHSO_2-O-O-H + H_2O + SO_2 \longrightarrow \underset{R^1}{\overset{R}{\diagdown}}CHSO_3H + H_2SO_4 \tag{14}$$

The insoluble alkyl sulfonic acid is then separated from the hot aqueous solution.

Currently, the only sulfoxidation processes in operation are in four plants in Western Europe, with a total capacity of 96000 tonnes per year (1991). Similar processes have also been developed by ATO Chemie, Enichem, and Nippon Mining.

3.3.3.2. Branched Higher Olefins

The oligomerization and co-oligomerization of light olefins such as propene, isobutene, and *n*-butenes are the favored routes for the manufacture of branched olefins with six or more carbon atoms. The production capacity (1991) for, for example, propylene dimers and oligomers (*i. e.,* nonenes and dodecenes, for example) of this type was about 0.55, 0.28, and 0.08×10^6 tonnes per year in the USA, Western Europe, and Japan, respectively.

The current feedstocks, reaction products and manufacturing processes are summarized in the following table:

manufacture of branched monoolefins by oligomerization and co-oligomerization of:

propene
isobutene
n-butenes

Table 3−6. Feedstocks and manufacture of branched olefins.

Feedstock	Type of process	Reaction product
Propene	Dimerization with AlR_3	2-Methyl-1-pentene
Propene	Dimerization with alkali metals	4-Methyl-1-pentene
Propene	Tri- and tetramerization with H_3PO_4	Isononenes, isododecenes
Propene and *n*-butenes	Co-dimerization	Isoheptenes
Isobutene	Dimerization preferably with H_3PO_4/support but also with ion-exchangers or with H_2SO_4	'Diisobutenes'

Olefin oligomerization processes can be divided into three main groups according to the type of catalyst employed:

1. Acidic catalysis with mineral acids such as H_3PO_4 or H_2SO_4, and with acidic ion-exchangers
2. Organometallic catalysis with Al-alkyls and possibly co-catalysts
3. Organometallic catalysis with alkali metals

three catalyst systems with marked product specficity:
1. proton catalysts
2. organo-Al catalysts
3. alkali metal catalysts

To 1:

The UOP process employing H_3PO_4/SiO_2 in a fixed bed is widely used industrially. In this process, propene-rich fractions are oligomerized at 170−220 °C and 40−60 bar with 90% propene conversion to yield a liquid reaction mixture consisting mainly of 'tri-' and 'tetrapropene'. Both products were once used mainly for the alkylation of benzene and phenol. However, sulfonates manufactured from alkyl derivatives have greatly diminished in importance as detergent bases due to their poor biodegradability (*cf.* Section 13.1.4). Increasing amounts of tetrapropene are undergoing hydroformylation, analogous to diisobutene (described below), and then hydrogenation to produce plasticizer alcohols.

1. processes involving proton catalysis:

1.1. H_3PO_4/SiO_2 gas-phase oligomerization of propene at 170−220 °C with UOP process

used as motor fuel additive (after hydrogenation) to increase octane rating

feed olefin for oxo alcohols (plasticizing alcohols) which are then esterified with phthalic anhydride and used as plasticizers

1.2. H_2SO_4 or ion exchange liquid-phase oligomerization process

example:
Bayer process for isobutene, also suitable for propene

uses of diisobutene:

precursor for
1. motor fuel additive
2. isononanol, a plasticizer alcohol

Isobutene oligomerization with H_2SO_4 or acid ion-exchangers (Bayer Process) has been mentioned previously (*cf.* Section 3.3.2). After hydrogenation, 'diisobutene' is used as an additive in motor gasoline. It is also a precursor in the manufacture of isononanol by hydroformylation and hydrogenation. The alcohol is employed in plasticizer manufacture.

Exxon Chemical is the world's largest producer of branched higher olefins using this technology.

To 2:

2. alkyl-Al-catalyzed processes:

2.1. without cocatalyst – high selectivity and reduced activity
2.2. with transition metal compounds or complexes as cocatalysts – reduced selectivity but higher activity

There are two basic parameters affected by the choice of catalyst in the dimerizations and co-dimerizations of lower olefins. When a trialkylaluminum alone is used, high selectivity can be attained. When it is used in combination with a transition metal cocatalyst such as a nickel salt, the activity of the system increases at the expense of the selectivity.

Industrial applications have been developed for both types of processes.

process example for 2.1:

Goodyear SD process for propene dimerization to 2 MP1, precursor for isoprene manufacture

In the Goodyear Scientific Design process, propene is dimerized to 2-methyl-1-pentene, an isoprene precursor (*cf.* Section 5.2.2), using tripropylaluminum at at 200 °C and 200 bar:

$$
\begin{array}{c}
\overset{\displaystyle CH_3}{\overset{|}{H_2C=CH}} + H_2C=CHCH_3 \xrightarrow{\ [R_3Al]\ } \\[2em]
\overset{\displaystyle CH_3}{\overset{|}{H_2C=C-CH_2CH_2CH_3}}
\end{array}
\qquad (15)
$$

process example for 2.2:

IFP Dimersol process favored for codimerization of C_3H_6 + C_4H_8 (see Alphabutol process for dimerization of C_2H_4, Section 3.3.2)

The other process was developed by IFP (Dimersol process). In this liquid-phase process, propene or butene can be homodimerized, or propene and butene can be codimerized. The Dimersol process is operated continuously with a trialkylaluminium/nickel salt catalyst at 60 °C and 18 bar with 50% selectivity, producing relatively few branched isoheptenes. The low selectivity is not due solely to the addition of cocatalyst, but also to the homodimerization of the two reaction partners, which takes place simultaneously with statistical probability. If only propene or *n*-butene is introduced into the dimerization, then isohexenes or isooctenes are formed with high selectivity (85–92% at 90% conversion), and are distinguished by their low degree of branching.

This dimerization process to C_6, C_7, and C_8 isoolefins is therefore a good source of intermediates for the production of high-octane gasoline as well as feedstocks for hydroformylation, where branching on both sides of the olefinic double bond would sharply reduce the rate of the 'oxo' reaction.

slightly branched C_6, C_7, C_8 isoolefins very suitable for hydroformylation

Since the first commercial operation of the Dimersol process in 1977, worldwide production capacity has grown to ca. 3×10^6 tonnes per year in 26 plants.

To 3:

The third possibility, dimerization of propene with an alkali metal catalyst, is of only limited industrial significance. The reaction is conducted in the liquid phase at ca. 150°C and 40 bar using Na/K_2CO_3 catalyst. Selectivity to 4-methyl-1-pentene (4MP1) can reach 80%:

3. processes using alkali metal catalysis:

C_3H_6 liquid-phase dimerization to 4MP1 with high selectivity but low activity, BP process

$$H_2C=CHCH_3 + \overset{\overset{\displaystyle CH_3}{|}}{HC}=CH_2 \xrightarrow{\text{[cat.]}}$$

$$H_2C=CHCH_2-\overset{\overset{\displaystyle CH_3}{|}}{CH}-CH_3 \quad (16)$$

4MP1 is produced by BP in a 25 000 tonne-per-year plant and is polymerized by ICI to high melting (240°C) transparent polymer known as TPX. In 1975 Mitsui Petrochemical began operation of a plant using the BP technology (current capacity, 15 000 tonnes per year) to produce monomer to be used in the manufacture of poly-4-methyl-1-pentene. Phillips Petroleum began smaller-scale production in 1986. 4MP1 is also used as a comonomer in polyethylene (LLDPE).

process example for 3:

Mitsui Petrochemical manufacture and polymerziation of 4MP1 to highly crystalline thermoplast; comonomer in LLDPE

3.4. Olefin Metathesis

Olefin metathesis is an exchange reaction between two olefins in which alkylidene groups are interchanged. Formally, this occurs *via* double bond cleavage, though the transition state has not yet been described explicitly. In the simplest example of a metathesis reaction, two moles of propene react to form one mole of ethylene and one of *n*-butene (mainly 2-butene):

principle of olefin metathesis:

reversible, catalytic exchange of alkylidene groups between two olefins

mechanism of metathesis explained using four-center transition state:

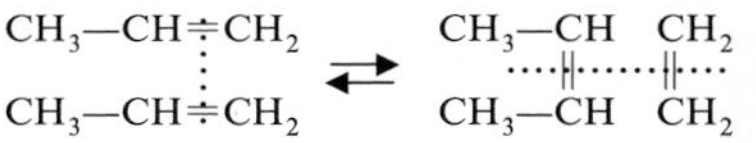

$$\begin{array}{c} CH_3-CH \doteq CH_2 \\ \vdots \\ CH_3-CH \doteq CH_2 \end{array} \rightleftharpoons \begin{array}{c} CH_3-CH \quad CH_2 \\ \cdots \| \cdots \cdots \| \cdots \\ CH_3-CH \quad CH_2 \end{array}$$

$$2\,CH_3CH=CH_2 \xrightleftharpoons{\text{[cat.]}}$$

$$H_2C=CH_2 + CH_3-CH=CH-CH_3 \quad (17)$$

disproportionation – alternative designation – covers possible metathesis reactions only to a limited extent

This transition of an olefin with a certain carbon number into two other olefins of lower and higher carbon numbers resulted in the term disproportion being applied to the reaction. However, as the metathesis is a reversible reaction, this is incorrect.

catalysts for olefin metathesis:

favored – Mo, W, Re

1. in homogeneous phase, as salts or carbonyls with organometallic compounds and O-containing promoter, e. g., $WCl_6/C_2H_5AlCl_2/C_2H_5OH$

2. in heterogeneous phase, e. g., as oxides on supports

metal oxides: WO_3, $CoO—MoO_3$, Re_2O_7
supports: Al_2O_3, SiO_2

Metathesis reactions generally use catalyst systems based on molybdenum, tungsten, or rhenium. In the liquid phase, soluble halides or carbonyls are reduced to lower valency states by the simultaneous presence of organometallic compounds and activated using oxygen-containing promoters. In the gas phase, the favored catalysts are the oxides, sulfides or carbonyls of the above-mentioned metals on supports with large surface areas.

While homogeneous catalysts are often effective at room temperature, heterogeneous catalysts require reaction temperatures of up to 500 °C.

The metathesis reaction can increase the flexibility of the olefin manufacturing processes. The ethylene share from naphtha cracking mixtures could, for example, be increased at the cost of the propene, simultaneously producing additional butene feedstock for dehydrogenation to butadiene.

advantages of olefin metathesis:

refinery flexibility can be increased as a function of raw material and cracking process to meet changing market requirements

Metathesis can also be used in combination with ethylene oligomerization in countries where natural gas cracking is favored to increase an otherwise inadequate supply of higher olefins.

Other opportunities for metathesis include synthesis of otherwise inaccessible diolefins and ring-opening polymerizations of cycloolefins.

first industrial applications of olefin metathesis:

Phillips Petroleum Triolefin process operated by Shawinigan in a 30 000 tonne-per-year plant for about six years

In the 1960s, Phillips Petroleum introduced the metathesis of propene as the Triolefin process. An industrial plant with a capacity of 30 000 tonnes butene per year was operated by Shawinigan in Canada between 1966 and 1972. It was shut down for economic reasons, due to an altered feedstock situation.

Triolefin: one olefin as educt
         two olefins as products

types of catalysts for Triolefin process in order of increasing activity:

1. WO_3/SiO_2
2. $CoO—MoO_3/Al_2O_3$
3. Re_2O_7/Al_2O_3

The reaction conditions depend on the catalyst employed. With Co-molybdate catalysts, temperatures of 120−210 °C at 25−30 bar are sufficient to obtain propene conversions of ca. 40%. Operating temperatures of 450−500 °C are required with WO_3/SiO_2; at 500 °C the propene conversion is about 42%. The propene feed must be free from acetylenes and diolefins , so they are removed by selective hydrogenation with a palladium catalyst.

A rhenium catalyst was developed by BP for metathesis processes. In the form of Re_2O_7/Al_2O_3, it is substantially more active than the two previously mentioned catalysts and is, in principle, already active at room temperature.

A recent need for additional propene in the USA led to another application of the Phillips process. Since 1985, Lyondell (Arco) has operated a 136 000 tonne-per-year plant where ethylene is first dimerized to 2-butene; the butene is then reacted with additional ethylene in a metathesis step to produce propene. This process is in use in two additional plants in the USA.

New use of Phillips process in the USA – propene manufacture from $H_2C=CH_2$ and $CH_3CH=CHCH_3$

Metathesis is also used commercially as a part of a combination process from Shell (SHOP = **S**hell **H**igher **O**lefins **P**rocess). The initial capacity in 1977 was 104 000 tonnes per year, but has since been expanded several times. Here ethylene is first oligomerized at $80-120\,°C$ and $70-140$ bar in the presence of a phosphine ligand (*e.g.*, $(C_6H_5)_2PCH_2COOK$) to a mixture of even-numbered linear α-olefins, from which $C_{10}-C_{18}$ olefins for use in detergents are isolated directly. Higher and lower olefins are subjected to a combination of double bond isomerization and metathesis. Isomerization will produce primarily olefins with internal double bonds, in a statistical mixture. If the subsequent metathesis is performed on these olefins alone, then a new mixture of internal olefins whose chain length distribution depends on the location of the original double bond will be produced. If, on the other hand, ethylene is added to the isomerized olefins, they can be cleaved to terminal olefins (ethenolysis):

Shell **H**igher **O**lefin **P**rocess (SHOP) operating since 1977 in 320 000 tonne-per-year plant in USA

process steps in SHOP:

1. ethylene oligomerization
2. double bond isomerization
3. metathesis (ethenolysis)

$$
\begin{array}{c}
CH_3\text{-}(CH_2)_{\overline{m}}CH{=}CH\text{-}(CH_2)_{\overline{n}}CH_3 \\
+ \\
H_2C{=}CH_2 \qquad \longrightarrow \qquad \text{(18)} \\[4pt]
CH_3\text{-}(CH_2)_{\overline{m}}CH \quad CH\text{-}(CH_2)_{\overline{n}}CH_3 \\
\| \qquad + \quad \| \\
H_2C \qquad\quad CH_2
\end{array}
$$

In this way addition olefins in the desired C range $(C_{11}-C_{14})$ can be obtained. These olefins can, for example, undergo hydroformylation to produce $C_{12}-C_{15}$ detergent alcohols. Undesired higher and lower olefins are recycled. Altogether, an n-α-olefin concentration of $94-97\%$ with a monoolefin content of more than 99.5% is obtained.

characteristics of ethenolysis:

internal even-numbered olefins catalytically cleavable with ethylene to terminal odd-numbered olefins

Since the first commercial application of the SHOP process in the USA, many plants have begun operation. By 1993 the worldwide capacity for α-olefins by the SHOP process had grown to about 10^6 tonnes per year.

Other areas of application could result from the series of reactions being studied by many firms. One example is the manufacture of neohexene (3,3-dimethyl-1-butene) from technical grade

Examples of applications of metathesis polymerization:

1. manufacture of neohexene by ethenolysis of diisobutene (Phillips)

'diisobutene' (approx. 75% 2,4,4-trimethyl-2-pentene) by ethenolysis. Isobutene is formed as coproduct:

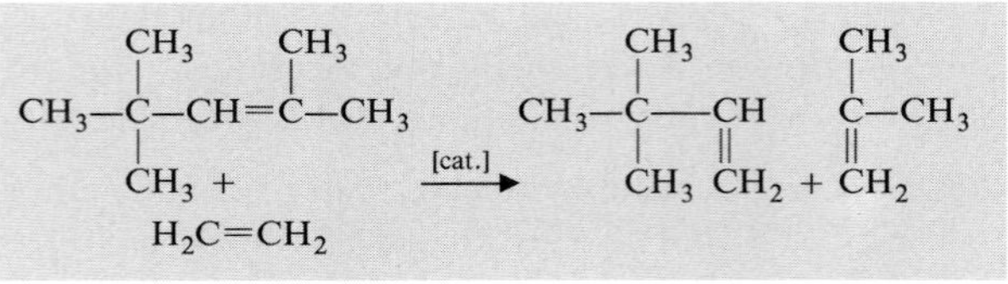

$$
\begin{array}{c}
\underset{\substack{|\\CH_3\ +}}{CH_3-\underset{\substack{|\\CH_3}}{C}-CH=\underset{\substack{|\\CH_3}}{C}-CH_3} \\
H_2C=CH_2
\end{array}
\xrightarrow{[cat.]}
\underset{\substack{|\\CH_3}}{CH_3-\underset{\substack{|\\CH_3}}{C}-\underset{\substack{\|\\CH_2}}{CH}}\quad \underset{\substack{\|\\CH_2}}{C-CH_3} \qquad (19)
$$

Phillips has begun operation of a smaller plant (initial capacity 1600 tonnes per year but expanded several times) for the production of neohexene, most of which is reacted further to produce fragrances.

2. manufacture of a,ω-diolefins by ring opening ethenolysis of cycloolefins (Shell)

Another example is the manufacture of a,ω-diolefins by ethenolysis of cycloolefins. Using this method, readily available cyclododeca-1,5-9-triene (*cf.* Section 10.1.2) can be initially selectively hydrogenated to cyclododecene, which can then undergo metathesis with ethylene to produce tetradeca-1,13-diene:

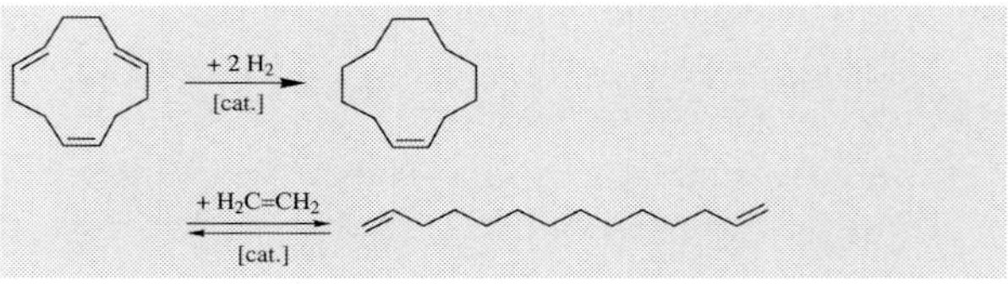

$$\qquad (20)$$

Other a,ω-diolefins, primarily 1,5-hexadiene and 1,9-decadiene, are produced by Shell in large quantities (total ca. 3000 tonnes per year) in a multipurpose facility.

a,ω-Dienes are of industrial interest as crosslinking agents in olefin polymerization, and for the manufacture of bifunctional compounds.

3. metathesis polymerization of cycloolefins to polyalkenamers – side reaction of cycloolefin ethenolysis

Cyclododecene can also be a starting material for polydodecenamers, which are formed in a parallel metathesis reaction. This is the main reaction in the absence of ethylene. The reaction starts with a ring expansion to form a cyclic diolefin with twice the number of C atoms:

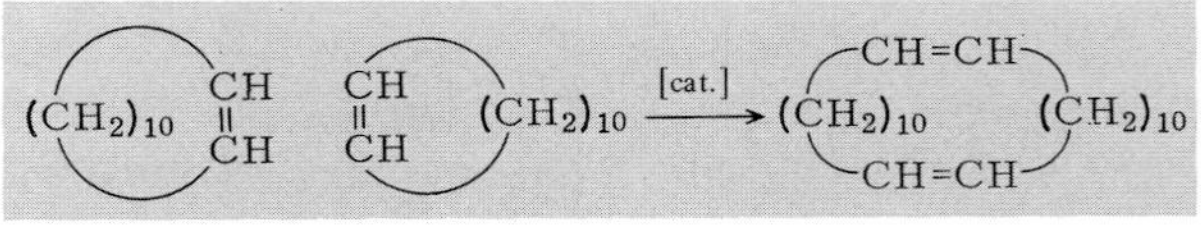

$$\qquad (21)$$

Open-chain structures are also present with the macrocyclic polydodecenamers. Catalyst systems containing tungsten, e. g., $WCl_6/C_2H_5OH/C_2H_5AlCl_2$, can be used.

A major industrial use of this reaction type is found in the metathesis polymerization of cyclooctene to polyoctenamers by Hüls (cap. in 1989 ca. 12000 tonnes per year). The feedstock cyclooctene is produced by the catalytic cyclodimerization of butadiene to *cis,cis*-1,5-cyclooctadiene and subsequent partial hydrolysis.

examples of applications of metathesis polymerization:

1. manufacture of polyoctenamers
$$\{CH-(CH_2)_6-CH\}_x$$
(Hüls, Vestenamer®)

Polyoctenamers are vulcanizable elastomers used in the rubber industry.

The oldest polyalkenamer is polynorbornene, produced since 1976 by CDF-Chemie in a 5000-tonne-per-year unit by the metathesis polymerization of norbornene with a W- or Rh-catalyst system:

2. manufacture of polynorbornene
(CDF-Chemie, Norsorex®)

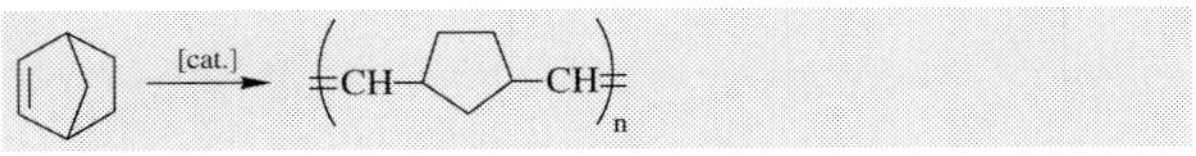

$$(22)$$

Polynorbornene is vulcanized to give a rubbery material used for damping vibration and noise.

Other industrially interesting, though not yet applied, examples of metathesis reactions include the production of isoprene by disproportionation of isobutene and 2-butene (*cf.* Section 5.2.2), and a styrene technology in which toluene is dimerized to stilbene, which then undergoes metathesis with ethylene to give the product (*cf.* Section 13.1.2.).

possible future metathesis for the manufacture of:

isoprene
styrene

4. Acetylene

4.1. Present Significance of Acetylene

Until the blossoming of the petrochemical industry in the USA in the 1940s and in Western Europe in the early 1950s, acetylene was one of the most important raw materials for industrial organic chemistry.

Olefins have replaced acetylene as a raw material during this development, especially for numerous monomers. The olefins are more available (and thus more economical) and easier to handle than acetylene.

By way of example, the large decrease in the fraction of important industrial chemicals derived from acetylene between 1965 and 1995 in the USA is shown in the following table:

'acetylene flow sheets' must be replaced by 'olefin flow sheets'

olefins are:
cheaper, mass produced, transportable by pipeline, safer but less reactive than acetylene

Table 4-1. Production of industrial chemicals based on acetylene and olefins in the USA (in 1000 tonnes).

	Acetylene based				Olefin based		
	1965	1974	1987	1995	1974	1987	1995
Vinyl chloride	159	59	73	36	2490	3657	6754
Acrylonitrile	91	–	–	–	640	1157	1455
Chloroprene	82	–	–	–	144	110	105[1]
Vinyl acetate	64	32	–	–	604	1137	1312
Miscellaneous (1,4-butanediol, THF, acrylic acid/esters, etc.)	68	91	60	87			

[1] 1993

In the USA, acetylene manufacture reached its peak of nearly 500000 tonnes per year in the 1960s. The worldwide capacity for acetylene production in 1995 was about 800000 tonnes per year, of which 260000, 190000, and 180000 tonnes per year were located in Western Europe, the USA, and Japan, respectively.

Production figures for acetylene in several countries are given in the adjacent table.

acetylene production[1] (in 1000 tonnes):

	1988	1991	1993
Germany	219	216	103
USA	173	175	157
Japan	91	84	73

[1] for chemical and other uses

earlier C_2H_2 base replaced today by:

C_2H_4 for: CH_3CHO
$\qquad$ $H_2C=CHCl$
$\qquad$ $H_2C=CHOAc$
$\qquad$ $Cl_2C=CHCl$
$\qquad$ $Cl_2C=CCl_2$
C_3H_6 for: $H_2C=CHCN$
$\qquad$ $H_2C=CHCOOH$
C_4H_6 for: $H_2C=C\!-\!CH=CH_2$
$\qquad\qquad\quad$ $|$
$\qquad\qquad\quad$ Cl

return to coal as basis for C_2H_2 with advantages such as:

1. C selectivity with C_2H_2 higher and
2. H_2 requirements lower than with C_2H_4, C_3H_6

but the disadvantage of higher specific energy requirements for C_2H_2 manufacture

coal-based chemical process for C_2H_2 favored in countries with inexpensive raw materials and low energy costs

other uses for CaC_2:

1. traditional $CaCN_2$ production limited by synthetic fertilizer
2. new use in desulfurization of pig iron increasing

arguments for present uses of acetylene:

1. capacity of an amortized plant for C_2H_2 secondary products covers demand, and conversion requires large investment

examples:

$H_2C=CHCOOH$ and HCl use in VCM manufacture from EDC cracking and addition to C_2H_2

Large-volume intermediates such as acetaldehyde, vinyl chloride, vinyl acetate and chlorinated solvents such as tri- and perchloroethylene (once a domain of acetylene chemistry) are now being manufactured from ethylene in modern plants operating industrially mature processes. Similarly, acrylonitrile and, to an increasing extent, acrylic acid are being produced from propene, and chloroprene is being manufactured from butadiene.

Retrospection on the large worldwide reserves of coal triggered by the oil crisis led to the development of new and improved processes for acetylene manufacture (cf. Section 4.2.2). However, because olefin manufacture is more energy efficient, advantages such as the higher carbon selectivity in many processes for acetylene manufacture from coal and the lower H_2 requirements are only economically useful when the coal price is sufficiently low.

This applies to countries which have very inexpensive hard coal — and thus electricity — such as Australia, India and South Africa. In these places, new plants for the manufacture of vinyl chloride, acetic acid and vinyl acetate based on acetylene are still being built. In many cases, the route chosen for production of acetylene goes through calcium carbide, an intermediate with numerous uses.

Calcium carbide, besides its accepted but decreasing use in the chemical industry for the manufacture of calcium cyanamide (presently ca. 50% of the CaC_2 production worldwide), has recently found an interesting application in the steel industry for the desulfurization of pig iron. $5-10$ kg CaC_2 per tonne of pig iron are required, opening up a new market which could markedly affect the carbide demand and thus influence the supply and cost of acetylene.

In other countries, there are basically four reasons for the further use of acetylene in chemical processes:

1. Large, tax-amortized plants exist whose capacities can cover the demand for one product. The transfer to another feedstock base would mean building new plants with high investment costs, and — for the most part — running technically difficult processes based on olefins.

Typical examples are the acrylic acid manufacture by carbonylation of acetylene (cf. Section 11.1.7.1) and the ethylene–acetylene combined process for the manufacture of vinyl chloride (cf. Section 9.1.1.2).

2. There are low-volume products which can be readily manufactured with C_2H_2 in which the C_2 part of the whole
   molecule is small and therefore insignificant with regard to
   material costs, as for example with the vinyl esters of higher
   carboxylic acids, vinyl ethers of higher alcohols (*cf.* Section
   9.2.2 and 9.2.3), and N-vinyl compounds (*cf.* Section 4.3).

   2. product is made in small quantities and
      has minor C_2 mole content

      examples:

      vinyl esters and ethers of higher alcohols,
      N-vinyl compounds

3. There are no suitable alternative processes, *i. e.* with comparable investment costs, inexpensive feedstocks, economic
   yields, and conventional technology.

   3. no economical alternative processes
      available:

      examples:

1,4-Butanediol (*cf.* Section 4.3) is one example; another is acetylene black which is required for special applications.

   1,4-butanediol, acetylene black for
   batteries (*e. g.*, Acetogen® Hoechst-
   Knapsack), or as extender for rubber

4. Since the oil supply is limited and can be influenced by
   political decisions, a partial independence from oil is being
   achieved through carbide-based acetylene, setting possible
   economic considerations aside.

   4. independence arising from coal-based
      C_2H_2 overshadows economic
      considerations

4.2. Manufacturing Processes for Acetylene

4.2.1. Manufacture Based on Calcium Carbide

The worldwide production of calcium carbide from the strongly
endothermic ($\Delta H = +111$ kcal or 465 kJ/mol) electrothermal
reaction between quicklime and coke at $2200-2300\,°C$ reached
a maximum of about 10^7 tonnes per year in 1960. Current
production worldwide is about 5×10^6 tonnes per year. Production figures for several nations can be seen in the accompanying
table. Despite the decline in production, new interest in modern
carbide/acetylene manufacture has been awakened in several
countries. This is shown, for example, by research at the US
Institute of Gas Technology on the use of solar energy for
CaC_2 synthesis.

classic manufacture of acetylene based on
calcium carbide of significantly decreasing
importance

calcium carbide production (in 1000 tonnes):

	1988	1990	1993
CIS	604	580	540
Japan	342	344[1)]	245
USA	225	236	222

[1)] 1989

Calcium carbide can be converted into acetylene and calcium
hydroxide in an exothermic reaction with H_2O:

main cost factors:

electrical energy (ca. 10 kwhr/kg C_2H_2)
and coal

$$CaC_2 + 2\,H_2O \rightarrow HC\equiv CH + Ca(OH)_2$$
$$\left(\Delta H = -\,\frac{31\ \text{kcal}}{130\ \text{kJ}}\big/\text{mol} \right) \qquad (1)$$

Industrially, acetylene is formed in one of two types of
generators. In the wet generator, an excess of water is used. In a
dry generator, an approximately stoichiometric amount of water
is used, and the calcium hydroxide is obtained as a pourable

hydrolysis of CaC_2 by two commercial
variations:

1. wet generator with H_2O excess, technically simpler, but wastewater generation

2. dry gasification with approximately stoichiometric amount of H_2O

purification of acetylene in three steps before use as feedstock:

1. H_2SO_4 washing (NH_3 from $CaCN_2$)
2. oxidation with $HOCl$ or H_2SO_4
3. NaOH washing

chlorine water should not contain any free Cl_2, as it can react violently with C_2H_2

manufacture of acetylene by pyrolysis of hydrocarbons

general process characteristics:

endothermic cracking process requires large energy supply at high temperature with short residence time of the cracked products

the equilibration resulting in formation of elements C and H_2 must be prevented by short residence time (quenching)

C_2H_2 isolation:

low C_2H_2 concentration in cracked gas requires washes with solvents

pyrolysis processes can be classified into three types according to heat source (allothermal = external heat source, endothermal = internal heat source) and transfer method:

1. electrothermal cracking

2. cracking with heat transfer agent

powder. The heat of reaction is removed by evaporation of a portion of the added water.

Continuously operating large-scale dry generators with an output up to 3750 m^3/h $C_2H_2 \triangleq 32\,000$ tonnes per year have been developed by Knapsack and Shawinigan. The byproduct calcium hydroxide is either recycled (up to 50%) for carbide manufacture or employed in construction, in the chemical industry, or in agriculture.

Before being used as a chemical feedstock, C_2H_2 is washed in three steps. After removing basic products such as NH_3 by means of a H_2SO_4 wash, impurities containing P and S are oxidized either with chlorine water or 98% H_2SO_4 and scrubbed. In the third step, traces of acid are removed with caustic soda solution.

4.2.2. Thermal Processes

Numerous processes for the manufacture of C_2H_2 are based on the uncatalyzed pyrolysis of hydrocarbons in the C-range from methane to light petrol to crude oil. With newer process developments, higher boiling fractions, residual oils, and even coal can also be used.

In principle, thermodynamics and kinetics have a deciding influence on the choice of reaction conditions.

Important for all processes are a rapid energy transfer at a high temperature level ($> 1400\,°C$), very short residence times of the feed or reaction products (10^{-2} to 10^{-3} s), low partial pressure of acetylene, and rapid quenching of the pyrolyzed gases. The C_2H_2 present in the cracked gas is relatively dilute, about $5-20$ vol%. It is extracted from the gas mixture by means of selective solvents such as N-methylpyrrolidone (NMP), dimethylformamide (DMF), kerosene, methanol or acetone and purified in further steps.

The individual acetylene processes differ mainly in the type of generation and transfer of the high-temperature heat required for the pyrolysis reaction. From basic principles, three processes can be differentiated:

1. Allothermal processes with direct heat transfer, usually with electrical heating.

2. Allothermal processes with indirect heat transfer by heat transfer agent.

3. Autothermal processes in which heat from a partial combustion of the feed serves to crack the remainder endothermally.

To 1:

The Hüls electric arc process belongs to this group. This process has been operated on an industrial scale since 1955 in a plant in Germany with an annual capacity (1993) of 120 000 tonnes C_2H_2 and ca. 50 000 tonnes C_2H_4, as well as in a smaller plant in Rumania. In this process, hydrocarbons with boiling points up to 200 °C are cracked in a meter-long stabilized arc with internal temperatures of up to 20 000 °C. At the end of the burner, the gas mixture has a working pressure of about 1.2 bar and a temperature of ca. 1800 °C, which is quickly lowered to 200 °C with a water spray. The residence time of the gas in the arc furnace is a few milliseconds. The yields of acetylene and ethylene attained are 1.0 and 0.42 tonnes per 1.8 tonnes of hydrocarbon feed, respectively. Hüls discontinued operation of this large plant in 1993 and, due to a smaller demand for this energy-intensive acetylene (*e.g.,* for 1,4-butanediol and tetrahydrofuran), started up a modern, environmentally friendly 40 000 tonne-per-year plant.

Another H_2 electric arc process called the plasma process (Hoechst-Hüls) was tested in two large-scale pilot plants. The heat-transfer agent, H_2, is initially heated by an electric arc to 3000–4000 °C, whereby 30–65 % is dissociated into atoms. In the coupled reactor, all types of hydrocarbons — from methane to crude oil — can be introduced into the plasma and cracked. The cracked gas is quickly quenched and separated. With light gasoline as feedstock, yields of acetylene and ethylene of ca. 80 wt% can be obtained if byproducts are recycled to the cracking process. The acetylene concentration in the cracked gas reaches almost 14 vol%.

Plasma processes have also been developed in the USA and the CIS, but none has been used industrially.

A modified electric arc process was developed by Du Pont; a plant with a production capacity of 25 000 tonnes per year was operated in the USA from 1963 to 1968.

Recently, the use of coal as a feedstock for plasma pyrolysis has been investigated on a pilot plant scale by Hüls in Germany and AVCO in the USA. Here — analogous to the process with hydrocarbons — powdered coal is introduced into a hydrogen plasma arc. With a residence time of a few milliseconds, and with

3. cracking by partial combustion of feed

1. electrothermal cracking processes:

1.1. Hüls electric arc process with 19 arc furnaces and preferred feedstocks of natural gas, refinery gases, LPG

process characteristics:

tangential gas injection stabilizes arc, electrode lifetimes of over 1000 hours, and high C_2H_2 and C_2H_4 yields (ca. 56% and 25% by weight, respectively)

1.2. HEAP (**H**ydrogen **E**lectric **A**rc **P**yrolysis) of Hoechst-Hüls with plasma consisting of 30 to 65% H atoms, *i.e.,* high enthalpy density

plasma production in Hoechst modification by electric arc operating with alternating current

Hüls electric arc operating with direct current

process characteristics:

wide range of hydrocarbons as feed, low soot formation, high yields (80 wt%) of C_2H_2/C_2H_4

other plasma processes:

UCC, Cyanamid, CIS

1.3. Du Pont electric arc

1.4. Hüls, AVCO plasma process with coal as feedstock

optimal temperature and pressure conditions, a C_2H_2 yield of 35 wt% can be attained, depending on the type of coal used. The cracked gas is quenched with water, and the unreacted solid is recycled. The separation of the mixture is more complicated than with a hydrocarbon feed, both because of the slag produced and because of the formation of compounds formed from the N, O, and S contained in the coal.

Due to low oil prices, commercial realization of this technology is not currently possible.

To 2:

2. cracking processes with heat transfer agent:
2.1. Wulff process according to Cowper principle, *i.e.*, two regenerator ovens operating alternatively hot and cold, not currently in commercial use due to intrinsic defects

The Wulff process, which operates a regeneration by alternate heating in fire-resistant lined ovens and subsequent cracking, belongs to this group. This process was developed mainly by UCC. However, the relatively large formation of soot, as compared to the electrothermal or partial oxidation processes, and the excessively long residence times for the acetylene could not be prevented. All such plants in the USA, Brazil and Europe were shut down at the end of the 1960s.

2.2. Kureha process as forerunner of a newer high-temperature cracking process:

H_2O as heat transfer agent at 2000 °C from H_2/CH_4 combustion leads to high C_2H_2 fraction

A novel high-temperature cracking process using superheated steam (ca. 2000 °C) from the combustion of H_2/CH_4 mixtures was developed by Kureha in Japan. The feedstock of crude oil or residual oils gives a 46% yield of C_2H_2/C_2H_4 in approximately equal amounts.

This process, first tested in a pilot plant in 1970, is the forerunner of the ACR process developed in conjunction with UCC and Chiyoda (*cf.* Section 3.3.1).

To 3:

3. direct autothermic cracking processes (autothermic = coupling of exothermic and endothermic reactions)
3.1. BASF (Sachsse-Bartholomé) for light petroleum

Two BASF processes belong to this group. The first, and that with the greatest application, is the Sachsse-Bartholomé process. Worldwide, thirteen plants with a total capacity of 400 000 tonnes per year C_2H_2 have used this technology. In 1991 only seven plants with a total capacity of 330 000 tonnes per year were still in operation. This process will be described in more detail below.

3.2. BASF (submerged-flame process) for crude oil

The second BASF process is the submerged-flame process in which an oil/O_2 flame conducts the cracking in the oil phase. Sisas plants in Italy use this technology to produce 110 000 tonnes per year of C_2H_2 and C_2H_4.

3.3. Montecatini and SBA/Kellogg with partial combustion of CH_4, natural gas, light gasoline

Two other commercial autothermal processes have been developed by Montecatini and Société Belge d'Azote (SBA)/Kellogg. They are very similar to the Sachsse-Bartholomé process in principle, differing mainly in the design of the burners and, in the Montecatini process, by the use of several bars excess pressure.

Methane, natural gas, and light gasoline can be used as feed-stocks.

The Hoechst HTP (**H**igh **T**emperature **P**yrolysis) process can be grouped with the autothermic processes since cracked gases are combusted for heat generation. A plant with a capacity of 85 000 tonnes per year C_2H_2 and C_2H_4 was operated in Frankfurt-Hoechst, Germany, until 1975 when it was closed for economic reasons. In Czechoslovakia, 25 000 tonnes each of C_2H_2 and C_2H_4 are still produced using this technology.

3.4. Hoechst (HTP process) for light gaso-line, cracking in hot combustion gases

The autothermal cracking process developed by BASF is suitable for feedstocks such as methane, liquid gas or light gasoline. The majority of plants all over the world are based on natural gas feedstock; only a few employ naphtha as starting material. In the industrial process CH_4 and O_2, for example, are separately preheated to $500-600\,°C$, then mixed and caused to react in a special burner with flame formation. The $O_2:CH_4$ ratio is adjusted to approximately $1:2$ so that only incomplete combustion can take place. The exothermic oxidation of part of the CH_4, as well as the endothermic dehydrodimerization of CH_4 to form C_2H_2 and H_2, takes place in the flame:

to 3.1. BASF process:

process characteristics:

combination of burner and reaction chamber for cracking of CH_4 and light hydrocarbons

high rate of gas flow in order to avoid flashing back of flame in mixing chamber

$$2\ CH_4 \rightarrow C_2H_2 + 3\ H_2 \left(\Delta H = +\frac{90\ \text{kcal}}{377\ \text{kJ}} \big/ \text{mol} \right) \qquad (2)$$

After a residence time of a few milliseconds, the reaction gas is quenched by injecting water or quench oil; otherwise the C_2H_2 would decompose to soot and H_2. The C_2H_2 separation is usually conducted with a solvent such as N-methylpyrrolidone or dimethylformamide. Fractional desorption and suitable rectify-ing steps are used to separate the codissolved substances. Up to 30% of the carbon present in CH_4 can be recovered as C_2H_2. The volume content of C_2H_2 in the cracked gas is about 8%. The main components are H_2 (57%) and CO (26%), *i.e.,* a readily utilizable synthesis gas. Five kilograms of soot are formed per 100 kg C_2H_2.

short residence time prevents equilibration to $C + H_2$

typical cracked gas composition from CH_4 (in vol%):

ca. 57 H_2
    26 CO
     8 C_2H_2
     4 CH_4
     3 CO_2

The largest plant using this process is in Ludwigshafen, Germany. Following expansion in 1972, it has a current produc-tion capacity for C_2H_2 of ca. 90 000 tonnes per year.

i.e., BASF simultaneously produces synthesis gas

Possibilities for obtaining acetylene as a byproduct of olefin production have increased simultaneously with better use of raw materials through more extreme cracking conditions and with the development of new cracking technologies.

improved acetylene supply through:

high-severity steam cracking at $1000\,°C$ or, *e.g.,*

advanced cracking reactor process at 2000 °C

Thus, for example, the trend to high-severity cracking conditions (to 1000 °C) of naphtha has led to C_2H_2 contents of the C_2 fraction of up to 2 wt%, and, in the Advanced Cracking Reactor (ACR) process (up to 2300 °C), even as high as 10 wt% relative to C_2H_4.

C_2H_2 must be removed from C_2H_4 before its further use. Two methods are available:

1. partial hydrogenation to C_2H_4
2. extraction with NMP or DMF

C_2H_2 has an adverse effect on the conventional processing, necessitating its conversion into C_2H_4 by partial hydrogenation. However, it is also possible to wash out the C_2H_2 with N-methyl-pyrrolidone or dimethylformamide before fractionating the C_2 mixture. By this means, the C_2H_2 can be concentrated to 50 − 90 vol%. This mixture, which has been employed as fuel gas, is also suitable for the manufacture of acetylene-based secondary products.

For instance, in the BASF process, C_2H_2 is selectively extracted with NMP from cracked gas at 1 bar and 20 °C and recovered with 98.5% purity. In 1990 this process was used worldwide in 18 plants with a total capacity of approximately 550 000 tonnes.

route 2 can be modified:

extract with 50 − 90 vol% C_2H_2 burnt up to now; BASF process converts it to 98.5% and Linde process converts it to 99.5% C_2H_2

analogous commercial processing also with UCC and other firms

In the Linde process, the C_2 fraction − consisting of around 84% C_2H_4, 14% C_2H_6, and 2% C_2H_2 − is cooled nearly to the dew point and washed with DMF. C_2H_2 dissolves in DMF, and can be obtained in 99.5% purity by fractional degassing. Other processes for the isolation of acetylene from ethylene manufacture have been developed by Stone and Webster using DMF as the solvent in the United States, for example, and used commercially there and in other countries.

Acetylene obtained as an unavoidable byproduct from cracking processes will probably most economically cover the acetylene demand for organic syntheses.

4.3. Utilization of Acetylene

The use of acetylene in the USA, in Western Europe and in Japan is outlined in the following table:

Table 4–2. Acetylene use (in %).

Product	USA			Western Europe			Japan	
	1982	1985	1995	1982	1987	1995	1992	1995
Vinyl chloride	43	31	27	33	31	–	–	–
1,4-Butanediol	28	47	64	12	13	34	18	17
Vinyl acetate	17	–	–	10	18	36	–	–
Acetylene black	8	14	5	} 45	} 38	3	26	25
Miscellaneous	4	8	4			27	56	58
(*e.g.,* trichloroethylene, acrylic acid, chloroprene only in Japan, etc.)								
Total use (in 1000 tonnes)	115	80	135	ca. 320	330	173	50	53

In Section 4.1, four basic criteria which must be fulfilled before acetylene can serve as feedstock for large-scale industrial processes were mentioned.

The manufacture of higher vinyl esters and ethers will be discussed in Sections 9.2.2 and 9.2.3 as examples of cases in which the C_2 unit forms a small part of the molecule.

The manufacture of 1,4-butanediol in the Reppe process as practiced by, *e.g.*, BASF, Du Pont, or GAF/Hüls is characteristic of the third point. Acetylene is still currently an economical base for the manufacture of 2-butyne-1,4-diol, a precursor of 1,4-butanediol.

In the manufacture of 1,4-butanediol, acetylene is first converted with $10-30\%$ aqueous formaldehyde at $100-110\,°C$ and $5-20$ bar in the presence of modified copper acetylide to produce 2-butyne-1,4-diol:

1. C_2H_2 also future economical reaction component for vinylation of higher acids and alcohols

2. economical basis for 1,4-butanediol

manufacture of 1,4-butanediol by the Reppe process, used commercially since 1929, proceeds in two steps:

$$HC\equiv CH + 2\ HCHO\ (aq.) \xrightarrow{\text{[cat.]}}$$

$$HOCH_2-C\equiv C-CH_2OH\ (aq.)\ \left(\Delta H = -\dfrac{24\ kcal}{100\ kJ}/mol\right) \qquad (3)$$

The reaction is conducted in a trickle-column reactor containing the copper acetylide catalyst with Bi as promoter on SiO_2 or magnesium silicate.

The intermediate propargyl alcohol ($HC\equiv C-CH_2OH$) is recycled, together with formaldehyde, to the reaction. The butynediol selectivity amounts to 80% (C_2H_2) and $>90\%$ (HCHO).

In the second step, 2-butyne-1,4-diol is hydrogenated to 1,4-butanediol:

1. ethynization of HCHO in aqueous solution using trickle-phase principle, catalyzed by solid CuC_2–Bi_2O_3/SiO_2 or Mg silicate

intermediate product ($HC\equiv C-CH_2OH$ from primary addition of HCHO to C_2H_2) can be isolated or recycled to reactor

$$HOCH_2-C\equiv C-CH_2OH + 2\ H_2 \xrightarrow{\text{[bat.]}}$$

$$HOCH_2(CH_2)_2CH_2OH\ \left(\Delta H = -\dfrac{60\ kcal}{251\ kJ}/mol\right) \qquad (4)$$

The hydrogenation can be conducted in the liquid phase at $70-100\,°C$ and $250-300$ bar in the presence of Raney nickel catalyst. Alternatively, the hydrogenation can take place in the trickle phase at $180-200\,°C$ and 200 bar employing Ni catalysts with Cu and Cr promoters. The selectivity to 1,4-butanediol reaches about 95% (based on 2-butyne-1,4-diol).

2. complete hydrogenation with Raney-Ni in the liquid phase at lower temperature or with Ni-Cu-Cr/SiO_2 in the trickle phase at higher temperature

producers of 1,4-butanediol:

W. Europe:	BASF
	GAF/Hüls
USA:	Arco
	BASF-Wyandotte
	Du Pont
	GAF (now ISP)

This acetylene-based manufacturing method profits more from the increasing worldwide interest in 1,4-butanediol than other processes based on C_4 feedstocks, especially in the USA and Western Europe.

In 1992, about 85% of the production capacity for 1,4-butanediol in the USA was based on acetylene.

based on C_2H_2 with increased capacity, against

| Japan: | Mitsubishi Chemical |
| | Toyo Soda |

based on C_4H_6

In Japan, on the other hand, the manufacture of 1,4-butanediol from butadiene by acetoxylation and chlorination (to be described later) has been developed and is used commercially.

1,4-butanediol production (in 1000 tonnes):

	1989	1991	1993
USA	191	216	244
W. Europe	165	145	131
Japan	26	39	48

The worldwide production capacity for 1,4-butanediol was about 620000 tonnes per year in 1996, with the USA, Western Europe, and Japan accounting for 310000, 220000, and 110000 tonnes per year, respectively. Production figures for these countries are given in the adjacent table.

variation of 2nd process step:

partial hydrogenation to 2-butene-1,4-diol
$HOCH_2CH=CHCH_2OH$

An additional industrially useful intermediate can be obtained from the acetylene process for 1,4-butanediol. If the hydrogenation of 2-butyne-1,4-diol is conducted in the presence of catalysts whose activity has been reduced either by their manufacturing process or by additives, then the reaction stops at the 2-butene-1,4-diol stage. Iron catalysts, nickel catalysts with iron additives and possibly amine inhibitors or palladium catalysts, usually partially poisoned with zinc, are employed industrially.

use of 2-butene-1,4-diol:

intermediate product, *e. g.*, Thiodan precursor

Thiodan:

Butenediol is a reactive trifunctional intermediate which, as the diacetate, takes part in the Diels–Alder reaction with hexachlorocyclopentadiene to form a precursor of Thiodan® insecticide.

The hydrogenation of maleic anhydride is another manufacturing process which has been used by various firms for several years (*cf.* Section 13.2.3.4). This route has now become sufficiently attractive that plants are being operated in, for example, South Korea, Japan and Western Europe.

alternative routes to 1,4-butanediol:

1. hydrogenation of

2. hydrolysis and hydrogenation of
$ClCH_2CH=CHCH_2Cl$

3. acetoxylation/AcOH addition to butadiene followed by hydrogenation of the butadiene derivative as well as hydrolysis with partial cyclization with loss of H_2O

Two other processes are industrially important. They are based on butadiene, a less expensive and more reactive starting material, and go through the intermediates 1,4-dichloro-2-butene (*cf.* Section 5.3) and 1,4-diacetoxy-2-butene.

Thus Mitsubishi Chemical has developed a technology of concerted manufacture of 1,4-butanediol and tetrahydrofuran from simultaneous acetoxylation/AcOH addition to butadiene, and has employed this technology commercially in a 20000 tonne-per-year plant since 1982.

In the first step, butadiene and acetic acid are converted to 1,4-diacetoxy-2-butene with more than 90% selectivity. The reaction is done in the liquid phase at about 70 °C and 70 bar over a Pd/C catalyst with promoters such as Sb, Bi, Se or (preferably) Te. This is then hydrogenated, also in the liquid phase, at about 60 °C and 50 bar over a conventional hydrogenation catalyst to produce 1,4-diacetoxybutane with more than 98% selectivity. This is then reacted on an acidic ion exchange resin. Depending on the temperature and residence time, different mixtures of 1,4-butanediol and tetrahydrofuran are obtained with 90% selectivity by hydrolysis and dehydration cyclization, respectively. The acetic acid is recycled to the first reaction step:

to 3, Mitsubishi process:

Pd, Te as an industrially used catalyst combination changed the view that Te is a poison for noble metal catalysts

$$H_2C=CH-CH=CH_2 + 2\ AcOH + 0.5\ O_2$$

$$\xrightarrow{[Pd/C]} AcOCH_2CH=CHCH_2OAc + H_2O$$

$$AcOCH_2CH=CHCH_2OAc + H_2 \qquad\qquad (5)$$

$$\xrightarrow{[cat.]} AcOCH_2(CH_2)_2CH_2OAc$$

$$AcOCH_2(CH_2)_2CH_2OAc$$

$$\xrightarrow[+H_2O]{[H^{\oplus}]} HOCH_2(CH_2)_2CH_2OH + \underset{O}{\bigcirc} + 2\ AcOH$$

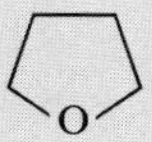

Another synthetic route to 1,4-butanediol consists of hydroformylation of allyl alcohol or allyl acetate followed by hydrogenation of the 4-hydroxy or 4-acetoxybutyraldehyde and, in the case of the monoester, the requisite hydrolysis. Numerous firms have investigated this method, and Arco (USA) has used it since 1991 for the simultaneous production of tetrahydrofuran and N-methylpyrrolidone in a 34000 tonne-per-year unit (*cf.* Section 11.2.2).

4. hydroformylation of $H_2C=CHCH_2OH/OAc$ and hydrogenation of the intermediate $OCH(CH_2)_2CH_2OH/OAc$ and hydrolysis of the monoester

The use of 1,4-butanediol differs by country; in the USA 51% is currently used for the production of tetrahydrofuran (THF), while in Western Europe this portion is only 34%. In Japan only 12% is presently used to make THF because of the coproduction of THF from butadiene through the intermediate 1,4-dichlorobutene or 1,4-diacetoxybutene.

utilization of 1,4-butanediol:

1. tetrahydrofuran by dehydration used as a solvent and monomer for ring-opening polymerization

The use of 1,4-butanediol in the USA, Western Europe and Japan is shown in Table 4–3.

The manufacture of THF from 1,4-butanediol is done by splitting off water in the presence of H_3PO_4, H_2SO_4, or acidic ion exchangers.

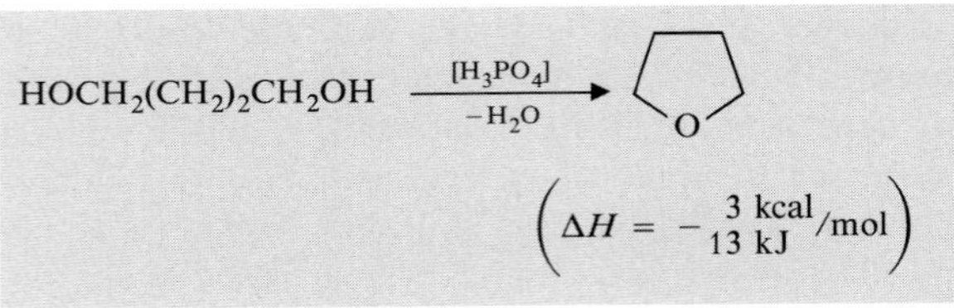

$$HOCH_2(CH_2)_2CH_2OH \xrightarrow[-H_2O]{[H_3PO_4]} \Box_O$$

$$\left(\Delta H = - \begin{matrix} 3 \text{ kcal} \\ 13 \text{ kJ} \end{matrix} / mol \right) \qquad (6)$$

Table 4–3. 1.4-Butanediol use (in %).

Product	USA 1986 1994		Western Europe 1986 1992		Japan 1986 1991	
THF	58	51	35	35	13	12
Acetylene chemicals[1]	16	20	20	16	3	–
Polybutene terephthalate	16	20		20	64	66
Polyurethanes	7	6	} 45	23	} 20	16
Other substances	3	3		6		6
Total use (in 1000 tonnes)	160	289	75	104	22	38

[1] vinyl esters/ethers, methyl vinyl ketone, propargyl alcohol, etc.

uses of THF (worldwide:

ca. 70% for polymerization
ca. 25% as solvent
remainder as intermediate for, *e. g.,*
  tetrahydrothiophene
  pyrrolidone
  γ-butyrolactone
  chlorinated products

HO-[(CH$_2$)$_4$O]$_n$-H goes by various
names:

1. polytetramethyleneglycol
2. polyoxytetramethyleneglycol
3. polytetramethyleneetherglycol
4. poly THF
5. α-hydro-ω-hydroxypoly(oxy-1,4-
   butanediyl)

In this reaction, 1,4-butanediol is heated with acid to 110–125 °C in a boiler and replenished with 1,4-butanediol as THF/H$_2$O is distilled over. The conversion to THF is virtually quantitative. 1,4-Butanediol and the dissolved mineral acid can also be fed together to a high-pressure tube reactor at 300 °C and 100 bar. The selectivity to THF is greater than 95%.

Another method for manufacturing THF, starting from 1,4-di-chloro-2-butene, is described in Section 5.3. A third method is the hydrogenation of maleic anhydride (*cf.* Section 13.2.3.4). The world production capacity for THF in 1994 was 186000 tonnes per year, and in the USA, Western Europe, and Japan it was 119000, 30000, and 33000 tonnes per year, respectively. THF is an important solvent for many high polymers such as PVC, rubber, Buna® S and others. It is increasingly used in the manufacture of polytetramethylene glycol (HO-[-(CH$_2$)$_4$—O-]$_n$H, mean molecular weight = 600–3000), a precursor for polyurethanes and Spandex® fibers. The world production capacity for polytetramethylene glycol is currently about 95000 tonnes per year. With a production capacity of 70000 tonnes per year (1995), Du Pont is the most important producer of polytetra-methyleneglycol.

THF is manufacured from 1,4-butanediol by Du Pont, BASF-Wyandotte, GAF and Arco (since 1990) in the USA, by BASF and GAF/Hüls in Western Europe, and by Mitsubishi Chemical and Toyo Soda in Japan.

Another product from 1,4-butanediol is γ-butyrolactone. In the USA, this use of the diol is second only to production of THF. It is formed by dehydrogenation cyclization, preferably in the gas phase, and generally over copper catalysts at 200–250 °C and at slightly higher than atmospheric pressure, with yields of up to 93%:

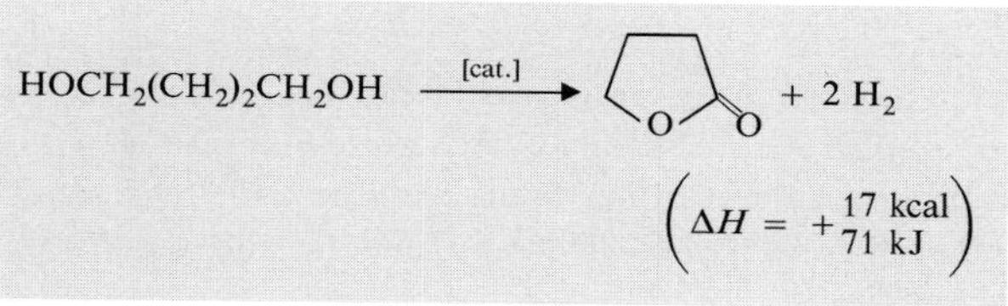

$$HOCH_2(CH_2)_2CH_2OH \xrightarrow{[cat.]} \text{⟨} + 2\,H_2 \qquad (7)$$

$$\left(\Delta H = +\frac{17\ \text{kcal}}{71\ \text{kJ}} \right)$$

Industrial processes are operated by BASF and GAF in USA and Western Europe.

An alternative process for the manufacture of γ-butyrolactone, the partial hydrogenation of maleic anhydride (*cf.* Section 13.2.3.4), is carried out in Japan by Mitsubishi Kasei. Due to the low price of maleic anhydride, use of this process is increasing in many countries.

γ-Butyrolactone is a solvent as well as a precursor of another industrially important solvent — N-methylpyrrolidone (NMP). About 27 000 tonnes of NMP are currently used annually worldwide. NMP is obtained by reacting γ-butyrolactone with methylamine:

$$\text{⟨} + CH_3NH_2 \longrightarrow \text{⟨} + H_2O \qquad (8)$$

γ-Butyrolactam, usually known as 2-pyrrolidone, results from the reaction between γ-butyrolactone and ammonia. This manufacturing method is, compared to several other processes, that used most often commercially.

A new route for the manufacture of 2-pyrrolidone has been developed by DSM. It is based on the hydrocyanization of acrylonitrile to succinonitrile at atmospheric pressure and 70 °C in the presence of triethylamine. The succinonitrile is then par-

2. γ-butyrolactone by cyclic dehydrogenation, used as:

solvent
intermediate for herbicides, pharmaceuticals and other uses

γ-butyrolactone production (in 1000 tonnes):

	1989
USA	43
WE	16
Japan	4

3. N-methylpyrrolidone from the reaction of γ-butyrolactone with CH_3NH_2, used for:

extraction of acetylene
extraction of butadiene
solvent for production of polyphenylene sulfide and substitute for chlorofluorocarbons

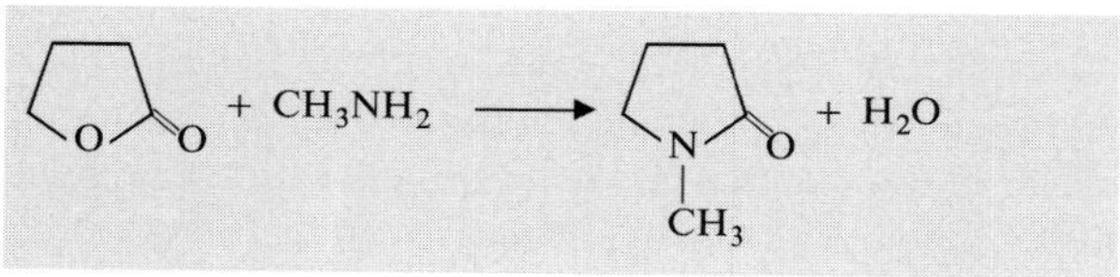

4. 2-pyrrolidone from reaction of γ-butyrolactone and ammonia, or possibly multistep based on $H_2C=CHCN$ in the future

tially hydrogenated over a nickel catalyst at $80-100\,°C$, and finally hydrolytically cyclized to 2-pyrrolidone at $210\,°C$ under pressure:

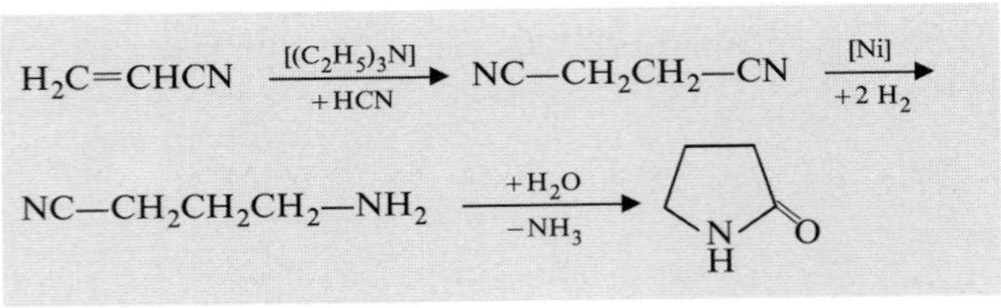

$$H_2C{=}CHCN \xrightarrow[{+\,HCN}]{[(C_2H_5)_3N]} NC{-}CH_2CH_2{-}CN \xrightarrow[{+\,2\,H_2}]{[Ni]}$$

$$NC{-}CH_2CH_2CH_2{-}NH_2 \xrightarrow[{-\,NH_3}]{+\,H_2O} \underset{\overset{|}{H}}{N}{\bigcirc}{=}O \qquad (9)$$

5. N-vinylpyrrolidone from γ-butyrolactone after reaction with NH_3 and vinylation with C_2H_2, used as:

monomer
co-monomer, *e.g.*, with vinyl acetate or styrene

The overall selectivity is greater than 80%. This method has yet to be applied commercially. The worldwide demand for 2-pyrrolidone is about 14000 tonnes per year, with the main use being the production of N-vinyl-2-pyrrolidone, and is increasing considerably.

The most important manufacturers are BASF and GAF.

2-Pyrrolidone can be converted into N-vinylpyrrolidone by vinylation with acetylene in the presence of basic catalysts:

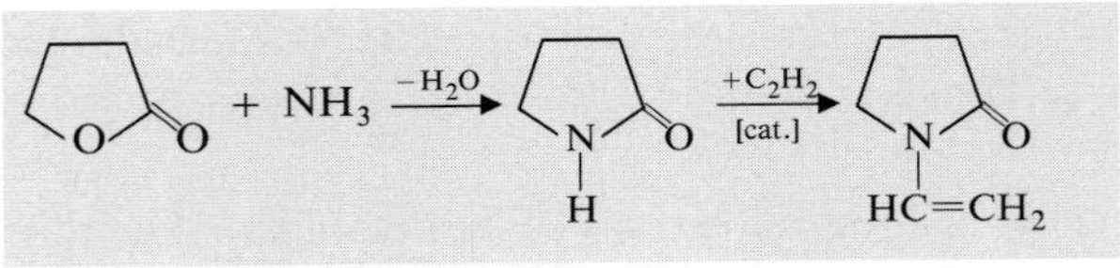

$$\bigcirc_{O}^{O} + NH_3 \xrightarrow{-\,H_2O} \underset{\overset{|}{H}}{N}{\bigcirc}{=}O \xrightarrow[{[cat.]}]{+\,C_2H_2} \underset{\overset{|}{HC{=}CH_2}}{N}{\bigcirc}{=}O \qquad (10)$$

It is used for the manufacture of specialty polymers which have applications in the cosmetic, medical and commercial sectors, for example, as binders, blood plasma substitutes (Periston®) or protective colloids. Next to GAF, BASF is the world's largest producer of polyvinylpyrrolidone and its secondary products.

6. polybutene terephthalate by polycondensation with terephthalic acid

PBT production (in 1000 tonnes):

	1990	1993	1995
Japan	51	54	66

In the last few years, 1,4-butanediol has become an important component in the manufacture of polyesters with terephthalic acid (*cf.* Section 14.2.4). In 1996, the annual production capacity of polybutene terephthalate (PBT) in the USA, Japan, and Western Europe was 144000, 76000, and 69000 tonnes per year, respectively. In 1989, world consumption was about 310000 tonnes. Monsanto is the largest producer of PBT worldwide.

PBT's main use is as an engineering plastic, but it is also used for fibers, films (*e.g.*, for laminated glass), and adhesives.

7. polyurethane by polyaddition to diiso-cyanates

1,4-Butanediol is also used in the production of polyurethanes, and as a feed material for specialty chemicals.

5. 1,3-Diolefins

In industrially important diolefins, or dienes, the C—C double bonds are conjugated, *i.e.*, in the 1,3 position to each other. Consequently, they have reaction properties different than compounds with isolated double bonds; in particular, the conjugated dienes are considerably more reactive. For this reason, free 1,3-dienes are not found in nature.

industrially important dienes have conjugated double bonds

compared to diolefins with isolated double bonds they possess:

1. different reaction behavior
2. higher reactivity

Industrially important 1,3-diolefins include the C_4 and C_5 dienes butadiene, chloroprene, isoprene and cyclopentadiene.

1,3-diolefins with

1. C_4-skeleton:

2. C_5-skeleton:

5.1. 1,3-Butadiene

1,3-Buadiene (generally referred to as simply butadiene) is the most industrially important of the aforementioned C_4 and C_5 dienes. Both World Wars provided the stimulus — especially in Germany — for intensive research and subsequent rapid development of manufacturing processes for the monomer and its further conversion into elastomers.

butadiene has become the quantitatively most important diene due to:

1. utilization as monomer and comonomer for elastomers, thermoplasts, dispersions

Moreover, the availability of butadiene improved continuously in the 1960s and 1970s, making it an attractive feedstock for chemical syntheses as well.

2. ready and economical availability with stimulus of expanding use as intermediate

More recently, an unfavorable development in the availability of butadiene has become apparent. Cracking processes for the manufacture of olefins have made the coproduct butadiene less expensive, and thus made the C_4 dehydrogenation process unprofitable. However, due to the growing use of natural gas and refinery waste gas as feedstocks for ethylene/propylene production, the butadiene share from the cracking process is decreasing. In the future, a more limited availability of butadiene is to be expected.

recent restricted availability of butadiene due to:

1. decrease in C_4 dehydrogenation
2. trend towards lighter feedstocks in cracking

Production figures for butadiene in several countries are summarized in the adjacent table. The worldwide production capacity for butadiene in 1993 was about 8.5×10^6 tonnes, of which the USA, Western Europe, and Japan accounted for 1.8, 2.3 and 0.9×10^6 tonnes, respectively.

butadiene production (in 10^6 tonnes):

	1990	1992	1995
W. Europe	1.91	1.88	2.00
USA	1.43	1.44	1.67
Japan	0.83	0.85	0.99

5.1.1. Traditional Syntheses of 1,3-Butadiene

traditional butadiene manufacture by synthesis of C_4-skeleton stepwise from

C_1: HCHO

C_2: HC≡CH (CH$_3$CHO)

  C_2H_5OH

three important older manufacturing routes:

1. four-step process with acetaldehyde (from acetylene or ethanol) with 1,3-butanediol as intermediate

The first industrial manufacturing processes for butadiene were based on coal conversion products such as acetylene, acetaldehyde, ethanol, and formaldehyde. There are basically three synthetic routes characterized by formation of the C_4 butadiene chain either from C_2 units or from C_2 and C_1 units, generally in multistep processes.

In former East Germany, a certain amount of butadiene is still being produced from acetylene in a four-step process. In this process, acetylene is initially converted into acetaldehyde and then aldolized to acetaldol. The acetaldol is reduced to 1,3-butanediol with a Ni catalyst at 110 °C and 300 bar. Finally, in the fourth step, the 1,3-butanediol is dehydrated in the gas phase at 270 °C using a Na polyphosphate catalyst:

$$2\ CH_3CHO \xrightarrow{[OH^{\ominus}]} CH_3\underset{\underset{OH}{|}}{CH}-CH_2CHO$$

$$\xrightarrow{+H_2} CH_3\underset{\underset{OH}{|}}{CH}-CH_2CH_2OH \xrightarrow{-2\,H_2O} H_2C=CH-CH=CH_2 \tag{1}$$

The selectivity to butadiene is about 70% (based on CH_3CHO).

One variation of the four-step process uses acetaldehyde from the dehydration of ethanol. This acetaldehyde is then converted over a Zr-/Ta-oxide/SiO$_2$ catalyst at 300–350 °C with an overall yield of about 70%. This process is used commercially in India and China (two plants with a total capacity of 85 000 tonnes per year in 1990).

2. Lebedew process – single-step direct synthesis from ethanol

Another method for butadiene manufacture based on ethanol is known as the Lebedew process. It was developed in the CIS, and is still employed commercially there, as well as in Poland and Brazil. In this process, ethanol is dehydrogenated, dehydrated, and dimerized in one step at 370–390 °C over a MgO—SiO$_2$ catalyst:

$$2\ C_2H_5OH \rightarrow H_2C=CH-CH=CH_2 + 2\ H_2O + H_2 \tag{2}$$

The selectivity to butadiene reaches as high as 70%. Today, this process could be of interest to countries not pocessing a petrochemical base but having access to inexpensive ethanol from fermentation.

In the third traditional method, the Reppe process, acetylene and formaldehyde are initially converted into 2-butyne-1,4-diol from which 1,4-butanediol is manufactured. This product is still of great industrial significance today (*cf.* Section 4.3). Subsequently, a direct twofold dehydration ensues, but due to technical considerations this is usually a two-step process with tetrahydrofuran as the intermediate product. The Reppe process is totally uneconomical today.

Modern industrial processes for butadiene are based exclusively on petrochemicals. C_4 cracking fractions or butane and butene mixtures from natural and refinery waste gases are economical feedstocks.

3. Reppe process using acetylene and formaldehyde with 1,4-butanediol as intermediate

two modern routes for producing butadiene:

1. isolation from C_4 steam cracking fractions
2. dehydrogenation and oxydehydrogenation of *n*-butane and *n*-butenes

5.1.2. 1,3-Butadiene from C_4 Cracking Fractions

C_4 fractions with an economically isolable butadiene content are available in countries where ethylene is manufactured by steam cracking of naphtha or higher petroleum fractions. The C_4 fraction amounts to about 9 wt% of the cracked product from conventional high-severity cracking of naphtha. The C_4 fraction contains 45–50 wt% of butadiene (*cf.* Section 3.3.2).

Western Europe and Japan are the main areas utilizing this raw material base for butadiene. In the USA, the usual cracking of natural and refinery gases supplies only very small amounts of butadiene compared to naphtha or gas oil cracking. The butadiene content obtained by cracking various feedstocks in ethylene plants at different cracking severities is summarized in the following table:

feedstocks for butadiene separation:

C_4 fraction from naphtha cracker (9 wt% of cracked product) with 45–50 wt% butadiene using high-severity cracking

C_4 fraction from cracking of natural or refinery gas contains relatively small amounts of butadiene

Table 5–1. Butadiene content (in kg per 100 kg ethylene) using various feedstocks.

Feedstock	Butadiene content
Ethane	1 – 2
Propane	4 – 7
n-Butane	7 – 11
Naphtha	12 – 15
Gas oil	18 – 24

However, naphtha and gas oil cracking has increased in importance in the USA. In 1976, 36% of the total butadiene capacity was based on isolation from these cracked products. During the 1990s, this has remained stable at about 45–50%. In Western

coproduction of butadiene in the naphtha/gas oil cracking increasingly displaces butane/butene dehydrogenation in the USA changing to lighter feedstocks for cracking in Western Europe is decreasing the butadiene supply

similar boiling points in C_4 fraction and azeotrope formation prevent distillative workup

	b.p. (°C)
isobutene	-6.90
1-butene	-6.26
trans-2-butene	$+0.88$
cis-2-butene	$+3.72$
1,3-butadiene	-4.41
vinylacetylene	$+5.10$

butadiene separation based on two principles:

1. chemical, by reversible complex formation with $[Cu(NH_3)_2]OAc$ (Exxon CAA (cuprous ammonium acetate) process) only of limited importance

2. physical, by extractive distillation – addition of solvent alters the relative volatility of the components to be separated

suitable solvents for purification of butadiene by extractive distillation:

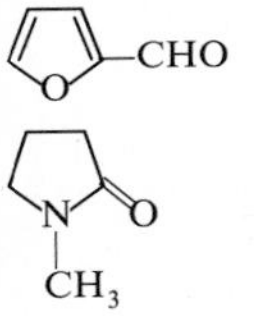

CH_3COCH_3
CH_3CN
$CH_3CON(CH_3)_2$
$HCON(CH_3)_2$

additional alkynes and allenes in C_4 fraction from modern high-severity cracking processes include

$HC{\equiv}C{-}CH_3$
$HC{\equiv}C{-}CH_2CH_3$
$HC{\equiv}C{-}CH{=}CH_2$
$H_2C{=}C{=}CH{-}CH_3$

these require beginning with hydrogenation of the C_4 fraction; some solvents separate C_4 alkynes during process, *e.g.*,

dimethylformamide
dimethylacetamide
N-methylpyrrolidone (NMP)

principle of solvent extraction of butadiene from a C_4 cracking fraction:

Europe, the opposite is occuring; *i. e.,* a shortage of butadiene is being caused by the increasing use of natural gas in cracking processes to manufacture olefins.

Separation of butadiene from a mixture of C_4 hydrocarbons is not possible by simple distillation as all components boil within a very close temperature range, and some form azeotropic mixtures. Consequently, two isolation processes based on a chemical and a physical separation have been developed:

1. The older, chemical separation process exploits the formation of a complex between butadiene and cuprous ammonium acetate, $[Cu(NH_3)_2]OAc$. This process was developed by Exxon as an extraction procedure for processing C_4 fractions with low butadiene content. These fractions can contain only small amounts of acetylenes, or the extraction process will be disturbed by foam formation. Another disadvantage is the relatively involved regeneration of the extractant.

2. All modern processes for butadiene isolation are based on the physical principle of extractive distillation. If selective organic solvents are added, the volatility of particular components of a mixture is lowered (in this case butadiene). They then remain with the solvent at the bottom of the distillation column, while the other impurities, previously inseparable by distillation, can be removed overhead.

Acetone, furfurol, acetonitrile, dimethylacetamide, dimethylformamide and N-methylpyrrolidone are the principal solvents employed in this extractive distillation.

Extractive distillations are particularly suitable for the presently available butadiene-rich C_4 cracking fractions with a relatively high share of alkynes such as methyl-, ethyl- and vinylacetylene, as well as methylallene (1,2-butadiene). In modern processes with solvents such as dimethylformamide (Nippon Zeon, VEB Leuna), dimethylacetamide (UCC), or N-methylpyrrolidone (BASF, ABB Lummus Crest), alkyne separation is a stage in the operation of the process. In the older processes developed and operated in the USA with solvents such as acetone, furfurol (Phillips Petroleum) or acetonitrile (Shell, UOP, Arco), removal of C_4 alkynes, for example by partial hydrogenation, prior to distillation was essential to avoid problems arising from resin formation.

The basic principle of solvent extraction of butadiene from a C_4 cracking fraction can be described as follows. The vaporized C_4

fraction is introduced at the bottom of an extraction column. The solvent (*e. g.*, dimethylformamide or N-methylpyrrolidone) flows through the gas mixture from the top, and on the way down becomes charged with the more readily soluble butadiene and small amounts of butenes. Pure butadiene is introduced at the bottom of the extraction column to drive out as much of the butenes as possible. The butenes leave the separating column overhead. In another column, the degasser, the butadiene is freed from solvents by boiling and is subsequently purified by distillation. In the BASF N-methylpyrrolidone process, butadiene is obtained in approximately 99.8% purity. The butadiene yield is 96% relative to the original butadiene content in the C_4 cracking fraction.

The BASF process for the extraction of butadiene with N-methylpyrrolidone (NMP) was first operated commercially in 1968 by EC-Dormagen. By 1990, there were 24 production facilities using the BASF process in operation or under construction worldwide with a combined capacity of 2.25×10^6 tonnes per year.

This is one of the leading butadiene extraction processes, exceeded only by the Nippon Zeon GPB process (**G**eon **P**rocess **B**utadiene) which was still used in more than 30 plants in 1985.

countercurrent extraction of butadiene with partial butadiene recycling to increase extraction selectivity

isolation of C_4 components:

butene overhead; butadiene after distillation of bottoms (solvent/butadiene)

example of butadiene extraction process:

BASF NMP process

first industrial application in 1968 by EC-Dormagen; in 1990, licensed capacity of 2 250 000 tonnes per year butadiene

5.1.3. 1,3-Butadiene from C_4 Alkanes and Alkenes

Butane and butene mixtures from natural gas and refinery waste gases are feedstocks for pure dehydrogenation or for dehydrogenation in the presence of oxygen. Historically they were particularly available in the USA, and for this reason, industrial processes were developed almost exclusively there. Today, butane/butene dehydrogenation has lost much of its former importance due to high costs. The last Petro-Tex dehydrogenation plant in the USA is currently idle, though several plants are still operating in the CIS. In Japan, the last Houdry dehydrogenation plant was shut down in 1967 after eight years of operation.

feedstocks for C_4 alkane and alkene dehydrogenation and oxidative dehydrogenation:

C_4 fractions from natural gas condensates and refinery waste gases

By 1985, worldwide dehydrogenation of *n*-butane and *n*-butenes accounted for less than 3% of the total butadiene volume.

Dehydrogenations of *n*-butane and *n*-butenes are endothermic processes requiring large amounts of energy:

general characteristics of dehydrogenation of C_4 alkanes and alkenes:

$$\left(\Delta H = +\frac{30 \text{ kcal}}{126 \text{ kJ}} \middle/ \text{mol} \right) \qquad (3)$$

$$+ \quad \underset{}{\overset{-H_2}{\rightleftarrows}} \quad \diagup\diagdown\diagup \quad \left(\Delta H = + \begin{matrix} 26 \text{ kcal} \\ 109 \text{ kJ} \end{matrix} / \text{mol} \right) \qquad (4)$$

1. endothermic processes with high energy demand at high temperatures

2. rapid equilibration with short residence time and selective catalysts necessary to limit secondary and side reactions

3. dehydrogenation favored by reduction of hydrocarbon partial pressure with H_2O (Le Chatelier–Braun principle) H_2O-sensitive catalysts require decrease in total pressure

C deposition decreased by water gas reaction:

$C + H_2O \rightleftharpoons CO + H_2$

Houdry butane dehydrogenation:

catalytic dehydrogenation with coke deposition, requiring an alternating oxidative catalyst regeneration

hydrogenation energy is supplied by heat of oxidation

Dow butene dehydrogenation:

alternating catalytic 'steam' dehydrogenation with reaction and regeneration in two parallel reactors

Relatively high temperatures (600–700 °C) are necessary to achieve economical conversions. To attain the same rate of reaction, n-butane requires a temperature roughly 130 °C higher than the n-butenes. At these temperatures, side reactions such as cracking and secondary reactions involving the unsaturated compounds become important. Therefore, a short residence time and a selective catalyst must be used.

Dehydrogenation involves an increase in the number of moles of gas, and so is favored by an addition of steam as in the case of steam cracking. The lowering of the partial pressure of the hydrocarbons achieved in this way decreases coke deposition, isomerization and polymerization.

If catalysts unstable in the presence of steam are employed, the dehydrogenation process is usually operated under reduced pressure. This will shift the equilibrium of dehydrogenation towards butadiene.

The Houdry single-step process for the dehydrogenation of butane (Catadiene process from ABB Lummus Crest used in 20 plants in 1993) is one of the most important processes commercially, and also one of the oldest. It is also the basis for the Catofin alkane dehydrogenation process (cf. Section 3.3.1). A H_2O-sensitive Cr-, Al-oxide catalyst is introduced at 600–620 °C and 0.2–0.4 bar. The catalyst must be regenerated after a few minutes by injecting air to burn off the coke layer. With a butane conversion of 30–40%, butadiene yields of up to 63% can be reached. Between the dehydrogenation and regeneration periods, the catalyst is evacuated to remove the reaction mixture. The three operations take place in cycle in separate reactors which are part of a single unit. The high energy requirement for the dehydrogenation is met by the oxidation of the coke layer on the catalyst (adiabatic method).

The Dow process is a butene dehydrogenation method which takes place with the addition of steam. It operates at 600–675 °C and 1 bar over a Ca-Ni-phosphate catalyst stabilized with Cr_2O_3. The heat of dehydrogenation is provided by the addition of superheated steam (H_2O: butene ratio of 20:1) to the reaction, analogous to the dehydrogenation of ethylbenzene to form styrene (cf. Section 13.1.2). The conversion of butene is about 50%, with a selectivity to butadiene of about 90%. After a reaction period of 15 minutes, the catalyst must be regenerated for 11 minutes. In practice, parallel reactors, alternately regenerated and utilized for hydrogenation, are employed. The butadiene is isolated from the reaction mixture by extractive distillation.

Similar processes have been developed by Shell using a Fe-Cr-oxide catalyst with K_2O additive, and by Phillips Petroleum with a Fe-oxide–bauxite catalyst.

Besides the dehydrogenation of C_4 hydrocarbons to butadiene, another dehydrogenation method (in the presence of oxygen) has gained in importance. In this process, known as oxidative dehydrogenation, the dehydrogenation equilibrium between butenes and butadiene is displaced by the addition of oxygen towards greater formation of butadiene. The oxygen not only removes H_2 by combustion, but also initiates dehydrogenation by abstracting hydrogen from the allyl position. At these high temperatures (up to 600 °C), oxygen also acts to oxidatively regenerate the catalyst.

In industrial operation, a sufficient quantity of oxygen (as air) is introduced so that the heat supplied by the exothermic water formation roughly equals the heat required for the endothermic dehydrogenation. In this way the butene conversion, the selectivity to butadiene, and the lifetime of the catalyst can be improved. By using an excess of air, the maximum temperature can be controlled by addition of steam. Mixed oxide catalysts based on Bi/Mo or Sn/Sb are most often used.

The Phillips O-X-D process (**ox**idative **d**ehydrogenation) for the manufacture of butadiene from *n*-butenes is an example of an industrially operated dehydrogenation process. *n*-Butenes, steam and air are reacted at 480–600 °C on a fixed-bed catalyst of unrevealed composition. With butene conversions between 75 and 80%, the butadiene selectivity reaches roughly 88–92%. This process was used by Phillips until 1976.

Petro-Tex also developed a process for the oxidative dehydrogenation of butenes (Oxo-D process) that was first used in the USA in 1965. The conversion with oxygen or air is performed at 550–600 °C over a heterogeneous catalyst (probably a ferrite with Zn, Mn or Mg). By adding steam to control the selectivity, a selectivity to butadiene of up to 93% (based on *n*-butenes) can be reached with a conversion of 65%.

A new oxidative dehydrogenation process for butane/butene has been developed and piloted by Nippon Zeon. Details of the

butadiene isolation by extractive distillation with solvents analogous to separation from C_4 cracked fractions (*cf.* Section 5.1.2)

other dehydrogenation processes:

Shell (Fe_2O_3 + Cr_2O_3 + K_2O)
Phillips (Fe_2O_3/bauxite)

newer dehydrogenation process:

oxidative dehydrogenation of butenes to butadiene

characteristics of oxidative dehydrogenation:

enthalpy of reaction for dehydrogenation is provided by the reaction enthalpy from H_2O formation. Furthermore, presence of O_2 facilitates the dehydrogenation by abstraction of allylic hydrogen. O_2 acts to regenerate catalyst

industrial application of a butene oxidative dehydrogenation:

Phillips O-X-D process characteristics:

adiabatic (*i. e.*, process supplies own heat requirement) catalyzed oxidative dehydrogenation of *n*-butenes with relatively high conversions and selectivities

Petro-Tex Oxo-D process

characteristics:

adiabatic, heterogeneously catalyzed autoregenerative oxidative dehydrogenation of *n*-butenes

catalyst composition and process conditions have not yet been disclosed.

modifications of oxidative dehydrogenation using halogens:

Shell 'Idas' process with I_2 for dehydrogenation, I_2 recovery by HI oxidation

Another method for removing H_2 from the dehydrogenation equilibrium involves reacting it with halogens to form a hydrogen halide, from which the halogen is later recovered by oxidation. For a time, Shell employed iodine as the hydrogen acceptor in the Idas process (France) for the dehydrogenation of butane to butadiene.

5.1.4. Utilization of 1,3-Butadiene

butadiene employed industrially for:

1. polymerization to form

1.1. homopolymers and
1.2. copolymers

2. synthesis of intermediate products by:

2.1. addition
2.3. sulfone formation
2.4. selective hydrogenation

to 1:

polymerization products:
elastomers, thermoplasts, and drying oils

elastomers are polymers which exhibit elastic properties after cross-linking (vulcanization)

numerous monomer combinations lead to broad spectrum of properties of the elastomers

natural and synthetic rubber consumption (in 10^6 tonnes):

	1989	1992	1995
World	15.6	14.3	15.1
USA	3.0	3.4	4.1
W. Europe	2.5	3.0	3.2
Japan	1.5	2.1	2.2

largest volume synthetic rubber type:
SBR with approximate composition:

$$+(CH_2{-}CH{=}CH{-}CH_2)_3(CH_2{-}CH)\!+_n$$

synthetic rubber production (in 10^6 tonnes):

	1990	1992	1994
World	10.6	10.6	9.10
USA	2.11	2.30	2.42
W. Europe	2.54	2.20	2.36
Japan	1.43	1.39	1.35

The industrial uses of butadiene are based on its ability to homopolymerize to polybutadiene and copolymerize with numerous unsaturated monomers.

Since it possesses several reactive centers, butadiene can take part in numerous addition and ring-formation reactions, leading to the syntheses of important intermediates.

The aforementioned polymerization products comprise a series of elastomers, *i.e.*, synthetic rubber. Depending on the polymer structure, various types of rubber are obtained with properties such as elasticity; resistance to abrasion, wear, cold and heat; and stability to oxidation, aging and solvents. Styrene and acrylonitrile are the main comonomers used for polymerization with butadiene. The two most important types of synthetic rubbers are SBR (**s**tyrene **b**utadiene **r**ubber; old IG Farben name, Buna S) and BR (**b**utadiene **r**ubber; 1,4-cis-polybutadiene).

A breakdown of the production of these and other types of synthetic rubber in several countries can be found in the following table (see next page).

The production figures for several countries are given in the accompanying table. Bayer is the world's largest producer of synthetic rubber and specialty rubber products.

The worldwide production of natural rubber was about 5.4×10^6 tonnes in 1994; natural rubber thus accounted for about 38% of the total rubber production of 14.5×10^6 tonnes.

Table 5–2. Pattern of synthetic rubber production (in wt%).

Product	World 1984	World 1994	USA 1986	USA 1994	Western Europe 1984	Western Europe 1994	Japan 1986	Japan 1995
Styrene butadiene rubber (SBR)	59	45	43	38	56	49	49	46
Polybutadiene rubber (BR)	15	18	19	20	14	22	19	21
Chloroprene rubber (CR)	5	4	6	5	5	5	7	6
Olefin rubber (EPR, EPTR[1])	6	9	7	12	8	8	9	12
Butyl rubber (IIR)	7	7	7	9	8	6	4	} [2]
Polyisoprene rubber (IR)	5	11	4	3	2	4	5	}
Nitrile rubber (NBR)	4	5	3	3	5	5	5	5
Others	1	1	11	10	2	1	2	10
Total production (in 10^6 tonnes)	9.0	9.1	2.0	2.5	1.8	1.7	1.2	1.5

[1] also abbreviated EPDM (ethylene propene diene monomer)
[2] included in "Others"

EP = ethylene propene
EPT = ethylene propene terpolymer
II   = isobutene isoprene
NB = acrylonitrile butadiene

ABS polymers, synthesized from acrylonitrile, butadiene, and styrene, belong to the group of terpolymers suitable for thermoplastic processing. They are characterized by a high impact strength that is maintained at low temperatures.

ABS plastics are acrylonitrile-butadiene-styrene terpolymers for thermoplastic processing – with high impact strength

The following breakdown of butadiene use in several countries shows that more than 70% of the total butadiene production was used for the manufacture of homo- and copolymers (Table 5-3).

Table 5-3. Use of butadiene (in %).

Product	World 1990	World 1994	USA 1981	USA 1994	Western Europe 1980	Western Europe 1994	Japan 1984	Japan 1992
Styrene butadiene rubber (SBR)	43	42	40	45	46	41	49	46
Polybutadiene rubber (BR)	23	22	21	23	24	14	27	32
Adiponitrile	8	6	12	11	6	8	–	–
Chloroprene	5	3	7	4	5	4	5	3
ABS polymers	11	8	6	5	8	6	11	12
Nitrile rubber(NBR)	4	4	3	3	6	4	6	6
Miscellaneous	6	15	11	9	5	23	2	1
Total usage (in 10^6 tonnes)	6.3	6.5	1.6	1.9	1.1	1.5	0.71	0.84

Butadiene has recently become increasingly important as an intermediate product. Chloroprene (*cf.* Section 5.3) is manufactured from 3,4-dichloro-1-butene – the addition product of

to 2.1:

chlorination followed by dehydrochlorination to form chloroprene

hydrocyanation to adiponitrile
multistep conversion to 1,4-butanediol

1. through 1,4-dichloro-2-butene
2. alternately through 3,4-epoxy-1-butene

butadiene and chlorine – by elimination of HCl. Adiponitrile (*cf.* Section 10.2.1.1) is obtained by hydrocyanation, *i. e.*, double addition of HCN. Butadiene can also be converted into 1,4-butanediol through various intermediate stages (*cf.* Section 5.3). The reaction described in Section 5.3 through 1,4-dichloro-2-butene can be run with substantially more favorable economics through a butadiene epoxidation developed recently by Eastman Chemical. In this process, butadiene is converted to 3,4-epoxy-1-butene with air over a silver catalyst. After a thermal rearrangement to 2,5dihydrofuran and hydrogenation to tetrahydrofuran, the final hydrolysis gives 1,4-butanediol:

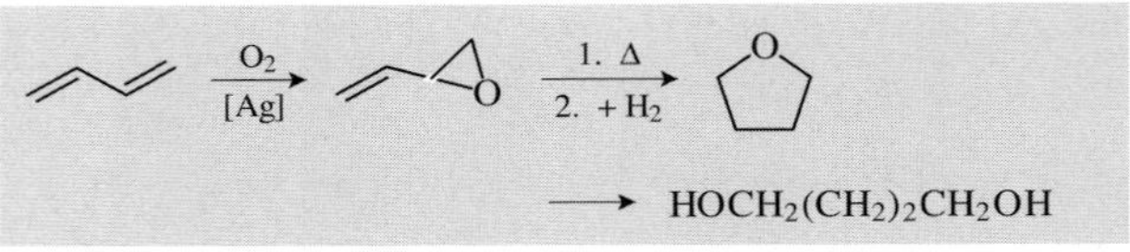

$$\text{(5)}$$

$$\longrightarrow \text{HOCH}_2(\text{CH}_2)_2\text{CH}_2\text{OH}$$

The intermediate steps are also raw materials for other compounds. Eastman plans to build a 140 000 tonne-per-year plant in Texas.

to 2.2:

cyclodimerization to 1,5-cyclooctadiene and secondary products cyclooctene and polyoctenamers as well as cyclotrimerization to 1,5,9-cyclododecatriene and secondary products (1,12-C_{12}-dicarboxylic acid, C_{12} lactam)

Cyclodimerization of butadiene in the presence of organometallic catalysts leads to 1,5-cyclooctadiene and trimerization to 1,5,9-cyclododecatriene. Both components are important precursors for higher polyamides (*cf.* Sections 10.1.2 and 10.3.2). Cyclooctadiene is partially hydrogenated to cyclooctene, which is a feedstock for the metathesis polymerization to form polyoctenamers (*cf.* Section 3.4).

to 2.3:

sulfolane synthesis and its application as selective solvent for extraction of aromatics or solvent for acidic gases (Sulfinol process)

Finally, butadiene reacts with SO_2 in a reversible 1,4-addition to form sulfolene which can be hydrogenated with hydrogen to give the very thermally stable sulfolane (tetrahydrothiophene dioxide):

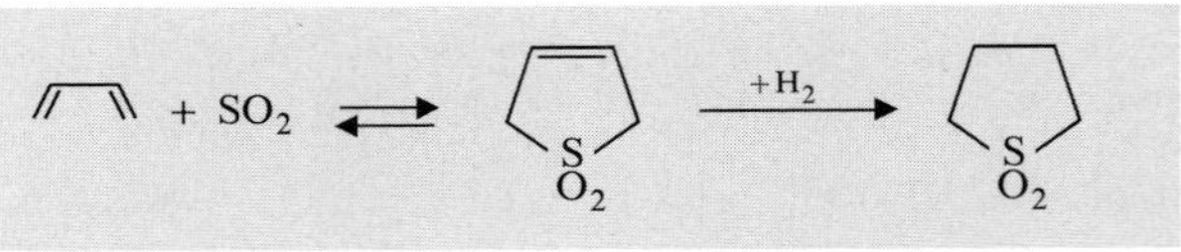

$$\text{(6)}$$

Sulfolane is a very stable aprotic industrial solvent used, for example, in the extractive distillation of aromatics (*cf.* Section 12.2.2.2) or together with diisopropanolamine in the Sulfinol process for gas purification (removal of acidic gases, *cf.* Section 2.1.2). Shell and Phillips both manufacture sulfolane.

A more recent use of butadiene, which has been made available by an overabundance of butadiene from increased olefin manufacture by naphtha cracking, is selective hydrogenation to *n*-butenes in the liquid phase with a Pd catalyst. Exxon has used this process since 1993 in a 35 000 tonne-per-year plant in England. *n*-Butene is a raw material for production of methyl ethyl ketone and higher olefins. Additional processes for selective hydrogenation of diolefins are being developed.

to 2.4:

selective hydrogenation to *n*-butene for further conversion to methyl ethyl ketone and higher olefins

5.2. Isoprene

Industrial interest in isoprene (2-methyl-1,3-butadiene) and its manufacturing processes increased enormously after the stereoselective polymerization of isoprene had been mastered. The isotactic 1,4-*cis*-polyisoprene thus obtained is virtually identical with natural rubber and is actually superior in terms of purity and uniformity.

isoprene, although a basic unit of numerous natural products (terpenes, natural rubber), was commercially insignificant before its polymerization to 1,4-*cis*-polyisoprene

isoprene production (in 1000 tonnes):

	1990	1992	1995
USA	227	139	145
Japan	93	76	100
W. Europe	18	15	22

The world production capacity for isoprene 1992 was 1.5×10^6 tonnes per year, with the USA, Japan and Western Europe accounting for 150 000, 111 000 and 20 000 tonnes per year, respectively.

The manufacture of isoprene takes place largely in processes analogous to those employed for butadiene, *i.e.*, by direct isolation from C_5 cracking fractions or by dehydrogenation of C_5 isoalkanes and isoalkenes. In contrast to butadiene, synthetic reactions using smaller units to build up the C_5 skeleton have continued to account for a substantial fraction of the total production. However, as worldwide use of steam crackers based on naphtha or gas oil increases — particularly in the USA — the proportion of isoprene from extraction will increase markedly.

methods for production of isoprene:

1. from C_5 cracking fractions

1.1. by direct isoprene isolation
1.2. by dehydrogenation of isoprene precursors

2. by synthesis of isoprene skeleton from smaller units

5.2.1. Isoprene from C_5 Cracking Fractions

The feedstocks for isoprene are the C_5 naphtha cracking fractions from which it can be isolated by extractive distillation or, more recently, by distillation as an azeotrope with *n*-pentane.

naphtha C_5 cracking fractions contain about 14 – 23 wt% (ca. 2 – 5% of ethylene capacity of a steam cracker)

The direct isolation of isoprene from corresponding fractions of naphtha cracked products has the advantage that isoprene can be obtained without additional synthetic steps. This method can be economical, especially since the isoprene concentration in a typical C_5 fraction from more severe cracking can comprise from 14 to 23 wt%. The other components of the C_5 fraction are pentanes, cyclopentadiene/dicyclopentadiene, piperylene and pentenes.

isolation of isoprene by two techniques:

1. extractive distillation, *e. g.*, with NMP, DMF, or acetonitrile

2. fractional distillation as *n*-pentane azeotrope employed directly for polymerization in Goodyear process

dehydrogenation can be conducted with mixtures of C_5 isoalkanes and isoalkenes, preseparation by H_2SO_4 extraction can also take place (Shell process)

characteristics of Shell Sinclair extraction:

primary isomerization to internal olefins, then reversible semi-esterification by addition of H_2SO_4, thereafter dehydrogenation to isoprene analogous to butadiene manufacture from butenes

Various butadiene isolation processes can be modified for isoprene, including extractive distillation with N-methylpyrrolidone (BASF), dimethylformamide (Nippon Zeon), or acetonitrile (Shell, Goodrich-Arco). Commercial units — chiefly in Japan and the United States — have isoprene capacities as large as 30000 tonnes per year.

Goodyear has developed a fractional distillation process using the C_5 fraction from a steam cracker in which isoprene is recovered as an azeotrope with *n*-pentane. Isoprene can be selectively reacted from this mixture, by polymerization for example. Commercial application of this technology, characterized by its low energy requirements, is not yet widespread.

Dehydrogenation processes with isopentane and isopentenes are used mainly in the USA. The feedstocks are the C_5 fractions from catalytic cracking processes which are introduced as mixtures to the dehydrogenation. Due to their greater reactivity, the alkenes are dehydrogenated before the alkanes. The subsequent purification, especially the separation of 1,3-pentadiene (piperylene), is very involved. Therefore, Shell first separates the 2-methylbutenes from the rest of the C_5 products in a so-called Sinclair extraction with 65% H_2SO_4. The acid initially isomerizes the mixture of the 2-methylbutenes mainly to 2-methyl-2-butene, and then adds to form the sulfuric acid ester:

$$\underset{H_3C}{\overset{H_3C}{\diagdown}} C=CHCH_3 + H_2SO_4 \; \rightleftarrows \; H_3C-\overset{\overset{\displaystyle CH_3}{|}}{\underset{\underset{\displaystyle OSO_3H}{|}}{C}}-CH_2CH_3 \qquad (7)$$

The ester is then split at 35 °C and the C_5 olefin is extracted. The required dehydrogenation to isoprene is similar to processes for butadiene (*cf.* Section 5.1.3). The Shell process employs a $Fe_2O_3-Cr_2O_3-K_2CO_3$ catalyst at 600 °C, and achieves yields of about 85%. Pure isoprene (99.0−99.5 wt%) is isolated by extractive distillation with acetonitrile. Isoprene plants using this process are operated by Shell, Arco and Exxon.

In the CIS, isoprene is manufactured in three plants (total capacity 180000 tonnes per year) by dehydrogenation of isopentene in a process which has not been disclosed in detail.

5.2.2. Isoprene from Synthetic Reactions

There are basically four different ways to construct the C_5 skeleton from smaller carbon units, two of which are still being used commercially:

1. Addition of acetone to acetylene to form 2-methyl-3-butyne-2-ol followed by partial hydrogenation and dehydration.
2. Dimerization of propene to isohexene followed by demethanation.
3. Double addition of formaldehyde to isobutene resulting in 4,4-dimethyl-1,3-dioxane followed by dehydration and cleavage of formaldehyde.
4. Dismutation of isobutene and 2-butene to form 2-methyl-2-butene followed by dehydrogenation.

isoprene synthesis by construction of C_5 skeleton in one of four routes:

1. $C_3 + C_2 \rightarrow C_5$
   (acetone-acetylene route)
2. $C_3 + C_3 \rightarrow C_5 + C_1$
   (isohexene route)
3. $2\,C_1 + C_4 \rightarrow C_5 + C_1$
   (*m*-dioxane route)
4. $2\,C_4 \rightarrow C_5 + C_3$
   (dismutation route)

The acetone–acetylene process was developed by Snamprogetti, and was used in a 30000 tonne-per-year plant in Italy until 1982. In the first step, acetone and acetylene are reacted together at $10-40\,°C$ and 20 bar in liquid ammonia, with KOH as the catalyst. The product methylbutynol is selectively hydrogenated to methylbutenol, which is dehydrated at $250-300\,°C$ on Al_2O_3 at atmospheric pressure to give isoprene:

to 1:

three-step Snamprogetti process consisting of:

1. ethynization of acetone
2. partial hydrogenation of alkynol
3. dehydration of alkenol

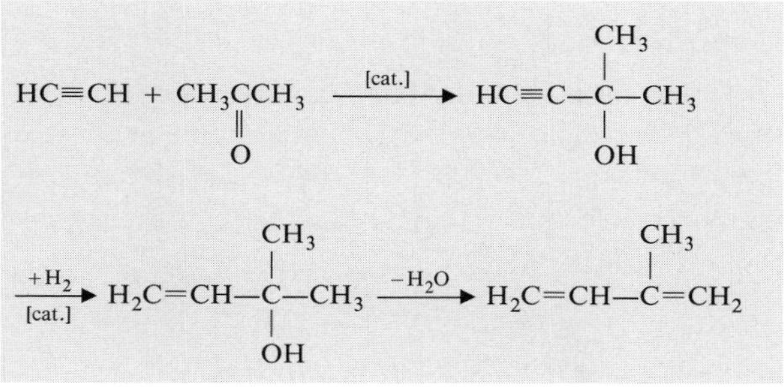

(8)

Overall selectivity to isoprene is 85% (based on CH_3COCH_3, C_2H_2).

The Goodyear–Scientific Design process operates according to the second principle, the isohexene route. The dimerization of propene on a Ziegler catalyst (*e. g.*, tri-*n*-propylaluminum) leads to the formation of 2-methyl-1-pentene (*cf.* Section 3.3.1), which is then isomerized to 2-methyl-2-pentene with a supported acidic catalyst. This can be cracked with superheated steam and catalytic amounts of HBr at $650-800\,°C$ to give isoprene and methane:

to 2:

three-step Goodyear-SD process consisting of:

1. propene dimerization
2. 2MP1 isomerization
3. 2MP2 demethanation

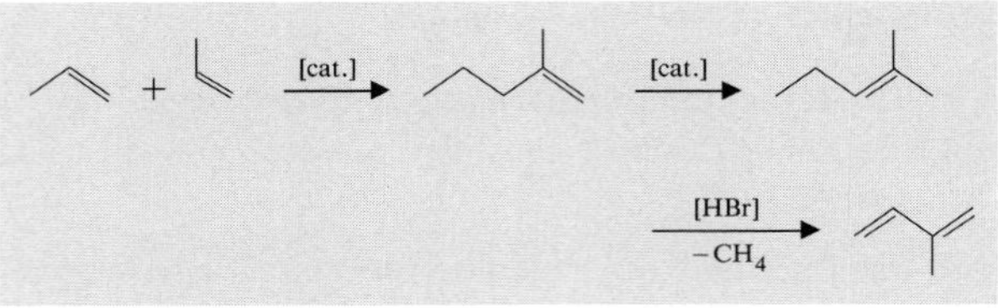

(9)

Isoprene selectivity is about 50% (based on propene).

Goodyear operated a plant in Texas for several years using this process. It was shut down for economic reasons.

commercial operation in Goodyear plant (capacity about 75 000 tonnes per year) has been stopped

The third route with *m*-dioxane as intermediate has been extensively studied and also developed into industrial processes by various companies (Bayer, IFP, Marathon Oil, Kuraray and in the CIS).

The first step is a Prins reaction between formaldehyde and isobutene. It takes place in the presence of a strong mineral acid such as H_2SO_4 or on acidic ion-exchangers at $70-95\,°C$ and about 20 bar. Aqueous formaldehyde reacts with the isobutene present in a butadiene-free C_4 fraction to form 4,4-dimethyl-1,3-dioxane:

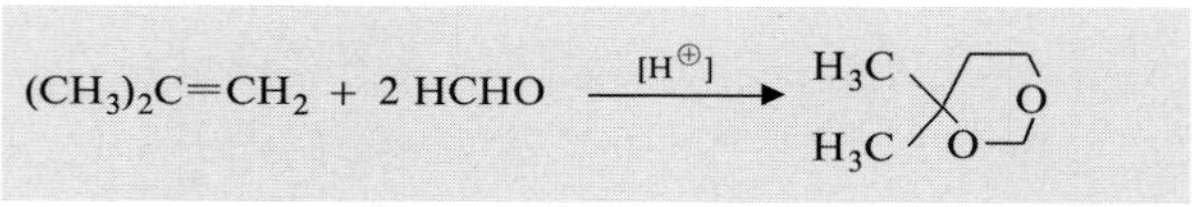

to 3:

two-step process of numerous companies consisting of:

1. double addition of HCHO to isobutene forming *m*-dioxane derivative

2. thermolytic cleavage into HCHO, H_2O, and isoprene

(10)

In the next step, the dioxane derivative is cracked at $240-400\,°C$ on a H_3PO_4/charcoal or $Ca_3(PO_4)_2$ catalyst in the presence of additional water:

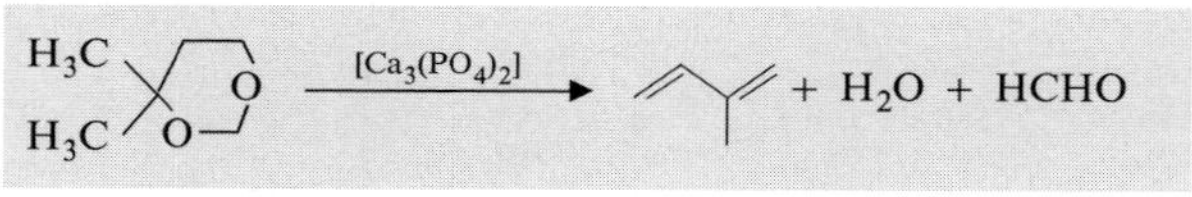

(11)

The total selectivity to isoprene is about 77% (based on isobutene).

m-dioxane derivative also starting material for new C_5 solvent with ether and alcohol functionality

Polyols are formed as byproducts in both steps, and the amounts formed strongly influence the economic viability. Kuraray operates a 30 000 tonne-per-year plant in Japan using this process. Kuraray also uses the 4,4-dimethyl-1,3-dioxane to manufacture 3-methyl-3-methoxybutanol, an economical solvent, by selective hydrogenolysis:

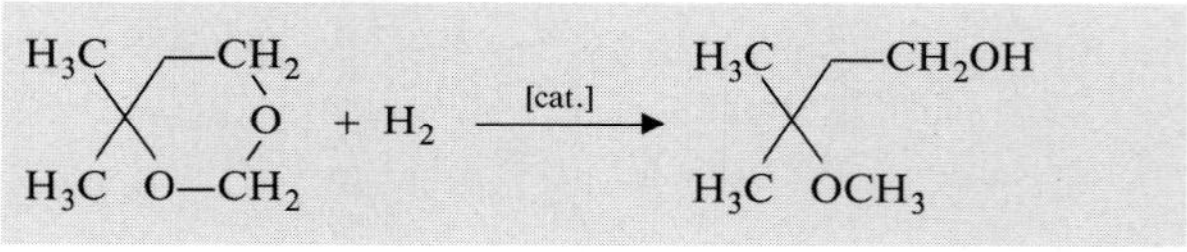

(12)

There are also several plants operating in the CIS and Eastern Europe. Recently, several firms have made proposals to simplify the process and improve the profitability of the *m*-dioxane route.

According to Takeda Chemical, isoprene can be manufactured from isobutene and formaldehyde in the gas phase at 300 °C and 1 bar in a single-step process with catalysts such as oxides of silicon, antimony or the rare earths. Instead of formaldehyde, Sun Oil proposes using methylal $H_2C(OCH_3)_2$ for the single-step reaction with isobutene, since it is more stable than formaldehyde and less decomposes into $CO + H_2$. Dioxolane, another formaldehyde source, has been suggested by Sumitomo Chemical.

Another improvement and simplification is also said to result from Sumitomo Chemical's single-step reaction of isobutene with methanol and O_2 which goes through a formaldehyde intermediate:

$$\text{(structure)} + CH_3OH + 0.5\,O_2 \xrightarrow{\text{[cat.]}} \text{(structure)} + 2\,H_2O \qquad (13)$$

H_3PO_4—MoO_3/SiO_2 and mixed oxide systems based on Mo—Bi—P—Si, Mo—Sb—P—Si, or H_3PO_4—V—Si are the catalysts used. At 250 °C, the isobutene conversion is 12%, with selectivity to isoprene of 60% (based on isobutene) and 40% (based on methanol). Most of the methanol is oxidized to formaldehyde.

The potential advantages of this process, such as less expensive C_1 components and the lower investment costs of a single-step process, are still to be confirmed in an industrial plant.

The fourth synthetic route to an isoprene precursor has not yet been employed industrially although it is the subject of great interest. This method is analogous to the dismutation of propene to 2-butene and ethylene developed by Phillips Petroleum and extended by other firms (*cf.* Section 3.4). In this case, a dismutation takes place between isobutene and 2-butene. These starting materials can be obtained from the butadiene raffinate (*cf.* Section 5.1.2):

process modifications of the *m*-dioxane route:

1. Takeda simplifies process with a single-step gas-phase reaction

2. in addition to single-step performance, increase in selectivity using HCHO in the form of methylal (Sun Oil) or dioxolane (1,3-dioxacyclopentane, Sumitomo)

3. in addition to single-step performance, Sumitomo uses less expensive feedstock $CH_3OH + O_2$ to replace HCHO

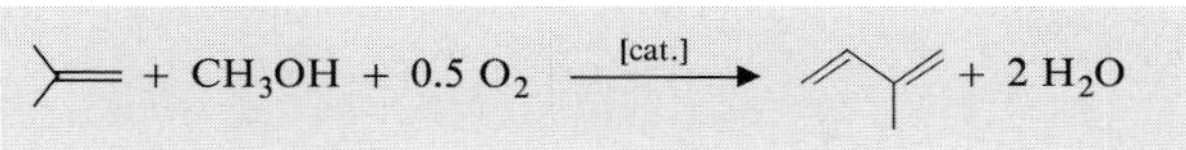

to 4:

two-step process consisting of:

1. catalytic dismutation of *n*-C_4 and iso-C_4 olefin
2. dehydrogenation of 2-methyl-2-butene

$$\begin{array}{c} CH_3 \\ | \\ CH_3C{=}CH_2 \\ + \\ CH_3CH{=}CHCH_3 \end{array} \underset{}{\overset{\text{[cat.]}}{\rightleftarrows}} \begin{array}{c} CH_3 \\ | \\ CH_3C \qquad CH_2 \\ \| \quad + \quad \| \\ CH_3CH \qquad CHCH_3 \end{array} \qquad (14)$$

The resulting 2-methyl-2-butene can be dehydrogenated to isoprene.

uses of isoprene:

1. minor significance as comonomer with isobutene in butyl rubber
2. main use in 1,4-*cis*-polyisoprene rubber

Isoprene was used solely as a comonomer with isobutene in the manufacture of butyl rubber for a long time. The demand was small because of the low isoprene content (2–5 wt%) in the copolymer. Isoprene first became important with the manufacture of 1,4-*cis*-polyisoprene rubber. This rubber has gained increasing commercial interest due to its high thermal stability and durability when employed in tire treads, especially in radial tires.

Goodyear is the greatest producer of isoprene rubber worldwide with a production capacity of 61 000 tonnes per year in the USA (1996).

Of the estimated world production of isoprene in 1995 (ca. 0.56 $\times$ 10^6 tonnes), in various countries 50–80% will be used for polyisoprene, 4–40% for butyl rubber, and increasing amounts for copolymers (*e. g.*, with styrene or acrylonitrile) and other applications.

5.3. Chloroprene

order of importance of butadiene and derivatives:

1. 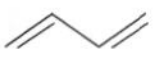
2. 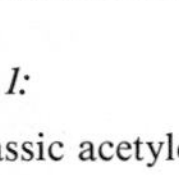
    CH$_3$
3.
    Cl

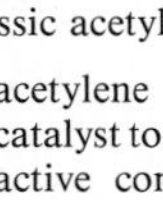

Chloroprene (2-chloro-1,3-butadiene) assumes third place in industrial importance behind butadiene and isoprene. Its main use is as a building block for synthetic rubbers.

manufacturing processes for chloroprene:

1. acetylene route of decreasing importance
2. butadiene route of increasing importance

Acetylene is used as a feedstock in the older manufacturing processes for chloroprene, while the more modern processes use butadiene. The change of raw materials took place in the early 1970s. By 1980, more than 80% of the worldwide production of chloroprene (ca. 0.4 $\times$ 10^6 tonnes) was manufactured from butadiene. In the USA and in Western Europe, chloroprene is produced exclusively from butadiene, while in Japan a small plant at Denki-Kagaku still uses acetylene.

to 1:

classic acetylene route is two-step:

1. acetylene dimerization with Nieuwland catalyst to form vinylacetylene (CuCl is an active component, NH$_4$Cl is added to increase CuCl solubility in H$_2$O)

The traditional chloroprene synthesis is done in two steps. First, acetylene is dimerized to vinylacetylene in an aqueous hydrochloric acid solution of CuCl and NH$_4$Cl at 80°C in a reaction tower:

$$2\ HC\equiv CH \xrightarrow{\text{[cat.]}} H_2C=CH-C\equiv CH$$

$$\left(\Delta H = -\frac{63\ \text{kcal}}{264\ \text{kJ}}\big/\text{mol}\right) \tag{15}$$

The substantial evolution of heat is controlled by vaporization of water. The C$_2$H$_2$ conversion reaches about 18%. The selectivity to vinylacetylene is as high as 90%, with the main byproduct being divinylacetylene. In the second stage, HCl is added at 60 °C to vinylacetylene, forming chloroprene:

2. HCl addition to vinylacetylene using the same catalyst system as step 1

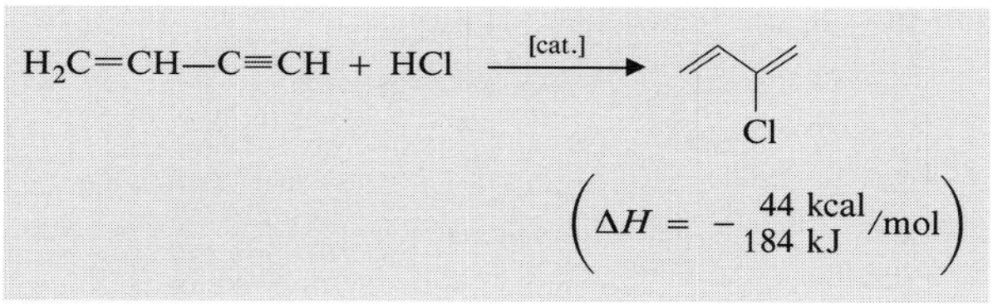

$$H_2C{=}CH{-}C{\equiv}CH + HCl \xrightarrow{\text{[cat.]}} \underset{Cl}{\diagup\!\!\diagup\!\!\diagdown\!\!\diagup\!\!\diagup}$$

$$\left(\Delta H = -\frac{44\ \text{kcal}}{184\ \text{kJ}}\big/\text{mol}\right) \qquad (16)$$

A solution of CuCl in hydrochloric acid can also serve as the catalyst. The selectivity to chloroprene is about 92%, based on vinylacetylene. The main byproducts are methyl vinyl ketone and 1,3-dichloro-2-butene.

byproducts result from secondary reactions involving chloroprene such as:

hydrolysis

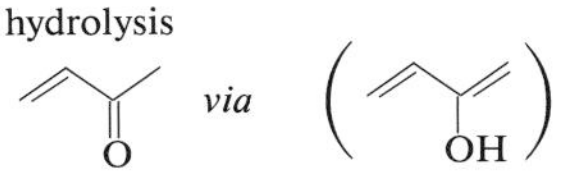

$$\underset{O}{\diagup\!\!\diagup\!\!\diagdown\!\!\diagup}\quad via \quad \left(\underset{OH}{\diagup\!\!\diagup\!\!\diagdown\!\!\diagup}\right)$$

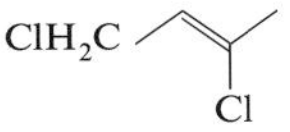

The more recent chloroprene processes are based on butadiene, an inexpensive feedstock. Even though the conversion to chloroprene appears to be relatively simple and was described early on, it was only after intensive development work by British Distillers and then BP that an economical process was established. This process was first operated by Distugil (50% BP) in a plant with 30000 tonne-per-year chloroprene capacity.

1,4-addition of HCl

$$ClH_2C\diagup\!\!\diagup\!\!\underset{Cl}{\diagdown\!\!\diagup}$$

Numerous other processes based on butadiene followed.

In the manufacture of chloroprene from butadiene, the initial step is a gas-phase free-radical chlorination with Cl_2 at 250 °C and 1–7 bar to give a mixture of *cis*- and *trans*-dichloro-2-butene as well as 3,4-dichloro-1-butene:

to 2:

three-step butadiene route:

1. noncatalytic radical addition of Cl_2, *i.e.,* gas-phase chlorination forming dichlorobutene mixtures (oxychlorination of butadiene industrially insignificant)

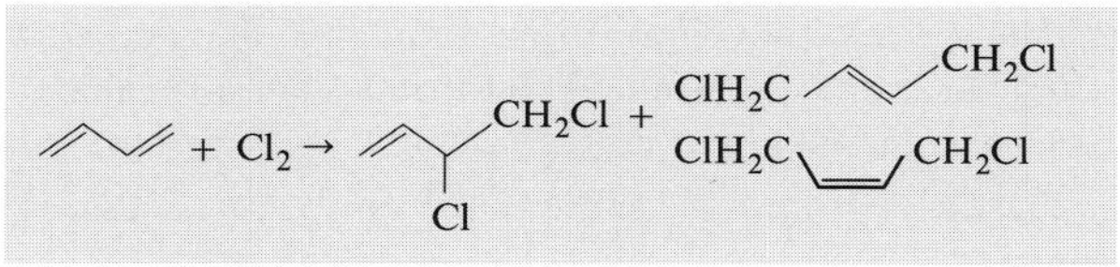

$$\diagup\!\!\diagup\!\!\diagdown\!\!\diagup + Cl_2 \rightarrow \underset{Cl}{\diagup\!\!\diagup\!\!\diagdown}\!\!\diagup^{CH_2Cl} + \begin{matrix}ClH_2C\diagup\!\!\diagdown^{CH_2Cl}\\ ClH_2C\diagdown\!\!\diagup_{CH_2Cl}\end{matrix} \qquad (17)$$

At butadiene conversions of 10–25%, the selectivity to this mixture of dichlorobutenes is 85–95% (based on butadiene).

The 1,4-adduct is unsuitable for chloroprene manufacture, but can be isomerized to the 1,2-adduct by heating with catalytic amounts of CuCl or with iron salts.

2. catalytic isomerization of 1,4- into 1,2-adduct in an allyl rearrangement

The equilibrium is continuously displaced in the desired direction by distilling off the 3,4-isomer (b. p. = 123 °C compared to 155 °C), allowing a selectivity of 95–98% to be attained.

Processes for the oxychlorination of butadiene have been developed by various firms (ICI, Monsanto, Shell), but there have been no industrial applications.

3. dehydrochlorination to chloroprene with short residence time and addition of stabilizers (sulfides or nitric oxides)

Chloroprene is obtained with a yield of 90–95% by dehydrochlorination with dilute alkaline solution at 85 °C:

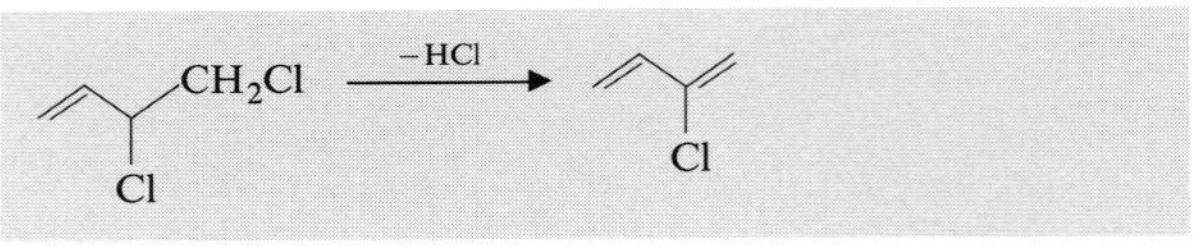

(18)

All recently constructed chloroprene capacity is based on this three-step process, since a direct conversion of butadiene into chloroprene is not economically feasible due to low selectivities.

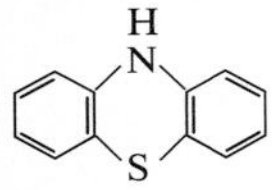

The autoxidation of chloroprene results in a peroxide which can initiate polymerization. Traces of oxygen must therefore be carefully excluded during the dehydrochlorination and purification steps. In addition, sulfur compounds, being very effective polymerization inhibitors, are introduced.

purification of chloroprene:

distillation with exclusion of O_2 and in presence of polymerization inhibitors, e. g., phenothiazine

utilization of chloroprene:

1. monomer for chloroprene rubber

Almost all chloroprene is polymerized to chloroprene rubber (Neoprene® or Baypren®). It is distinguished from other rubber types by its stability towards sunlight and oil.

In 1994, the production capacity for polychloroprene was 163 000, 133 000, and 88 000 tonnes per year in the USA, Western Europe, and Japan, respectively.

2. 1,4-dichloro-2-butene as chloroprene precursor also intermediate for adiponitrile, 1,4-butanediol, tetrahydrofuran

1,4-Dichloro-2-butene, an intermediate in chloroprene manufacture, is employed to a limited extent as starting material for adiponitrile (cf. Section 10.2.1.1), 1,4-butanediol and tetrahydrofuran. In a process developed by Toyo Soda for the manufacture of 1,4-butanediol, 1,4-dichlorobutene is hydrolyzed at approximately 110 °C in the presence of an excess of sodium formate:

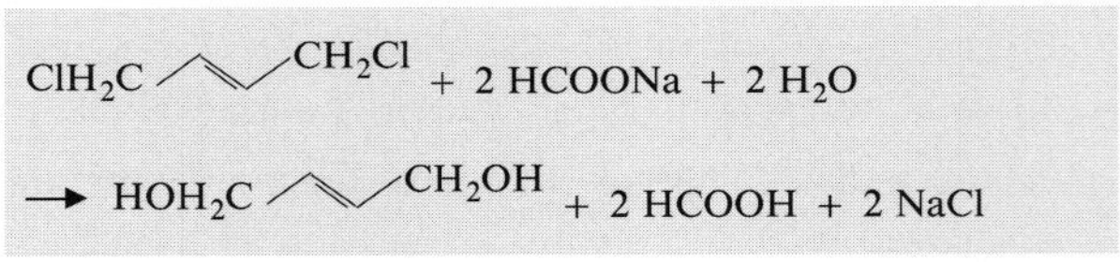

(19)

characteristics of 1,4-butanediol manufacture from 1,4-dichlorobutene:

1. hydrolysis with butene-1,4-diformate as intermediate

After the hydrolysis, the free formic acid is neutralized with NaOH. The 1,4-dichlorobutene conversion is almost 100% with a 1,4-butenediol selectivity of more than 90%.

Direct hydrolysis would lead to noticeably lower selectivity due to the formation of polycondensation products and 3-butene-1,2-diol.

The aqueous solution of 1,4-butenediol is then hydrogenated to 1,4-butanediol in the presence of a Ni/Al catalyst at ca. 100 °C and 270 bar.

2. heterogeneous catalytic (Ni/Al) liquid-phase hydrogenation to 1,4-butanediol

Toyo Soda operates two plants with a total capacity of about 7000 tonnes per year 1,4-butanediol and 3000 tonnes per year tetrahydrofuran (obtained by the cleavage of water from 1,4-butanediol) with this technology.

industrial applications:

Toyo Soda process with coupled manufacture of chloroprene
1,4-butanediol
tetrahydrofuran

Other chloroprene producers include Bayer, Denka Chemical, Denki Kagaku, Distugil, Du Pont, and Showa.

5.4. Cyclopentadiene

Cyclopentadiene is another commercial 1,3-diolefin obtained in limited amounts from coal tar, as well as in higher concentrations (between 15 and 25 wt%) in the C_5 fractions from naphtha cracking mixtures. Its share, relative to the ethylene production, is about 2–4 wt%. With the construction of large steam crackers for ethylene production, cyclopentadiene has become an inexpensive feedstock. However, due to lack of demand, it usually remains in the C_5 fraction which, after removal of the diolefins and hydrogenation, is incorporated in motor gasoline as a high-octane component.

cyclopentadiene starting materials:

1. coal tar
2. C_5 cracking fractions

At the present time, the sole Western European producers of cyclopentadiene based on petrochemicals are Shell and Dow in the Netherlands. Chemical Exchange, Dow, Exxon, Lyondell and Shell produce cyclopentadiene in the USA.

Nowadays, cyclopentadiene is generally obtained from the first runnings of pyrolysis gasolines where its tendency to dimerize – a special case of Diels–Alder reaction – is exploited. The C_5 fraction is either heated to 140–150 °C under pressure or left for several hours at 100 °C. Under these conditions, cyclopentadiene dimerizes to dicyclopentadiene (endo form favored). This boils about 130 °C higher than its monomer, so the remaining C_5 components can be readily separated by distillation under reduced pressure.

cyclopentadiene isolation from C_5 cracking fraction via thermal processes:

1. dimerization to dicyclopentadiene and separation
2. cleavage of dicyclopentadiene to regenerate cyclopentadiene (reverse Diels–Alder reaction)

The residue is split at an industrially suitable rate at temperatures over 200 °C (for example: 300 °C, tubular reactor, 80–85% yield) to regenerate the monomer, which is distilled off as pure cyclopentadiene.

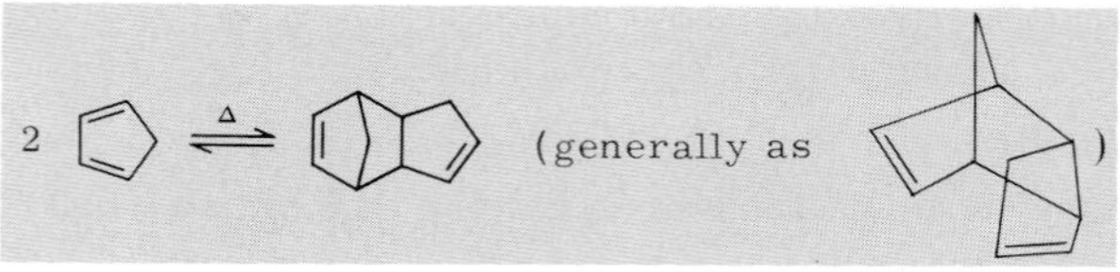

It readily reverts to the dimer with considerable evolution of heat. It is transported and stored in this form.

utilization of cyclopentadiene:

1. as monomer and comonomer in manufacture of polycyclopentadiene and C_5 hydrocarbon resins for the printing ink and adhesive sectors

Cyclopentadiene is finding increasing use in the manufacture of hydrocarbon resins. It is either copolymerized with the other components of the C_5 cracking fraction (pentanes, pentenes, isoprene, piperylene) to C_5 fraction resin, or pure cyclopentadiene is thermally polymerized to polycyclopentadiene in an aromatic solvent at 250–280 °C. If necessary, it can be stabilized by subsequent hydrogenation. It can also be cationically copolymerized. The polymers are a constituent of thermoplastic and contact adhesives as well as of printing ink resins. In Europe, Exxon and ICI are the main producers of C_5 hydrocarbon resin.

2. as diene for Diels-Alder reactions, e. g., two-step manufacture of ethylidenenorbornene from cyclopentadiene and butadiene; i. e., (4 + 2)-cycloaddition + double bond isomerization

Cyclopentadiene is also used in other Diels–Alder reactions; for example, 5-ethylidenenorbornene can be obtained from a two-step reaction between cyclopentadiene and butadiene. 5-Vinylbicyclo-[2.2.1]hept-2-ene is formed initially, and then isomerized to 5-ethylidenebicyclo-[2.2.1]hept-2-ene (5-ethylidenenorbornene) with alkali metal catalysts:

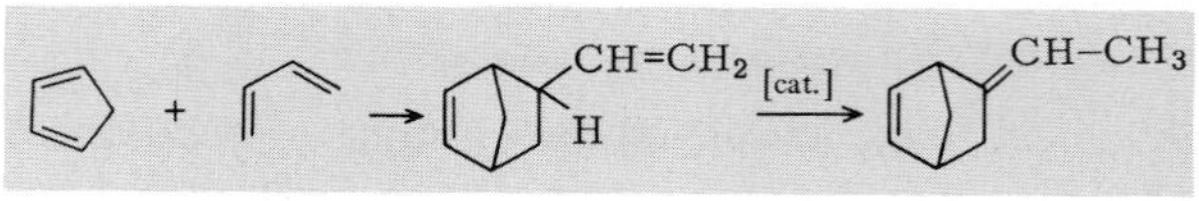

5-Ethylidenenorbornene is added to the ethylene–propene polymerization in small amounts as a third component.

ethylidenenorbornene use for EPTR ('ter' = three-component)

The resulting terpolymer can be vulcanized under the usual conditions to EPTR (ethylene propene terpolymer rubber).

3. as hexachlorinated diene component

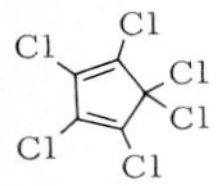

precursor for insecticides and flame retardants

Other industrial examples of Diels-Alder reactions start from the important intermediate hexachlorocyclopentadiene obtained from multistep chlorination of cyclopentadiene.

6. Syntheses involving Carbon Monoxide

The economic significance of carbon monoxide as a synthetic component, alone or with hydrogen, has already been considered in Chapter 2, 'Basic Products of Industrial Synthesis', where the manufacture of methanol, its quantitatively most important secondary product, was discussed. Other industrial processes involving CO as feedstock, for example the reaction with acetylene to yield acrylic acid (*cf.* Section 11.1.7.1), with methanol in the synthesis of acetic acid (*cf.* Section 7.4.1.3), or with propene resulting in butanol (*cf.* Section 8.1.3) are so specific for the product involved that they will be dealt with separately. The same goes for products of 'C₁ chemistry' — that is, for products from the synthesis of intermediates from synthesis gas or their simple derivatives — such as methanol and formaldehyde. In this section, CO reactions such as hydroformylation, carbonylation, and the Koch reaction which can be used to convert a wide range of olefins are discussed.

CO alone or together with H_2 is an economical synthetic unit

besides application in individual industrial processes for:

methanol (synthesis gas)
acrylic acid (C_2H_2 + CO + H_2O)
acetic acid (CH_3OH + CO)
butanol (C_3H_6 + 3 CO + 2 H_2O)

and recently 'C₁ chemistry'

also general processes such as:

hydroformylation
carbonylation
Koch reaction

6.1. Hydroformylation of Olefins

Hydroformylation or 'oxo' synthesis is an industrial process for the manufacture of aldehydes from olefins, carbon monoxide, and hydrogen.

principle of hydroformylation:

catalytic addition of H_2 and CO to olefins resulting in chain extension by one C atom with formation of an aldehyde

The basic reaction was discovered by O. Roelen of Ruhrchemie in 1938. He observed the formation of propionaldehyde when ethylene was reacted with CO and H_2 using cobalt–thorium catalysts at elevated pressures and temperatures.

In the years following, the hydroformylation reaction was quickly developed to an industrial process for the manufacture of detergent alcohols with chain lengths between C_{12} and C_{14}. In 1945, the first 10 000 tonne-per-year 'oxo' plant went into operation. (The current detergent alcohol situation is discussed in Section 8.2).

first industrial manufacture:

$C_{12}-C_{14}$ detergent alcohols by hydrogenation of resulting primary aldehydes

The hydroformylation reaction has become very important throughout the world. By 1995, the world capacity for all hydroformylated products amounted to more than

today the most important 'oxo' synthesis products are:

n-butanol and 2-ethylhexanol

7 × 10⁶ tonnes per year of which the USA, Western Europe and Japan accounted for roughly 2.1, 2.5, and 0.73 × 10⁶ tonnes per year, respectively. The most important olefin feedstock is propene, and *n*-butanol and 2-ethylhexanol are the most significant final products. Ethylene is also hydroformylated to propionaldehyde, which is then oxidized in the most important commercial route (ca. 50% of worldwide production) to propionic acid (*cf.* Section 6.2). Propionaldehyde is also used for the production of methacrylic acid (*cf.* Section 11.1.4) in a new process developed by BASF, increasing the range of used and demand for propionaldehyde.

6.1.1. The Chemical Basis of Hydroformylation

Hydroformylation can be conducted with a multitude of straight-chain and branched olefins with terminal or internal double bonds. Olefins with 2 to about 20 carbon atoms are significant industrial feedstocks. Typically − except in the case of ethylene, where propionaldehyde is the only product − mixtures of isomeric aldehydes resulting from attachment of the formyl group at either end of the double bond are formed:

range of application of olefin hydroformylation:

C_2 to C_{20} olefins, branched or unbranched, with terminal or internal C—C double bond

reaction products of hydroformylation:

mixtures of aldehydes with higher content of *n*-aldehydes (higher *n*-/iso ratio)

$$RCH{=}CH_2 + CO + H_2 \xrightarrow{\text{[cat.]}} \begin{array}{l} RCH_2CH_2CHO \\[2mm] RCHCH_3 \\ \quad | \\ \quad CHO \end{array} \qquad (1)$$

The formation of the *n*-aldehyde is favored over the isoaldehyde.

effects of olefin feedstock:

1. olefin structure preserved in aldehyde, no chain isomerization
2. double bond isomerization with internal olefins can lead to aldehyde mixtures

   internal double bonds generally migrate to end of chain as terminal π-complex is most stable

The structure and size of the olefin molecule also influence the course of the hydroformylation. Although the carbon skeleton of the olefin is retained in the 'oxo' aldehyde, the double bond can migrate from an internal toward a terminal position resulting in more components in the aldehyde mixture. If sufficient residence time and temperature allow the isomerization taking place at the hydroformylation catalyst to go to completion (terminal double bond), then double-bond isomers of an olefin will lead to the same 'oxo' aldehydes.

3. unbranched α-olefins are readily hydroformylated in contrast to the scarcely reactive sterically hindered internal olefins

Moreover, the rate of reaction is reduced by branching in the olefin, particularly when branching is at an olefinic carbon atom.

example of steric hindrance:

H₃C   CH₃
    \   /
     C=C
    /   \
H₃C   CH₃

Branching on both sides of the double bond can markedly impede the hydroformylation, as in the case of 2,3-dimethyl-2-butene.

Hydroformylation is a homogeneously catalyzed reaction which takes place at high pressures and temperatures, usually 200 to 450 bar and 100 to 200 °C.

The Co, Rh or Ru compounds or complexes which are employed as catalysts can be modified with amine or phosphine containing ligands to influence their activity and selectivity. Although the addition of promoters is reported in the literature, they are not industrially significant.

Until recently, Co compounds were preferentially employed in industrial processes because of their comparatively low cost and high activity. Today, most new plants use rhodium, and more and more older plants are changing over. In 1985, 60% of the propene hydroformylation capacity was still based on Co catalysis, while 40% was already based on Rh catalysis. When Co is used, the active form of the catalyst, tetracarbonylhydrocobalt ($HCo(CO)_4$), is formed in equilibrium with dicobalt octacarbonyl ($Co_2(CO)_8$) under the conditions of the reaction.

During the course of the reaction, tetracarbonylhydrocobalt – after transformation into tricarbonylhydrocobalt – possesses a free coordination position which is occupied by the olefin with formation of a π-complex:

catalysis principle of the hydroformylation: homogeneous catalysis with hydrido-carbonyl complexes possessing following structure:

possible central atoms:

Co, Rh, Ru

possible ligands besides CO:

$-NR_2$, and $-PR_3$;
R = alkyl or aryl

industrially favored catalyst:

Co fed as metal or compound, then in active form

$$2\ H-Co(CO)_4 \rightleftarrows Co_2(CO)_8 + H_2$$

mechanism of catalysis:

simultaneous, formally multistep, reversible process with irreversible hydrogenolysis step

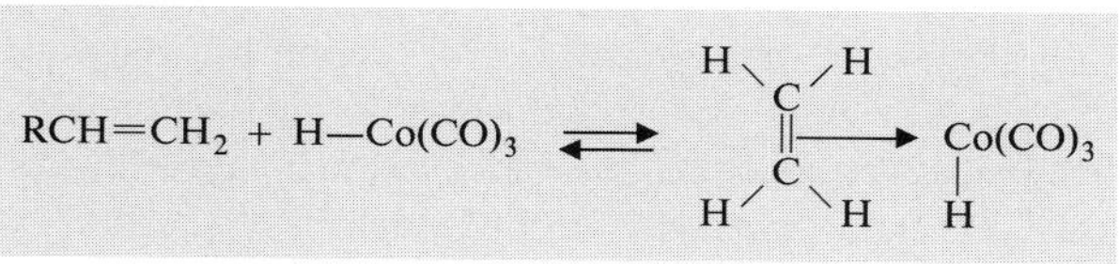

$$RCH{=}CH_2 + H-Co(CO)_3 \rightleftarrows \quad \overset{H}{\underset{H}{{}}}\overset{}{\underset{}{C}}{=}\overset{}{\underset{H}{C}}\overset{H}{\underset{H}{{}}} \longrightarrow Co(CO)_3 \qquad (2)$$

The π-complex rearranges with formation of a C–Co bond to a σ-complex which is saturated with CO to give an alkyl tetracarbonylcobalt:

1. $HCo(CO)_4$ forms free coordination position by release of CO

   $$H-Co(CO)_4 \rightleftarrows H-Co(CO)_3 + CO$$
2. carbonyl species adds electrophilically to olefin, forming a π-complex
3. π-σ-rearrangement with stabilization by uptake of CO

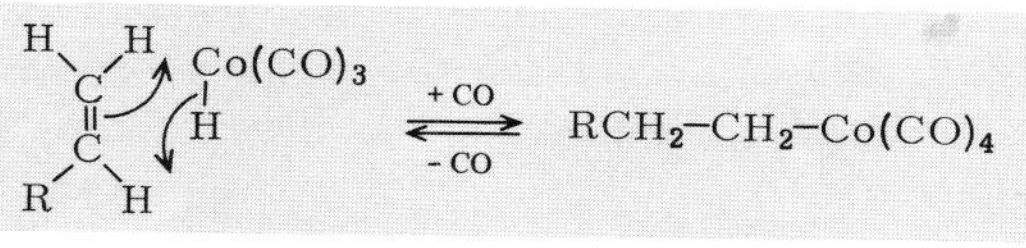

$$\qquad \overset{+\ CO}{\underset{-\ CO}{\rightleftarrows}} \quad RCH_2{-}CH_2{-}Co(CO)_4 \qquad (3)$$

During the next step, the alkyl tetracarbonylcobalt complex is transformed into an acyl tricarbonylcobalt complex by expulsion of a CO ligand from the coordination sphere of the cobalt. This undergoes hydrogenative cleavage to produce the aldehyde and re-form the tricarbonylhydrocobalt:

4. CO insertion leads to acyl complex, recent theories also concern carbenium ion migration
5. hydrogenolysis of the C–Co bond

$$RCH_2CH_2{-}Co(CO)_4 \rightleftarrows RCH_2CH_2{-}\overset{\overset{\textstyle O}{\|}}{C}{-}Co(CO)_3 \xrightarrow{+H_2}$$

$$RCH_2CH_2CHO \ + \ HCo(CO)_3$$

$$(4)$$

catalysis with coordination compounds also allows an internal Co–C complex (*cf.* eq 3) leading to the isoaldehyde

In the course of the reaction, these partial reactions take place simultaneously and are inseparable from one another. In addition, isomer formation also occurs, as can be appreciated from equation 3 in which the cobalt–carbon bond can form at either the terminal or the penultimate carbon atom.

sterically favored terminal Co–C complex leads to higher proportion of *n*-aldehyde (higher *n*-/iso ratio)

Although the *n*-aldehyde dominates in the isomeric mixture, because it is of much greater industrial importance than the iso-aldehyde, all commercial processes strive to increase its share by catalyst modifications and suitable process conditions.

possibilities for regulating the *n*:iso ratio:
1. by reaction conditions such as temperature, CO partial pressure
2. by catalyst modifications such as variation of ligands and central atom

Although the *n*:iso ratio can be influenced to a certain extent by temperature and CO partial pressure (*cf.* Section 6.1.2) the strongest effect is obtained by modifying the catalyst system.

With metal complexes as catalysts, modification is possible by exchanging ligands and the central atom. Both are used industrially and lead essentially to the following conclusions:

effect of ligands:

large volume ligands increase *n*-aldehyde share but lower olefin conversion with simultaneous increase of hydrogenation activity:

olefin → paraffin
aldehyde → alcohol

1. Complex-forming additives such as *tert*-amines, phosphites or, in particular, phosphines – for example tributyl or triphenylphosphine – cause an increase in the fraction of *n*-aldehyde formed, but also cause a reduction in rate of hydroformylation and lower selectivity to aldehydes.

effect of central atom:

Rh instead of Co reduces *n*-aldehyde share, increases olefin conversion and *n*:iso aldehyde selectivity

2. Although using rhodium instead of cobalt catalysts leads to a reduction in the fraction of *n*-aldehydes, it also causes an increase in the rate of hydroformylation and in selectivity to total aldehyde (*cf.* Section 6.1.3).

The effect of ligand on the *n*:iso ratio can be explained by a reduction in electron density at the central atom along with the steric effect of the catalyst molecule. Large-volume phosphine ligands hinder the attack at an internal carbon atom, and thus increase the fraction of *n*-aldehydes in the reaction product.

practical application of ligand and central atom effects:

combination of rhodium with phosphine ligands leads to maximum proportion of *n*-aldehyde

With the appropriate combination of rhodium as central atom with phosphine ligands, the positive effects can be optimized to give a higher *n*:iso ratio. With a propene feed, the ratio of butyraldehyde and isobutyraldehyde can be varied between 8:1 and 16:1 (in practice a ratio of about 10:1 is generally employed) compared to the 8:2 ratio with unmodified Co catalysts. Other advantages such as lower reaction pressures and simplified workup of the reaction products are set against disadvantages such as lower catalyst activity and a high rhodium cost.

Besides the formation of undesired isoaldehydes as outlined in the principles of the hydroformylation reaction and the olefinic double bond isomerization already mentioned, there are several other side and secondary reactions which, depending on catalyst type and reaction conditions, affect *n*-aldehyde selectivity.

side reactions of hydroformylation:

1. alkene + H_2 → alkane
2. internal alkene → terminal alkene

Hydrogenation of the olefin feedstock to a saturated hydrocarbon is one of the side reactions. Further hydrogenation of *n*- and isoaldehyde to the corresponding alcohols, aldol condensations, formation of formic acid esters by aldehyde hydroformylation, and acetal formation are all examples of secondary reactions. Thus, there is a whole series of optimization problems to be solved if the selectivity to a linear aldehyde is to be maximized.

secondary reactions in hydroformylation:

1. *n*/isoaldehyde + H_2 → *n*/isoalcohol
2. aldol condensation
3. aldehyde hydroformylation to formic acid esters
4. aldehyde/alcohol conversion to acetals

6.1.2. Industrial Operation of Hydroformylation

Today, the main industrial processes employed in the hydroformylation of olefins use cobalt compounds as catalysts without phosphine or phosphite additives (*cf.* Catalyst Modifications, Section 6.1.3).

The individual steps of the industrial process will now be explained using a cobalt catalyst in the hydroformylation of the quantitatively most important olefin feedstock − propene. The process can be divided into three stages:

1. Hydroformylation (including catalyst preparation)
2. Catalyst separation (including workup)
3. Isolation of the reaction products

To 1st process step:

Cobalt, either as metallic powder, hydroxide or as a salt, is fed into the stainless steel high pressure reactor. It reacts with the oxo gas (H_2 + CO) in the liquid phase (propene + 'oxo' products) under hydroformylation conditions (250−300 bar, 140−180 °C) with sufficient speed to form cobalt hydrocarbonyl. Although the olefin reaction product generally serves as solvent, alkane mixtures can also be used for this purpose.

1. catalyst preparation can be combined with hydroformylation, as active species $HCo(CO)_4$ is formed quickly enough from various Co precursors

Propene is then converted with H_2 and CO into a mixture of butyraldehyde and isobutyraldehyde:

principle of reaction:

exothermic, homogeneously catalyzed liquid-phase reaction with different solubilities for gaseous partners CO, H_2, and olefin

$$CH_3CH=CH_2 + CO + H_2 \xrightarrow{\text{[cat.]}} \begin{array}{l} CH_3CH_2CH_2CHO \\ CH_3CHCH_3 \\ \qquad | \\ \qquad CHO \end{array} \qquad (5)$$

characteristic composition of condensate (in wt%):

C$_4$ aldehydes	80
C$_4$ alcohols butyl formates }	10−14
various	6−10

n:iso from 75:25 to 80:20

The heat of reaction, about 28–35 kcal (118–147 kJ)/mol olefin, is removed by a tubular heat exchanger.

The condensable crude product consists of ca. 80 wt% butyraldehydes, 10–14% butanols and butyl formates and 6–10% various compounds such as high boiling products. The n:iso butyraldehyde ratio varies from about 75:25 to 80:20. Up to 90% of the CO/H$_2$ mixture is converted into isolable aldehydes and alcohols. The remainder is discharged with inerts and combusted.

The selectivity to C$_4$ products is 82–85% (C$_3$H$_6$). 15–17% of the converted propene is present in the higher boiling substances or as propane in the waste gas.

numerous process variables influence the rate of aldehyde formation and composition of 'oxo' products

The conversion and selectivities depend in a complex manner on numerous process variables. A few simplified concentration effects based on a rate equation derived by G. Natta are presented here. An industrially desirable high rate of formation of aldehyde is dependent on a large value for the concentration or partial pressure quotient:

$$\frac{d[\text{aldehyde}]}{dt} = k \cdot \frac{[\text{olefin}] \cdot [\text{Co}] \cdot p\text{H}_2}{p\text{CO}}$$

This can, in principle, be achieved in two ways:

1. by low CO partial pressure
2. by high olefin and cobalt concentrations and a high H$_2$ partial pressure

To 1:

low CO partial pressure favors hydroformylation

stabilization of the catalyst by complex formation, *e.g.*, with phosphine ligands, possible

A low CO partial pressure causes the rate of hydroformylation to increase. However, a minimum CO pressure depending on reaction temperature must be maintained to ensure stability and consequently activity of the HCo(CO)$_4$ catalyst.

To 2:

increase of H$_2$ partial pressure and catalyst concentration increases the C$_3$H$_6$ conversion, but lowers selectivity, *i.e.*, favors byproduct formation (*n*-butanol and propane)

If, for example, the H$_2$ partial pressure in the 'oxo' gas or the catalyst concentration is increased, then the rate of formation (*i.e.*, propene conversion to butyraldehydes) becomes greater.

However, the *n*-butyraldehyde selectivity is simultaneously lowered due to further hydrogenation to the alcohol, and propane formation increases. Therefore, in order to attain high selectivity, the propene conversion should be decreased. This precaution soon reaches its limit because of an uneconomical space–time yield. Optimization is further complicated by other process variables interacting with one another.

One unsolved problem in the hydroformylation of propene is the resulting isobutyraldehyde, which cannot always be used economically. Therefore process modifications of the hydroformylation, mainly involving other catalyst systems, were developed primarily to increase the *n*-butyraldehyde selectivity.

thermal cleavage of isobutyraldehyde to regenerate propene, CO and H_2 is technologically possible, but not yet applied

To 2nd process step:

Two basic procedures, with variations by each 'oxo' producer, have been developed for the separation of the cobalt hydrocarbonyl from the liquid reaction products. In one case, the reaction mixture is heated after reducing the pressure to about 20 bar. A cobalt sludge results which is separated, regenerated, and recycled to the reactor (*e. g.*, in the Ruhrchemie process).

Another cobalt separation is used mainly with the lower aldehydes. The catalyst is treated with aqueous acid (*e. g.*, CH_3COOH in the BASF process, long-chain carboxylic acids in the Mitsubishi process, and H_2SO_4/CH_3COOH in the UCC process) in the presence of air or O_2 and the cobalt recovered as an aqueous Co salt solution or precipitated with alkali hydroxide as $Co(OH)_2$. Alternatively, it is extracted with $NaHCO_3$ as a hydrocarbonyl (*e. g.*, in the Kuhlmann process) and, after acidification and extraction with the olefin feedstock or auxiliary agents, recycled to the process.

two approaches to $HCo(CO)_4$ separation

1. thermolysis results in metallic Co sludge. H_2O addition hinders Co precipitation on reactor walls

2. recovery as Co compound:

2.1. by conversion of Co hydrocarbonyl in aqueous Co salt solutions with mineral or carboxylic acids
2.2. by extraction of Co hydrocarbonyl with $NaHCO_3$ and subsequent steps

To 3rd process step:

The Co-free reaction product is separated by distillation at normal pressure.

A mixture of *n*-butyraldehyde and isobutyraldehyde is isolated in the first column. Because of the small difference in boiling points (10 °C), this must be separated into its pure components on a second column. The residue from the aldehyde separation contains *n*-butanol and isobutanol from the hydrogenation of the aldehydes during the oxo reaction, as well as other byproducts such as formates, acetals, and the so-called heavy oils. The residual mixture is hydrogenated either directly or after pretreatment (*e. g.*, hydrolysis), and worked up to butanols.

workup of reaction product by two-step distillation:

1st column: *n*- plus isoaldehyde separated from high boiling substances as overhead

2nd column: isoaldehyde separation overhead, with *n*-aldehyde at bottom

simplified workup of 'oxo' products of higher olefins:

without *n*/iso separation, directly hydrogenated to *n*/isoalcohols

If olefins higher than propene are hydroformylated, the aldehydes are not ordinarily isolated, but the crude product is hydrogenated to a mixture of *n*-alcohols and isoalcohols immediately after removal of the cobalt catalyst.

6.1.3. Catalyst Modifications in Hydroformylation

catalyst variants in hydroformylation:

1. Co phosphine complexes
2. Rh phosphine complexes

To minimize the amount of isobutyraldehyde formed, Shell employs catalysts of the type $HCo(CO)_3 \cdot P(n\text{-}C_4H_9)_3$ at the low pressures of $50-100$ bar and $180-200\,°C$ in several 'oxo' plants.

industrial examples of Co phosphine catalysis:

Shell low-pressure process in USA and United Kingdom

characteristics of Co phosphine catalysis:

higher *n*-aldehyde share
lower reaction pressure
increased side reactions, alkene → alkane, aldehyde → alcohol

However, due to the relatively high phosphine content, the catalyst exhibits a lower activity (ca. 20%) and selectivity than the unmodified cobalt hydrocarbonyl. Despite the more favorable *n*:iso ratio (approx. 90% unbranched products), the total aldehyde selectivity is decreased, since side and secondary reactions such as hydrogenation of the olefin feed to a saturated hydrocarbon and, above all, of the aldehydes to the corresponding alcohols occur to a greater extent.

The Shell process is therefore especially suitable in cases where the alcohols are the desired 'oxo' products.

industrial examples of Rh phosphine catalysis:

UCC LPO process (low pressure Oxo) for hydroformylation of ethylene and propene in industrial plants in USA, Sweden

The most important catalyst modification in hydroformylation is the use of rhodium carbonyls, alone or together with phosphines, instead of the conventional cobalt catalysts. Although numerous firms had been engaged in pilot-stage development of Rh-catalyzed hydroformylation for many years, the initial industrial breakthrough first occurred in 1975; in this year, UCC, in cooperation with Johnson Matthey and Davy McKee (formerly Davy Powergas), started operation of a plant with a capacity of 70000 tonnes per year propionaldehyde based on rhodium triphenylphosophine catalysis. In 1976, they started up a second plant with a capacity of 136000 tonnes per year *n*-butyraldehyde. With this, the advantages of a high selectivity ratio for the *n*/isobutyraldehydes in the range $8-16:1$, a low reaction pressure between 7 and 25 bar at $90-120\,°C$, and a simplified workup of the reaction products − due to the higher stability of the modified rhodium carbonyls and the absence of C_4 alcohols in the product mixture, which would otherwise lead to acetal formation − were first industrially exploited (*cf.* Section 6.1.1). Of course, the high cost of rhodium requires − as in the Monsanto methanol carbonylation process − that the catalyst be recovered as completely as possible.

characteristics of Rh phosphine catalysis:

higher *n*-aldehyde content
lower reaction pressure
simplified workup

The mechanism of rhodium triphenylphosphine-catalyzed hydroformylation is similar to that for cobalt catalysis; the starting compound in the catalytic scheme is taken to be a rhodium carbonyl phosphine complex from which, after exchange of a triphenylphosphine ligand for an olefin molecule, an alkyl rhodium compound is formed:

mechanism of catalysis:

1. $HRhCO(PPh_3)_3$ forms free coordination sites by ligand dissociation
2. Rh-species adds to the olefin electrophilically to form π-complex
3. π-σ-rearrangement to form Rh—C species
4. CO insertion to acyl complex
5. hydrogenolysis of the C—Rh bond
$$\overset{\|}{O}$$

$$HRhCO(PPh_3)_3 \xrightarrow[-PPh_3]{+C_3H_6} \begin{array}{c} HRhCO(PPh_3)_2 \\ \uparrow \\ H_2C=CHCH_3 \end{array} \qquad (6)$$
$$\longrightarrow C_3H_7RhCO(PPh_3)_2$$

Insertion of carbon monoxide in the C—Rh bond is followed by hydrogenolysis to complete the catalytic cycle:

$$C_3H_7RhCO(PPh_3)_2 \xrightarrow{+CO} C_3H_7{-}\overset{\|}{\underset{O}{C}}{-}RhCO(PPh_3)_2$$

$$\xrightarrow[+PPh_3]{+H_2} C_3H_7\overset{\|}{\underset{O}{C}}H + HRhCO(PPh_3)_3 \qquad (7)$$

Other companies have built plants based on the UCC/Johnson Matthey/Davy McKee rhodium technology, and by 1993 55% of the worldwide production of butyraldehyde was based on the LPO process.

Ruhrchemie/Rhône-Poulenc have improved the LPO process by sulfonating the triphenylphosphine ligand in the *m*-position of the phenyl groups. The resulting water-soluble catalyst can be removed from the organic-phase reaction product by a simple phase separation. Furthermore, at 50–130 °C and 10–100 bar, a higher activity (98% propene conversion) and a better *n*/iso ratio (95/5) for the butyraldehyde are obtained.

additional process development by, *e.g.*, Ruhrchemie/Rhône-Poulenc by modification of the Rh ligand led to improved two-phase processing; *i.e.*, simple, low-energy catalyst separation with minimal loss of Rh and increased activity and selectivity to *n*-aldehyde

The first 100 000 tonne-per-year plant began production in 1984 at Ruhrchemie. By 1996, two additional plants with a combined capacity of 400 000 tonnes per year were in operation.

A further modification of the rhodium catalyst with diorganophosphite ligands was developed by UCC. This modification led to a notable improvement in activity, even in the conversion of sterically hindered olefins such as 2-methylpropene, while increasing the stability of the ligands.

new diorganophosphite ligands from UCC, *e.g.*,

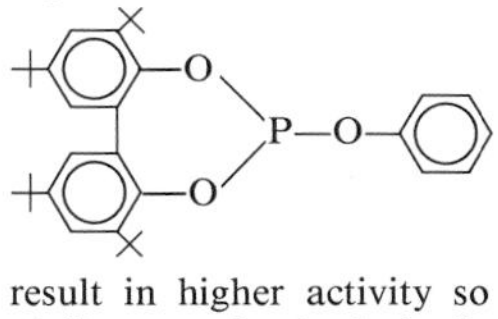

result in higher activity so that branched olefins can also be hydroformlated in high yields

6.1.4. Utilization of 'Oxo' Products

As aldehydes are the primary products of hydroformylation, this synthetic route is known as the 'oxo' reaction and all the aldehydes and their secondary products are termed 'oxo' products.

utilization of 'oxo' aldehydes:

primary products insignificant, but important intermediary products for:

1. alcohols
2. carboxylic acids
3. aldol condensation products
4. primary amines

'Oxo' aldehydes have virtually no importance as final products. However, they are important reactive intermediates for the manufacture of 'oxo' alcohols, carboxylic acids, and aldol condensation products. 'Oxo' aldehydes are also converted, to a limited extent, into primary amines by reductive amination:

$$RCHO + H_2 + NH_3 \xrightarrow{[cat.]} RCH_2NH_2 + H_2O \qquad (8)$$

6.1.4.1. 'Oxo' Alcohols

basis for 'oxo' alcohols:

1. 'oxo' aldehydes
2. aldols and enals from 'oxo' aldehydes

The group of 'oxo' alcohols includes hydrogenation products formed directly from the 'oxo' aldehydes, as well as those from their primary aldolization or aldol condensation products (*cf.* Section 6.1.4.3).

The worldwide production capacity for 'oxo' alcohols in 1991 was 5.1×10^6 tonnes per year, with 1.7, 1.4 and 0.56×10^6 tonnes per year in Western Europe, the USA and Japan, respectively. In 1995, BASF had the largest production capacity for oxo alcohols worldwide (ca. 880000 tonnes per year).

manufacture of 'oxo' alcohols:

catalytic hydrogenation of 'oxo' aldehydes usually according to two variants:

1. gas phase with Ni or Cu catalysts
2. liquid phase with Ni catalysts

At higher temperatures, the catalysts used for hydroformylation are in principle also suitable for the further hydrogenation of the 'oxo' aldehydes; however, in most cases more selective Ni or Cu catalysts are preferred. The cobalt-free aldehyde distillates are usually fed to the hydrogenation, though cobalt-free crude 'oxo' products can also be used.

The hydrogenation can be conducted either in the gas phase with Ni catalysts at $2-3$ bar and $115\,°C$ or with Cu catalysts at a higher temperature of $130-160\,°C$ and $30-50$ bar, or in the liquid phase with Ni catalysts at 80 bar and $115\,°C$:

$$
\begin{array}{l}
RCH_2CH_2CHO \\[4pt]
RCHCH_3 \\
\;\;| \\
\;\;CHO \\
R = H,\ alkyl
\end{array}
\quad \xrightarrow[\text{[cat.]}]{+H_2} \quad
\begin{array}{l}
RCH_2CH_2CH_2OH \\[4pt]
RCHCH_3 \\
\;\;| \\
\;\;CH_2OH
\end{array}
\qquad (9)
$$

If the hydrogenation conditions are made more severe, *e. g.*, 200 °C and 280 bar, then crude 'oxo' products containing, for example, butyl formates and butyraldehyde dibutylacetals can also be produced. Esters and acetals, which must normally be saponified separately, can supply further amounts of butanols by hydrogenolysis:

crude 'oxo' products require more severe hydrogenation conditions than the pure aldehyde for the hydrogenative cracking of byproducts

$$RCH_2-O\!-\!\!\overset{\displaystyle ||}{\underset{\displaystyle O}{C}}\!-\!H + 2\,H_2 \xrightarrow{\text{[cat.]}} RCH_2OH + CH_3OH \quad (10)$$

$$RC\overset{\displaystyle O-CH_2R}{\underset{\displaystyle O-CH_2R}{\big\langle}}\!\!H + H_2 \xrightarrow{\text{[cat.]}} RCH_2-O-CH_2R + HOCH_2R \quad (11)$$

$$R = CH_3CH_2CH_2-, \quad \overset{\displaystyle H_3C}{\underset{\displaystyle H_3C}{\big\rangle}}CH-$$

'Oxo' alcohols with a chain length of C_4-C_6 are mainly used, directly or after esterification with carboxylic acids (*e. g.*, acetic acid), as solvents for the paint and plastics industry.

The C_8-C_{13} 'oxo' alcohols obtained from olefin oligomers (*e. g.*, isoheptenes, diisobutenes, tripropenes) and from cracked olefins are generally esterified with dicarboxylic acids or their anhydrides (phthalic anhydride) and employed as plasticizers (*cf.* Section 14.1.3).

The industrially available branched and unbranched higher olefins are important feedstocks for the manufacture of surfactants and textile auxiliaries after hydroformylation and hydrogenation to the $C_{12}-C_{19}$ alcohols.

The largest volume lower 'oxo' alcohol is *n*-butanol. In the table opposite, *n*-butanol production figures are summarized for several countries. For the total production and production capacity of all butanols, refer to Section 8.1.3.

About half of all butanol is used, either directly or after esterification with carboxylic acids such as acetic acid, butyric acid, valeric acid, glycolic acid, or lactic acid, as a solvent for fats, oils, waxes, natural resins and plastics. *n*-Butylesters also have other applications, *e. g.*, *n*-butyl acrylate together with other comonomers is used to manufacture dispersions. For a long time **di-*n*-butyl phthalate** (DBP) was a standard plasticizer for PVC, but its importance relative to DOP (*cf.* Section 14.1.3) has decreased

utilization of 'oxo' alcohols:

main uses depend on chain length

1. C_4-C_6 alcohols, directly or as ester for solvents

2. C_8-C_{13} alcohol esters (*e. g.*, phthalates) as plasticizers

3. $C_{12}-C_{19}$ alcohols after conversion (*e. g.*, to RCH_2OSO_3H) as raw material for detergents or textile auxiliaries

n-butanol production (in 1000 tonnes):

	1990	1992	1994
USA	576	573	670
W. Europe	473	442	564
Japan	118	174	190

uses of *n*-butanol:

1. directly, or esterified with carboxylic acids, as solvent
2. butyl acrylate as comonomer for dispersions
3. di-*n*-butyl phthalate (DBP) as plasticizer

DBP-share of phthalate plasticizers
(in wt%)

	1985	1991
USA	2.0	1.4
W. Europe	8.7	7.2
Japan	4.7	5.2

uses of isobutanol:

1. directly, or esterified with carboxylic acids, as solvent
2. diisobutyl phthalate (DIBP) as plasticizer

substantially. The share of phthalate plasticizer production held by DBP in the USA, Western Europe and Japan is given in the adjacent table.

As in the case of *n*-butanol, a considerable share of isobutanol production is used in the solvent sector, because it possesses solvent properties similar to *n*-butanol. **Diisobutyl phthalate** (DIBP), like DBP, has also been used as a plasticizer.

6.1.4.2. 'Oxo' Carboxylic Acids

manufacture of 'oxo' carboxylic acids:

oxidation of 'oxo' aldehydes with air (O$_2$)

principle of aldehyde oxidation:

noncatalyzed or homogeneously-catalyzed (redox system) liquid-phase oxidation with percarboxylic acid as intermediate

$$R—CH \rightarrow R—C—O—O—H$$
$$\quad\;\; \| \qquad\qquad \|$$
$$\quad\;\; O \qquad\qquad O$$

The 'oxo' aldehydes can be oxidized to carboxylic acids by mild oxidizing agents, in the simplest case with air. This can be done catalytically in the presence of metal salts, or in the absence of catalysts at temperatures of up to about 100 °C and pressures of up to 7 bar:

$$
\begin{array}{ccc}
RCH_2CH_2CHO & & RCH_2CH_2COOH \\
 & \xrightarrow{+O_2} & \\
RCHCHO & & RCHCOOH \\
\;\;| & & \;\;| \\
CH_3 & & CH_3
\end{array}
\qquad (12)
$$

R = H, alkyl

In particular, metals capable of a valency change such as Cu, Fe, Co, Mn, etc., are used.

The resulting carboxylic acids are generally processed to esters, which are employed to a large extent as solvents. In addition, esters of branched, polyhydric alcohols such as neopentylglycol, trimethylolpropane or pentaerythritol are gaining importance as synthetic oils.

applications of 'oxo' carboxylic acids:

1. as ester for solvents or plasticizers
2. as acids for modifying alkyd resins
3. as salts for siccatives
4. as vinyl esters for manufacture of copolymer dispersions

The following are a few characteristic applications of the 'oxo' carboxylic acids: butyric acid is used in the manufacture of cellulose acetobutyrate, a mixed ester which can be processed to light-, heat-, and moisture-resistant coatings. Isooctanoic and isononanoic acids are suitable for modifying alkyd resins and, after esterification with ethylene glycols, as plasticizers for PVC. Their Co, Mn, Pb, Zn, and Ca salts serve as drying accelerators (siccatives) for paints, while their vinyl esters are starting materials for dispersions.

6.1.4.3. Aldol and Condensation Products of the 'Oxo' Aldehydes

A third type of secondary reaction of the 'oxo' aldehydes used industrially is aldolization. This is done by reacting the 'oxo' aldehydes in the liquid phase in the presence of basic catalysts. Diols can be obtained from the aldol, the primary reaction product, by a subsequent hydrogenation of the aldehyde group. Higher primary alcohols possessing twice the number of carbon atoms as the aldehyde starting material can be synthesized by cleavage of water from the aldol followed by hydrogenation of the resulting double bond and remaining aldehyde group. Both options are used industrially:

manufacture of aldolization products of 'oxo' aldehydes:

base-catalyzed homogeneous liquid-phase reaction (*e. g.*, NaOH or basic ion exchange resins)

further reactions of the aldol products:

1. hydrogenation to dihydric alcohols (*e. g.*, Pd catalyst)
2. dehydration (aldol condensation) and hydrogenation to monohydric alcohols (*e. g.*, Ni or Cu catalyst)

$$
RCH_2CHO + \underset{\underset{R}{|}}{CH_2CHO} \xrightarrow{[cat.]} RCH_2\underset{\underset{OH}{|}}{CH}-\underset{\underset{R}{|}}{CH}CHO \tag{13}
$$

$$
RCH_2\underset{\underset{OH}{|}}{CH}-\underset{\underset{R}{|}}{CH}CHO
\begin{cases}
\xrightarrow[{[cat.]}]{+H_2} RCH_2\underset{\underset{OH}{|}}{CH}-\underset{\underset{R}{|}}{CH}CH_2OH \\[2ex]
\xrightarrow{-H_2O} RCH_2CH\!=\!\underset{\underset{R}{|}}{C}-CHO
\end{cases} \tag{14}
$$

R = alkyl

$$
\xrightarrow[{[cat.]}]{+H_2} RCH_2CH_2-\underset{\underset{R}{|}}{CH}CH_2OH
$$

2,2,4-Trimethylpentane-1,3-diol, which can be synthesized from isobutyraldehyde, belongs to the first group, the diols. It is only of limited industrial importance.

2-Ethylhexanol belongs to the second group of primary monohydric alcohols and, with *n*-butanol, is one of the largest volume 'oxo' products. In 1994, the world production capacity of 2-ethylhexanol was about 2.3×10^6 tonnes per year, of which 0.85, 0.37, and 0.34×10^6 tonnes per year were located in Western Europe, Japan and the USA, respectively. Production figures for these countries are given in the adjacent table.

2-Ethylhexanol is obtained from hydrogenation of 2-ethylhexenal, the product of the aldol condensation of *n*-butyraldehyde:

example of 1:

$$
\underset{H_3C}{\overset{H_3C}{>}}CHCH-\underset{\underset{CH_3}{|}}{\overset{CH_3}{\underset{|}{C}}}-CH_2OH \text{ from iso-}C_3H_7CHO
$$

example of 2:

2-ethylhexanol (2 EH or isooctanol) from *n*-C$_3$H$_7$CHO

2-ethylhexanol production (in 1000 tonnes):

	1990	1992	1994
W. Europe	771	694	800
Japan	283	327	305
USA	295	314	332

$$2\ CH_3CH_2CH_2CHO \xrightarrow[\text{[OH}^\ominus]]{-H_2O} CH_3(CH_2)_2CH{=}\underset{\underset{C_2H_5}{|}}{C}{-}CHO$$

$$\xrightarrow[\text{[cat.]}]{+2\ H_2} CH_3(CH_2)_3\underset{\underset{C_2H_5}{|}}{C}HCH_2OH \tag{15}$$

manufacture of 2-ethylhexanol:

1. OH-catalyzed simultaneous aldolization and dehydration (aldol condensation) of *n*-butyraldehyde
2. Ni- or Cu-catalyzed gas-phase hydrogenation of double bond and CHO group of 2-ethylhexenal

n-Butyraldehyde is converted almost quantitatively into 2-ethylhexenal in the presence of a sodium hydroxide solution or a basic ion-exchanger at $80-100\,°C$.

This is followed by a gas-phase hydrogenation to 2-ethylhexanol at a slight excess pressure up to 5 bar and temperatures of $100-150\,°C$ with Ni, or $135-170\,°C$ with Cu, fixed-bed catalysts. If required, a further hydrogenation can be conducted in the liquid phase.

The pure product is obtained from a three-step distillation. The selectivity is about 95% (based on *n*-butyraldehyde).

process modification for 2-ethylhexanol:

single-step, 'oxo' reaction modified by co-catalysts combined with aldol condensation and hydrogenation (Aldox process)

Another process variation, used so far only by Shell in the USA and the United Kingdom and by Exxon in the USA and Japan, consists of the combination of the **ald**olization and **'oxo'** reactions into a single step (Aldox process). By adding cocatalysts, such as compounds of Zn, Sn, Ti, Al, or Cu or KOH, to the original 'oxo' catalyst, the three essential steps to 2-ethylhexanol — *i.e.*, propene hydroformylation, aldol condensation and hydrogenation — can take place simultaneously. Besides using KOH as a cocatalyst, Shell also uses a ligand-modified hydroformylation catalyst, $HCo(CO)_3P(R)_3$, in their Aldox process.

Today, the manufacture of 2-ethylhexanol from acetaldehyde is almost insignificant due to the long reaction path (*cf.* Section 7.4.3). However, this process is still used to a limited extent in Brazil. A return to this method would be possible in the future if inexpensive ethanol from fermentation or the homologation of methanol (that is, based totally on coal) were to become available.

uses of 2 EH:

mainly as ester component, with phthalic acid to 'di**o**ctyl **p**hthalate' (DOP)

Of all the higher alcohols, 2-ethylhexanol (2EH) is the most important economically. It has been used as a 'softening alcohol' since the middle of the 1930s. It is mainly employed in the manufacture of esters with dicarboxylic acids such as phthalic acid (*cf.* Section 14.1.3) or adipic acid.

DOP production (in 1000 tonnes):

	1990	1992	1993
Japan	295	297	279
Germany	223	231	259
USA	141	122	115

The worldwide production capacity for di-2-ethylhexyl phthalate, generally known as di**o**ctyl **p**hthalate (DOP), was about 6.1×10^6 tonnes in 1994. Production figures for several countries are given in the adjacent table.

Dioctyl phthalate, obtained by reacting 2-ethylhexanol with phthalic anhydride, is an excellent, physiologically harmless, standard plasticizer used by the plastics industry. More recently, DOP has been used as a dielectric liquid for capacitors, where it replaces toxic polychlorinated diphenyls.

Other esters of 2-ethylhexanol, in particular with aliphatic dicarboxylic acids, are employed as hydraulic oils or as components of synthetic lubricants.

2-Ethylhexanol is, in addition, oxidized to 2-ethylhexanoic acid. This acid can also be manufactured by oxidation of 2-ethylhexanal produced by selective hydrogenation of 2-ethylhexenal with Pd catalysts. It is used for modifying alkyd resins.

esters also with:

adipic acid
sebacic acid
phosphoric acid
trimellitic acid

main uses of the 2EH ester:

DOP as plasticizer (*e. g.*, PVC) with special properties (low volatility; heat-, cold-, H_2O resistant; nonpoisonous)

hydraulic fluids
synthetic lubricants

6.2. Carbonylation of Olefins

The carbonylation of olefins with CO and a nucleophilic reaction partner possessing a labile H atom results in the formation of carboxylic acids or their derivatives such as esters, thioesters, amides, anhydrides, etc., in the presence of metal carbonyls:

olefin reaction with CO and nucleophilic partners in presence of metal carbonyl leads to formation of carboxylic acids and derivatives (Reppe carbonylation)

$$H_2C{=}CH_2 + CO + HX \longrightarrow H_3C{-}CH_2{-}\underset{\underset{O}{\|}}{C}{-}X$$

$$X = \text{-OH, -OR, -SR, -NHR,} \quad {-}O{-}\underset{\underset{O}{\|}}{C}{-}R \quad \text{etc.}$$

(16)

Thus this type of carbonylation reaction, employing metal carbonyls as catalysts (*i. e.*, carbonyls of Ni, Co, Fe, Rh, Ru, and Pd), belongs to the wide range of Reppe reactions.

Analogous to hydroformylation, the conversion of olefins with CO and H_2O to carboxylic acids is designated hydrocarboxylation.

carbonylation with insertion of COOH in olefins termed hydrocarboxylation

The Koch reaction, described in Section 6.3, is a similar hydrocarboxylation which starts with the same materials (such as olefin, CO and H_2O), but which, by using proton catalysts and milder reaction conditions, leads mainly to tertiary carboxylic acids.

Koch and Reppe hydrocarboxylation differ in the following ways:

1. catalyst
2. reaction conditions
3. reaction products

The reaction of ethylene with CO and H_2O to form propionic acid is an example of an industrially operated hydrocarboxylation under Reppe conditions. The reaction takes place at 200–240 bar and 270–320 °C in a liquid phase consisting of the components and the catalyst Ni propionate dissolved in crude propionic acid:

characteristics of Reppe hydrocarboxylation:

homogeneously catalyzed liquid-phase reaction, at raised pressure and temperature, metal carbonyls used, preferably $Ni(CO)_4$ which forms *in situ* from Ni salts

$$H_2C=CH_2 + CO + H_2O \xrightarrow{\text{[cat.]}} CH_3CH_2COOH$$

$$\left(\Delta H = -\frac{38 \text{ kcal}}{159 \text{ kJ}} \middle/ \text{mol} \right) \qquad (17)$$

industrial example of a Reppe olefin hydocarboxylation:

BASF process for manufacture of propionic acid from ethylene, CO, H_2O

$Ni(CO)_4$, the actual catalyst, is formed *in situ*. Other carbonyls, such as Fe or Co carbonyls, are also catalytically active. After the reaction, the $Ni(CO)_4$ present in the crude product is reconverted into the propionate by air oxidation in the presence of propionic acid and recycled to the reactor. The propionic acid yield amounts to 95% (based on C_2H_4). The byproducts are CO_2, ethane and higher carboxylic acids, which, due to the synthetic route, possess an odd number of carbon atoms, *e. g.*, valeric acid. If, instead of H_2O, another reactant with an active hydrogen is used, then carboxylic acid derivatives can be formed directly. For example, propionic acid esters can be manufactured directly by replacing water with alcohols.

broadening of carbonylation with other active-H reactants possible to give esters, thioesters, amides, etc.

For example, BASF has run this reaction since 1952 in an industrial plant whose capacity was increased from 60000 to 80000 tonnes per year in 1995.

Reppe catalyst activation by halogen/phosphine ligands

One process improvement − a more active Reppe catalyst − has been developed by Halcon and also by Eastman. By adding halogen (*e. g.*, iodine) and phosphine ligands to the standard Ni/Mo catalyst, reaction conditions could be reduced to 10−35 bar and 175−225 °C without loss of yield.

production of propionic acid
(in 1000 tonnes):

	1988	1990	1992
USA	55	73	79

Propionic acid is also a byproduct of the oxidation of light distillate fuel to give acetic acid (*cf.* Section 7.4.1.2). However, the most important production possibility is the hydroformylation of ethylene followed by oxidation (*cf.* Section 6.1). In 1991, about 93000 of the 102000 tonne-per-year capacity in the USA was based on this route. In 1993, the production capacity for propionic acid in the USA and Western Europe was about 125000 and 92000 tonnes per year, respectively. Production figures for the USA are given in the adjacent table.

use of propionic acid:

1. Ca and Na salt for preservation
2. ester as solvent and plasticizer, *e. g.*, glycerine tripropionate
3. vinyl ester as an important comonomer
4. starting product for herbicides

Propionic acid is mainly employed as a preservative in the food sector and in the manufacture of animal fodder. Another application is the manufacture of esters, *e. g.*, amyl propionate, which is used as a solvent for resins and cellulose derivatives, and vinylpropionate, a comonomer. Another important use is in the manufacture of herbicides, vitamins, and pharmaceuticals.

When higher olefins are used in the Reppe hydrocarboxylation, double bond isomerization causes mixtures of carboxylic acids to be formed. This manufacturing route is therefore limited to the reaction with ethylene.

extension of hydrocarboxylation to higher olefins possible, however partial isomerization leads to carboxylic acid mixtures

Union Oil recently developed an oxidative carbonylation of ethylene to acrylic acid. This process is discussed in Section 11.1.7.3.

6.3. The Koch Carboxylic Acid Synthesis

Besides being catalyzed with metal carbonyls (*cf.* Section 6.2), the reaction of olefins with carbon monoxide and hydrogen to form carboxylic acids can also be conducted using proton catalysts. Mineral acids such as H_2SO_4, HF, and H_3PO_4, alone or in combination with BF_3 or SbF_5, *e.g.*, in $HF \cdot SbF_5$, are particularly suitable.

the Koch carbonylation (also called the Koch–Haaf reaction) is the synthesis of predominantly tertiary carboxylic acids from olefins

in contrast to Reppe carbonylation, proton catalysis also leads to double bond and structural isomerization

In the initial step of the reaction, a proton adds to the olefin to form a secondary carbenium cation, which is stabilized by isomerization of the double bond and rearrangement of the carbon skeleton:

principles of the Koch reaction:

1. formation of most stable (*i.e.*, tertiary) carbenium ion by isomerization
2. CO addition to acylium cation
3. further reaction with H_2O or ROH to acid or ester

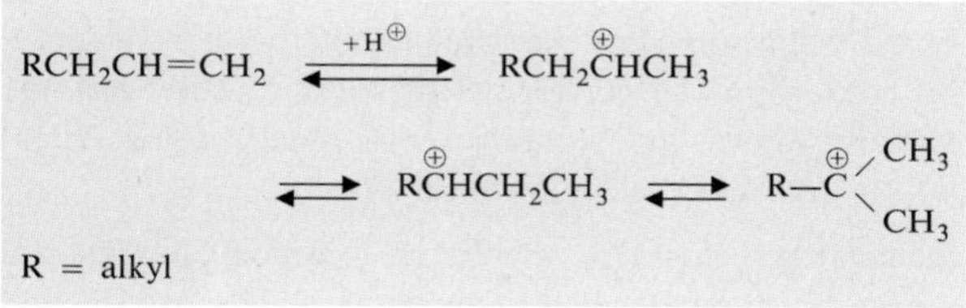

$$RCH_2CH=CH_2 \xrightarrow{+H^{\oplus}} RCH_2\overset{\oplus}{C}HCH_3$$

$$\rightleftharpoons R\overset{\oplus}{C}HCH_2CH_3 \rightleftharpoons R-\overset{\oplus}{C}\begin{smallmatrix}CH_3\\CH_3\end{smallmatrix} \qquad (18)$$

R = alkyl

CO then adds to the tertiary carbenium ion to form an acylium cation, which goes on to react with either water (to form a carboxylic acid) or with an alcohol (to form an ester directly):

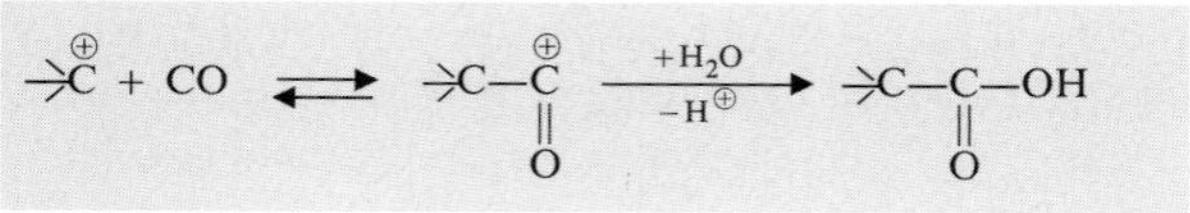

$$\overset{\oplus}{>}C + CO \rightleftharpoons >C-\overset{\oplus}{\underset{\underset{O}{\|}}{C}} \xrightarrow[-H^{\oplus}]{+H_2O} >C-\underset{\underset{O}{\|}}{C}-OH \qquad (19)$$

Thus, mixtures of isomeric branched carboxylic acids are obtained in which the fraction of tertiary carboxylic acids depends on the reaction conditions. For example, at 80 °C, 20 – 100 bar and longer reaction periods, all butene isomers can be converted into pivalic acid (trimethylacetic acid). In practice, isobutene is normally used for the manufacture of this acid:

reaction conditions determine equilibration and therefore the proportion of tertiary carboxylic acids

$$\begin{array}{c} H_3C \\ \diagdown \\ H_3C \diagup \end{array} C=CH_2 + CO + H_2O \xrightarrow{[H^{\oplus}]} \begin{array}{c} CH_3 \\ | \\ H_3C-C-COOH \\ | \\ CH_3 \end{array} \quad (20)$$

The carbenium ion intermediate can also be obtained from precursors other than the olefin. Isobutanol and 2-chlorobutane can also be used for pivalic acid manufacture, since they are subject to loss of H_2O or HCl, respectively, under the conditions of the reaction.

industrial operation of the Koch reaction:

1. H_3PO_4/BF_3-catalyzed CO addition to olefin
2. hydrolysis resulting in two-phase system

In the industrial process, CO is added to the olefin in the presence of catalyst in a multistage stirred-tank reactor at 20–80 °C and 20–100 bar. Water is introduced in the second step. H_3PO_4/BF_3 is the preferred proton-supplying catalyst, because a separation of product and catalyst phase occurs when H_2O is added. H_3PO_4/BF_3 is recycled to the process. In a more recent process development from BASF, a heterogeneous zeolite catalyst (pentasil type) is used at 250–300 °C and 300 bar.

With high olefin conversion, the selectivity to tertiary carboxylic acids is 80–100%. The byproducts are carboxylic acids from the dimerized olefins.

examples of industrial processes:

Shell ('Versatics')
Exxon ('Neo Acids')
Du Pont
Kuhlmann ('CeKanoic Acids')

characteristic properties of the Koch acids:

$$\begin{array}{c} C \\ | \\ neopentyl\ structure\ C-C-COOH \\ | \\ C \end{array}$$

leads to low reactivity due to steric hindrance, *i.e.*, acids are difficult to esterify, and esters are:

thermally
oxidatively   } stable
hydrolytically

important Koch acid derivatives:

vinyl ester   RCOOCH=CH$_2$
glycidyl ester   RCOOCH$_2$CH—CH$_2$
                              \O/

Shell, Exxon, Kuhlmann and Du Pont operate industrial processes based on the Koch reaction. Besides the manufacture of pivalic acid from isobutene, branched C_6-C_{11} carboxylic acids are produced from the corresponding olefins. The acids are commercially known as 'Versatic Acids' (Shell), 'Neo Acids' (Exxon), or 'CeKanoic Acids' (Kuhlmann).

The chemical properties of the tertiary carboxylic acids are largely determined by the alkyl branching at the α-position next to the carboxyl group. This branching creates strong steric hindrance, as demonstrated by their unusual thermal stability and the difficult saponification of the ester. They are therefore suitable components of synthetic oils, for example. The acids are also processed to resins and paints while the vinyl esters (*cf.* Section 9.2.2) are used as comonomers for the manufacture of dispersions or in VC copolymerization for internal plasticizing.

Glycidyl esters of the Koch acids are used to modify alkyd resins (*cf.* Section 11.2.3).

Metal salts of highly branched tertiary carboxylic acids are drying accelerators.

Worldwide production of these Koch acids and their derivatives is currently about 150 000 tonnes per year.

7. Oxidation Products of Ethylene

Ethylene oxide (oxirane) and acetaldehyde are the simplest partial oxidation products of ethylene. They are isomeric. Due to their reactivity and their numerous industrially important secondary products, they are among the major intermediate products based on ethylene. However, as can be appreciated from the following table, they both lie far behind polyethylene, and — in Western Europe and Japan — even behind vinyl chloride in their consumption of ethylene:

ethylene oxide (EO) and acetaldehyde are the most significant partial oxidation products of ethylene

however, largest volume application of ethylene (worldwide) is in the manufacture of low- and high-density polyethylene

Table 7-1. Ethylene use (in %).

Product	World		USA		Western Europe		Japan	
	1988	1993	1984	1993	1984	1994	1985	1993
Polyethylene (LDPE and HDPE)	54	56	48	53	54	57	52	49
Vinyl chloride	16	15	14	15	19	16	17	17
Ethylene oxide and secondary products	13	12	17	14	10	10	11	10
Acetaldehyde and secondary products	3	2	3	1	3	3	5	4
Ethylbenzene/styrene	7	7	8	7	7	7	9	11
Others (e. g., ethanol, vinyl acetate, 1,2-dibromoethane, ethyl chloride, ethylenimine, propionaldehyde, etc.)	7	8	10	10	7	7	6	9
Total use (in 10^6 tonnes)	52.6	62.3	13.9	19.2	12.5	17.8	4.1	5.7

7.1. Ethylene Oxide

Ethylene oxide has experienced a dramatic expansion in production since its discovery in 1859 by A. Wurtz and the first industrial process (UCC) in 1925. Production figures for several countries are summarized in the adjacent table. In 1995, the world capacity for ethylene oxide was about 11.2×10^6 tonnes per year; 34% (3.8×10^6 tonnes per year) of this was in the USA, with UCC as the largest producer (capacity almost 10^6 tonnes per year in 1992). The dominant technology in 1993, accounting for ca. 4×10^6 tonnes per year capacity worldwide, is based on a process from Scientific Design. In 1995, capacities in Western Europe and Japan were about 2.2 and 0.88×10^6 tonnes per year, respectively.

Parallel to the growth in capacity, the process changed from the costly traditional two-step process (with ethylene chlorohydrin as

UCC was first and is currently the largest producer of EO worldwide (ca. 10^6 tonnes per year capacity)

ethylene oxide production (in 10^6 tonnes):

	1990	1992	1995
USA	2.43	2.64	3.46
W. Europe	1.66	1.61	1.86
Japan	0.72	0.72	0.80

industrial syntheses of EO:

1. indirect oxidation with intermediate ethylene chlorohydrin (classical process)
2. direct oxidation of ethylene

an intermediate) to a more economical direct oxidation of ethylene in large plants with capacities of up to 360000 tonnes per year. Thus while about half of the ethylene oxide produced in the USA in the mid 1950s was still based on ethylene chlorohydrin, since 1975 EO has been manufactured exclusively by the direct oxidation of ethylene.

7.1.1. Ethylene Oxide by the Chlorohydrin Process

principles of the chlorohydrin process:

two-step epoxidation:
primary product $HOCH_2CH_2Cl$ is dehydrochlorinated with $Ca(OH)_2$ to EO

The two-step chlorohydrin process for the epoxidation of lower olefins is rarely used with ethylene today, though it is still used with propene (cf. Section 11.1.1.1). In this process, the non-isolated ethylene chlorohydrin intermediate was converted into ethylene oxide by heating with lime water:

$$H_2C=CH_2 + Cl_2 + H_2O \longrightarrow HOCH_2CH_2Cl + HCl \qquad (1)$$

$$2\ HOCH_2CH_2Cl + Ca(OH)_2$$

$$\longrightarrow 2\ H_2C\!\!-\!\!CH_2 + CaCl_2 + 2\ H_2O \qquad (2)$$
$$\underset{O}{\diagdown\diagup}$$

reasons for abandonment of chlorohydrin route:

1. high Cl_2 consumption
2. high salt load
3. resulting byproducts

The selectivity to ethylene oxide was about 80% (based on C_2H_4), and was therefore satisfactory. However, the chlorine was virtually lost. 10–15 kg 1,2-dichloroethane, 7–9 kg 2,2'-dichlorodiethyl ether and 300–350 kg $CaCl_2$ were formed per 100 kg ethylene oxide.

The main reasons for replacing this process with a direct oxidation were therefore the high chemical feedstock requirements, in particular the cost of chlorine (a significant factor), as well as the considerable effluent load.

7.1.2. Ethylene Oxide by Direct Oxidation

7.1.2.1. Chemical Principles

principles of direct oxidation process:

gas-phase reaction of ethylene, air or oxygen on supported Ag catalysts

Ag essential basis of all EO catalysts, on which three characteristic exothermic reactions take place:

1. partial oxidation, C_2H_4 to EO
2. secondary reaction, EO to $CO_2 + H_2O$
3. total oxidation, C_2H_4 to $CO_2 + H_2O$

In 1931, T. E. Lefort first oxidized ethylene directly to ethylene oxide. This was transformed into an industrial process in 1937 by UCC. Other companies participated in the further development, which led to fundamental progress in catalyst manufacture. However, silver maintained its position as the most active and selective catalyst component.

Partial oxidation of ethylene with silver catalysts is an exothermic reaction:

$$H_2C{=}CH_2 + 0.5\ O_2 \longrightarrow H_2C\underset{O}{\diagdown\diagup}CH_2$$

$$\left(\Delta H = -\genfrac{}{}{0pt}{}{25\ \text{kcal}}{105\ \text{kJ}}\Big/\text{mol}\right) \qquad (3)$$

It is normally accompanied by two even more exothermic side or secondary reactions. These are the complete combustion of ethylene, which is the main source of CO_2, and the further oxidation of ethylene oxide:

total oxidation of EO usually after isomerization to CH_3CHO

$$H_2C{=}CH_2 + 3\ O_2 \longrightarrow 2\ CO_2 + 2\ H_2O$$

$$\left(\Delta H = -\genfrac{}{}{0pt}{}{317\ \text{kcal}}{1327\ \text{kJ}}\Big/\text{mol}\right) \qquad (4)$$

$$H_2C\underset{O}{\diagdown\diagup}CH_2 + 2.5\ O_2 \longrightarrow 2\ CO_2 + 2\ H_2O$$

$$\left(\Delta H = -\genfrac{}{}{0pt}{}{292\ \text{kcal}}{1223\ \text{kJ}}\Big/\text{mol}\right) \qquad (5)$$

Industrial processes attain a selectivity of 65–75% ethylene oxide (air process), or 70–80% (O_2 process), with a total heat of reaction of 85–130 kcal (357–546 kJ)/mol ethylene. According to the reaction mechanism (*cf.* eq 6-8), the maximum possible ethylene oxide selectivity is only 80%.

relatively low EO selectivity means a large contribution by side reactions to heat of reaction causing heat removal problems

The specific O_2 activation on the metal surface of the silver is the fundamental reason for its catalytic acitivity. Initially, oxygen is adsorbed molecularly on the silver and reacts in this form with ethylene to yield ethylene oxide. The atomic oxygen generated cannot produce more ethylene oxide; it oxidizes ethylene or ethylene oxide to CO and H_2O:

connection between selective EO formation and nonselective combustion can be understood by different O_2 activation:

chemisorbed O_2 effects the partial, O the total, C_2H_4 oxidation – thus theoretical limit of selectivity is about 80%

$$[\text{Ag}] + O_2 \longrightarrow [\text{Ag}]\cdot O_{2\,\text{ads.}} \qquad (6)$$

$$\xrightarrow{\ H_2C{=}CH_2\ } H_2C\underset{O}{\diagdown\diagup}CH_2 + [\text{Ag}]\cdot O_{\text{ads.}}$$

$$4\ [\text{Ag}]\cdot O_{\text{ads.}} + H_2C{=}CH_2 \longrightarrow 2\ CO + 2\ H_2O + 4\ [\text{Ag}] \qquad (7)$$

$$[\text{Ag}]\cdot O_{2\,\text{ads.}} + 2\ CO \longrightarrow 2\ CO_2 + [\text{Ag}] \qquad (8)$$

[Ag] = metal surface

important parameters of Ag catalysts for regulating activity and selectivity:

1. impregnation method
2. silver salt reduction
3. support material
4. physical properties of support (texture)
5. cocatalyst or promoter

oxidation inhibitors such as 1,2-dichloro-ethane lead to an increase in selectivity

action explicable by partial reversible poisoning of Ag surface for atomic chemi-absorption of oxygen

Industrial catalysts generally contain up to 15 wt% Ag as a finely divided layer on a support. All other specific catalyst characteristics are proprietary and, together with technological differences, account for the various processes in existence today. The activity and selectivity of the catalyst is primarily influenced by the manufacturing process, the type of support and its physical properties, and any promoters or activators.

In all processes, inhibitors are used to prevent total oxidation. Several ppm of 1,2-dichloroethane are often introduced to the reaction mixture for this purpose. The chemiabsorbed atomic chlorine − from the oxidation of hydrogen chloride obtained in the dehydrochlorination of 1,2-dichloroethane − hinders the dissociative chemiabsorption of atomic oxygen and therefore the combustion of ethylene to CO_2 and H_2O.

7.1.2.2. Process Operation

process characteristics:

1. careful removal of reaction heat and limitation of ethylene conversion to <10% avoid local hot spots which cause loss in activity due to sintering of Ag (particles become larger)

2. special supports such as alundum (α-Al_2O_3) make catalyst robust, *i.e.*, increase its lifetime

3. special arrangement of catalyst in fixed-bed in bundled tube reactors

One of the most important tasks in large-scale units is the effective removal of the considerable heat of reaction. Overheating the catalyst could lead to changes in the Ag distribution on the support, causing a decrease in its activity and lifetime. Safe temperatures in the reactor are maintained by limiting the ethylene conversion to less than 10% to limit the amount of heat evolved. Extensive development has also led to highly resistant catalysts. For example, the Shell silver catalyst, used in numerous plants, shows only an insignificant drop in selectivity and space–time yield after many years of operation. In the present conventional ethylene oxide reactor, the catalyst is packed in a tubular bundle consisting of several thousand tubes through which the reaction mixture is cycled. A boiling liquid − *e. g.*, kerosene, tetralin or, for safety reasons, water − circulates between the tubes and functions as a heat transfer agent. The heat of reaction is normally used to generate medium pressure steam.

alternative arrangement of catalyst in fluidized bed not used industrially because of two problems:

1. Ag removal from support (erosion)
2. backmixing of reaction products results in long residence time and hence more total oxidation

Fluidized-bed processes are not industrially important. Among the main reasons are insufficient catalyst lifetime and low selectivities. However, a new fluidized-bed technology is in use in a 2000 tonne-per-year unit in the CIS. Here the Ag/Al_2O_3 catalyst is carried upward with the feed gas stream in the inner tube of a double-pipe reactor. It then falls in the outer tube back toward the gas inlet. This gives a good heat transfer with little catalyst abrasion. At 270–290 °C and 30 bar, a selectivity to ethylene oxide of 75% and a space–time yield of 600–700 g EO·(L catalyst)$^{-1}$·h^{-1} are obtained.

Air was employed as the oxidizing agent in the first industrial plants of UCC and Scientific Design in the USA, Distillers in the United Kingdom, and IG Farben in their plant in Ludwigshafen, Germany. However, the N_2 content interfered with the gas cycle and led to ethylene losses either by discharging the nitrogen after a single pass of the ethylene/air mixture through the reactor, or by following with a second reactor where more severe temperatures conditions are used and lower selectivity to ethylene oxide is obtained.

variation of fluidized-bed reactor − *i.e.,* double pipe reactor with catalyst transport by rising gas stream − in commercial use

two basic variants of industrial process:

1. air process
2. O_2 process (currently favored)

to 1:

advantage:
no air fractionating plant necessary

Therefore, oxygen is used for the oxidation of ethylene in almost all new units. Despite the investment and operating costs for an air fractionating plant, the total ethylene oxide manufacturing costs are lower than those resulting from an air feed.

However, in the (oxygen) oxidation process a constant amount (about 50%) of a gas such as methane, ethane, or CO_2 is employed as inert gas (trap for free radicals) in a closed cycle. Shell introduced this process in the mid-1950s. The main advantage of the process is the smaller amount of waste gas produced − only about 2% of the waste gas of the air process − which reduces ethylene losses considerably.

disadvantages:

major C_2H_4 loss with inert gas (N_2) removal, higher costs for gas cycle

Another deviation from the older processes consists of the removal of CO_2. In the air process, it is removed with the exhaust gas, while in the O_2 process it is washed out with a hot potash solution.

Some firms can operate their processes using either oxygen or air.

to 2:

advantages:
small amount of waste gas, constant proportion of inert gas

disadvantages:
air fractionation plants for O_2 production (>99% purity) and CO_2 washes necessary

isolation of EO from reaction gas:

EO absorption in H_2O to effect separation from C_2H_4, O_2 and inert gas (CH_4)

Typical conditions for the O_2 process are $10-20$ bar and $250-300\,°C$. The oxygen content in the reaction mixture is adjusted to $6-8$ vol% (C_2H_4 $20-30$ vol%) − outside the flammability limits for ethylene/O_2 mixtures. Selectivity to ethylene oxide is $65-75\%$ (air process) or $70-80\%$ (O_2 process) at an ethylene conversion of $8-10\%$.

The workup of the reaction gases begins in an absorption column, where the ethylene oxide is washed out with water. It is then driven out of aqueous solution with steam in a stripping column (desorber) and fractionated in distillation columns. The ethylene oxide/H_2O mixtures are also, in part, directly converted into glycol.

workup of aqueous EO solutions:

1. desorption with steam and distillation to remove, *e.g.,* CO_2 and N_2
2. direct conversion to glycol

7.1.2.3. Potential Developments in Ethylene Oxide Manufacture

improvements in EO process mainly concentrated on EO selectivity, as C_2H_4 price is decisive cost factor in EO manufacture

The economics of ethylene oxide manufacture is mainly determined by the ethylene price. The total (destructive) oxidation of 20–30 mol% of the converted ethylene is one of the most important cost factors. Before the first oil crisis in autumn 1973, the ethylene share of the total costs was 60–70%; today, it lies at 70–80%.

Thus improvements in the technology, and above all in catalyst selectivity, are necessary because of high raw material costs.

increase in selectivity has dual effect:

1. higher EO yield
2. lower ΔH, due to less combustion, *i.e.*, longer catalyst lifetime, or C_2H_4 conversion can be increased without danger, leading to greater STY (**S**pace-**T**ime **Y**ield)

Experiments to increase selectivity are therefore particularly worthy of attention since an increase in selectivity has a favorable effect in two respects: not only is a higher effective ethylene oxide yield obtained, but less heat is generated due to decreased total oxidation. Lower heat of reaction mainly increases the catalyst lifetime, but can also allow a risk-free increase in ethylene conversion leading to enhanced plant capacity.

increase in EO selectivity only with Ag as basis metal and cocatalysts mainly from the alkali metals

Numerous experiments to increase the selectivity of the ethylene oxide catalyst have shown that silver, due to its high fundamental selectivity, cannot be replaced by any other metal. However, both the specific surface of the support – generally Al_2O_3 or Al silicate – and the generated silver crystallites are important factors.

examples of Ag modification:

Shell: Cs
Halcon: Ba/Cs
ICI: K/Rb

In addition, many firms have proposed using primarily alkali metal salt cocatalysts with silver. The selectivity to ethylene oxide is reported to increase to over 80%, and, in the case of Rb/Cs, to 94%. Although this improvement in the selectivity does not at first glance seem to agree with the proposed mechanistic theory, it can be explained by recombination of atomic oxygen initiated by the cocatalyst (*cf.* Section 7.1.2.1).

new Ag catalyst regeneration without removal from the tube reactor and thus without loss by 'in situ impregnation' with Cs salt solution

One new variation on the process technology uses the effect of alkali metal salts to increase catalyst activity for the regeneration of silver catalyst. After several years in operation, a small amount of cesium salts is impregnated into the catalyst as a methanolic solution. This technique – already used many times industrially – averts an expensive catalyst change in the tube reactor, can be used repeatedly, and gives a selectivity increase of up to 8% depending on the type and history of the catalyst.

Of the ethylene oxide produced worldwide, 40–60% is used for the manufacture of glycol. Thus, besides improvements in the ethylene oxide technology, processes for glycol that avoid the relatively expensive intermediate ethylene oxide have a bright future (*cf.* Section 7.2.1.1).

7.2. Secondary Products of Ethylene Oxide

Ethylene oxide is used directly only to a very limited extent, *e. g.*, as a fumigant in grain storage and as an agent for sterilization and fermentation inhibition.

utilization of EO:

1. minor importance of its own (diluted with CO_2 or CF_2Cl_2 as insecticide, for sterilization or fermentation inhibitor)

Its great importance stems from the reactivity of the oxirane ring, making it a key substance for a multitude of other intermediates and final products.

2. outstanding importance as intermediate for ethoxylation, *i.e.*, introduction of hydroxyethyl group $—CH_2CH_2OH$ or polyethyleneoxide group $+CH_2CH_2O\rightarrow_n H$

The secondary reactions of ethylene oxide are based on the exothermic opening of the three-membered ring by a nucleophilic partner such as water, alcohols, ammonia, amines, carboxylic acids, phenols, or mercaptans to form ethoxylated products. In general, the aqueous solubility of the organic reaction partner will be increased by the addition of the hydroxyethyl group. The rate of reaction can be increased by either alkaline or acid catalysis. Acidic catalysts such as mineral acids or acidic ion-exchangers are favored when working with a catalyst at higher temperatures and under pressure.

increased rate of ethoxylation by catalysis: $H^{\oplus}$-catalysis by S_N1 through oxonium complex:

$$\begin{array}{c} H_2C\!-\!\!-\!CH_2 \\ \diagdown \overset{\oplus}{O} \diagup \\ | \\ H \end{array}$$

$OH^{\ominus}$-catalysis by S_N2

Since primary products of the addition possess a reactive hydroxyl group, they can add a further ethylene oxide molecule. In this way, there is a stepwise formation of di-, tri- and poly-ethoxylated products.

characteristics of ethoxylation:

reactivity of primary products lowers selectivity due to secondary reactions

If only the monoethoxylated product is to be produced, less than a stoichiometric amount of ethylene oxide must be used.

The most significant, and increasingly important, reaction partners of ethylene oxide are summarized in Table 7–2 with their main reaction products and other secondary products.

Table 7–2. Ethylene oxide – secondary products.

Reaction partners	Reaction products	Secondary products
Water	Ethylene glycol	Glyoxal, Dioxolane
	Diethylene glycol	Dioxane
	Polyethylene glycols	
Alkylphenols	Polyethoxylates	
Fatty alcohols		
Fatty acids		
Fatty amines		
Ammonia	Monoethanolamine	Ethylenimine
	Diethanolamine	Morpholine
	Triethanolamine	
Alcohols RCH_2OH	Glycol monoalkyl ethers	Glycol dialkyl ethers
$R = H, CH_3, n\text{-}C_3H_7$	Diglycol monoalkyl ethers	Esters of glycol monoalkyl ethers
Carbon dioxide	Ethylene carbonate	Carbamates
Synthesis gas	1,3-Propanediol	Polyesters

The relative importance of ethylene oxide secondary products can be readily appreciated from the breakdown of ethylene oxide applications in several countries:

Table 7-3. Utilization of ethylene oxide (in %).

Product	World		USA		Western Europe		Japan	
	1985	1994	1981	1992	1981	1995	1981	1992
Ethylene glycol	59	58	58	61	45	42	62	55
Nonionic surfactants	13	18	12	13	23	30	17	24
Ethanolamines	6	6	7	8	10	9	7	6
Glycol ethers	6	4	8	6	10	8	5	5
Miscellaneous (*e. g.*, higher ethylene glycols, urethane polyols, etc.)	16	14	15	12	12	11	9	10
Total use (in 10^6 tonnes)	6.5	9.0	2.3	2.7	1.3	1.9	0.48	0.76

7.2.1. Ethylene Glycol and Higher Ethylene Glycols

main secondary product of EO – ethylene glycol (glycol) – consumes 40–60% of EO production, depending on country

ethylene glycol production (in 10^6 tonnes):

	1990	1992	1995
USA	2.28	2.32	2.37
W. Europe	0.86	0.85	1.10
Japan	0.59	0.56	0.71

Ethylene glycol, usually known simply as glycol, is the most important secondary product of ethylene oxide. In 1995 the world capacity for ethylene glycol was about 9.7×10^6 tonnes per year, of which about 3.1, 0.99, and 0.83×10^6 tonnes per year were in the USA, Western Europe and Japan, respectively. The production figures for ethylene glycol in these countries are summarized in the adjacent table.

Ethylene glycol results from the addition of H_2O to ethylene oxide:

$$H_2C\!-\!CH_2 + H_2O \longrightarrow HOCH_2CH_2OH$$
$$\left(\Delta H = -\begin{smallmatrix}19\ \text{kcal}\\80\ \text{kJ}\end{smallmatrix}\Big/\text{mol}\right) \qquad (9)$$

EO hydration – industrial operation:

EO:H_2O = 1:10, two types:

1. proton catalysis, without presssure at low temperature
2. no catalyst, medium pressure, higher temperature

workup of aqueous crude glycol solution:

concentration by multistep vacuum distillation

In the commercial process, ethylene oxide is reacted with an approximately tenfold molar excess of water, either in the liquid phase in the presence of an acidic catalyst (*e. g.*, 0.5–1.0 wt% H_2SO_4) at normal pressure and 50–70 °C, or without a catalyst at 140–230 °C and 20–40 bar. The ethylene glycol manufacture takes place almost exclusively in a coupled reactor following direct oxidation of ethylene. The resulting aqueous crude glycol solution is concentrated to 70% by evaporation and fractionally distilled in several vacuum columns. Despite the large excess of water, monoethylene glycol selectivity is only about 90%.

Roughly 9% diglycol, 1% triglycol, and higher ethylene glycols are also formed. The total yield is 95 – 96%. Purity requirements vary depending on the application of the glycols; a particularly high quality is required for the manufacture of polyesters (99.9 wt% purity).

A well-known noncatalytic process developed by Scientific Design was used in more than 45 plants (1993).

If the proportion of water is lowered in the hydration of ethylene oxide, di-, tri-, and polyethylene glycols are formed in a stepwise manner:

alternatives for manufacture of higher ethylene glycols:

1. by EO hydration with less than stoichiometric amounts of H_2O

$$HOCH_2CH_2OH + H_2C\underset{\displaystyle O}{-}CH_2 \longrightarrow HOC_2H_4OC_2H_4OH$$

$$\xrightarrow{+n \cdot EO} HO(C_2H_4O)_{n+2}H \tag{10}$$

In another process, ethylene oxide is fed to ethylene glycols (glycolysis of ethylene oxide). Usual conditions are 120 – 150 °C and slight excess pressure, generally in the presence of an alkaline catalyst. With increasing molecular weight, the polyethylene glycols become viscous liquids and finally wax-like products; these do, however, remain water soluble.

2. feeding EO to ethylene glycols, usually catalyzed by alkali (base catalysis promotes glycolysis, acid catalysis promotes hydrolysis)

production of diethylene glycol (in 1000 tonnes):

	1988	1990	1992
USA	224	250	239

7.2.1.1. Potential Developments in Ethylene Glycol Manufacture

The current basis for the manufacture of ethylene glycol is ethylene. In future developments, the possible use of synthesis gas will have to be considered.

improved routes to ethylene glycol from:

1. ethylene
2. C_1 chemistry

Furthermore, in the predominant process based on ethylene, optimization of the epoxidation and following hydration and minimization of the energy requirements for the isolation of ethylene glycol from the dilute aqueous solution will be important tasks.

improved selectivity in the classical route by avoidance of:

total oxidation
byproducts
energy costs

However, other indirect routes based on ethylene with intermediates such as ethylene glycol acetate or ethylene carbonate are also of commercial interest.

new routes based on ethylene by:

1. ethylene acetoxylation, *i.e.*, oxidative C_2H_4 reaction followed by hydrolysis of glycol mono- and diacetates and recycling of AcOH

The limitation in the ethylene oxide selectivity shown by the epoxidation method led to the development of processes for the oxidation in the presence of acetic acid (acetoxylation) by many firms. Ethylene is converted to glycol mono- and diacetate with up to 98% selectivity in the first step, and hydrolyzed to ethylene glycol in the second step:

$$H_2C=CH_2 + 2\ AcOH + 0.5\ O_2 \xrightarrow{[cat.]} AcOCH_2CH_2OAc$$

$$+ H_2O \quad \left(\Delta H = -\begin{smallmatrix}30\ kcal\\124\ kJ\end{smallmatrix}/mol\right) \quad (11)$$

$$AcOCH_2CH_2OAc + 2\ H_2O \xrightarrow{[cat.]} HOCH_2CH_2OH$$

$$+ 2\ HOAc \quad (12)$$

$$\left(\Delta H = -\begin{smallmatrix}4\ kcal\\17\ kJ\end{smallmatrix}/mol\right)$$

TeO_2/HBr catalysis with following elementary steps:

$Br_2 + H_2C=CH_2 \rightarrow BrCH_2CH_2Br$
$\xrightarrow{+2\ AcOH} AcOCH_2CH_2OAc + 2\ HBr$
$2\ HBr + 0.5\ O_2 \rightarrow Br_2 + H_2O$

industrial use by Oxirane in 360 000 tonne-per-year unit stopped due to HBr/HOAc corrosion

2. indirect EO hydration with intermediate product ethylene carbonate (1,2-glycol carbonate)

$$\begin{array}{ccc} H_2C & \!\!\!-\!\!\!- & CH_2 \\ | & & | \\ O & & O \\ & \diagdown \ \diagup & \\ & C & \\ & \| & \\ & O & \end{array}$$

with subsequent hydrolysis or methanolysis

$(CH_3O)_2C=O$, coproduct of methanolysis, an important replacement for $COCl_2$ and $(CH_3O)_2SO_2$

new routes based on C_1 chemistry by:

1. direct CO/H_2 reaction
2. indirect CO/H_2 use by using C_1 product

to 1:

single-step Co-, Rh-, Ru-catalyzed synthetic reaction with little chance of industrial use

Halcon, in cooperation with Arco (Oxirane Chemical Co.), was the first firm to use this acetoxylation of ethylene. They started up a commercial unit using a TeO_2/HBr catalyst in 1978, but were forced to shut down in 1979 due to corrosion.

Another area pursued by many firms is indirect hydrolysis, *i.e.*, ethylene oxide is reacted with CO_2 to form the intermediate ethylene carbonate, which is then hydrolyzed to ethylene glycol and CO_2. Apart from the additional process step, it has the advantages of a nearly quantitative hydrolysis of the carbonate to ethylene glycol with only a small excess of water.

In a new development, Texaco has replaced the hydrolysis of ethylene carbonate with a methanolysis, leading to ethylene glycol and dimethyl carbonate. Dimethyl carbonate can then be hydrolyzed to methanol and CO_2, but it is increasingly used for carbonylation and methylation, for example, for the manufacture of isocyanates (*cf.* Section 13.3.3) and polycarbonates (*cf.* Section 13.2.1.3).

Increasing interest in synthesis gas as a basis for ethylene glycol paralleled the advancement of C_1 chemistry. Direct processes of CO hydrogenation to ethylene glycol as well as indirect methods based on synthesis gas dependent intermediates like methanol, methyl formate (*cf.* Section 2.3.3) or formaldehyde have been developed.

By the end of the 1940s, Du Pont had already been able to show that CO hydrogenation in aqueous cobalt salt solutions leads to ethylene glycol. In the 1970s, UCC investigated synthesis gas conversion using homogeneous rhodium carbonyl catalyst systems with numerous salt promoters and nitrogen-containing Lewis bases. Ethylene glycol, 1,2-propanediol, and glycerine can be produced in a high-pressure (1400–3400 bar) reaction at 125–350 °C with a total selectivity of up to 70%. Both the extreme reaction conditions and the low catalyst efficiency are obstacles to practical application of this technology.

Further research, including investigations by other firms of other catalyst systems (*e. g.*, ruthenium catalysts), has resulted in little or no progress.

Indirect processes based on synthesis gas products such as methanol and formaldehyde (*cf.* Section 2.3.2.2) are being more intensively investigated. Generally, other intermediates such as glycol aldehydes, glycolic acid or oxalic acid ester are formed by hydrogenative, oxidative or simple carbonylation, respectively. These can then be converted into ethylene glycol in a final hydrogenation.

Recently, much attention has been given to the oxidative carbonylation of methanol to dimethyloxalate – pursued mainly by Ube – and a following hydrogenolysis to ethylene glycol and methanol:

to 2:

hydroformylation, oxidative carbonylation or carbonylation of CH_3OH or $HCHO$ to form intermediates for the following hydrogenation

oxalic ester manufacture used commercially by Ube since 1978, with further development of hydrogenation step by UCC, has highest economic appeal

$$2\,CH_3OH + 2\,CO + 0.5\,O_2 \xrightarrow{\text{[cat.]}} CH_3OCCOCH_3 + H_2O \qquad (13)$$
$$\qquad\qquad\qquad\qquad\qquad\quad \underset{OO}{\overset{\|\ \|}{}}$$

$$2\,CH_3OCCOCH_3 + 4\,H_2 \xrightarrow{\text{[cat.]}} HOCH_2CH_2OH + 2\,CH_3OH \quad (14)$$
$$\underset{OO}{\overset{\|\ \|}{}}$$

The first step gives a yield of 97% when run at 110 °C and 90 bar in the presence of a Pd catalyst and 70% HNO_3. A commercial unit has been in operation since 1978. The hydrogenation can be done, *e. g.*, on ruthenium oxide with about 90% yield and the reformation of methanol. Other alcohols such as *n*-butanol can be used in place of methanol.

HNO_3 reacts to form methylnitrite (CH_3ONO) as reactive intermediate for carbonylation

The second step is being piloted by UCC, giving this oxalate route to ethylene glycol the greatest chance of commercial realization.

7.2.1.2. Uses of Ethylene Glycol

Ethylene glycol is predominantly used in two areas: in antifreeze mixtures in automobile cooling systems, and as a diol in polyester manufacture. Polyethylene terephthalate (PET), the most important product, is mainly used for fiber manufacture, but is also employed for films and resins. The breakdown of ethylene glycol usage for these two purposes differs greatly from country to country. In the USA, more than 50% has long been used in antifreeze, but because of the trend to smaller motors and to longer

uses of glycol:

1. component of polyester
   most important product:
   PET (polyethylene terephthalate) for fibers, films, resins, and recently bottles

$$\{CH_2CH_2\!-\!OC\!-\!\langle\bigcirc\rangle\!-\!CO\}_n$$
$$\qquad\quad \underset{O}{\overset{\|}{}}\qquad\quad \underset{O}{\overset{\|}{}}$$

2. antifreeze agents (ca. 95% glycol)

intervals between antifreeze changes, this use is decreasing (see following table):

Table 7-4. Use of ethylene glycol (in %).

Product	World		USA		Western Europe		Japan	
	1991	1995	1981	1991	1981	1991	1981	1991
Antifreeze	23	18	54	45	34	28	14	11
Polyester fibers	41	55	} 40	27	} 52	32	} 71	43
Polyester films, resins	24	17		18		8		23
Remaining uses	12	10	6	10	14	32	15	23
Total use (in 10^6 tonnes)	6.2	7.5	1.79	1.84	0.70	1.20	0.46	0.66

In other industrialized countries, the use for polyester dominates. A smaller share of the ethylene glycol is used for several secondary products, as described in the following section. Recently, PET has also been used for nonreturnable bottles for beverages (*cf.* Section 14.2.4).

uses of polyglycol:

as brake fluid, plasticizer, lubricant, as well as in the cosmetic and pharmaceutical industries

reaction component for esters, polyurethanes, ester resins

Depending on molecular weight, polyethylene glycols are employed, alone or after esterification, as brake fluids, plasticizers, or lubricants. They are also used in the manufacture of polyurethanes and ester resins (*cf.* Section 14.2.4).

7.2.1.3. Secondary Products – Glyoxal, Dioxolane, 1,4-Dioxane

secondary products of glycol:

dialdehyde (glyoxal)
cyclic formal (dioxolane)
cyclic diether (1,4-dioxane)

Glyoxal, dioxolane (1,3-dioxacyclopentane), and 1,4-dioxane are the industrially most important secondary products of ethylene glycol.

manufacture of glyoxal from ethylene glycol analogous to oxidative dehydrogenation of CH_3OH to HCHO with Ag or Cu metal

Glyoxal is manufactured from glycol with yields of 70–80% in a gas-phase oxidation with air. The reaction takes place at over 300 °C in the presence of Ag or Cu catalysts and small amounts of halogen compounds to prevent total oxidation:

$$HOCH_2CH_2OH + O_2 \xrightarrow{[cat.]} OHC-CHO + 2\,H_2O \qquad (15)$$

alternative manufacture by oxidation of CH_3CHO with HNO_3 and $Cu(NO_3)_2$-catalysis

numerous acids as byproducts, *e.g.*, OHC-COOH of industrial importance

Another possibility for the synthesis is the oxidation of acetaldehyde with HNO_3, either noncatalytically or with metal salt catalysts. Acetaldehyde is reacted with 69% HNO_3 under mild conditions at about 40 °C with a maximum yield of about 70%. In addition to the byproducts glycolic acid, oxalic acid, acetic acid and formic acid, about 10% glyoxylic acid is also formed. By raising the temperature and the HNO_3 concentration, the oxidation can be directed to favor glyoxylic acid, which is needed for large-scale manufacture of vanillin, ethyl vanillin, and allantoin.

The aqueous glyoxal solution is purified by ion exchange. Since glyoxal is unstable in its pure form, it is used either as a 30–40% aqueous solution or as a solid hydrate with an 80% glyoxal content. Glyoxal is frequently used for condensation and cross-linking reactions because of the reactivity of its two aldehyde groups towards polyfunctional compounds with hydroxy or amino groups, *e. g.,* with urea and its derivatives, or with starch, cellulose, cotton, casein, or animal glue. It is also used in textile and paper processing. The largest producer of glyoxal world-wide is BASF, with capacities (1995) of 20 000 and 25 000 tonnes per year in the USA and Germany, respectively.

properties and uses of glyoxal:

simplest dialdehyde stable only as hydrate $(HO)_2 CHCH(OH)_2$

bifunctional character exploited for cross-linking of cellulose and cotton, *i. e.,* for surface finishing (textile and paper industry)

A new glyoxylic acid process was adopted in 1989 by Chemie Linz in a 25 000 tonne-per-year plant in Austria. In this process, glyoxylic acid is obtained in 95% yield by the ozonolysis of maleic acid dimethyl ester.

oxidative cleavage of maleic acid dimethyl ester with ozone and subsequent hydrogenation/hydrolysis to $HOOC-CHO$

The world demand for glyoxylic acid is currently between 8000 and 10 000 tonnes per year.

Dioxolane (1,3-dioxacyclopentane) is another glycol derivative employed industrially. It can be prepared by the proton-catalyzed reaction of glycol with aqueous formaldehyde:

manufacture of dioxolane:

proton-catalyzed cyclocondensation of glycol with formaldehyde

$$HOCH_2CH_2OH + HCHO \underset{}{\overset{[H^\oplus]}{\rightleftarrows}} \quad \text{O}\bigcirc\text{O} + H_2O \qquad (16)$$

The H_2O/dioxolane azeotrope is distilled out of the reaction mixture and the dioxolane is isolated from the aqueous phase by extraction with, for example, methylene chloride.

dioxolane isolation is made difficult because of:

1. azeotrope formation with H_2O
2. H_2O solubility

therefore extraction (CH_2Cl_2) after distillation

Analogous to ethylene oxide, dioxolane is used as a comonomer in the polymerization of trioxane to polyoxymethylene copolymers.

Furthermore, dioxolane is a powerful solvent with properties similar to tetrahydrofuran (THF). The cyclic formal (acetal of formaldehyde) has less tendency to form a peroxide than does the cyclic ether, THF; however, this advantage is balanced by the lability of the formal bond in an acidic aqueous environment.

uses of dioxolane:

1. as monomer for cationic polymerization
2. as solvent with properties similar to THF (*e. g.,* for polyvinylidene chloride)

1,4-Dioxane is manufactured by the cyclic dehydration of glycol or diglycol:

$$\text{O}\bigcirc\overset{OH}{\underset{OH}{}} \longrightarrow \text{O}\bigcirc\text{O} + H_2O \qquad (17)$$

manufacture of dioxane:

from glycol, diglycol or chlorine substituted products by inter- or intramolecular dehydration or dehydrochlorination

or by EO dimerization

Dilute H_2SO_2 (or another strong acid) is allowed to react with the starting material at $150-160\,°C$, and the dioxane is distilled off as it forms. Two other processes, which take place with elimination of HCl and ring closure, are the reactions of chlorohydrin and 2,2-dichlorodiethyl ether with NaOH.

1,4-Dioxane can also be prepared by heating ethylene oxide with dilute H_2SO_4 or H_3PO_4:

$$2\ \triangleright\!O \xrightarrow{[H^{\oplus}]} O\bigcirc O \tag{18}$$

uses of dioxane:

as cyclic ether, good solvent and complex component

Dioxane is a valuable solvent for cellulose esters and ethers as well as for oils and resins. Since dioxane is a cyclic ether, it forms oxonium salts and complexes e.g., with Br_2 or SO_3; these are used in preparative chemistry.

7.2.2. Polyethoxylates

polyethoxylate types:

1. general, from multiple EO addition to components with mobile H

2. special (ethoxylates in strict sense), from multiple EO addition to:

2.1. alkyl phenols
  (e.g., octyl, nonyl, dodecyl)

2.2. fatty alcohols
  fatty acids
  fatty amines

degree of ethoxylation of starting materials determines the extent of hydrophilic character

manufacture of ethoxylates:

alkaline-catalyzed EO addition

uses:

nonionic surfactants, special detergents, wetting agents, and emulsifiers

The second most important use for ethylene oxide is for multi-ethoxylated alkyl phenols, fatty alcohols, fatty acids, and fatty amines.

In the reaction of these starting products with $10-30$ moles ethylene oxide per mole, they lose their hydrophobic character and form widely used industrial products whose hydrophilic character can be regulated by the number of ethylene oxide units. An inadequate water solubility with an ethylene oxide content below five units can be improved by esterifying the terminal hydroxyl groups with H_2SO_4 to form ether sulfates.

Industrially, the addition of ethylene oxide is generally conducted at several bar pressure in the presence of alkaline catalysts such as NaOAc or NaOH at $120-220\,°C$. The ethoxylation is usually a batch process run in a stirred vessel or recycle reactor.

The ethoxylated products are nonionic surfactants causing little foam formation and are widely used as raw materials for detergents, and as wetting agents, emulsifiers, and dispersants. They are also reacted with isocyanates in the production of polyurethanes, e.g., of polyurethane foams.

In 1990, with a worldwide capacity of nearly 3.5×10^6 tonnes per year, about 2.54×10^6 tonnes of linear alcohol ethoxylates (polyether alcohols) were produced for industrial and household use.

An additional important factor when evaluating these products is their biodegradability. It is known that alkyl phenol ethoxylates are more difficult to degrade than the fatty alcohol ethoxylates. An interesting feature is that the hydrophobic part of the fatty alcohol ethoxylate is very easily degraded, while degradation of the hydrophilic polyester residue is noticeably more difficult, becoming more difficult with increasing chain length.

ecological behavior:

ethoxylates from alkyl phenols difficult, from fatty alcohols easy, to biodegrade

7.2.3. Ethanolamines and Secondary Products

Ethylene oxide reacts exothermically with $20-30\%$ aqueous ammonia at $60-150\,°C$ and $30-150$ bar to form a mixture of the three theoretically possible ethanolamines with high selectivity:

manufacture of ethanolamines:

three reactive H atoms in NH_3 allow threefold ethoxylation

$$H_2C\!-\!CH_2 + NH_3 \longrightarrow H_2NC_2H_4OH + HN(C_2H_4OH)_2$$
$$\underset{O}{\diagup\diagdown}$$
$$+ N(C_2H_4OH)_3 \qquad (19)$$

When run under pressure at about $100\,°C$ and with a triethanolamine salt, the ethoxylation can go one step further to form the quaternary base. The composition of the reaction product can be influenced by temperature and pressure, but is especially sensitive to the ratio of NH_3 to ethylene oxide; the greater the excess of NH_3, the higher the monoethanolamine content:

EO/NH_3 ratio determines extent of ethoxylation, e. g.,

$EO:NH_3 \ll 1$ favors monoethoxylation, high reactivity of secondary reaction does not preclude triple ethoxylation

Table 7-5. Ethanolamine from NH_3 and ethylene oxide at $30-40\,°C$ and 1.5 bar.

Mol ratio $NH_3:EO$	Mono-,	Selectivity ratios Di-,	Triethanolamine
10:1	75	21	4
1:1	12	23	65

The high triethanolamine content resulting from an equimolar mixture of reactants indicates that the primary reaction with NH_3 is slower than the secondary reaction. Ethoxylated byproducts can also result from the reaction between ethylene oxide and the OH groups of triethanolamine.

The world production capacity for ethanolamines in 1993 was 0.84×10^6 tonnes per year, with about 450000, 240000, and 70000 tonnes per year in the USA, Western Europe and Japan, respectively.

Ethanolamines are industrially valuable products whose main use is for the manufacture of detergents by reaction with fatty acids. At $140-160\,°C$, the fatty acid can react with the amino

uses of the ethanolamines:

1. as fatty acid derivatives for detergents and emulsifiers, e. g., in the form of:

RCONHCH$_2$CH$_2$OH
RCOOCH$_2$CH$_2$NH$_2$
RCOOCH$_2$CH$_2$NHCOR

2. for gas purification (H$_2$S, HCl, CO$_2$)

3. as soap and cosmetic cream component, e. g., in the form of:

RCOO$^{\ominus}$H$_3$N$^{\oplus}$CH$_2$CH$_2$OH

4. as starting material in organic syntheses (heterocycles)

ethanolamine production (in 1000 tonnes):

	1989	1991	1993
USA	287	291	320
W. Europe	155	158	164
Japan	63	77	47

secondary products of ethanolamines:

1. morpholine
2. ethylenimine
3. ethylenediamine

manufacture of morpholine:

1. acid-catalyzed dehydration of HN(C$_2$H$_4$OH)$_2$

2. ring-forming amination of O(C$_2$H$_4$OH)$_2$ under hydrogenation conditions

manufacture of ethylenimine (aziridine):

indirect two-step chemical dehydration with monoester of sulfuric acid as intermediate

group to form a fatty acid ethanolamide, with the hydroxyl group to give a fatty acid aminoethyl ester, or with both groups to form a di-fatty acid ethanolamide ester.

Ethanolamine can also be used directly as a weak base in industrial gas purification for removal of acidic gases such as H$_2$S and CO$_2$. Because of their mild alkaline properties, ethanolamines are frequently used as constituents of soaps and cosmetic creams. In contrast to detergents, only fatty acid salts are used here. In addition, they are used as auxiliary agents in the cement industry and as cooling lubricants for drilling and cutting oils. They are also used in the synthesis of heterocyclic organic compounds. The production figures for several countries are summarized in the adjacent table. Worldwide, the largest producer of ethanolamines is UCC, with a capacity of 0.28×10^6 tonnes per year. With a capacity of 0.14×10^6 tonnes per year, BASF is the largest producer in Western Europe (1997).

As the principal secondary products of ethanolamine, morpholine, ethylenimine, and ethylenediamine have become commercially important.

Morpholine, a solvent and intermediate for, e. g., optical brighteners and rubber chemicals, is obtained from diethanolamine by dehydration with 70% H$_2$SO$_4$ to close the ring:

(20)

A new manufacturing route has completely eclipsed the cyclization of diethanolamine described above, e. g., in the USA. In this method, diethylene glycol, NH$_3$ and a little H$_2$ are reacted at 150–400 °C and 30–400 bar over a catalyst containing Ni, Cu, Cr, or Co to give morpholine. More specific details of this process as operated by Jefferson Chemical Company (Texaco) are not known.

Ethylenimine is another commercially important intermediate which can be manufactured from monoethanolamine. Most of the industrial processes operated today have two steps. In the first step, ethanolamine is esterified with 95% H$_2$SO$_4$ to β-aminoethyl sulfuric acid:

$$H_2NCH_2CH_2OH + H_2SO_4 \longrightarrow$$
$$H_2NCH_2CH_2OSO_3H + H_2O \quad (21)$$

In the second step, the ester is heated with a stoichiometric amount of NaOH at $220-250\,^\circ$C and $50-80$ bar. The acid is cleaved from the ester and the ring closes to form aziridine:

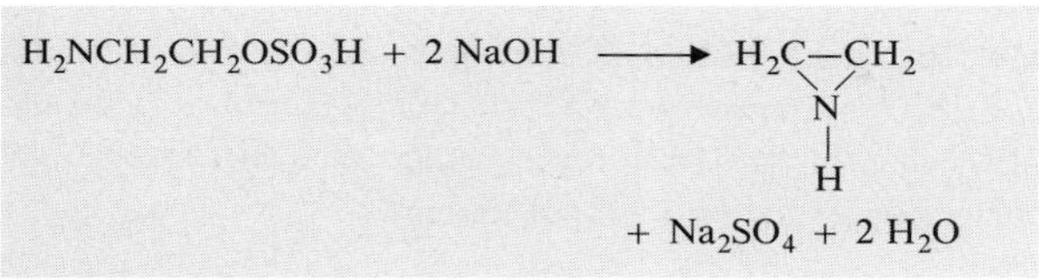

$$H_2NCH_2CH_2OSO_3H + 2\,NaOH \longrightarrow \underset{\underset{H}{|}}{\underset{N}{\diagdown\diagup}} H_2C\!-\!CH_2$$
$$+ Na_2SO_4 + 2\,H_2O \qquad (22)$$

If the imine formation is conducted in a flow tube, residence times of $4-10$ s can be chosen to suppress secondary reactions such as the polymerization of ethylenimine. The imine selectivity then increases to $80-85\%$ (based on ethanolamine). Processes for ethylenimine manufacture based on this route have been developed by BASF and Hoechst.

In a process from Dow, ethylenimine is synthesized by reacting 1,2-dichloroethane with NH_3 in the presence of CaO at about $100\,^\circ$C:

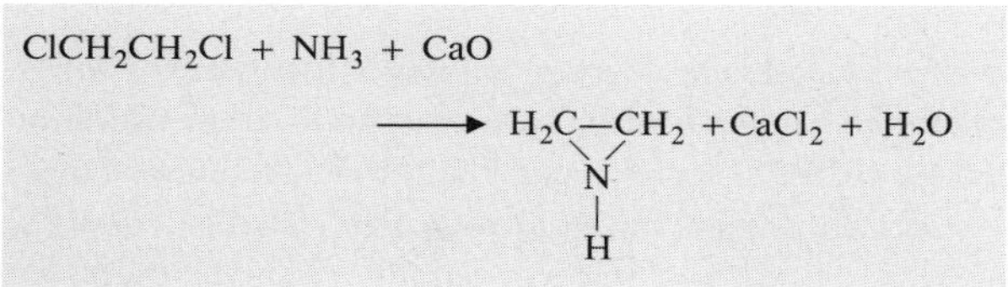

alternative ethylenimine process (Dow) achieves cyclization of dichloroethane with NH_3 and HCl-acceptor CaO

$$ClCH_2CH_2Cl + NH_3 + CaO$$
$$\longrightarrow \underset{\underset{H}{|}}{\underset{N}{\diagdown\diagup}} H_2C\!-\!CH_2 + CaCl_2 + H_2O \qquad (23)$$

Along with the ready polymerizability of ethylenimine, which can occur spontaneously in the presence of small amounts of acidic substances, its high toxicity must also be taken into account in its manufacture, storage, and processing.

Most ethylenimine is converted into polyethylenimine, a processing aid in the paper industry. It also serves as an intermediate in syntheses, e.g., ureas are formed with isocyanates.

uses of ethylenimine:

1. polymerization to $-(CH_2CH_2NH)_n$ – aid in paper processing
2. intermediate product for syntheses

Ethylenediamine is obtained from the reaction of monoethanolamine with ammonia:

manufacture of ethylenediamine:

Ni-catalyzed ammonolysis of ethanolamine in the gas phase

$$H_2NCH_2CH_2OH + NH_3 \xrightarrow[\text{H}_2]{\text{[cat.]}} H_2NCH_2CH_2NH_2 + H_2O \qquad (24)$$

excess of NH_3 determines selectivity

important byproducts:

$(H_2NCH_2CH_2)_2NH$

$$HN\diagup\diagdown NH$$

This gas-phase reaction is run at temperatures less than $300\,^\circ$C and pressures under 250 bar in the presence of water and nickel catalysts with various promoters such as Co, Fe, and Re. As in the conversion of 1,2-dichloroethane to ethylenediamine with ammonia (*cf.* Section 9.1.1.4), and dependent on the excess of ammonia, secondary reactions can give higher amines such as piperazine. A particular advantage of using ethanolamine rather than 1,2-dichloroethane is the absence of the coproduct NaCl.

In 1969, BASF in Wesern Europe became the first firm to produce ethylenediamine commercially (7000 tonnes per year capacity) using this salt-free process. Since then, there have been similar developments by other companies.

The uses of ethylenediamine are discussed in Section 9.1.1.4.

7.2.4. Ethylene Glycol Ethers

principle of monoalkyl ether formation of glycols:

ethoxylation of alcohols = alcoholysis of ethlene oxide

Another important use of ethylene oxide comes from its reaction with alcohols to form glycol monoalkyl ethers:

$$\text{ROH} + \text{H}_2\text{C}\underset{\text{O}}{-}\text{CH}_2 \xrightarrow{\text{[cat.]}} \text{ROCH}_2\text{CH}_2\text{OH}$$
$$+ \text{RO}\text{-}(\text{CH}_2\text{CH}_2\text{---O})_n\text{H} \tag{25}$$

characteristics of industrial process:

for desired monoethoxylation, large excess of alcohol, catalyzed homogeneously or heterogeneously with Al_2O_3

preferred alkyls in monoethers:

$R = $ -CH_3, -C_2H_5, -n-C_4H_9

The most frequently employed alcohols are methanol, ethanol, and *n*-butanol. The industrial process is similar to the hydration of ethylene oxide, the difference being that alkaline catalysts (such as alkali hydroxide or the corresponding alkali alcoholate) or Al_2O_3 are usually employed. Despite a large excess of alcohol, the secondary products di-, tri- and higher ethylene glycol monoalkyl ethers are generally still formed. Ethylene glycol monoethyl ether is manufactured at $170-190\,°C$ and $10-15$ bar.

uses of the monoethers:

1. of monoethylene glycol:
favored paint solvent, starting material for diethers and ether esters

2. of higher ethylene glycols:
components for brake fluids, solvent and compounding agent, starting material for diethers

ethylene glycol ether production (in 1000 tonnes):

	1990	1992	1993
W. Europe	306	301	309
USA	184	228	247
Japan	76	74	75

The monoalkyl ethers of ethylene glycol are commercially known as Cellosolve®, Carbitol®, and Dowanol®. These substances have many uses: as solvents in the paint and lacquer sector, components for adjusting viscosity in brake fluids, emulsifiers for mineral and vegetable oils, and for ball point pen and printing inks. In 1993, the production capacity for ethylene glycol ethers in the USA, Western Europe, and Japan was about 0.58, 0.57, and 0.14×10^6 tonnes per year, respectively. The production figures for glycol ethers in these countries are summarized in the adjacent table. The glycol ether production includes the monomethyl, monoethyl, and monobutyl ethers of monoethylene glycol.

Other industrially important inert aprotic solvents are obtained by etherification or esterification of the remaining OH group in the glycol monoalkyl ethers. Dimethyl glycols such as dimethyl glycol (Glyme®) and dimethyl diglycol (Diglyme®) are the most important members of this group.

Ethylene glycol dialkyl ethers, exemplified by the dimethyl ether, are generally manufactured in two steps. First, the monomethyl ether is converted into the alcoholate by NaOH; the water of reaction is removed by distillation:

two-step manufacture of dialkyl ethers of glycols (Williamson synthesis)

1. conversion into Na alcoholate

$$CH_3 (OCH_2CH_2)_n\text{—}OH + NaOH \quad (26)$$
$$\longrightarrow CH_3 (OCH_2CH_2)_n\text{—}ONa + H_2O$$

In the second step, the sodium alcoholate is etherified with an alkyl chloride (usually methyl chloride) or dimethyl sulfate:

2. alkylation with alkyl chloride or sulfate, preferably methylation with CH_3Cl or $(CH_3O)_2SO_2$

$$CH_3 (OCH_2CH_2)_n\text{—}ONa + CH_3Cl \quad (27)$$
$$\longrightarrow CH_3 (OCH_2CH_2)_n\text{—}OCH_3 + NaCl$$

As this method is burdened by the stoichiometric consumption of chlorine and alkali, Hoechst has developed a new process for ethylene glycol dimethylether in which dimethyl ether is reacted with ethylene oxide in the presence of boron trifluoride:

new manufacture of dimethylethylene glycols:

Friedel–Crafts-catalyzed insertion of ethylene oxide into dimethylether

$$CH_3OCH_3 + H_2C\text{—}CH_2 \xrightarrow{\text{[cat.]}}$$
$$\underset{O}{\qquad}$$
$$CH_3OCH_2CH_2OCH_3 + CH_3O(CH_2CH_2O)_nCH_3 \quad (28)$$

This process is used commercially.

Ethylene glycol dialkyl ethers are widely used as inert, aprotic reaction media for organometallic reactions (*e.g.*, for Grignard reactions or in boron chemistry), as specialized solvents for resins, plastics, latex paints and varnishes (*e.g.*, for the manufacture of polyurethane coatings), as performance liquids (*e.g.*, for brake and hydraulic fluids as well as heating and cooling media), and as extractants. Mixtures of polyethylene glycol dimethyl ethers, $CH_3(OCH_2CH_2)_nOCH_3$ with 4 to 7 ethylene oxide units, are used as absorption agents in Allied's "Selexol process" in which natural gas and synthesis gas are freed from acidic components such as H_2S, CO_2 or SO_2 by pressurized washing. The acidic gases are then released one at a time by reducing the pressure stepwise.

uses of glycol dialkyl ethers:

1. aprotic reaction media
2. specialized solvents
3. functional liquids
4. extraction media
5. tetra- to heptaglycol dimethyl ether mixtures as Selexol® for purification of natural and synthesis gas using the principle of pressure-dependent reversible absorption of acidic gases

manufacture of ether esters of ethylene glycol:

1. esterification of monoalkyl ethers of glycols, generally with AcOH
2. Friedel–Crafts-catalyzed insertion of ethylene oxide into ethyl acetate

To esterify the free OH groups, ethylene glycol monoalkyl ethers are reacted with carboxylic acids, especially acetic acid, using proton catalysts:

$$C_2H_5OC_2H_4OH + CH_3COOH \xrightarrow{[cat.]} C_2H_5OC_2H_4OCCH_3 + H_2O \qquad (29)$$
$$\overset{\|}{O}$$

Ethyl glycol acetate (2-ethoxyethyl acetate) is also produced (*e.g.*, by Nisso) from ethyl acetate by reaction with ethylene oxide:

$$CH_3COOC_2H_5 + H_2C\!-\!CH_2 \xrightarrow{[cat.]} CH_3COOCH_2CH_2OC_2H_5 \qquad (30)$$
$$\underset{O}{\diagdown\diagup}$$

Friedel–Crafts systems with, *e.g.*, N- or P-containing ligands such as $AlCl_3 \cdot N(C_2H_5)_3$ are used as catalysts.

glycol ether ester, *e.g.*,

$CH_3OC_2H_4OAc$ and $C_2H_5OC_2H_4OAc$

are favored solvents for cellulose derivatives

Ethyl glycol acetate is an excellent solvent for nitrocellulose and cellulose ether used in the manufacture and processing of varnish.

7.2.5. Additional products from ethylene oxide

EO products with limited production:

1. ethylene carbonate
2. 1,3-propanediol

Ethylene carbonate and 1,3-propanediol are two products made from ethylene oxide which have not yet reached the production volumes of those already listed.

manufacture of ethylene carbonate:

traditionally by reaction of glycol with phosgene, today by catalytic addition of CO_2 to EO to give cyclic carbonate

Ethylene carbonate, also known as glycol carbonate or 1,3-dioxolan-2-one, was originally only obtainable from glycol and phosgene by double elimination of HCl. In 1943, IG Farben discovered a new route using ethylene oxide and CO_2:

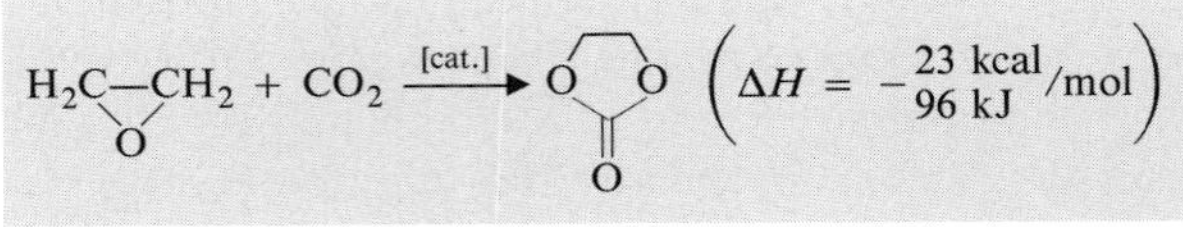

$$H_2C\!-\!CH_2 + CO_2 \xrightarrow{[cat.]} \left(\Delta H = -\frac{23\ kcal}{96\ kJ}/mol \right) \qquad (31)$$

Hüls process operated in plants in West Germany and Rumania

Hüls developed this process to industrial maturity and started up a smaller production unit in West Germany. There are two plants operating in Rumania using the same process with a total capacity of 8000 tonnes per year. Tertiary amines, quaternary ammonium salts or active carbon impregnated with NaOH are all suitable catalysts. The process is conducted at 160–200 °C and 70–100 bar. A 97–98% yield is obtained. Texaco and Dow produce ethylene carbonate commercially in the USA.

further production in USA

Ethylene carbonate is an excellent solvent for many polymers and resins; this particular property is utilized industrially, for example, in Rumania in the manufacture of polyacrylonitrile fibers. It is also an intermediate in organic syntheses; *e.g.*, carbamates can be obtained from the reaction with NH_3 or amines. In some cases, ethylene carbonate can replace ethylene oxide in ethoxylations. More recently, ethylene carbonate has been used as an intermediate in the conversion of ethylene oxide to ethylene glycol (*cf.* Section 7.2.1.1).

uses of ethylene carbonate:

1. high boiling solvent (b.p. = 238 °C) for polymers and resins
2. intermediate for, *e.g.*, carbamates, pyrimidines, purines, etc.
3. ethoxylation agent for special applications

A second, growing product is 1,3-propanediol, which can be manufactured from acrolein (*cf.* Section 11.1.6) as well as from ethylene oxide. Several companies have developed processes based on ethylene oxide; Shell's commercial manufacturing route was the first to be used openly.

production of 1,3-propanediol:

1. combined hydration and hydrogenation of acrolein
2. combined hydroformylation and hydrogenation of ethylene oxide

Ethylene oxide can be converted to the diol in two steps through hydroformylation followed by hydrogenation of the intermediate 3-hydroxypropionaldehyde, or in a single step using an alternate catalyst system:

$$H_2C\!-\!CH_2 + CO/H_2 \xrightarrow{\text{[cat.]}} HOCH_2CH_2CHO$$
$$\underset{O}{}$$
$$HOCH_2CH_2CHO + H_2 \xrightarrow{\text{[cat.]}} HOCH_2CH_2CH_2OH \tag{32}$$

Technical details have not yet been disclosed. The use of 1,3-propanediol is outlined in Section 11.1.6.

7.3. Acetaldehyde

The main raw materials for the manufacture of acetaldehyde have been ethanol, acetylene, and hydrocarbon fractions. The choice between these has varied from country to country and has been strongly influenced by economic and historical factors.

manufacturing processes for acetaldehyde:

traditional processes:

1. dehydrogenation and oxydehydrogenation of ethanol
2. hydration of acetylene
3. gas-phase oxidation of C_3/C_4 alkanes

Until the mid 1960s, the process with ethylene as feedstock and ethanol as intermediate still played in important role in the USA, Great Britain, and France. In Germany and Italy, the high price of ethanol set by fiscal policy caused the hydration of acetylene to dominate.

The oxidation of C_3/C_4 alkanes as an additional route to acetaldehyde was then developed and applied commercially, especially in the USA.

With the supply of inexpensive ethylene from the cracking of natural gas and naphtha on the one hand and the development of industrial direct oxidation processes by Wacker–Hoechst on

current acetaldehyde situation:

traditional raw materials of only limited, geographically specific importance

the other, the older processes have been increasingly replaced in recent years, especially in Western Europe and Japan.

Today, the manufacture of acetaldehyde appears as follows: the ethanol-based process has been able to hold its own in only a few countries in Western Europe, and there it accounts for only a small share of the production.

The acetylene-based process is only practiced in a few Eastern European countries and in countries like Switzerland and Italy where low-cost acetylene is available.

The oxidation of propane/butane has also become less important; since it is an unselective process, it is only profitable in large plants where the coproducts can be recovered.

direct oxidation of ethylene diminishing following worldwide growth due to alternative routes for acetaldehyde secondary products

larger drop in acetaldehyde production due to expansion of the Monsanto acetic acid process (carbonylation of methanol)

Another development was the direct oxidation of ethylene. In the mid 1970s, this reached a maximum production capacity of about 2.6×10^6 tonnes per year worldwide. The cause of the decline in the following years was the increase in the manufacture of acetic acid — the most important product made from acetaldehyde — by the carbonylation of methanol (*cf.* Section 7.4.1).

future development:

more products now based on acetaldehyde, *e. g.*, AcOH, to be made using C_1 chemistry

In the future, new processes for chemicals such as acetic anhydride (*cf.* Section 7.4.2) and alkyl amines — chemicals formerly made from acetaldehyde — will cause a further decrease in its importance.

With the growing interest in synthesis gas as the foremost raw material, the one-step conversion of CO/H_2 mixtures to acetaldehyde and other acidic C_2 products (*cf.* Section 7.4.1.4) may also limit the remaining need.

acetaldehyde production (in 10^6 tonnes):

	1988	1990	1993
W. Europe	0.66	0.58	0.61
Japan	0.33	0.38	0.35
USA	0.32	0.31	0.16

Production figures for acetaldehyde in several countries are summarized in the adjacent table. In 1993, the worldwide capacity for acetaldehyde was about 2.4×10^6 tonnes per year, of which 0.65, 0.42, and 0.23×10^6 tonnes per year were in Western Europe, Japan, and the USA, respectively.

7.3.1. Acetaldehyde *via* Oxidation of Ethylene

7.3.1.1. Chemical Basis

manufacture of acetaldehyde from ethylene:

by partial oxidation on noble metal catalysts combined with a redox system

The principle of the process currently in general use — the partial oxidation of ethylene to acetaldehyde — is based on the observation made by F. C. Phillips back in 1894 that platinum metal salts stoichiometrically oxidize ethylene selectively to acetaldehyde while themselves being reduced to the metal. However, industrial application was first possible only after the discovery by Wacker of a catalytic process using a redox system and the development of a commercial process by Wacker and Hoechst.

The total process, developed by Wacker and Hoechst between 1957 and 1959, can be depicted as an exothermic catalytic direct oxidation:

$$H_2C{=}CH_2 + 0.5\ O_2 \xrightarrow{\text{[Pd—Cu—cat.]}} CH_3CHO$$

$$\left(\Delta H = -\frac{58\ \text{kcal}}{243\ \text{kJ}}/\text{mol}\right)$$

(33)

The catalyst is a two-component system consisting of $PdCl_2$ and $CuCl_2$. $PdCl_2$ functions as the actual catalyst in a process involving ethylene complex formation and ligand exchange. The important elementary steps in the mechanism are seen as the formation of a π charge-transfer complex, rearrangement to a σ complex, and its decomposition into the final products:

principle of Wacker–Hoechst direct oxidation process:

homogeneously catalyzed, reacting through π-σ-$Pd^{2\oplus}$ complexes, very selective oxidation of $H_2C{=}CH_2$ to CH_3CHO

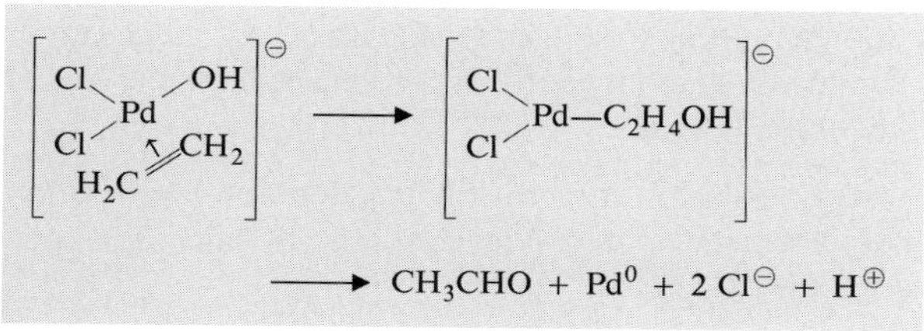

$$\begin{bmatrix} \begin{array}{c} Cl \\ Cl \end{array}\!\! \begin{array}{c} OH \\ Pd \\ CH_2 \\ H_2C \end{array} \end{bmatrix}^{\ominus} \longrightarrow \begin{bmatrix} \begin{array}{c} Cl \\ Cl \end{array}\!\! Pd{-}C_2H_4OH \end{bmatrix}^{\ominus}$$

(34)

$$\longrightarrow CH_3CHO + Pd^0 + 2\ Cl^{\ominus} + H^{\oplus}$$

$CuCl_2$ reoxidizes the nonvalent palladium to the divalent state. Although numerous other oxidizing agents can also convert Pd^0 into $Pd^{2\oplus}$, the copper redox system has the advantage that $Cu^{\oplus}$ can be easily reoxidized to $Cu^{2\oplus}$ with O_2. Recently, a new catalyst development using a phosphorous–molybdenum–vanadium–polyoxoanion system for the reoxidation of Pd^0 has been disclosed by Catalytica in the USA. There are several advantages, including a higher selectivity in the absence of chlorinated coproducts. This new catalyst has been demonstrated commercially.

characteristics of oxidation mechanism:

intramolecular transfer of OH ligand from Pd to ethylenic C, *i. e.*, total aldehyde oxygen derived from aqueous medium

New Pd-reoxidation one- or two-step using alkali salt of a P, Mo, V-polyoxoanion

$CuCl_2$ reoxidizes Pd^0 to $PdCl_2$ and is itself regenerated to the divalent oxidation state by O_2 (air)

The previous net equation (eq 33) summarizes the various reactions taking place. They can be formally divided into the rapid olefin oxidation:

$$H_2C{=}CH_2 + PdCl_2 + H_2O \longrightarrow CH_3CHO + Pd + 2\ HCl$$ (35)

and the rate determining regeneration:

$$Pd + 2\ CuCl_2 \rightleftharpoons PdCl_2 + 2\ CuCl \qquad (36)$$

$$2\ CuCl + 0.5\ O_2 + 2\ HCl \longrightarrow 2\ CuCl_2 + H_2O \qquad (37)$$

The relative rates of the partial reactions can be determined by adjusting the HCl content, the regeneration being accelerated by a higher HCl concentration.

Thus the quantity of palladium salt required for the selective oxidation of ethylene can be limited to catalytic amounts by using a large excess of $CuCl_2$ (cf. Section 9.2.1.2).

7.3.1.2. Process Operation

industrial operation of two-phase reaction with corrosive metal salt solutions at pH 0.8–3 preferably employs bubble column constructed with corrosion-resistant material

The large-scale manufacture of acetaldehyde takes place in a two-phase, i.e., gas/liquid, system. The gaseous reaction components – ethylene, and air or O_2 – react with the acidic (HCl) aqueous catalyst solution in a titanium or lined bubble column reactor.

two versions are possible:

1. single-step process, reaction and regeneration in same reactor, i.e., O_2 and C_2H_4 keep catalyst in a stationary, active state

2. two-step process, reaction and regeneration in separate reactors with exchange of catalyst charge (active and inactive form)

Two versions of the process were developed at the same time:

1. Single-step process – in which the reaction and regeneration are conducted simultaneously in the same reactor. O_2 is used as the oxidizing agent.

2. Two-step process – in which the reaction and regeneration take place separately in two reactors. In this case, air can be used for the oxidation.

characteristics of single-step process:

incomplete C_2H_4 conversion requires recycling of C_2H_4, i.e., C_2H_4 and O_2 must be largely free of inerts

In the single-step process, ethylene and O_2 are fed into the catalyst solution at 3 bar and $120-130\,°C$, where $35-45\%$ of the ethylene is converted. The resulting heat of reaction is utilized to distill off acetaldehyde and water from the catalyst solution, which must be recycled to the reactor. In this way, around $2.5-3.0\ m^3\ H_2O$ per tonne acetaldehyde are recycled. It is necessary to use a pure O_2 and ethylene (99.9 vol%) feed to avoid ethylene losses which would otherwise occur on discharging the accumulated inert gas.

characteristics of two-step process:

total C_2H_4 conversion allows higher share of inert components, regeneration with air saves air fractionation plant, but catalyst circulation consumes more energy than gas recycling method

In the two-step process, ethylene is almost completely converted with the catalyst solution at $105-110\,°C$ and 10 bar. After reducing the pressure and distilling off an acetaldehyde/H_2O mixture, the catalyst solution is regenerated with air at $100\,°C$ and 10 bar in the oxidation reactor and then returned to the reactor. Since the O_2 in the air is largely removed, a residual gas with a high N_2 content is obtained which can be used as an inert gas. The advantages of total ethylene conversion and the

use of air contrast with the disadvantages of a greater investment arising from the double reactor system at higher pressure
and the catalyst circulation.

In both processes the aqueous crude aldehyde is concentrated
and byproducts such as acetic acid, crotonaldehyde and chlorine-
containing compounds are removed in a two-step distillation.
The selectivities are almost equal (94%).

isolation of acetaldehyde in two-step purification:

1. separation of low-boiling substances
   (CH_3Cl, C_2H_5Cl, CO_2, C_2H_4)
2. separation of high-boiling substances
   (CH_3COOH, $CH_3CH=CHCHO$,
   $ClCH_2CHO$)

Currently, the Wacker–Hoechst process accounts for 85% of
the worldwide production capacity for acetaldehyde.

7.3.2. Acetaldehyde from Ethanol

Acetaldehyde can be obtained by catalytic dehydrogenation of
ethanol:

acetaldehyde manufacture from ethanol
according to dehydrogenation principle with
two process variants:

$$C_2H_5OH \xrightarrow{\text{[cat.]}} CH_3CHO + H_2 \quad \left(\Delta H = +\dfrac{20 \text{ kcal}}{84 \text{ kJ}}/\text{mol} \right) \quad (38)$$

This is analogous to the manufacture of formaldehyde from
methanol (*cf.* Section 2.3.2.1). However, in contrast to methanol
manufacture, oxidation is not used as an alternative. Instead, two
modifications of the dehydrogenation are customary:

1. Dehydrogenation on silver, or preferably copper, catalysts

2. Oxidative dehydrogenation with silver catalysts in the presence
   of oxygen.

1. endothermic dehydrogenation favored
   over Cu metal with H_2 formation
2. exothermic oxidative dehydrogenation
   over Ag metal with H_2O formation

To 1:

The dehydrogenation of ethanol is usually done over Cu-
catalysts activated with Zn, Co, or Cr. One frequently applied
process originates from the Carbide & Carbon Corporation.

to 1:

ethanol dehydrogenation with process
characteristic:

limitation of conversion by temperature
regulation, unpressurized

The temperature is regulated to 270–300 °C so that the ethanol
conversion is limited to 30–50%. A selectivity to acetaldehyde
of 90–95% is attained. Byproducts include crotonaldehyde,
ethyl acetate, higher alcohols, and ethylene. The hydrogen
formed is pure enough to be used directly for hydrogenation.

To 2:

If the dehydrogenation of ethanol is conducted in the presence
of air or oxygen (*e. g.*, the Veba process), the concomitant
combustion of the hydrogen formed supplies the necessary heat
of dehydrogenation (oxidative dehydrogenation or autothermal
dehydrogenation):

to 2:

oxidative dehydrogenation with process
characteristic:

dehydrogenation as primary step, oxidation
as secondary step

$$C_2H_5OH + 0.5 \; O_2 \xrightarrow{\text{[cat.]}} CH_3CHO + H_2O$$

$$\left(\Delta H = - \frac{43 \; \text{kcal}}{180 \; \text{kJ}} \Big/ \text{mol} \right) \quad (39)$$

most economical operation:

thermoneutral, *i. e.*, measured amount of air brings heat evolution and consumption into equilibrium

In the industrial process, silver catalysts in the form of wire gauzes or bulk crystals are preferred. Ethanol vapors mixed with air at 3 bar and at $450 - 550 \,^\circ\text{C}$ are passed over the catalyst. The reaction temperature is contingent upon the amount of air used; a temperature is reached at which the heat of oxidation and heat consumption of the dehydrogenation compensate one another. Depending on the reaction temperature, $30 - 50\%$ of the ethanol is converted per pass with a selectivity of $85 - 95\%$. The by-products are acetic acid, formic acid, ethyl acetate, CO, and CO_2.

both process routes characterized by incomplete ethanol conversion $(30 - 50\%)$ and thus by involved distillation and recycle operations

In both process modifications, the acetaldehyde is separated from the unreacted alcohol and byproducts and purified in various washes and distillations. The recovered ethanol is recycled to the reaction.

In 1994 only about 13% of the acetaldehyde production capacity in Western Europe was based on this route. The last plant in the USA was shut down in 1983. In Japan, acetaldehyde is only produced by the direct oxidation of ethylene. A second plant for dehydrogenation of fermentation ethanol from sugar cane molasses was started up in Indonesia in 1995. They now have a total acetic acid capacity of 33 000 tonnes per year based on the oxidation of acetaldehyde to acetic acid.

7.3.3. Acetaldehyde by C_3/C_4 Alkane Oxidation

less important, no longer practiced acetaldehyde manufacture from $C_3 - C_4$ alkanes with process characteristics:

radical, unselective oxidative degradation leads to numerous oxidation products

In a process developed by Celanese and operated in the USA from 1943 to 1980, propane or propane/butane mixtures are oxidized in the gas phase to produce mixtures containing acetaldehyde. The noncatalytic, free-radical reaction takes place at $425 - 460 \,^\circ\text{C}$ and $7 - 20$ bar. One version of the Celanese butane oxidation process is conducted in the liquid phase (*cf.* Section 7.4.1.2). Either oxygen or air can be used as the oxidizing agent. About $15 - 20\%$ of the hydrocarbon is completely oxidized. The remaining complex reaction mixture contains, besides acetaldehyde, mainly formaldehyde, methanol, acetic acid, *n*-propanol, methyl ethyl ketone, acetone, and numerous other oxidation products.

complex product separation involves high investment and operating costs, essentially determining the profitability

After decomposing any peroxide, the separation of the oxidized products takes place in an involved, and therefore costly, combination of extractions and distillations.

7.4. Secondary Products of Acetaldehyde

Acetaldehyde is an important precursor for the manufacture of many major organic chemicals, e.g., acetic acid, peracetic acid, acetic anhydride, ketene/diketene, ethyl acetate, crotonaldehyde, 1-butanol, 2-ethylhexanol, pentaerythritol, chloral, pyridines, and many others.

From the following table, it is clear that the manufacture of acetic acid and its anhydride accounts for the largest share of the acetaldehyde production, although this percentage has been decreasing since the beginning of the 1980s. It can also be seen that the manufacture of 1-butanol and 2-ethylhexanol from acetaldehyde has been superceded by propene hydroformylation (cf. Section 6.1).

uses of acetaldehyde:

almost no direct use, only secondary products such as

CH_3COOH, $(CH_3CO)_2O$, $H_2C=C=O$, $CH_3COOC_2H_5$, $CH_3CH=CHCHO$, $n\text{-}C_4H_9OH$, $C_4H_9CH(C_2H_5)CH_2OH$, etc.

trends in acetaldehyde use:

decrease in fraction for $AcOH/Ac_2O$: instead, relative increase of remaining uses

Table 7-6. Acetaldehyde use (in %).

Product	Western Europe		USA		Japan	
	1980	1995	1980	1993	1978	1993
Acetic acid	60	52	} 63	–	} 69	29
Acetic anhydride	11	8		–		–
1-Butanol	0.2	–	–	–	2	–
2-Ethylhexanol	0.8	–	–	–	–	–
Ethyl acetate[1]	18	25	–	22	16	45
Miscellaneous (e.g., pyridines, pentaerythritol, peracetic acid)	10	15	37	78	13	26
Total use (in 10^6 tonnes)	0.64	0.52	0.36	0.15	0.59	0.35

[1] + small amount isobutyl acetate

The fraction of acetaldehyde used for each product differs from country to country. For example, in Western Europe, the USA, and Japan, ethyl acetate is still made from acetaldehyde. In Japan and the USA, a fraction of the acetaldehyde is used to make peracetic acid (cf. Section 7.4.1.1). Acetaldehyde is also used in Japan as paraldehyde for the co-oxidation of p-xylene to terephthalic acid (cf. Section 14.2.2).

acetaldehyde use dependent on country:

Japan: ethyl acetate
            peracetic acid
            co-oxidation of p-xylene

W. Europe   ethyl acetate
            acetic anhydride

USA: ethyl acetate
            peracetic acid

7.4.1. Acetic Acid

Acetic acid, one of the most important aliphatic intermediates, ranked eighth in the number of tonnes produced in the USA in 1992. It was also the first carboxylic acid used by man.

In 1994, the worldwide production capacity for acetic acid was about 6.0×10^6 tonnes per year, of which 1.7, 1.5, 0.80, and 0.33×10^6 tonnes per year were located in the USA, Western Europe, Japan, and the CIS, respectively.

AcOH importance, e.g., in USA 1992 behind

$ClCH_2CH_2Cl$, $H_2C=CHCl$,

CH_3OH, $HCHO$, $H_2C\overset{\diagdown}{\underset{O}{-}}CH_2$,

$HOCH_2CH_2OH$, $(CH_3)_3COCH_3$

in 8th place among the aliphatic intermediates (in tonnes produced)

acetic acid production (in 10^6 tonnes):

	1990	1992	1995
USA	1.71	1.63	2.12
W. Europe	1.26	1.16	1.47
Japan	0.46	0.45	0.57

AcOH named according to origin:

synthetic acetic acid
spirit vinegar
wood vinegar

AcOH manufacture by three basic methods:

1. unchanged C skeleton by CH_3CHO oxidation

2. longer chain C skeleton by oxidative degradation

2.1. *n*-butane (Celanese, Hüls, UCC)
2.2. *n*-butenes (Bayer, Hüls)
2.3 light petrol (BP, British Distillers)

3. C_1 + C_1 by CH_3OH carbonylation

3.1. Co catalysis, high pressure (BASF, Celanese)
3.2. Rh catalysis, normal or low pressure (Monsanto)

acetic acid production capacity worldwide (in %):

	1988	1989	1994
CH_3OH carbonylation	47	50	58
CH_3CHO oxidation	27	27	23
C_2H_5OH dehydrogenation/ oxidation	6	7	4
butane/naphtha oxidation	7	12	9
other methods	13	4	6

Production figures for acetic acid in these countries are listed in the adjacent table.

These figures refer only to synthetic acetic acid. In some countries, small amounts of acetic acid are still obtained from the fermentation of ethanol-containing substrates (spirit vinegar, generally used in food) and by wood distillation (wood vinegar). In addition, there is another manufacturing route to acetic acid based on sugar cane molasses (*cf.* Section 7.3.2). In the following, acetic acid will refer only to synthetic acetic acid.

For many years synthetic acetic acid was mainly manufactured from acetaldehyde. At the beginning of World War I, commercial oxidation processes were operated by Hoechst, Wacker, and Shawinigan.

Thus acetic acid was closely coupled to acetaldehyde manufacturing processes, and underwent the same change in feedstock from acetylene to ethylene.

The economic necessity of utilizing the lower hydrocarbons led to the development of oxidation processes for light paraffins in the USA, England, and Germany by BP, Celanese, British Distillers, Hüls, and UCC. The utilization of C_4 olefins was studied by Bayer and Hüls. However, only a few of these processes were ever used commercially.

Although the first attempts to carbonylate methanol were made in the 1920s, an industrial process was first proposed by BASF much later. A few years ago, this method was given new momentum by Monsanto's Rh-catalyzed process.

Acetic acid production in Western Europe can be taken as a good example of competing feedstocks. In 1979, 62% of the acetic acid came from acetaldehyde. This percentage had decreased to 28% by 1995 under pressure from methanol carbonylation, which accounted for 55% of the production and continues to grow. The basic processes listed in the adjacent table currently contribute to the world production capacity of acetic acid.

7.4.1.1. Acetic Acid by Oxidation of Acetaldehyde

Oxidation of acetaldehyde with air or O_2 to acetic acid takes place by a radical mechanism with peracetic acid as an intermediate.

The acetyl radical, formed in the initiation step, reacts with O_2 to make a peroxide radical which leads to peracetic acid. Although peracetic acid can form acetic acid by homolysis of the peroxy group, it is assumed that the peracetic acid preferentially reacts with acetaldehyde to give α-hydroxyethyl peracetate, which then decomposes through a cyclic transition state to two moles of acetic acid:

(40)

2 CH_3COOH

Peracetic acid can also be obtained as the main product when the oxidation is conducted under mild conditions; that is, preferably without catalyst in a solvent such as ethyl acetate at −15 to 40 °C and 25–40 bar, and with air.

Commercial plants using this technology for the manufacture of peracetic acid are operated by UCC in the USA, by Daicel in Japan, and by British Celanese in England.

An industrial method for the production of peracetic acid from acetic acid and H_2O_2 is described in Section 11.1.1.2.

If a redox catalyst is used for the oxidation of acetaldehyde to acetic acid, it not only serves to generate acetyl radicals initiating the oxidation but also to accelerate the decomposition of peracetic acid. The resulting acetoxy radical causes chain branching.

The usual catalysts are solutions of Co or Mn acetates in low concentration (up to 0.5 wt% of the reactant mixture).

Today, the oxidation is usually done with oxygen; one example is the Hoechst process, which operates continuously at 50–70 °C in oxidation towers made of stainless steel (bubble columns) with acetic acid as solvent. Temperatures of at least 50 °C are necessary to achieve an adequate decomposition of peroxide and thus a sufficient rate of oxidation. The heat of reaction is removed by circulating the oxidation mixture through a cooling system. Careful temperature control limits the oxidative decomposition of acetic acid to formic acid, CO_2, and small amounts of CO and H_2O. Acetic acid selectivity reaches 95–97% (based on CH_3CHO).

mechanism of peracetic acid formation:

principle of CH_3CHO oxidation:

radical reaction with peracetic acid as intermediate in the presence of redox catalysts

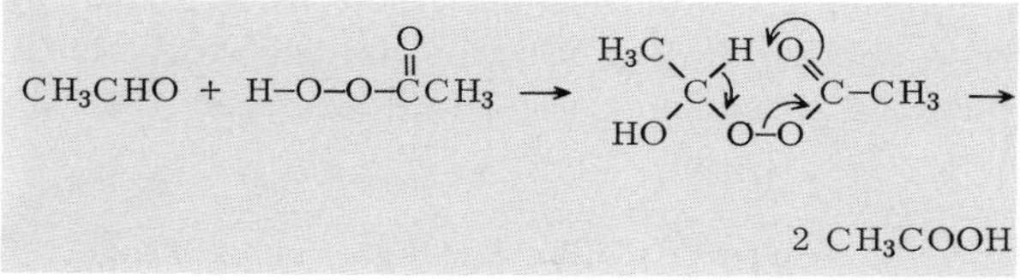

mode of operation of redox catalysts:

1. formation of acetyl radicals

2. cleavage of peracetic acid

two modifications of CH_3CHO oxidation process:

1. O_2 as oxidizing agent with higher O_2 material costs, due to air fractionation, but advantage of absence of inerts

2. air as oxidizing agent with cost advantages over O_2, but disadvantage of off-gas washings to remove CH_3CHO/CH_3COOH in air effluent

both processes have similar selectivities and the same byproducts

workup of AcOH:

distillation with byproducts as entraining agents leads directly to anhydrous AcOH

As an alternative to acetaldehyde oxidation with oxygen, Rhône-Poulenc/Melle Bezons developed a process using air as the oxidizing agent. Selectivities similar to those with pure oxygen were obtained. However, the greater amount of inert gas in the air oxidation is disadvantageous since it contains acetaldehyde and acetic acid which must be removed by washing.

Byproducts of both oxidation processes are very similar. Besides CO_2 and formic acid, they include methyl acetate, methanol, methyl formate, and formaldehyde. These are separated by distillation. Anhydrous acetic acid is obtained directly, since the byproducts act as entraining agents to remove water.

7.4.1.2. Acetic Acid by Oxidation of Alkanes and Alkenes

common characteristics of nonselective hydrocarbon oxidation to acetic acid:

radical, strongly exothermic degradation of $C_4 - C_8$ hydrocarbons to C_4, C_3, C_2, and C_1 fragments

C_4 to C_8 hydrocarbons are the favored feedstocks for the manufacture of acetic acid by oxidative degradation. They can be separated into the following groups and process modifications:

1. *n*-Butane (Hoechst Celanese, Hüls, UCC)
2. *n*-Butenes (Bayer, with *sec*-butyl acetate as intermediate; Hüls directly)
3. Light gasoline (BP, British Distillers)

To 1:

process examples for *n*-butane oxidation:

Celanese and UCC in USA, Hüls in W. Germany operated for many years, but have now shut down; Hoechst Celanese was back on line in 1989

In 1982, about 31% of the acetic acid capacity in the USA was still based on the liquid-phase oxidation of *n*-butane as practiced by UCC and Celanese. The UCC plant has been shut down, and following an accident the Hoechst Celanese plant resumed operation in 1989. In 1991, their production of ca. 230000 tonnes accounted for about 14% of the total acetic acid production in the USA. Formerly, the UCC plant in Texas produced about 225000 tonnes acetic acid, 36000 tonnes methyl ethyl ketone, 23000 tonnes formic acid, and 18000 tonnes propionic acid annually by the uncatalyzed oxidation of *n*-butane with oxygen in a bubble column at $15-20$ bar and $180\,°C$ using the liquid oxidation products as reaction medium. More extensive oxidative degradation was controlled by limiting the conversion to $10-20\%$.

characteristics of Celanese LPO process for *n*-butane:

catalytic liquid-phase oxidation with simplified workup due to recycling of byproducts

The Celanese LPO process (**L**iquid-**P**hase **O**xidation) operates at $175°C$ and 54 bar with Co acetate as catalyst. After separating the acetic acid, part of the byproduct mixture is recycled to the process and oxidized either to acetic acid or completely. This simplifies the product workup. In addition to acetic acid, acetaldehyde, acetone, methyl ethyl ketone, ethyl acetate, formic acid, propionic acid, butyric acid, and methanol are isolated. On demand, amounts of methyl ethyl ketone up to 17% of the plant capacity can be manufactured, though at the cost of acetic acid production.

From 1965 to 1984, a plant in The Netherlands with an acetic acid capacity of 130 000 tonnes per year (after several expansions) used the Celanese LPO process.

For several years, Hüls also carried out *n*-butane oxidation — without a catalyst — in a commercial plant with a capacity of 20 000 tonnes per year. The oxidation took place at 60–80 bar and 170–200 °C with air or O_2 enriched air (ca. 30% O_2) in a liquid phase consisting of crude acetic acid. *n*-Butane conversion was limited to about 2% to prevent secondary reactions. The main product was in fact acetic acid (about 60% selectivity), but numerous byproducts such as acetone, methyl ethyl ketone, methyl and ethyl acetate, and smaller amounts of formic and propionic acid, were also formed. Processing required a multistep distillation unit with, for example, 14 columns at normal pressure and others at reduced or excess pressure.

characteristics of Hüls butane process:

uncatalyzed, nonselective liquid-phase oxidation with low butane conversion, necessitating a high butane recycle

large number of byproducts requires multistep distillation and high technology with added corrosiveness of product

To 2:

Hüls developed yet another acetic acid process based on the oxidation of *n*-butenes. This was operated for a long time in a pilot plant, though an industrial-scale unit was never built. In this method, butene is oxidized at slight excess pressure at 200 °C in a liquid phase consisting essentially of crude acetic acid. Titanium and tin vanadate are employed as catalysts. For various reasons, including the extremely narrow explosion limits, this oxidation takes place in the presence of a large amount of steam. The very dilute acetic acid produced must be concentrated to a crude acid consisting of 95% acetic acid, which requires a great deal of energy. The crude acid selectivity reaches 73% at 75% butene conversion.

characteristics of Hüls *n*-butene process:

catalytic liquid-phase oxidation of butene in the presence of large amounts of H_2O, requires energy-consuming AcOH concentration

Bayer uses a different two-stage process for its liquid-phase oxidation of *n*-butenes to acetic acid. After removing butadiene and isobutene from the C_4 cracking fraction (*cf.* Section 3.3.2), a mixture of 1-butene and *cis*- and *trans*-2-butene remains. This is converted into 2-acetoxybutane, *i. e., sec*-butyl acetate:

characteristics of Bayer *n*-butene process:

two-step process with $C_2H_5CH(OAc)CH_3$ as intermediate

1st step:

proton-catalyzed AcOH addition to *n*-butene mixtures to form 2-acetoxybutane

$$CH_3CH_2CH=CH_2$$

$$CH_3CH=CHCH_3 \xrightarrow[{[H^{\oplus}]}]{+ AcOH} \underset{OAc}{CH_3CH_2CHCH_3} \qquad (41)$$

(*cis* and *trans*)

acidic ion-exchanger with two catalytic functions:

isomerization of 1-butene to 2-butene and addition of AcOH to 2-butene

2nd step:

uncatalyzed oxidative cleavage of C_4 acetate to AcOH

The addition of acetic acid takes place at $100-120\,°C$ and $15-25$ bar over acidic ion-exchange resins with sulfonic acid groups. Due to the simultaneous isomerization of the *n*-butenes, only 2-acetoxybutane is formed. In the second stage, this is oxidized with air to acetic acid in a noncatalytic liquid-phase reaction at $200\,°C$ and 60 bar:

$$CH_3CH_2-\underset{\underset{O}{\overset{\|}{O}}{\overset{|}{O}CCH_3}}{CHCH_3} + 2\,O_2 \longrightarrow 3\,CH_3COOH \qquad (42)$$

isolation of AcOH by azeotropic and normal distillation and partial recycle to 1st process step

After working up the reaction mixture by an azeotropic and a normal distillation, a portion of the acetic acid is recycled to the *sec*-butyl acetate manufacture.

Selectivity to acetic acid reaches about 60%. The main byproducts are formic acid and CO_2. To date, no large-scale plant has been constructed.

To 3:

characteristics of Distillers-BP light gasoline oxidation:

uncatalyzed, nonselective liquid-phase oxidation of inexpensive C_4-C_8 feedstock to form C_1 to C_4 carboxylic acids and other byproducts

isolated products important for economics of process:

CH_3COOH
$HCOOH$
C_2H_5COOH
CH_3COCH_3

A process for the oxidation of crude oil distillates in the boiling range $15-95\,°C$ (roughly corresponding to a light gasoline in the C_4-C_8 region) was developed by British distillers in England. The uncatalyzed air oxidation is done in the liquid phase in a stainless steel reactor at $160-200\,°C$ and $40-50$ bar, and proceeds by a free-radical mechanism. The product from the oxdiation is separated by a two-stage distillation into starting material, more volatile byproducts, and an aqueous acid mixture. The acids and a small amount of water are first extracted from the mixture with a low boiling solvent such as isoamyl acetate. The organic phase is then separated into its individual components by distillation. Besides acetic acid, formic acid, propionic acid, and small amounts of succinic acid are formed. Enough acetone is produced to warrant its isolation. All byproducts contribute to the profitability of the process. Depending on the process conditions, 0.35 to 0.75 tonnes of byproducts can be obtained per tonne of acetic acid.

Distillers BP process with focal point in England; formerly however also plants in

France	(30 000 tonnes per year)
Japan	(15 000 tonnes per year)
CIS	(35 000 tonnes per year)

BP developed the Distillers process further and brought it to maturity, particularly in England. In 1996 the capacity there was 220 000 tonnes per year acetic acid and about 18 000 tonnes per year propionic acid. This is the last plant using the BP process.

7.4.1.3. Carbonylation of Methanol to Acetic Acid

BASF prepared the a way for new acetic acid process with their work on the catalytic conversion of CO and H_2. It was discovered around 1913 that methanol, the primary reaction product from synthesis gas, could be carbonylated to acetic acid. This route became economically feasible after 1920 when methanol was available in commercial quantities. Other firms, including British Celanese after 1925, then began working intensively with the carbonylation reaction, which takes place by the following equation:

carbonylation of methanol to acetic acid – principle of manufacture:

$C_1 + C_1$ synthesis

two important prerequisites for industrial development:

1. inexpensive synthetic CH_3OH available in sufficient amounts

$$CH_3OH + CO \xrightarrow{\text{[cat.]}} CH_3COOH \left(\Delta H = - \frac{33 \text{ kcal}}{138 \text{ kJ}} \big/ \text{mol} \right) \quad (43)$$

Corrosion problems, present from the beginning, were only solved at the end of the 1950s with the use of newly developed, highly resistant Mo-Ni alloys (Hastelloy®). In 1960, the first small plant was brought on line by BASF. In the industrial process (BASF), methanol – alone or mixed with dimethyl ether and a small amount of H_2O – is reacted with CO in the presence of CoI_2 in the liquid phase at 250 °C and 680 bar. The cation and anion act independently of one another in the reaction mechanism. It is assumed that cobalt iodide initially reacts to form tetracarbonylhydridocobalt and hydrogen iodide, which is then converted into methyl iodide with methanol:

2. control of corrosion problems and development of economical technology

first industrial plant (BASF) using process principle:

CoI_2-catalyzed liquid-phase high pressure carbonylation of CH_3OH or mixtures with $(CH_3)_2O$

$$2 \text{ CoI}_2 + 3 \text{ H}_2\text{O} + 11 \text{ CO} \rightarrow 2 \text{ HCo(CO)}_4 + 4 \text{ HI} + 3 \text{ CO}_2 \quad (44)$$

$$\text{HI} + \text{CH}_3\text{OH} \rightleftarrows \text{CH}_3\text{I} + \text{H}_2\text{O} \quad (45)$$

Tetracarbonylhydridocobalt and methyl iodide react to form the important intermediate $CH_3Co(CO)_4$ which, after CO insertion, hydrolyzes to form acetic acid and regenerate tetracarbonylhydridocobalt:

catalytic mechanism:

CoI_2 supplies the components for the precursors $HCo(CO)_4$ and CH_3I, which synthesize the essential intermediate $CH_3Co(CO)_4$

$$\text{CH}_3\text{I} + \text{HCo(CO)}_4 \rightleftarrows \text{CH}_3\text{Co(CO)}_4 + \text{HI} \quad (46)$$

$$\text{CH}_3\text{Co(CO)}_4 \xrightarrow{+\text{CO}} \text{CH}_3\text{COCo(CO)}_4 \xrightarrow{+\text{H}_2\text{O}} \text{CH}_3\text{COOH} \quad (47)$$
$$+ \text{HCo(CO)}_4$$

process characteristics:

virtual no-loss catalyst processing, high AcOH selectivity with CO_2 formation as consequence of process, and numerous by-products present in small amounts, *e.g.*, C_2H_5OH, C_2H_5COOH, C_2H_5CHO, C_3H_7CHO, C_4H_9OH.

Thus both catalyst components are available for a new reaction sequence. In the industrial process, cobalt and iodine can be almost completely recovered. The selectivities to acetic acid are 90% (based on CH_3OH) and 70% (based on CO). The by-products (4 kg per 100 kg acetic acid) include a multitude of chemicals. CO_2 is regarded as a coproduct (*cf.* eq 44). After a five-column distillation of the crude product, 99.8% acetic acid is obtained.

used industrially in two plants (W. Germany, USA)

In 1983, there were two plants using the BASF process: one in West Germany, with a capacity of 50 000 tonnes per year, and a Borden plant in the USA (since shut down) with a capacity of 65 000 tonnes per year.

modification of $C_1 + C_1$ synthesis for AcOH (Monsanto development):

Rh/I_2-catalyzed liquid-phase carbonylation of CH_3OH at normal or slight excess pressure

Around the mid 1960s, Monsanto discovered that rhodium combined with iodine was a considerably more active catalyst system for methanol carbonylation than cobalt iodide. As with CoI_2 it is assumed that the active species is a metal carbonyl complex with methyl ligands, in the form of $[CH_3-Rh(CO)_2I_3]^{\ominus}$. By insertion of CO into the CH_3-Rh bond, an acetylrhodium complex is formed. This can go on to react, for example, by methanolysis to form acetic acid and regenerate the initial complex:

$$
[CH_3\!-\!Rh(CO)_2I_3]^{\ominus} \xrightarrow{\;+CO\;} [CH_3\overset{\displaystyle O}{\underset{\displaystyle \|}{C}}\!-\!Rh(CO)_2I_3]^{\ominus}
$$

$$
\xrightarrow{\;+CH_3OH\;} CH_3\overset{\displaystyle O}{\underset{\displaystyle \|}{C}}OH \; + \; [CH_3Rh(CO)_2I_3]^{\ominus} \tag{48}
$$

catalytic mechanism:

analogous to CoI_2 high pressure process, in this case CH_3–Rh–CO complex is the active species.

In 1970, the first industrial plant went on stream in Texas City, with a capacity of 150 000 tonnes per year acetic acid.

In the years following, the Monsanto process (now BP) was preferred for new acetic acid plants, so that by 1991 about 55% of the acetic acid capacity worldwide was based on this technology. In Japan, the first plant to use Monsanto technology — a 150 000 tonne-per-year unit by Daicel Chemical — started operation in 1980; in Western Europe, the first was a 225 000 tonne-per-year unit brought up by Rhône-Poulenc in 1981.

process characteristics:

high selectivities with $CO_2 + H_2$ as main byproducts from CO + H_2O conversion, as well as special precautions to ensure virtually no-loss catalyst recycle due to high Rh price

In the industrial process, methanol and CO react continously in the liquid phase at 150–200 °C under a slight pressure of up to 30 bar to form acetic acid with selectivities of 99% (based on CH_3OH) and over 90% (based on CO). The main byproducts are CO_2 and H_2 from the water gas shift reaction. In a modern commerical unit, the fully automated process control system includes production and regeneration of the catalyst system, since a low-loss rhodium recycle is very important to the profitability of the process.

A new process concept has been employed by BP in a plant in England since 1989. By using a rhodium-catalyzed carbonylation of methanol/methyl acetate mixtures, yields of acetic acid and acetic anhydride (*cf.* Section 7.4.2) of between 40/60 and 60/40 (which correspond to market demand) can be produced.

7.4.1.4. Potential Developments in Acetic Acid Manufacture

The availability and pricing of raw materials for the manufacture of acetic acid have − as shown in the case of acetaldehyde, a possible precursor − been subjected to considerable change in the last few years.

changes in the raw materials favor AcOH synthesis from C_1 components CH_3OH and CO from fossil and, in the future, renewable sources

Monsanto methanol carbonylation has proven to be the most successful process on economic grounds, mainly because of the price increases for ethylene following both oil crises. Low-loss recycling of rhodium is necessary due to its high price and its established use in hydroformylation (*cf.* Section 6.1.3) and in automobile exhaust catalysts.

Rh broadly useful catalyst, but with limited availability (world production ca. 8 tonnes Rh compared to about ten times as much Pt or Pd) compels most loss-free Rh use possible

For this reason, research on other catalyst systems for methanol carbonylation is very relevant.

alternate catalyst developments for carbonylation based on other metals, *e. g.,* Ni, Co, Mn with promoters of Pt metals, remain relevant

Thus BP has developed a new catalyst based on iridium acetate with a promoter system that has a higher space–time yield. Details have not yet been disclosed. BP is planning the first plants in Korea and China.

The increasing interest in C_1 chemistry worldwide, *i. e.*, the use of synthesis gas for making valuable intermediates from the versatile CO/H_2 gases of different origins, has also provided new possibilities for the manufacture of acetic acid.

A direct, heterogeneously catalyzed conversion of synthesis gas with a rhodium system to a mixture of oxygenated C_2 compounds such as acetic acid, acetaldehyde, and ethanol was introduced by UCC in the mid 1970s.

$AcOH/CH_3CHO/C_2H_5OH$ manufacture from H_2/CO on Rh/carrier by UCC and Hoechst; a more direct path than the CH_3OH/CO route of economic interest

Hoechst also developed a fixed-bed catalyzed direct synthesis of the three oxygenated C_2 compounds from synthesis gas. For example, at 290 °C and 50 bar, a total selectivity to the $O-C_2$ compounds of up to 80 % with a space time yield of 200 g/L·h can be attained with a $Rh-Yb-Li/SiO_2$ catalyst. With an additional oxidation of the crude product, and after extraction and separation of propionic and butyric acids, acetic acid is obtained as the main product.

This process has been piloted. It could have economic advantages over the Monsanto process, which first must go through the energy- and capital-intensive manufacture of methanol.

HCOOCH$_3$ as third C$_1$-based route for AcOH with established consumption structure for HCOOH derivatives of interest

Methyl formate is another product derived from synthesis gas which could, with its isomerization to acetic acid, be of commercial interest if there were clear economic advantages over the Monsanto carbonylation route (*cf.* Section 2.3.3), or when a company-specific use for derivatives of formic acid exists.

AcOH manufacture from inexpensive lower alkanes with unselective oxidation and extensive workup is strongly influenced by the value of byproducts

As always, inexpensive C$_4$–C$_8$ fractions and butane from natural gas appear to be reasonable raw materials for acetic acid manufacture. However, capital costs for processing the product mixture in the production plants are high, and the great variety of byproducts leads to a stronger dependence on market demands. Thus it is understandable that several firms operate acetic acid plants using different processes – *i.e.,* different feedstocks – side by side.

7.4.1.5. Use of Acetic Acid

uses of AcOH:

1. in acetic acid esters as monomers (vinyl acetate), for artificial silk manufacture (cellulose acetates) and as solvent for paints and resins (methyl, ethyl, isopropyl, butyl acetate)

As can be clearly seen from the breakdown of consumption figures worldwide and in the USA, Western Europe, and Japan, most acetic acid is used for the manufacture of various acetic acid esters (see Table 7–7):

Table 7-7. Use of acetic acid (in wt%).

	World		USA		Western Europe		Japan	
	1988	1995	1984	1993	1985	1994	1984	1993
Vinyl acetate	40	43	52	59	35	33	26	26
Cellulose acetate	4	} 13	16	–	8	6	18	17
Acetic acid esters	9		11	11	13	11	5	7
Acetic anhydride, acetanilide, acetyl chloride, acetamide	} 11	} 8	} 4	} 15	} 21	} 20	} 4	} 3
Solvent for terephthalic acid and terephthalic dimethyl ester manufacture	8	12	8	10	6	8	11	16
Chloroacetic acids	4	3	2	1	11	10	5	4
Others (*e.g.*, Al, NH$_4$, alkali metal acetates, etc.)	23	21	6	4	4	12	20	27
Total use (in 10^6 tonnes)	4.36	5.67	1.41	15.6	0.94	1.25	0.36	0.48

In most countries, the order of the esters is similar. Vinyl acetate is the most important, usually followed by cellulose acetate.

n-Butyl, isobutyl, and methyl acetate are also important esters which, like ethyl acetate, are preferred solvents for paints and resins.

Cellulose acetate can have a acetate content of $52-62.5$ wt%, depending on the degree of acetylation of the three esterifiable OH groups on each $C_6H_{10}O_5$ unit. Cellulose acetate is used in the manufacture of fibers, films, and paints.

The salts of acetic acid, for instance Na, Pb, Al, and Zn acetate, are auxiliary agents in the textile, dye, and leather industries, and in medicine.

2. in inorganic salts, as aid in dye and clothing industries, medicine

Chloroacetic acids are also important acetic acid derivatives. The most significant of these is monochloroacetic acid. Production capacities for monochloroacetic acid in 1991 in Western Europe, Japan, and the USA were 243 000, 37 000 and 36 000 tonnes per year, respectively. Production figures are listed in the adjacent table. Chloroacetic acid is manufactured by the chlorination of acetic acid in the liquid phase at a minimum of 85 °C and up to 6 bar, usually without catalyst but often with addition of initiators such as acetic anhydride or acetyl chloride. Of less importance is the hydrolysis of trichloroethylene to monochloroacetic acid (*cf.* Section 9.1.4). Monochloroacetic acid is an important starting material for the manufacture of carboxymethylcellulose and an intermediate for pesticides, dyes, and pharmaceuticals. The Na salt of trichloroacetic acid is used as a herbicide.

3. as chloroacetic acids for organic syntheses

most important chloroacetic acid:

$ClCH_2COOH$ by liquid-phase chlorination of CH_3COOH used for carboxymethylcellulose

H   H
\|/   +NaOH   \|/
Y   ───────→   Y
/   −NaCl   /
CH_2OH CH_2OCH_2COOH

monochloroacetic acid production (in 1000 tonnes):

	1987	1989	1991
W. Europe	199	214	218
Japan	35	34	31
USA	17	30	30

Industrially significant secondary products of acetic acid which are also intermediates (such as acetic anhydride and ketene) will be dealt with in the following sections.

4. precursor for anhydrides:

    ketene and acetic anhydride

Acetic acid is also used as a solvent for liquid-phase oxidations, as for example in the oxidation of *p*-xylene to terephthalic acid or dimethyl terephthalate (*cf.* Section 14.2.2).

5. solvent for liquid-phase oxidations

In the future, acetic acid could be used as methyl acetate for the production of ethanol in a new Halcon process. Methyl acetate is converted to methanol and ethanol in a gas-phase hydrogenolysis with high conversion and selectivity. The methanol is used either for carbonylation to acetic acid, or ester formation. Thus methyl acetate is also interesting as an intermediate for methanol homologation (*cf.* Section 8.1.1) and for carbonylation to acetic anhydride (*cf.* Section 7.4.2).

6. potentially as intermediate for CH_3OH homologation by CH_3OAc hydrogenolysis

7.4.2. Acetic Anhydride and Ketene

acetic acid forms two anhydrides:

1. intermolecular:
   acetic anhydride (Ac_2O) of greater importance, and
2. intramolecular:
   ketene (diketene)

Ac_2O production (in 1000 tonnes):

	1987	1989	1990
USA	740	778	830
W. Europe	315	325	330
Japan	152	145	144

Ac_2O manufacture by three routes:

1. oxidative dehydration of CH_3CHO
2. 'acetylation' of AcOH with ketene
3. carbonylation of CH_3OCH_3 or CH_3OAc

to 1:

process principle:
homogeneously catalyzed liquid-phase oxidation of CH_3CHO with Cu/Co acetate with formation of AcOH as coproduct

Acetic anhydride, the intermolecular, and ketene, the intramolecular, anhydride of acetic acid are very closely related to acetic acid in their manufacture, use, and importance.

Production figures for acetic anhydride in several countries are summarized in the adjacent table. In 1994, the world production capacity was about 1.9×10^6 tonnes per year, with about 890000, 520000, and 160000 tonnes per year in the USA, Western Europe, and Japan, respectively.

Ketene and its dimer (diketene) are, apart from their conversion to acetic anhydride, of little importance. In 1995, the production capacity for ketene/diketene in the USA and Western Europe was about 340000 and 210000 tonnes per year, respectively.

Two different approaches have been used for manufacturing acetic anhydride. One process is a modified acetaldehyde oxidation; the other involves ketene as an intermediate. The ketene is usually obtained from the dehydration of acetic acid, and then reacted with acetic acid. Alternatively, ketene can be manufactured from acetone. In Western Europe in 1995, about 21% of acetic anhydride production was based on acetaldehyde oxidation, and 60% was based on the ketene/acetic acid reaction. In Japan the only production process in use is the ketene/acetic acid reaction.

New routes include the carbonylation of dimethyl ether and − already used commercially − of methyl acetate. In 1995 in the USA, about 45% of the total acetic anhydride produced was by acetaldehyde oxidation, with 22% from the ketene/acetic acid reaction and 10% from butane oxidation.

The modified acetaldehyde oxidation with air or O_2 was developed by Hoechst–Knapsack. A variation was developed by Shawinigan, in which a mixture of Cu and Co acetate was used as catalyst instead of Mn acetate. The reaction is run in the liquid phase at $50\,°C$ and 3–4 bar.

The primary acetyl radical is formed from acetaldehyde by abstraction of hydrogen (*cf.* Section 7.4.1.1). This is oxidized by Cu to the acetyl cation, which reacts with acetic acid to form acetic anhydride:

$$CH_3CHO \xrightarrow[-HX]{+X\cdot} CH_3\overset{\bullet}{\underset{\displaystyle O}{C}} \xrightarrow[-Cu^{\oplus}]{+Cu^{2\oplus}} CH_3\overset{\oplus}{\underset{\displaystyle O}{C}}$$

$$CH_3\underset{\displaystyle O}{C}^{\oplus} + CH_3COOH \longrightarrow \begin{array}{c} CH_3C\diagup\!\!\diagdown^O_{\diagdown O} \\ CH_3C\diagdown\!\!\diagup_O \end{array} + H^{\oplus} \qquad (49)$$

In a parallel reaction, peracetic acid is formed by addition of O_2 to the acetyl radical with subsequent abstraction of hydrogen from acetaldehyde. The acid serves to reoxidize $Cu^{\oplus}$ to $Cu^{2\oplus}$:

redox system regeneration by peracetic acid and creation of possibly explosive amounts of peracid

$$CH_3\underset{\displaystyle O}{C}-O-O-H + Cu^{\oplus} \longrightarrow Cu^{2\oplus} + CH_3\underset{\displaystyle O}{C}O^{\bullet} + OH^{\ominus} \qquad (50)$$

Acetic acid formed from this acetoxy radical is, therefore, an inevitable coproduct.

The water formed can now initiate the secondary reaction of acetic anhydride to acetic acid if it is not quickly removed from the equilibrium by distillation with an entraining agent such as ethyl acetate. With 95% conversion of acetaldehyde an optimal ratio (56:44) of anhydride to acid is attained.

process characteristics:

azeotropic H_2O removal hinders Ac_2O hydrolysis

examples of process:

Hoechst–Knapsack
Shawinigan

In the acetic anhydride manufacturing process involving the dehydration of acetic acid *via* ketene (Wacker process), acetic acid is first thermally dissociated into ketene and H_2O. The reaction takes place in the presence of triethyl phosphate at $700-750\,°C$ and reduced pressure:

to 2:

principles of two-step process:

1. catalytically regulated ketene equilibrium, back reaction hindered by:

$$CH_3COOH \underset{\text{[cat.]}}{\overset{}{\rightleftharpoons}} H_2C=C=O + H_2O$$

$$\left(\Delta H = +\begin{array}{c} 35 \text{ kcal} \\ 147 \text{ kJ} \end{array}/\text{mol} \right) \qquad (51)$$

To freeze the equilibrium, the resulting H_3PO_4 is neutralized with NH_3 or pyridine while still in the gas phase, and the cracked gas is quickly cooled.

1.1. neutralization of catalyst
1.2. quenching (lowering of temperature) of cracked gases
1.3. short residence time (0.2–3 s)

The higher boiling components (acetic anhydride, acetic acid, and water) are separated from the gaseous ketene in a system of graduated coolers. After removing the water, they are recycled to the cracking stage. The conversion of acetic acid is about 80%. The ketene selectivity exceeds 90% (based on CH_3COOH) at an acetic acid conversion of $70-90\%$. The ketene purified in this way is fed directly into acetic acid (*e.g.,* in the Wacker process,

2. Ac_2O manufacture by uncatalyzed exothermic AcOH addition to ketene *e.g.,* in a liquid-seal pump with two functions:
2.1. production of reduced pressure for ketene manufacture
2.2. reactor for siphoned AcOH with ketene

in a liquid-seal pump), and converted at $45-55\,°C$ and reduced pressures of $0.05-0.2$ bar into acetic anhydride:

$$H_2C{=}C{=}O + CH_3COOH \longrightarrow (CH_3CO)_2O$$

$$\left(\Delta H = -\frac{15 \text{ kcal}}{63 \text{ kJ}} \Big/ \text{mol} \right) \qquad (52)$$

In this stage, selectivity approaches 100%. This acetic anhydride process has the advantage that ketene can be obtained on demand as an intermediary product. Furthermore, acetic acid from the most economical process can be used, including acetic acid from acetylation reactions with acetic anhydride.

advantages of process:

1. isolable ketene intermediate
2. inexpensive AcOH utilizable *e. g.*, from other processes
3. recovered acetic acid from acetylation with Ac₂O usable

alternative ketene manufacture based on:

homogeneously-catalyzed irreversible de-methanation of $(CH_3)_2CO$

Ketene can also be obtained in a process independent of acetic acid from acetone by thermolysis at $600-700\,°C$ in the presence of a small amount of CS_2:

$$CH_3COCH_3 \xrightarrow{\;[\text{cat.}]\;} H_2C{=}C{=}O + CH_4 \qquad (53)$$

A ketene selectivity of 70–80% is attainable with an acetone conversion of 25%. Acetone cracking is only of minor industrial importance. A plant was once operated by Hofmann La Roche in the USA, but today ketene is only manufactured from acetone in Switzerland and the CIS.

to 3:

process principles:

homogeneous, Rh-salt-catalyzed carbonylation of CH_3OCH_3 or CH_3OAc

advantages of process:

manufacture solely from CO/H_2, and therefore on low-cost C-basis, possible

New routes for the manufacture of acetic anhydride are the homogeneously catalyzed carbonylation of dimethyl ether and methyl acetate as developed by Hoechst and Halcon, respectively. Starting from methanol or its further carbonylation to acetic acid, both processes are based fundamentally on synthesis gas, and thus also on coal, as a feedstock. The carbonylation of methyl acetate has been developed further industrially, and is preferably run at $150-220\,°C$ and $25-75$ bar in the presence of a rhodium salt catalyst system with promoters such as chromium hexacarbonyl/picoline:

$$\underset{\substack{\|\\ O}}{CH_3OCCH_3} + CO \xrightarrow{\;[\text{cat.}]\;} \underset{\substack{\|\ \|\\ O\ O}}{CH_3COCCH_3} \qquad (54)$$

industrial use:

Tennessee Eastman in Kingsport, USA, from CO/H_2 from Texaco coal gasification

In 1983, Tennessee Eastman started operation of a unit in the USA using the Halcon technology, which has since been expanded from its original capacity of 230 000 tonnes per year to 300 000 tonnes per year.

The main application of acetic anhydride is as an acetylating agent, above all in the manufacture of cellulose acetate, and in the manufacture of pharmaceuticals (*e. g.*, acetylsalicylic acid, acetanilide, etc.) and of intermediate products. Ketene is also used as an acetylating agent; the special case of acetylation of acetic acid to acetic anhydride is the most important example.

Ketene is also used in addition reactions (including the dimerization to diketene).

Diketene is obtained by ketene dimerization in trickle towers, into which a liquid stream of the diketene is introduced counter-current to ketene at $35-40\,°C$:

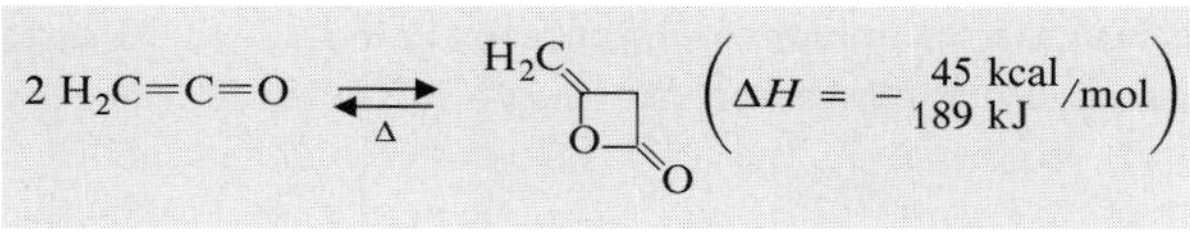

$$2\ H_2C{=}C{=}O \underset{\Delta}{\rightleftharpoons} \begin{array}{c} H_2C \\ \end{array}\ \left(\Delta H = -\ \begin{array}{c} 45\ \text{kcal} \\ 189\ \text{kJ} \end{array}/\text{mol}\right) \qquad (55)$$

The reaction goes almost to completion; small amounts of ketene are washed out of the off-gas with dilute acetic acid. The diketene must be redistilled with particular care since it is polymerized by both acids and bases.

Diketene is an important starting material for the manufacture of acetoacetic acid derivatives from the acid- or base-catalyzed addition − *i.e.*, opening of the β-lactone ring − of alcohols, ammonia, amines, hydrazines, etc.

Acetoacetic acid anilides with substitution on the phenyl ring are widely used as dye and pharmaceutical intermediates as are the pyrazolone derivatives obtained from diketene and substituted phenyl hydrazines.

Monomeric ketene forms β-lactones with aldehydes, *e. g.*, propiolactone is obtained from formaldehyde in the presence of $AlCl_3$:

$$HCHO\ +\ H_2C{=}C{=}O \xrightarrow{[AlCl_3]} \begin{array}{c} \square \\ O \quad O \end{array} \qquad (56)$$

Ketene reacts with crotonaldehyde to form a labile β-lactone which can stabilize itself by polyester formation (*cf.* Section 7.4.3).

use of Ac$_2$O and ketene:

1. for acetylation, *e. g.*, Ac$_2$O for acetylcellulose, for manufacture of pharmaceuticals, dyes, etc.; ketene for Ac$_2$O

Ac$_2$O demand for acetylcellulose (in %):

	USA	W. Europe	Japan
1991	80	71	81

2. for diketene formation

diketene also used for manufacture of acetoacetic acid derivatives (important intermediates for pharmaceuticals (pyrazolone, cumarine), dyes, insecticides)

3. for addition reactions, *e. g.*,
   with HCHO to propiolactone
   with CH$_3$CH=CHCHO to sorbic acid

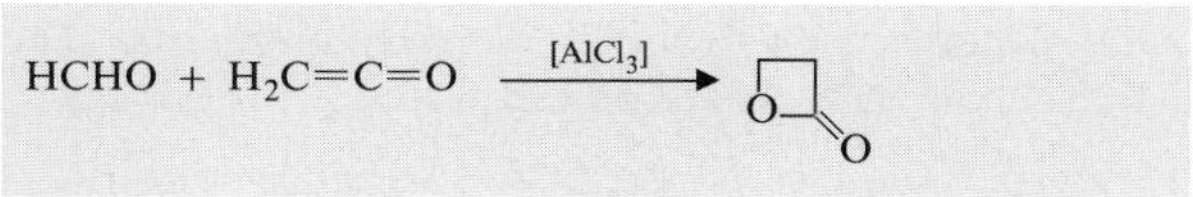

acetaldehyde 'homo-aldolization' to
β-hydroxybutyraldehyde (acetaldol)

also 'co-aldolization' *e.g.*, with HCHO to
pentaerythritol (*cf.* Section 8.3.1)

7.4.3. Aldol Condensation of Acetaldehyde and Secondary Products

Acetaldehyde, an aldehyde with active α-H atoms, can react in a characteristic way to form a dimer (acetaldol):

$$2 \; CH_3CHO \xrightarrow{\text{[cat.]}} CH_3CH(OH)CH_2CHO$$

$$\left(\Delta H = -\frac{27 \; kcal}{113 \; kJ}\middle/ mol\right) \tag{57}$$

principle of acetaldol manufacture:

selectivity of alkali-catalyzed CH_3CHO aldolization increased by limitation of conversion (neutralization with acid)

Analogous to an old IG Farben process, acetaldehyde is reacted at 20–25 °C in a tubular flow reactor with a residence time of several hours in the presence of dilute caustic soda. Acetaldehyde conversion in the aldolization reaction is restricted to 50–60% to limit resin formation and other side and secondary reactions. The reaction is stopped by adding acetic acid or H_3PO_4. After evaporating the unreacted aldehyde, acetaldol is obtained as a 73% aqueous solution. The selectivity to aldol approaches 85%. Crotonaldehyde is virtually the only byproduct.

two uses of acetaldol:

1. 1,3-butanediol
2. crotonaldehyde

Acetaldol can be converted into 1,3-butanediol by mild hydrogenation, avoiding the ready cleavage of water. Esters of 1,3-butanediol with long-chain carboxylic acids are used as plasticizers. In 1986, the world capacity for 1,3-butanediol was about 24000 tonnes per year.

to 1:

aldol is thermally unstable, therefore mild hydrogenation to
$CH_3CH(OH)CH_2CH_2OH$

to 2:

proton-catalyzed aldol dehydration forming reactive intermediate (crotonaldehyde)

However, most acetaldol is used in the manufacture of crotonaldehyde. The dehydration occurs readily in the presence of a small amount of acetic acid, which also prevents condensation reactions. Water is distilled off at 90–110 °C:

$$CH_3CH(OH)CH_2CHO \xrightarrow{\text{[H}^{\oplus}]} CH_3CH=CHCHO + H_2O \tag{58}$$

The crotonaldehyde is purified in a two-stage distillation. The selectivity is about 95% (based on CH_3CHO).

crotonaldehyde possesses various applications due to active double bond and aldehyde group, in secondary reactions such as:

1. selective hydrogenation of double bond to *n*-butyraldehyde, further hydrogenation to 1-butanol
2. additions to the double bond
3. oxidation of aldehyde group to crotonic acid

Crotonaldehyde was once important for partial hydrogenation to *n*-butyraldehyde, and further hydrogenation to 1-butanol. In larger industrialized countries today, both products are manufactured solely by hydroformylation of propene (*cf.* Section 6.1).

Only in countries like Brazil, which have plentiful ethanol based on agricultural products, can the acetaldehyde process for *n*-butanol still be maintained. The alcohol is obtained by hydrogenating crotonaldehyde with Cu/Cr catalysts at 170–180 °C or with nickel catalysts at a lower temperature.

2-Ethylhexanol is another secondary product of *n*-butyraldehyde. The alcohol is obtained from an aldol condensation followed by hydrogenation (*cf.* Section 6.1.4.3).

This multistep reaction sequence to 2-ethylhexanol through acetaldehyde and *n*-butyraldehyde is still used in countries like Brazil.

Apart from this, crotonaldehyde is gaining importance in the manufacture of *trans,trans*-2,4-hexadienoic acid (also known as sorbic acid). Crotonaldehyde reacts with ketene in an inert solvent (*e. g.*, toluene) at 30–60 °C to form a polyester. Soluble Zn or Cd salts of the higher carboxylic acids serve as catalysts. The polyester is then depolymerized in the second stage by thermolysis or proton-catalyzed hydrolysis to form sorbic acid:

to 1:

n-butyraldehyde/*n*-butanol manufacture, and secondary product 2-ethylhexanol, only still in countries with, *e. g.*, inexpensive ethanol from agriculture

to 2:

2.1. addition of ketene to crotonaldehyde to form sorbic acid

principle of two-step process:

Zn or Cd homogeneously catalyzed polyester formation

thermal or $H^{\oplus}$-catalyzed depolymerization to sorbic acid

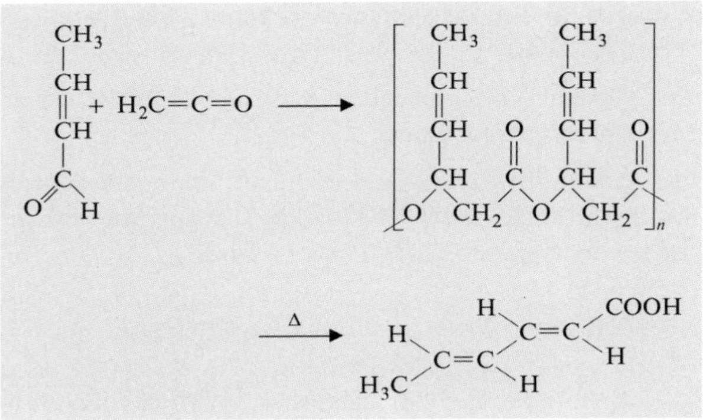

(59)

Sorbic acid and its potassium and calcium salts are of great importance in food preservation. It is manufactured in Germany by Hoechst (capacity in 1993, 7000 tonnes per year), and by several manufacturers in Japan. After the only industrial plant for sorbic acid in the USA was shut down by UCC in 1970, Monsanto began production (1977; capacity 4500 tonnes per year) using the crotonaldehyde/ketene technology. The UCC process was based on sorbic aldehyde (2,4-hexadienal), which was converted into sorbic acid by a silver-catalyzed oxidation. Sorbic aldehyde was manufactured by the aldol condensation of acetaldehyde in the presence of secondary amine salts. The world demand for sorbic acid is estimated at 19 000 tonnes per year, while the production capacity today is about 25 000 tonnes per year.

use of sorbic acid:

preservation of normal and luxury foods

industrial sorbic acid manufacture:

Germany: Hoechst
USA: Monsanto
Japan: several producers

2.2. addition of CH_3OH to crotonaldehyde
with subsequent hydrogenation to
3-methoxybutanol

principles of process:

base-catalyzed CH_3OH addition, with high
pressure, bubble-column hydrogenation to
stable alcohol

Another addition product of crotonaldehyde is obtained from the reaction with methanol. As the resulting 3-methoxybutanal readily cleaves to regenerate the starting materials, it is immediately hydrogenated to 3-methoxybutanol. In the manufacturing process, an excess of methanol in the presence of caustic soda at temperatures below 5 °C is used for the addition to crotonaldehyde. 3-Methoxybutanal is hydrogenated (without isolation) in the liquid phase in the presence of Ni or preferably Cu catalysts:

$$CH_3CH=CHCHO + CH_3OH \xrightarrow{[OH^{\ominus}]} CH_3CH-CH_2CHO$$
$$\underset{OCH_3}{|}$$

$$\xrightarrow[\text{[Ni]}]{+H_2} CH_3CH-CH_2CH_2OH$$
$$\underset{OCH_3}{|}$$

(60)

use of 3-methoxybutanol:

1. for hydraulic fluids
2. acetate as paint solvent

The selectivity to 3-methoxybutanol is about 90% (based on $CH_3CH=CHCHO$).

The ether–alcohol is a component of hydraulic fluids. Its acetate is an excellent solvent for paints.

to 3:

crotonaldehyde oxidation to crotonic acid

Crotonic acid, another secondary product of crotonaldehyde, is obtained by the liquid-phase oxidation of the aldehyde with air or O_2 at low temperature (20 °C) and 3−5 bar:

$$CH_3CH=CHCHO + 0.5 O_2 \rightarrow CH_3CH=CHCOOH$$
$$\left(\Delta H = -\frac{64 \text{ kcal}}{268 \text{ kJ}}/mol \right)$$

(61)

process principles:

uncatalyzed liquid-phase oxidation with per-
crotonic acid intermediate

use of crotonic acid:

comonomer, leveling agent for alkyd resins

The more stable *trans*-crotonic acid is formed preferentially, with a selectivity of about 60% (based on $CH_3CH=CHCHO$). It is produced by Hoechst in Germany, Eastman Kodak in the USA, and by several firms in Japan. While its main use is as a component in copolymerizations, it is also used in the manufacture of alkyd resins, where it improves the flow characteristics of raw materials for paints.

7.4.4. Ethyl Acetate

ethyl acetate ('acetic ester') is manufactured
by one of three routes, depending on
country:

In 1991, the worldwide production capacity for ethyl acetate was more than 700 000 tonnes per year, with about 300 000, 160 000, and 120 000 tonnes per year in Western Europe, Japan, and the

USA, respectively. Of the possible synthetic routes to ethyl acetate, only two have been developed into industrial processes. The feedstock, which varies from country to country, is either ethanol or acetaldehyde.

In places where inexpensive ethanol is available, it is esterified with acetic acid using an acidic catalyst:

$$C_2H_5OH + CH_3COOH \; \underset{}{\overset{[H^\oplus]}{\rightleftharpoons}} \; CH_3COOC_2H_5 + H_2O \quad (62)$$

1. ethanol esterification with AcOH, proton-catalyzed

2. formation as byproduct in butane oxidation

If the esterification is run continuously in a column, a yield of 99% can be attained.

Ethyl acetate is also formed in *n*-butane oxidation (*cf.* Section 7.4.1.2) along with numerous other products and can be isolated economically (*e. g.*, UCC in USA since 1983).

The Tishchenko reaction with acetaldehyde is the favored process in other countries where acetaldehyde is present in sufficient quantities as in Japan and Germany, or where the price of ethanol is artificially high:

3. Tishchenko reaction with acetaldehyde

$$2 \; CH_3CHO \; \overset{[cat.]}{\longrightarrow} \; CH_3COOC_2H_5 \quad (63)$$

The catalyst is a solution of Al ethylate in an ethanol/ethyl acetate mixture, with zinc and chloride ions as promoters. The exothermic conversion of acetaldehyde takes place at $0-5\,°C$ (with cooling) in this solution. At 95% conversion the selectivity is roughly 96% (based on CH_3CHO). The byproduct, acetaldol, is easily dehydrated; the resulting water hydrolyzes the ethylate to cause a rapid deactivation of the catalyst. Therefore another process, until now only described in patents, could possibly become important; this is the addition of acetic acid to ethylene:

process principles of 2:

homogeneously catalyzed dismutation, *i. e.*, conversion of aldehyde into the higher (acid) and lower (alcohol) oxidation state (oxidation/reduction)

disadvantage of process:

OH groups displace OC_2H_5 ligands on Al and deactivate the catalyst

potential alternative ethyl acetate manufacture:

addition of AcOH to $H_2C=CH_2$, not yet used industrially

$$CH_3COOH + H_2C=CH_2 \; \overset{[cat.]}{\longrightarrow} \; CH_3COOC_2H_5 \quad (64)$$

Ethyl acetate is an important solvent which is used mainly in the paint industry. It is also used as an extraction solvent, for example in the manufacture of antibiotics. Production figures for ethyl acetate in several countries are summarized in the adjacent table.

use of ethyl acetate:

solvent in paint industry
extractant

ethyl acetate production (in 1000 tonnes):

	1988	1990	1992
W. Europe	213	220	270
Japan	147	158	165
USA	123	124	135

7.4.5. Pyridine and Alkylpyridines

industrially important pyridines:

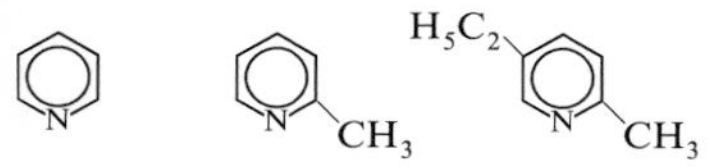

(2-picoline)   (MEP)

production methods:

pyridine, by extraction from coal tar
2-picoline and MEP, by synthesis based on
CH_3CHO/NH_3

synthetic principle:

C_2 aldehyde supplies alkylpyridines with
even C number
mixtures of C_2 and C_1 aldehydes supply
pyridine and alkylpyridines with odd C
number

two different processes with same feedstocks
(CH_3CHO or paraldehyde, NH_3):

1. liquid phase
2. gas phase

The industrially significant pyridine bases are pyridine itself,
2-methylpyridine (2-picoline) and 2-methyl-5-ethylpyridine
(MEP); 3- and 4-picoline are of limited use. Today much pyridine
is still isolated from coal tar, where it occurs with other low-
boiling derivatives at 0.1 wt%. However, synthetic routes —
especially to the alkylpyridines — are increasing in importance
due to growing demand.

In 1995, the world capacity of synthetic pyridines was about
95 000 tonnes per year, with about 46 000, 31 000, and 15 000
tonnes per year in Western Europe, USA, and Japan, respectively. The largest manufacturer, Reilly Chemicals in Belgium,
currently produces synthetic pyridines in a unit with a capacity
of 17 000 tonnes per year.

Of the numerous processes known, those based on acetaldehyde, alone or together with formaldehyde, and ammonia
have made the greatest impact (*cf.* Section 11.1.6).

Due to the economical selectivity and simplicity of the reaction,
2-methyl-5-ethylpyridine was the first to be industrially manufactured.

Acetaldehyde (directly or in the form of paraldehyde) is reacted
with an aqueous 30–40% ammonia solution in a continuous
process. The conversion takes place in the liquid phase at
220–280 °C and 100–200 bar in the presence of ammonium
acetate catalyst:

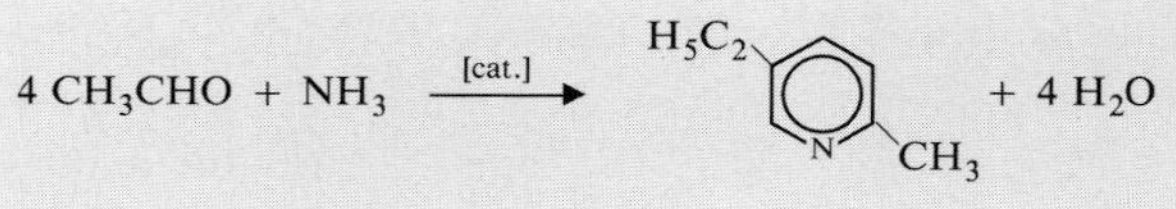

$$4 \ CH_3CHO + NH_3 \xrightarrow{\text{[cat.]}} \text{MEP} + 4 \ H_2O \qquad (65)$$

to 1:

homogeneously catalyzed liquid-phase reaction under pressure
(favored for MEP)

to 2:

heterogeneously catalyzed gas-phase reaction at normal pressure (favored for 2- and
4-picoline)

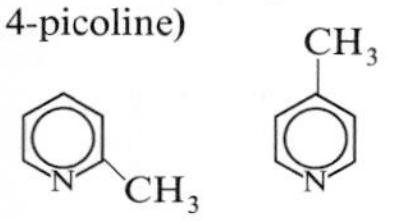

The reaction product is two-phase. The aqueous phase is largely
recycled to the reaction, and the organic phase is worked up by
azeotropic and vacuum distillations.

The selectivity to 2-methyl-5-ethylpyridine reaches 70% (based
on acetaldehyde). Byproducts are 2- and 4-picoline (in the ratio
3 : 1) and higher pyridine bases.

If the reaction is carried out in the gas phase at 350–500 °C and
atmospheric pressure over Al_2O_3 or $Al_2O_3 \cdot SiO_2$ catalyst, with
or without promoters, then 2- and 4-picoline are formed preferentially (in roughly the same ratio) from acetaldehyde and
ammonia.

When a mixture of acetaldehyde and formaldehyde is reacted with NH_3 in the gas phase, pyridine and 3-picoline are formed:

pyridine manufacture from CH_3CHO, $HCHO$, NH_3 used industrially

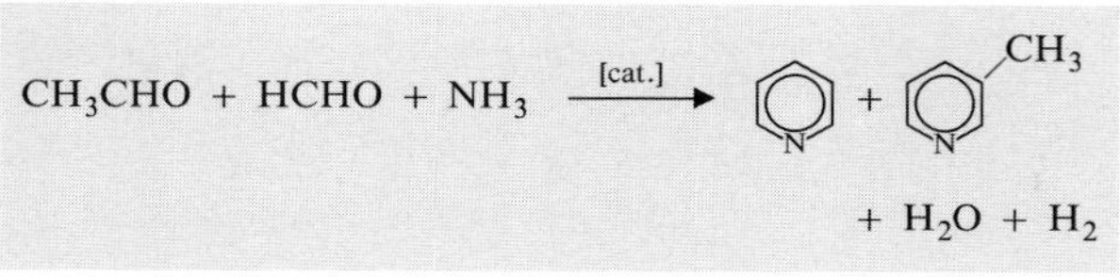

$$CH_3CHO + HCHO + NH_3 \xrightarrow{[cat.]} \bigcirc\!\!\!\!_N + \bigcirc\!\!\!\!_N{}^{CH_3}$$

$$+ \ H_2O + H_2 \qquad (66)$$

The ratio of the aldehydes in the feed determines the ratio of the reaction products. This route is used commercially.

alternative multistep 2-picoline manufacture:

1. cyanoethylation of acetone
2. ring-forming dehydration with dehydrogenation

In 1977, DSM began production of 2-picoline in a 5000 tonne-per-year (1991) unit using a new multistep selective pathway. In the first stage, acetone is reacted with acrylonitrile over a basic catalyst (isopropylamine) to form 5-oxohexanenitrile with a selectivity of more than 80% (based on acetone and acrylonitrile). This intermediate is then cyclized in the gas phase over supported metal catalysts (*e. g.*, Ni/SiO_2 or Pd/Al_2O_3) in the presence of hydrogen to eliminate water and form 2-picoline or its hydrogenated products:

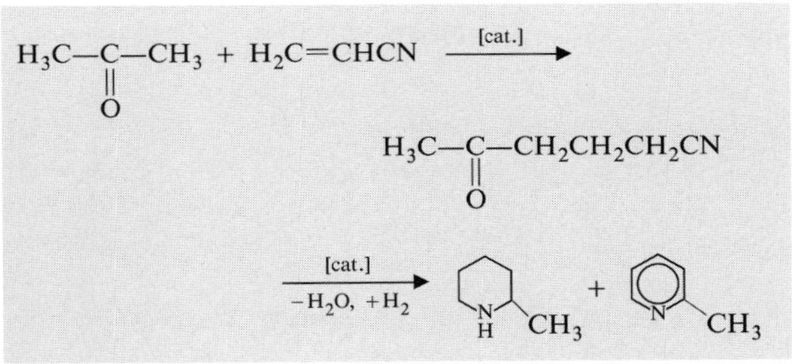

used commercially by DSM

$$H_3C\!-\!\underset{\underset{O}{\|}}{C}\!-\!CH_3 + H_2C\!=\!CHCN \xrightarrow{[cat.]}$$

$$H_3C\!-\!\underset{\underset{O}{\|}}{C}\!-\!CH_2CH_2CH_2CN \qquad (67)$$

$$\xrightarrow[-H_2O, +H_2]{[cat.]} \underset{\overset{|}{H}}{\bigcirc}\!\!\!\!_N{}^{CH_3} + \bigcirc\!\!\!\!_N{}^{CH_3}$$

A newer process for 2-picoline and 2-methyl-5-ethylpyridine introduced by the Nippon Steel Chemical Company has not yet been used commercially. Unlike all other processes, ethylene is used directly. It is reacted with NH_3 in the presence of an ammoniacal palladium salt solution and a Cu redox system at $100-300\,°C$ and $30-100$ bar to form the two alkylpyridines. The total selectivity is 80% (based on ethylene):

potential alternative alkylpyridine manufacture by homogeneously Pd-catalyzed $H_2C=CH_2/NH_3$ reaction not yet used industrially

reaction principle can be explained as intermediate CH_3CHO formation (Wacker–Hoechst)

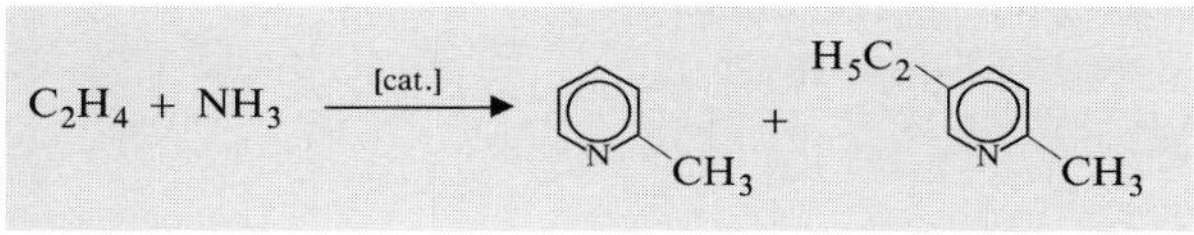

$$C_2H_4 + NH_3 \xrightarrow{[cat.]} \bigcirc\!\!\!\!_N{}^{CH_3} + \underset{}{\overset{H_5C_2}{\bigcirc}}\!\!\!\!_N{}^{CH_3} \qquad (68)$$

The main use of 2-methyl-5-ethylpyridine is in the manufacture of nicotinic acid (3-pyridinecarboxylic acid, Niacin®). The dialkylpyridine is subjected to oxidation with nitric acid followed

use of alkylpyridines:

MEP for nicotinic acid as vitamin precursor for nicotinamide (antipellagra vitamin B_3 and enzyme building block) by selective decarboxylation of the carboxyl group in the 2-position:

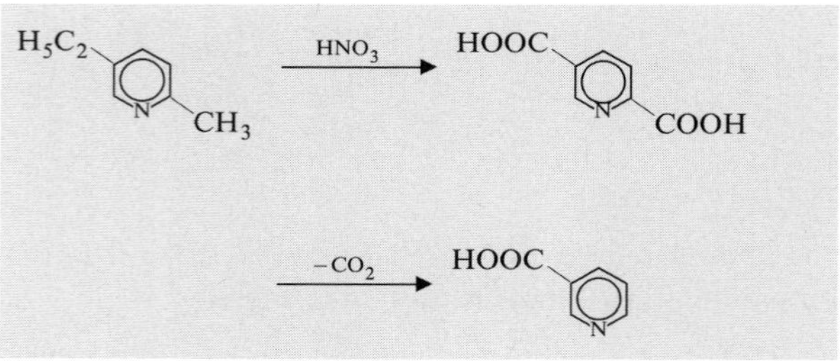

$$(69)$$

characteristics of nicotinic acid manufacture:

25% C loss by oxidation/decarboxylation, but high selectivity in 2nd step, since CO_2 is cleaved from positions 2, 3, 4 with increasing ease

2-picoline for 2-vinylpicoline with intermediate 2-hydroxyethylpyridine:

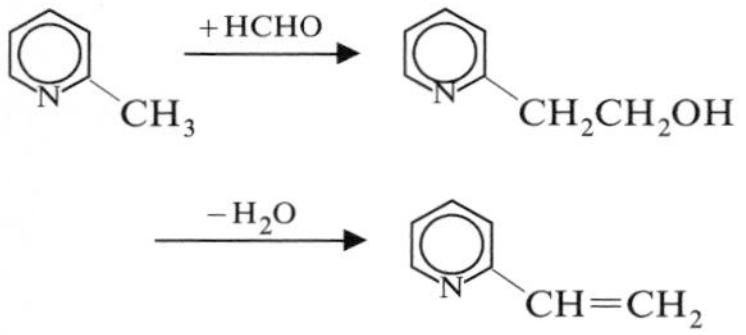

Nicotinic acid and its derivatives, *e. g.*, nicotinic acid amide, are B-complex vitamins. They are most important as pharmaceuticals, and are also additives in food and animals feeds. Nicotinic acid is manufactured from nicotinic acid esters by reaction with NH_3. Another route is the much-studied ammonoxidation of 3-picoline to nicotinic acid nitrile and, after partial hydrolysis, to nicotinic acid amide:

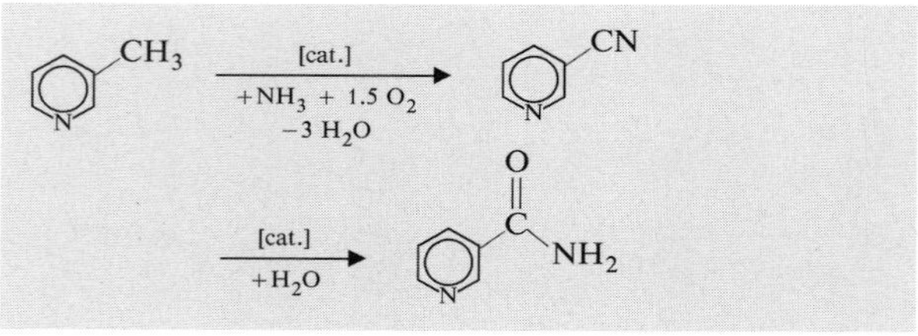

$$(70)$$

The first commercial unit using this process was brought into operation in 1983 in Antwerp by Degussa. The world demand for nicotinic acid and its derivatives is estimated at 10000 tonnes per year, with about 45% of this demand in the USA, 30% in Western Europe, and 10% in Japan.

2-Picoline is a starting material for 2-vinylpyridine, a comonomer with butadiene and styrene in copolymers to improve adhesion between synthetic fibers and rubbers in the tire industry. Pyridine derivatives are also used in the synthesis of herbicides and many pharmaceuticals.

8. Alcohols

8.1. Lower Alcohols

Methanol, ethanol, isopropanol, and the butanols are among the most commercially important lower alcohols. Allyl alcohol also belongs to this group due to its growing industrial importance. However, as it is an unsaturated compound with an allyl structure, it will be discussed in Section 11.2.2. Amyl alcohols have a more limited, but increasing, field of application.

In all industrialized countries, methanol is the highest volume alcohol. It has already been discussed in Section 2.3.1 as a C_1 product manufactured solely from synthesis gas. The production capacity for isopropanol, which ranks second in importance to methanol, is considerably less. The third position amongst the lower alcohols is taken by ethanol in countries such as the USA or England, where the price is determined by supply and demand.

In several Western European countries, especially Germany, and in other countries such as Japan, taxation to protect alcohol produced from the fermentation of agricultural products has inhibited the commercialization of synthetic alcohol. In these countries, the use of ethanol and its price are not sensitive to market pressures. However, since 1978 marketing of synthetic alcohol has been somewhat facilitated.

industrially important lower alcohols:

CH_3OH

C_2H_5OH

iso-C_3H_7OH

C_4H_9OH (n-, iso-, sec-, tert-)
$CH_3(CH_2)_3CH_2OH$
$CH_3CH_2CHCH_2OH$
$\qquad\quad |$
$\qquad\quad CH_3$
$(CH_3)_2CHCH_2CH_2OH$

production sequence of lower alcohols:

USA	W. Europe
methanol	methanol
isopropanol	butanols
ethanol	isopropanol
butanols	ethanol

in several countries, ethanol cannot develop freely due to:

1. beverage alcohol monopoly with C_2H_5OH tax to protect fermentation alcohol
2. prohibition of use of synthetic ethanol for food applications in EC countries

8.1.1. Ethanol

In 1994, the world production capacity for synthetic ethanol was about 2.6×10^6 tonnes per year, of which ca. 0.65, 0.62, and 0.11×10^6 tonnes per year were located in the USA, Western Europe, and Japan, respectively. In these same countries, the production capacity of ethanol from fermentation was about 4.3, 0.74, and 0.14×10^6 tonnes per year, respectively.

Currently, about 12.9×10^6 tonnes per year of ethanol are produced by fermentation of agricultural products such as sugar cane molasses and corn starch, or from wood hydrolysis pro-

synthetic ethanol production (in 1000 tonnes):

	1986	1988	1993
USA	232	255	308
W. Europe	410	383	499
Japan	72	84	77

C_2H_5OH manufacture:

1. indirect through sulfuric acid ester and subsequent hydrolysis
2. direct by proton-catalyzed hydration of ethylene

characteristics of H_2SO_4 process:

1.1. stepwise (possibly catalyzed) C_2H_4 absorption in conc. H_2SO_4 under pressure in bubble-tray or bubble-column reactors

1.2. hydrolysis after dilution (otherwise back reaction to C_2H_4) at raised temperature

disadvantages of H_2SO_4 method:

1. corrosion problems
2. concentration of H_2SO_4 from 70% to >90% costly
3. submerged burner produces SO_2 in off-gas

ducts and sulfite liquors (paper industry). The majority of plants are found in Brazil, India, and the USA. Other producing countries are Japan, Mexico, South Africa, Indonesia and several countries in Western Europe. The last few years have seen an increasing use of agricultural products as renewable raw materials for different processes, e.g., higher alcohols based on oil and fats (cf. Section 8.2). Production figures for synthetic ethanol can be found in the adjacent table.

Most ethanol is manufactured from ethylene in one of two processes:

1. By indirect hydration by addition of H_2SO_4 and subsequent saponification of the sulfuric acid ester.

2. By direct catalytic hydration.

The indirect hydration has been used industrially since 1930. Ethylene-containing gases, in which the ethylene content may vary between 35 and 95%, are reacted with $94-98\%$ H_2SO_4 in a system consisting of several absorption columns at $55-80\,°C$ and $10-35$ bar. Mono- and diethyl sulfate are formed exothermically in this step, which can be catalyzed by Ag_2SO_4:

$$H_2C{=}CH_2 + H_2SO_4 \rightarrow C_2H_5O{-}SO_3H$$

$$C_2H_5O{-}SO_3H + H_2C{=}CH_2 \rightarrow (C_2H_5O)_2SO_2 \qquad (1)$$

$$\left(\Delta H = -\frac{58\ \text{kcal}}{243\ \text{kJ}}\Big/\text{mol}\right)$$

After adjusting the H_2SO_4 concentration to $45-60$ wt%, both sulfuric acid esters are hydrolyzed to ethanol in acid-resistant, lined columns at temperatures between 70 and 100 °C. Diethyl ether is formed as a byproduct, particularly at higher temperatures:

$$C_2H_5O{-}SO_3H \text{ or } (C_2H_5O)_2SO_2 + H_2O \qquad (2)$$

$$\rightarrow C_2H_5OH + (C_2H_5)_2O$$

The dilute H_2SO_4 is concentrated using submerged burners, which evaporate the water in the sulfuric acid with an open flame. Small amounts of SO_2 are also produced.

Ethanol selectivity is about 86% (based on C_2H_4).

Due to economic considerations, this ethyl sulfate process has not been used in the United States since 1974. The last operating plant in Western Europe (France) was also closed down in the mid 1980s.

A substantial improvement in the treatment of dilute sulfuric acid — also from other production processes — can be realized through the Bertrams/Bayer process.

In this process, the aqueous sulfuric acid is concentrated in a forced circulation concentrator made of glass-lined steel in the last step. If necessary, organic contaminants are removed by addition of an oxidizing agent (generally 65% HNO_3) without the production of undesired waste gases.

There are many industrial applications of this technology.

The catalytic hydration of ethylene was first used commercially by Shell in 1947. The addition of water is carried out in the gas phase, generally over acidic catalysts:

$$H_2C{=}CH_2 + H_2O \xrightleftharpoons{[H^{\oplus}]} C_2H_5OH\left(\Delta H = -\frac{11 \text{ kcal}}{46 \text{ kJ}}\Big/\text{mol}\right) \quad (3)$$

H_3PO_4/SiO_2 catalysts have proven to be particularly useful in several different processes. Typical reaction parameters are 300 °C, 70 bar, and a short residence time to limit the formation of byproducts such as diethyl ether and ethylene oligomers. Under these conditions of temperature and pressure, only about 30% of the equilibrium ethanol concentration is obtained. The partial pressure of steam is limited, since it lowers catalyst activity and shortens catalyst lifetime by loss of phosphoric acid. Thus, the mole ratio of water to ethylene is limited to 0.6. Ethylene conversion is only about 4%. Since the ethylene must be recycled many times to use it economically, either it must be very pure or a larger portion must be vented to avoid building up inert gases in the recycle gas. The gas flow from the reactor is cooled to separate the condensable products, and the ethylene (for recycling) is once again brought to reaction temperature. The aqueous crude alcohol is concentrated and purified by extractive distillation. The selectivity to ethanol is 97%. Well-known processes of this type have been developed by such companies as BP, Shell, UCC, USI, and Veba (now Hüls). Single units can have a production capacity as high as 380 000 tonnes per year.

Basic disadvantages of this otherwise elegant catalytic process are the high ethylene purity necessary and the low conversion. These are also the reasons why the older H_2SO_4 process can still be competitive today.

The biggest factor in determining the most economical ethanol process in the future will not be wastewater problems in the H_2SO_4 process or process costs of a low-conversion catalytic

improvement of H_2SO_4 method:

two-step concentration and possible purification of H_2SO_4 from 20% to 96%:

1. falling film evaporator 20 → 78%
2. forced circulation concentrator 78 → 96%

characteristics of catalytic hydration:

heterogeneous gas-phase process generally with H_3PO_4/SiO_2 as catalyst, the first step is the protonation of ethylene to an ethyl carbenium ion

as exothermic, volume-reducing reaction, favored by high pressure (esp. H_2/partial pressure) and low temperature, though limited by lifetime and working temperature of the catalyst

C_2H_5OH isolation from dilute aqueous solution:

1. H_2O azeotrope distillation with 95% C_2H_5OH in distillate
2. water removal by addition of entraining agent (e. g., benzene)

disadvantages of catalytic hydration:

1. low C_2H_4 conversion
2. high C_2H_4 purity necessary
3. continuous loss of H_3PO_4
4. high energy consumption

direct coupling of C_2H_5OH manufacturing costs to C_2H_4 price increases interest in alternative manufacture

CH₃OH homologation with CO/H₂ to CH₃CHO or, with in situ hydrogenation, directly to C₂H₅OH thoroughly researched, not yet used industrially

traditional alcoholic fermentation (basis carbohydrates, sulfite waste liquors) with newer aspects:

1. expansion of feedstock base (also other biomass or domestic wastes)
2. development of process technology (fluidic optimized fermentation in continuous process)
3. increase in rate of fermentation (simultaneous removal of C₂H₅OH)

use of C₂H₅OH:

still basis for CH₃CHO, H₂C=CH₂
solvent
ester component
gasoline additive
C-source for single cell proteins (SCP)

---

system, but pricing of ethylene and its sources, crude petroleum and natural gas.

Most newly developed technology is based on synthesis gas, with its various feedstocks. One route is so-called homologation, in which methanol in either the liquid or the gas phase is reacted with CO/H₂ over Rh- or Co-containing multicomponent catalysts. Depending on the reaction conditions and the catalyst used, either acetaldehyde or ethanol can be obtained preferentially. This technology has not yet been used commercially (cf. Section 2.3.1.2).

It is therefore logical that in addition to the current alcoholic fermentation of carbohydrates and sulfite waste liquors, the value of all biological materials suitable for fermentation will be more strongly considered. Of particular interest are process developments for continuous processes, which lead to increased economic efficiency of fermentation. The space-time yield of ethanol can be increased if the inhibiting effect of increasing ethanol concentration on the fermentation can be lowered, for example by distillation of the ethanol at lowered pressure, by selective separation of ethanol by membranes, or by continuous extraction with solvents such as dodecanol.

Ethanol productivity can also be significantly improved by the use of loop-type bubble reactors with intensive mixing of the three-phase system, with a higher yeast concentration, and by continuous process management.

In some countries such as the USA, England, and Germany, ethanol has been an important precursor for acetaldehyde. However, this use for ethanol has decreased greatly (e.g., the last plant in the USA was shut down in 1983), since acetic acid – the principal product of acetaldehyde – is made more economically by methanol carbonylation. In countries such as Brazil, India, Pakistan, and Peru fermentation ethanol is still dehydrated to ethylene to complement the ethylene from petrochemical sources. The dehydration is done over activated alumina, aluminum silicate, or H₃PO₄-impregnated catalysts in fixed-bed reactors at 300–360 °C. Ethanol is also a solvent and ester component, and is increasingly used as a component in gasoline. Brazil was the first country to use fermentation ethanol to a great extent in gasoline, with roughly 4×10^6 automobiles fueled by ethanol in 1993. By 1985, the USA was already meeting about 5% of its gasoline demand with ethanol under the Gasohol project (ethanol use in gasoline). In 1995, the portion of total ethanol use for gasoline in the USA was 73%, while in Western Europe it was only 28%. Like methanol,

ethanol can be employed in the synthesis of SCP (*cf.* Section 2.3.1.2). In Czechoslovakia 4000 tonnes per year SCP are produced based on ethanol, and in the USA a 7000 tonne-per-year SCP plant (Amoco) for manufacturing Torula yeast (with 53% protein content) from extremely pure ethanol is in operation.

Ethyl chloride and ethyl acetate (*cf.* Section 7.4.4) are the most important esters of ethanol. In 1991, the production capacity for ethyl chloride was 111 000 tonnes per year in the USA, 112 000 tonnes per year in Western Europe, and 5000 tonnes per year in Japan. The esterification of ethanol with hydrogen chloride or concentrated hydrochloric acid can, as with methanol, be conducted in the liquid or gas phase. Catalysts include mineral acids and Lewis acids such as $ZnCl_2$, $FeCl_3$, $BiCl_3$, $AlCl_3$, and $SbCl_5$. As in the hydrochlorination of methanol, Al_2O_3 is also suitable.

industrially important esters of ethanol:

ethyl chloride
ethyl acetate

manufacture of ethyl chloride by three routes:

1. ethanol esterification with HCl
2. ethane chlorination
3. ethylene hydrochlorination of greatest industrial importance

Today most ethyl chloride is manufactured by one of two other processes: ethane chlorination and hydrochlorination of ethylene. Both will be dealt with briefly.

principle of ethanol esterification:

homogeneously catalyzed liquid- or gas-phase reaction with HCl in presence of Friedel–Crafts catalysts

principle of ethane chlorination:

The chlorination of ethane is, like methane chlorination, conducted purely thermally at 300–450 °C with slight excess pressure. Kinetics favor monochlorination over multiple chlorination so that with an additional over-stoichiometrical ethane/chlorine ratio of 3–5 : 1, a high ethyl chloride selectivity can be obtained.

catalyst-free gas-phase chlorination with favored monochlorination, as C_2H_6 is more rapidly chlorinated than C_2H_5Cl

As almost no byproducts arise from the addition of hydrogen chloride to ethylene, this reaction can be run very economically in combination with ethane chlorination (by using the hydrogen chloride produced) in the so-called integrated process. Since addition of chlorine to ethylene is minimal at 400 °C, ethane can be chlorinated in the presence of ethylene and, after separation of ethyl chloride, ethylene in the gas mixture can be catalytically hydrochlorinated in a second reactor:

utilization of HCl resulting from ethane chlorination:

1. oxychlorination of ethane possible but not used commercially
2. HCl used industrially hydrochlorination of ethylene
   characteristics of
   C_2H_6/C_2H_4—Cl_2/HCl combined process:

no $ClCH_2CH_2Cl$ (EDC) formation from $H_2C=CH_2$ and Cl_2 at higher temperature (400 °C)

$$\begin{matrix} H_2C=CH_2 \\ + Cl_2 \\ H_3C-CH_3 \end{matrix} \xrightarrow{400\,°C} \begin{matrix} H_2C=CH_2 \\ + HCl \\ H_3C-CH_2Cl \end{matrix} \xrightarrow[\text{[cat.]}]{150-250\,°C}$$

$$2\ H_3C-CH_2Cl$$

(4)

A combined process of this type was developed by Shell.

Hydrochlorination of ethylene is by far the preferred manufacturing process for ethyl chloride. Since 1979, no other process has been used in the United States. It can either be conducted in the liquid phase at 30–90 °C and 3–5 bar with Friedel–Crafts

principles of ethylene hydrochlorination:

alternative homogeneous liquid-phase or heterogeneous gas-phase HCl addition to ethylene

catalysts such as $AlCl_3$ or $FeCl_3$ or, as in new plants, in the gas phase with supported catalysts containing the aforementioned metal chlorides as active components at $130-250\,°C$ and $5-15$ bar. Selectivities of $98-99\%$ (based on C_2H_4, HCl) and conversions of 50% C_2H_4 and HCl are obtained.

use of ethyl chloride:

tetraethyllead (decreasing)
ethyl cellulose
alkylation agent
solvent
extractant
local anesthetic

ethyl chloride production (in 1000 tonnes):

	1986	1988	1990
USA	74	69	73

Most ethyl chloride (about 85% in the USA and 65% in Western Europe) is used for the manufacture of the antiknock agent tetraethyllead (*cf.* Section 2.3.6.1). However, the production figures for tetraethyllead (and tetramethyllead) are decreasing, since many countries have drastically restricted addition to gasoline for environmental reasons. This trend is illustrated by data from the USA given in the adjacent table. Most of the remaining ethyl chloride is used as an ethylating agent (e. g., for cellulose) or as a solvent or extractant. It is also used in medicine as a local anesthetic.

8.1.2. Isopropanol

general principle of H_2O addition to olefins: hydration of α-olefins (except for C_2H_4) leads to *sec*- and *tert*-alcohols (Markovnikov addition)

isopropanol production (in 1000 tonnes):

	1990	1992	1994
USA	660	563	658
W. Europe	587	555	570
Japan	122	122	130

The first commercial production of isopropanol (IPA, or 2-propanol) by the addition of water to propene was done in 1930 by Standard Oil of New Jersey (USA). This was also the first example of the manufacture of a petrochemical from a refinery product.

The world production capacity for isopropanol in 1993 was about 2.3×10^6 tonnes per year, of which 0.86, 0.79, and 0.14×10^6 tonnes per year were in the USA, Western Europe, and Japan, respectively. Production figures for isopropanol in several countries are summarized in the adjacent table.

three processes for iso-C_3H_7OH manufacture by propene hydration:

1. indirect, two-step through sulfuric acid monoester (liquid phase)
2. direct, single-step, heterogeneously catalyzed (either gas, trickle, or liquid phase)
3. direct, single-step, homogeneously catalyzed (liquid phase)

In the classical process, propene hydration with H_2SO_4 took place in the liquid phase. Along with this process, which is still operated today, other processes based on propene have also been established involving, however, a single-stage catalytic hydration in the gas, trickle, or liquid phase. Depending on the mode of operation, various acidic catalysts can be employed, such as:

1. supported heteropoly acids or mineral acids in the gas phase,

2. acidic ion-exchangers in the trickle phase,

3. water-soluble heteropoly acids containing tungsten.

industrial use of indirect propene hydration:

BP process in England and Japan

Texaco process in Germany until 1986 (60000 tonne-per-year capacity), then conversion of the process to direct hydration over an ion-exchange catalyst

The older, two-step sulfuric acid process is still operated by BP and Shell. A Deutsche Texaco plant used this technology until 1986, when it was converted to direct hydration over an acidic ion exchanger. The H_2O addition to propene in the H_2SO_4 process takes place indirectly *via* the sulfuric acid monoester. In the second step, the acid content is lowered to less than 40% by dilution with steam or water, and the ester is hydrolyzed:

$$CH_3CH=CH_2 + H_2SO_4 \rightarrow (CH_3)_2CH-O-SO_3H$$

$$\xrightarrow{+H_2O} (CH_3)_2CHOH + H_2SO_4$$

(5)

The resulting dilute H_2SO_4 is then concentrated. At the same time, the higher boiling organic byproducts are burned, usually with the addition of small amounts of HNO_3. In commercial implementation, two modifications are common. The strong acid process, which is two-step, has separate reactors for absorption and hydrolysis. The absorption is done with 94% H_2SO_4 at 10 – 12 bar and 20 °C. The weak acid process is single stage, and is carried out in a bubble reactor with only 70% H_2SO_4; the pressure and temperature must be increased to 25 bar and 60 – 65 °C. The isopropanol selectivity reaches more than 90% and the byproducts are diisopropyl ether and acetone.

The catalytic gas-phase hydration of propene takes place in a manner similar to ethanol manufacture from ethylene:

$$CH_3CH=CH_2 + H_2O \xrightleftharpoons{[cat.]} (CH_3)_2CHOH$$

$$\left(\Delta H = -\frac{12 \text{ kcal}}{50 \text{ kJ}} \middle/ \text{mol} \right)$$

(6)

In contrast to ethylene, the protonization of propene occurs much more readily during the first stage of the reaction as the resulting secondary propyl carbenium ion is more stable than the primary ethyl carbenium ion. Therefore, higher conversions are attainable with propene than with ethylene (*cf.* below).

In this exothermic reaction, the equilibrium is displaced towards the desired product by high pressure and low temperatures. However, the catalyst requires a certain minimum temperature to be effective, so it is not possible to benefit fully from the thermodynamic advantages of low temperature.

Suitable catalysts include WO_3/SiO_2 combinations (heteropoly acids), which have been used by IG Farben. Isopropanol was manufactured from 1951 until the end of the 1970s by ICI in a unit with a capacity of 48 000 tonnes per year, using WO_3/SiO_2 with ZnO as promoter. In Germany, Hüls (formerly Veba) uses H_3PO_4 on a SiO_2 support in a 110 000 tonne-per-year (1993) plant. Typical operating conditions are 270 °C and 250 bar for

process characteristics of indirect propene hydration:

use of dilute propene possible, however, disadvantages include corrosion, wastewater and outgoing air problems

process scheme analogous to C_2H_5OH manufacture from C_2H_4 according to concentration-dilution principle has two variants:

1. two-step strong acid process
2. single-step weak acid process

process characteristics of the direct hydration:

high pressure and low temperature are favorable, as reaction is exothermic and takes place with reduction in number of moles (Le Chatelier principle)

difference from ethylene hydration:

greater stability of secondary propyl carbenium ion results in higher propene conversions

catalyst for gas-phase hydration:

$WO_3 \cdot ZnO/SiO_2$ in ICI high-pressure process

H_3PO_4/SiO_2 in Veba (now Hüls) medium-pressure process

characteristics of gas-phase hydration:

high selectivity but incomplete C_3H_6 conversion together with high investment and operating costs; H_3PO_4 discharge causes corrosion

variant of direct hydration:

trickle-phase hydration with ion-exchanger in Deutsche Texaco process

principles of process:

trickle-phase is a three-phase system:

fixed-bed catalyst, over which the reactants water (liquid phase) and propene (gas phase) are passed in co- or countercurrent

characteristics of process:

high $H_2O:C_3H_6$ ratio avoids C_3H_6 oligomerization (fouling)

high pressure favors C_3H_6 solubility in liquid film, since only dissolved C_3H_6 reacts

saturation of liquid phase with C_3H_6 by repeated introduction to reactor intermediate bottom

isopropanol (IPA) purification:

H_2O/IPA azeotrope freed from water using IPA/H_2O/C_6H_6 azeotrope

direct hydration in the liquid phase:

Tokuyama process

characteristics of liquid-phase hydration:

very active heteropoly acid allows high and selective propene conversion

use of isopropanol:

in the USA and Western Europe with decreasing trend to $(CH_3)_2C{=}O$

also as frost protection additive for gasoline (carburetor icing), solvents and extracting agents, intermediate

the ICI process, and 170–190 °C and 25–45 bar in the Hüls process.

In the gas-phase process, isopropanol selectivity is 97% and therefore higher than in the H_2SO_4 liquid-phase process. The workup of the aqueous solutions is conducted in a manner similar to that described below. The disadvantages of both gas-phase processes are low propene conversion (5–6%) and high plant costs due to pressurized operation and gas recycles.

The trickle-phase variant of the direct hydration was developed by Deutsche Texaco. It avoids the disadvantages of the aforementioned processes to a great extent by employing a strongly acidic ion-exchanger in the trickle phase. The process is characterized by the introduction of liquid H_2O and gaseous propene (molar ratio from 12 to 15) at the head of the reactor, where they are then passed over a solid sulfonic acid ion-exchanger. The liquid and gas mix thoroughly at the acidic catalyst and react at 130–160 °C and 80–100 bar to form aqueous isopropanol. A 75% propene conversion is obtained.

Selectivity to isopropanol is 92–94%, with 2–4% diisopropyl ether as well as alcohols of the C_3H_6 oligomers formed as by-products. The catalyst life is at least 8 months, and is essentially determined by the hydrolytic degradation of the SO_3H groups of the ion-exchanger.

After removal of the low boiling substances, an isopropanol/H_2O azeotrope is distilled from the aqueous reaction product which originally contained 12–15% isopropanol. Benzene is added to the distillate, and further distillation gives anhydrous alcohol.

Since 1972, Deutsche Texaco (now Condea) has been operating a plant with this technology. The capacity was increased to 140 000 tonnes per year in 1986. Additional plants have been built and more are being planned.

Further development of gas-phase catalysts containing tungsten led Tokuyama new, very active, water-soluble silicotungsten acids, which can be used as the acid or its salts to convert propene to isopropanol in the liquid phase at 270–280 °C and 200 bar. The aqueous catalyst is recycled after distilling off the alcohol/water azeotrope. Propene conversion reaches 60–70% with a selectivity to isopropanol of about 99%. The first commercial unit has a production capacity of 30 000 tonnes per year, and began operation in 1972.

A significant fraction of isopropanol was previously used for the production of acetone (*cf.* Section 11.1.3) in Western Europe and in the USA, but this had decreased to 39% and 7%, respectively, by 1992. It is interesting to note that earlier, the opposite was the case: acetone – which was available from fermentation processes – was hydrogenated to isopropanol. Isopropanol is also used as a gasoline additive to prevent carburetor icing. The bulk is used as solvent, extractant, and ethanol substitute in the cosmetic and pharmaceutical industries.

Isopropyl acetate is an important derivative obtainable either by esterification of isopropanol with acetic acid or by the recently developed direct catalyzed (ion-exchange) addition of acetic acid to propene (*cf.* Section 11.1.7.3). In addition to its use as a solvent, a mixture of isopropyl acetate, ethyl acetate and water glass is used for soil stabilization. The acetic acid resulting from hydrolysis precipitates silicic acid from the alkali silicate which cross-links the soil.

manufacture of isopropyl acetate:
1. homogeneously catalyzed esterification of IPA with AcOH
2. heterogeneously catalyzed addition of AcOH to propene

Other commercially important derivatives are isopropylamine as an intermediate for the synthesis of such things as dyes, rubber chemicals, insecticides and pharmaceuticals, and isopropyl oleate and myristate as components of cosmetics.

8.1.3. Butanols

Saturated monohydric C_4 alcohols, or butanols, occur as four structural isomers which are named as follows:

$CH_3CH_2CH_2CH_2OH$ 1-butanol (*n*-butanol)

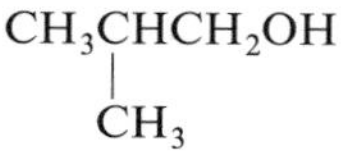

CH_3CHCH_2OH 2-methyl-1-propanol (isobutanol) 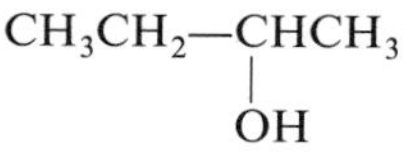
  |
  CH_3

$CH_3CH_2—CHCH_3$ 2-butanol (*sec*-butanol) 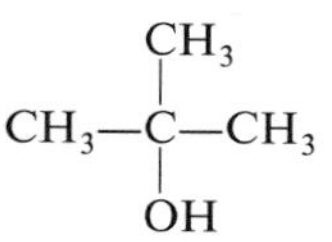
        |
        OH

    CH_3
     |
$CH_3—C—CH_3$ 2-methyl-2-propanol (*tert*-butanol)
     |
    OH

C_4 alcohols:
$CH_3(CH_2)_2CH_2OH$
$(CH_3)_2CHCH_2OH$
$C_2H_5CH(OH)CH_3$
$(CH_3)_3COH$

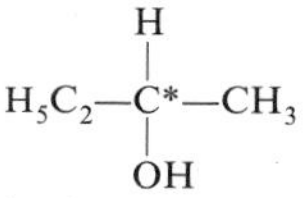

      H
      |
$H_5C_2—C^*—CH_3$
      |
     OH
is simplest alcohol with an asymmetric C atom

In addition, 2-butanol is present as an enantiomeric mixture (racemate of two optical isomers) due to its asymmetric C atom.

In 1994, the manufacturing capacity for butanols in Western Europe, the USA, and Japan was 0.74, 0.72, and 0.22×10^6 tonnes per year, respectively.

total butanol production (in 1000 tonnes):

	1989	1991	1993
USA	537	639	665
W. Europe	420	530	475
Japan	291	334	388

In many industrialized countries, butanols are second only to methanol as the highest-volume alcohol. An exception is the USA; here butanols are fourth in line after methanol, isopropanol, and ethanol. Production figures are listed in the adjacent table.

four routes for manufacture of n-butanol

1. through C_3H_7CHO from propene hydro- formylation
2. through $CH_3CH=CHCHO$
3. fermentation of sugar or starch
4. 'Reppe carbonylation', i. e., hydrocarbonylation of propene

The butanols can be manufactured in various ways. Two routes to n-butanol have already been discussed:

1. Hydroformylation of propene with subsequent hydrogenation (cf. Section 6.1.4.1); isobutanol can also be produced in this process.
2. Aldol condensation of acetaldehyde with subsequent hydro- genation of the crotonaldehyde (cf. Section 7.4.3).

A third production method is based on the fermentation of sugar or starch. Once pursued in such countries as the USA and South Africa, it is now only practiced in the CIS (cf. Section 11.1.3).

A fourth possibility is the Reppe process, i. e., the reaction of propene with CO and water in the presence of a modified iron pentacarbonyl catalyst:

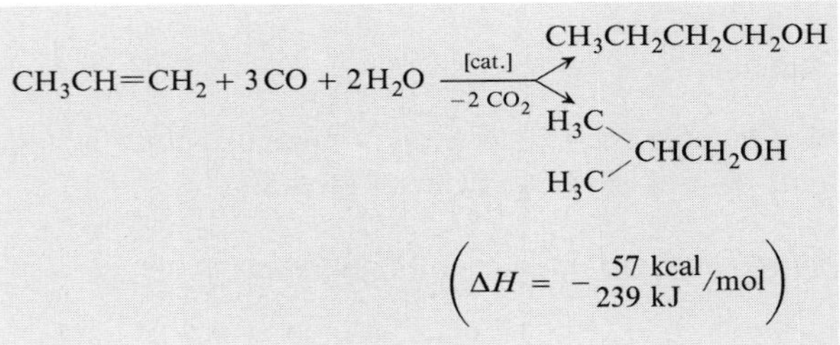

$$CH_3CH=CH_2 + 3\,CO + 2\,H_2O \xrightarrow[-2\,CO_2]{[cat.]} \begin{array}{l} CH_3CH_2CH_2CH_2OH \\[4pt] \begin{array}{c} H_3C \\ \diagdown \\ H_3C \diagup \end{array} CHCH_2OH \end{array} \qquad (7)$$

$$\left(\Delta H = -\dfrac{57\ \text{kcal}}{239\ \text{kJ}}/\text{mol} \right)$$

principle of Reppe hydrocarbonylation:

CO and H_2 (from $Fe(CO)_5 + H_2O$) are transferred to propene from an intermediate Fe-CO-H-complex

In this hydrocarbonylation reaction an iron hydrocarbonyl amine complex is formed in situ from $Fe(CO)_5$ and a tertiary amine such as N-butylpyrrolidine:

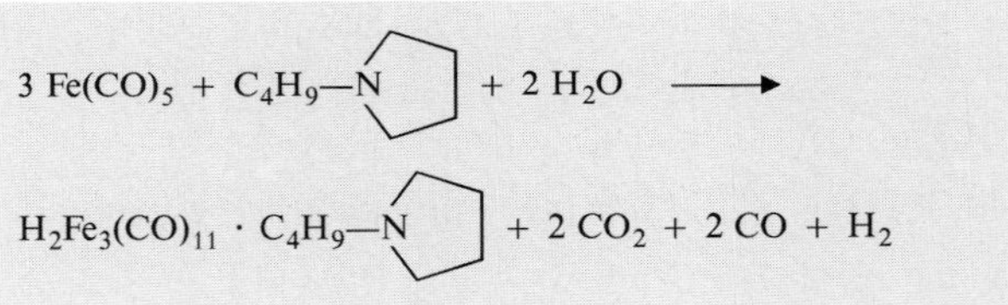

$$3\,Fe(CO)_5 + C_4H_9{-}N\!\!\bigcirc + 2\,H_2O \longrightarrow$$

$$H_2Fe_3(CO)_{11} \cdot C_4H_9{-}N\!\!\bigcirc + 2\,CO_2 + 2\,CO + H_2 \qquad (8)$$

This active species is involved both in hydrogenation and CO addition.

At 90–110 °C and 10–15 bar, about 85% *n*-butanol and 15% isobutanol are formed with a total selectivity of about 90% (based on C_3H_6 and CO), similar to the 'oxo' reaction. The catalyst separates as a discrete phase from the two-phase reaction mixture and is recycled to the process.

process characteristics of Reppe butanol manufacture:

analogous to hydroformylation – homogeneously catalyzed liquid-phase reaction under pressure yields mixture of butanols, but differs in that alcohols are manufactured directly under milder reaction conditions

BASF developed the Reppe route to an industrial process that was used until 1984 by Japan Butanol Company in a 30 000 tonne-per-year unit.

commercial use of Reppe butanol technology:

30 000 tonne-per-year plant in Japan until 1984

Sec- and *tert*-butanol can, as with ethanol and isopropanol, be manufactured in the H_2SO_4 process by indirect hydration at 20–40 °C. In accordance with Markownikoff's rule, 1-butene and the 2-butenes give rise to the same alcohol:

manufacture of *sec*-butanol (2-butanol): indirect hydration of *n*-butenes through

$$CH_3CH_2\underset{\underset{OSO_3H}{|}}{C}HCH_3$$

with subsequent hydrolysis

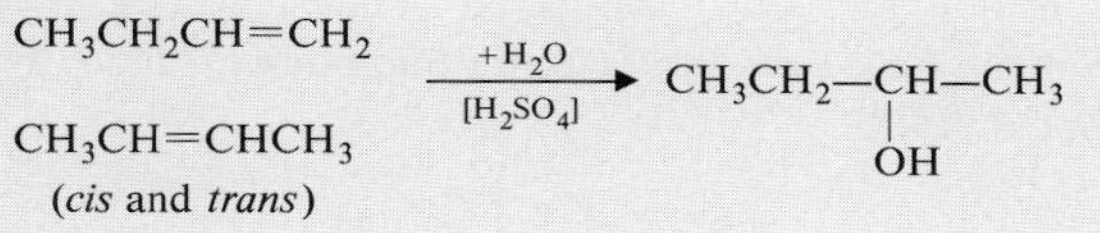

$$
\begin{array}{l}
CH_3CH_2CH=CH_2 \\
\\
CH_3CH=CHCH_3 \\
(cis \text{ and } trans)
\end{array}
\quad \xrightarrow[\text{[H}_2\text{SO}_4\text{]}]{+H_2O} \quad
CH_3CH_2-\underset{\underset{OH}{|}}{C}H-CH_3
\qquad (9)
$$

For the *n*-butenes, the H_2SO_4 concentration must be around 75–80%. However, 50–60% H_2SO_4 is sufficient for isobutene conversion:

manufacture of *tert*-butanol (2-methyl-2-propanol):

indirect isobutene hydration through $(CH_3)_3COSO_3H$ with subsequent hydrolysis

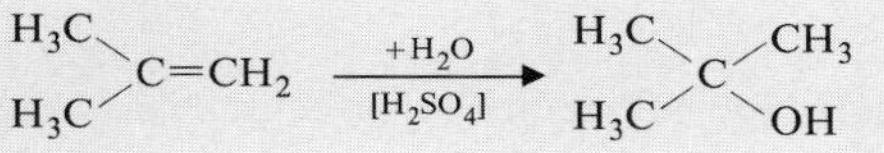

$$
\begin{array}{c}
H_3C \\ \\ H_3C
\end{array}\!\!C=CH_2
\quad \xrightarrow[\text{[H}_2\text{SO}_4\text{]}]{+H_2O} \quad
\begin{array}{c}
H_3C \\ \\ H_3C
\end{array}\!\!C\!\!
\begin{array}{c}
CH_3 \\ \\ OH
\end{array}
\qquad (10)
$$

The greater reactivity of isobutene can be used to separate it from the *n*-butenes. A *n*-butene/isobutene mixture is treated at 0 °C with 50–60% H_2SO_4 to convert isobutene to *tert*-butyl hydrogensulfate $(CH_3)_3C-OSO_3H$, which is soluble in H_2SO_4. The *n*-butenes are separated, and converted with 75–80% H_2SO_4 at 40–50 °C to *sec*-butyl hydrogensulfate. The esters are then diluted with water and saponified to the corresponding alcohols by heating.

principles of *n*-butene/isobutene separation by monoester formation with H_2SO_4:

higher reactivity of isobutene leads to esterification at lower temperatures and H_2SO_4 concentrations than with *n*-butenes

Several other hydration catalysts (including WO_3 and Al_2O_3) have been investigated, but are not yet used commercially.

In a new development by Hüls, the hydration is performed in a noncorrosive environment by using acidic ion-exchange resins as catalysts.

The manufacture of *tert*-butanol has already been discussed as a step in the separation of *n*-butenes and isobutene from C_4 fractions (*cf.* Section 3.3.2).

manufacturing variants for butanols:

1. direct butene hydration – commercially insignificant
2. isobutane oxidation through *tert*-butyl hydroperoxide → *tert*-butanol

Commercial application of a direct acid-catalyzed hydration of 1-butene and 2-butene, analogous to the manufacture of ethanol or isopropanol from ethylene or propene, respectively, was not achieved for a long time. The first unit — a 60000 tonne-per-year plant belonging to Deutsche Texaco — started operation in 1984. Acidic ion-exchange resins can be used as catalysts for the direct hydration of *n*-butenes, similar to the manufacturing process for isopropanol (*cf.* Section 8.1.2).

Tert-butanol is present as a cooxidized substance during the manufacture of propylene oxide (*cf.* Section 11.1.1.2).

uses of *sec*- and *tert*-butanol:

solvent (generally in paint industry)
de-icing agent
antiknock compound

Sec- and *tert*-butanol and their esters are frequently used as solvents in place of *n*-butanol and its esters. *tert*-Butanol is also used as a motor fuel additive (to prevent icing of the carburetor) and as an antiknock agent.

intermediate products, *e. g.*, dehydrogenation of *sec*-butanol to methyl ethyl ketone, introduction of *tert*-butyl groups in aromatics with *tert*-butanol

Sec- and *tert*-butanol are also chemical intermediates. The greatest demand for *sec*-butanol is for the manufacture of methyl ethyl ketone (MEK):

$$CH_3CH_2\underset{\underset{OH}{|}}{C}HCH_3 \xrightarrow{\text{[cat.]}} C_2H_5\underset{\underset{O}{\|}}{C}CH_3 + H_2 \qquad (11)$$

$$\left(\Delta H = -\begin{array}{c} 12 \text{ kcal} \\ 51 \text{ kJ} \end{array}/\text{mol}\right)$$

principle of MEK manufacture:

catalytic dehydrogenation of *sec*-butanol in

1. gas phase over ZnO or Cu/Zn, Cu/Cr or Pt systems
2. liquid phase over Raney Ni or CuO · Cr₂O₃

uses of MEK:

as solvent for nitro- and acetylcellulose and for manufacture of 'MEK hydroperoxide' (initiator for polymerizations)

main components in hydroperoxide mixture:

Similar to the manufacture of acetone from isopropanol, this process is generally a gas-phase dehydrogenation at 400–500°C with ZnO or Cu–Zn catalysts (*cf.* Sect. 11.1.3.2). One of the first plants used the IFP process of liquid-phase dehydrogenation in the presence of finely divided Raney nickel or Cu chromite in a high-boiling solvent. A selectivity to MEK of over 95% is obtained with an 80–95% *sec*-butanol conversion. MEK is also a byproduct in the butane oxidation to acetic acid as carried out, for example, in a Hoechst Celanese plant, which was brought back into operation after an accident in 1989 (*cf.* Section 7.4.1.2). After acetone, methyl ethyl ketone is the most important ketone industrially. It is used principally as a solvent for paints and resins, and also as a dewaxing agent for lubricating oils. The isomeric mixture ('methyl ethyl ketone peroxide') which results from the reaction of methyl ethyl ketone with H_2O_2 belongs, with dibenzoyl peroxide, to the highest-volume peroxides. Its main application is in the curing of unsaturated polyester resins.

$$HO-O-\underset{\underset{C_2H_5}{|}}{\overset{\overset{CH_3}{|}}{C}}-O-O-\underset{\underset{C_2H_5}{|}}{\overset{\overset{CH_3}{|}}{C}}-O-OH \quad \sim 45\%$$

$$\underset{H_5C_2}{\overset{H_3C}{>}}\underset{\underset{O}{}}{\overset{O-O}{C}}\underset{\underset{O}{}}{\overset{CH_3}{<}}_{C_2H_5} \quad \sim 25\%$$

$$\underset{H_3C}{\overset{C}{<}}{>}_{C_2H_5}$$

The production capacity for MEK in 1989 in the USA, Western Europe, and Japan was ca. 280000, 360000, and 180000 tonnes, respectively. Production figures for these countries are given in the adjacent table.

MEK production (in 1000 tonnes):

	1990	1992	1994
USA	211	217	272
W. Europe	275	n.a.	n.a.
Japan	178	178	187

n.a. = not available

8.1.4. Amyl Alcohols

The group of C_5 alcohols is also increasing in industrial importance, especially mixtures obtained from the hydroformylation and subsequent hydrogenation of *n*-butenes:

industrially most important C_5 alcohols:

1. $CH_3(CH_2)_3CH_2OH$

$CH_3CH_2CHCH_2OH$ 'amyl alcohol mixture'
  $|$
  CH_3

2. $(CH_3)_2CHCH_2CH_2OH$ = 3-methyl-1-butanol (isoamyl alcohol)

$CH_3(CH_2)_3CH_2OH$ 1-pentanol (*n*-pentanol)
+
$CH_3CH_2CHCH_2OH$ 2-methyl-1-butanol
  $|$
  CH_3

and isoamyl alcohol, which is formed from isobutene using the same process:

$(CH_3)_2CHCH_2CH_2OH$ 3-methyl-1-butanol (isoamyl alcohol)

manufacture of C_5 alcohols:

hydroformylation followed by hydrogenation of:

The mixture of amyl alcohols from *n*-butenes is used as a solvent for fats, oils, and many natural and synthetic resins. Their esters are used as perfumes and extractants, while isoamyl alcohol and its esters are employed as solvents.

1. *n*-butenes
2. isobutene

uses of C_5 alcohols and esters: solvents, perfumes, extractants

8.2. Higher Alcohols

Higher monohydric alcohols in the range C_6-C_{18} are of particular industrial significance. Commercial interest encompasses the whole group of primary and secondary, branched and unbranched, even- and odd-numbered alcohols.

primary and secondary, higher mono-alcohols above C_6 have many industrial applications
tertiary higher alcohols insignificant

The C_6-C_{11} and $C_{12}-C_{18}$ alcohols have earned the titles 'plasticizer alcohols' and 'detergent alcohols', respectively, by virtue of their major end use. The other conventional name, 'fatty alcohols', refers to the group of predominantly unbranched primary alcohols C_8 or higher that were previously only available as natural products, but are now also synthesized from petrochemical products. In the past few years, fats and oils

Common nomenclature:

C_6-C_{11} = plasticizer alcohols
$C_{12}-C_{18}$ = detergent alcohols
$\geq C_8$ = fatty alcohols

agricultural products used increasingly as renewable raw materials for "oleochemistry", e.g., for higher alcohols

from renewable resources such as rapeseed, sunflower seed and flaxseed have been used increasingly as raw materials for alcohol production.

In 1990, the world production capacity for C_{12+} alcohols was more than 1.4×10^6 tonnes per year. Of this, about 43% was derived from natural products.

manufacture of higher alcohols in four industrially important processes:

The four routes generally used for the manufacture of higher alcohols differ in feedstocks and in the basic nature of the process, so that all specific alcohol types can be synthesized.

1. hydrogenation of fatty acids manufactured from fats and oils (tallow, palm oil, coconut oil, etc.), from paraffins, or from ricinoleic or oleic acid

1. The hydrogenation of fatty acids from fats and oils, or of fatty acids from the catalytic oxidation of n-paraffins or two special processes (oxidative alkaline cleavage of ricinoleic acid and ozonolysis of oleic acid), yields linear primary alcohols. The natural fatty alcohols from fatty glycerides formed by synthesis or degradation over activated acetic acid (acetic acid thioester of coenzyme A) have an even number of carbon atoms.

2. hydroformylation of olefins

2. The hydroformylation of olefins followed by hydrogenation of the aldehydes yields straight chain and branched primary alcohols (n- and isoalcohols) from linear olefin mixtures.

3. partial oxidation of paraffins

3. The partial oxidation of linear paraffins yields linear secondary alcohols.

4. Ziegler growth reaction (Alfol synthesis)

4. The Alfol process leads to linear even-numbered primary alcohols.

To 1:

principle of fatty acid hydrogenation:

CuO/Cr$_2$O$_3$-catalyzed pressure hydrogenation (at high temperature) of fatty acids or their esters

four methods for manufacture of fatty acids and fatty acid esters:

1. hydrolysis or alcoholysis of fatty acid triglycerides from fats and oils, or
2. paraffin oxidation with subsequent esterification
3. ricinoleic acid thermolysis and oxidation
4. oleic acid ozonolysis and oxidation

Fatty acids or their methyl esters from the catalytic saponification (usually ZnO) or methanolysis of the fatty acid triglycerides (from fats or oils) are hydrogenated at about 200–300 bar and 250–350 °C in the presence of Cu–Cr-oxide catalysts (Adkins catalysts) in fixed-bed reactors to give fatty alcohols:

$$
\begin{array}{l}
\text{H}_2\text{COCOR} \\
| \\
\text{HCOCOR} \\
| \\
\text{H}_2\text{COCOR}
\end{array}
\xrightarrow[\text{[cat.]}]{\text{H}_2\text{O (CH}_3\text{OH)}}
\begin{array}{c}
\text{HOCH}_2\text{CHOHCH}_2\text{OH} \\
+ \\
\text{RCOOH} \\
(\text{RCOOCH}_3)
\end{array}
$$

(12)

$$\downarrow \; +\text{H}_2 \quad \text{[cat.]}$$

$$\text{RCH}_2\text{OH} + \text{H}_2\text{O}$$
$$(\text{CH}_3\text{OH})$$

A mixture of linear primary C_8 to C_{20} alcohols is obtained, for example, from coconut oil which in terms of quantity is the fifth most important vegetable oil after soybean oil, palm oil, rapeseed oil, and sunflower oil. Although considerable amounts of fatty acids are obtained from this source, their growth cannot compete with that of the Alfols, which have a similar structure. The share of natural fatty alcohols in the total capacity of natural and synthetic fatty alcohols varies considerably from country to country (cf. Table 8–1):

distribution of fatty alcohols in coconut oil (in wt%):

$CH_3(CH_2)_6CH_2OH$	8–9
$CH_3(CH_2)_8CH_2OH$	7–10
$CH_3(CH_2)_{10}CH_2OH$	45–51
$CH_3(CH_2)_{12}CH_2OH$	16–18
$CH_3(CH_2)_{14}CH_2OH$	7–10
$CH_3(CH_2)_{16}CH_2OH$	1–3
$CH_3(CH_2)_{18}CH_2OH$	6–11

Table 8-1. Fatty alcohol capacity.

	Total capacity (in 1000 tons)			Percentage of natural fatty alcohols		
	1981	1985	1991	1981	1985	1991
Western Europe	400	415	540	45	59	58
USA	433	570	524	17	26	21
Former Eastern Bloc	75	145	104	60	17	13
Japan	60	105	125	24	38	31
World	1100	1300	1332	30	40	40

worldwide vegetable oil production (in 10^6 tonnes):

	1990
Soybean oil	16.9
Palm oil	10.6
Rapeseed oil	8.1
Sunflower oil	8.0
Coconut oil	3.1

A recently developed catalyst (Henkel) allows the direct hydrogenation of fats and oils to fatty alcohols and 1,2-propanediol. Esterification can be avoided by using a Cu–Cr-oxide/spinel catalyst at 200°C, 250 bar, and a large excess of hydrogen.

Henkel fat/oil hydrogenation (hydrogenolysis) without esterification and simultaneous dehydration/hydrogenation of glycerol to

$$CH_3-\underset{\underset{OH}{|}}{C}HCH_2OH$$

A whole spectrum of C_1–C_{30} acids is obtained from the catalytic liquid-phase oxidation of linear C_{20}–C_{30} paraffins with Mn salts at about 130°C and slight pressure, with a relatively high proportion in the range C_{12}–C_{18}. After esterification with methanol or n-butanol, they are hydrogenated to the alcohols over Cu-Cr-oxide or Cu-Zn-oxide catalysts at about 200°C and 200 bar. The first production plant for synthetic fatty acids based on the oxidation of paraffins was erected jointly by Imhausen and Henkel in 1936. Today, paraffin oxidation and the subsequent hydrogenation to alcohols is mainly used in the former Eastern Bloc countries, China, Germany (ca. 40000 tonnes per year), and the CIS (300000 tonnes per year

commercial fatty acid manufacture via paraffin oxidation:

China, former E. Germany, the CIS, (temporarily in W. Europe)

fatty acid production (in 10^6 tonnes)

	1989	1991	1994
W. Europe	862	850	902
USA	723	755	916
Japan	307	309	262

paraffin oxidation products). In Italy, Liquichimia also produced fatty alcohols from linear paraffins for a short time in a 120 000 tonne-per-year plant. Here pure paraffins (C_8-C_{20}) from crude oil fractions were first dehydrogenated to internal olefins. These were converted into aldehydes in an 'oxo' reaction, hydrogenated to alcohols, and oxidized with air to give carboxylic acids, a large fraction of which are branched.

other processes for fatty acid manufacture:

thermolysis and ozonolysis of natural products

There are two other processes available in Western Europe for the production of fatty alcohols from natural products. ATO has a plant in France for the thermal cracking of ricinoleic acid into sebacic acid and heptanal (*cf.* Section 10.2.2). The second method — used by Unilever Emery in a single plant — is the ozonolysis of oleic acid in a two-step oxidation at temperatures under 100 °C to give pelargonic acid and azelaic acid:

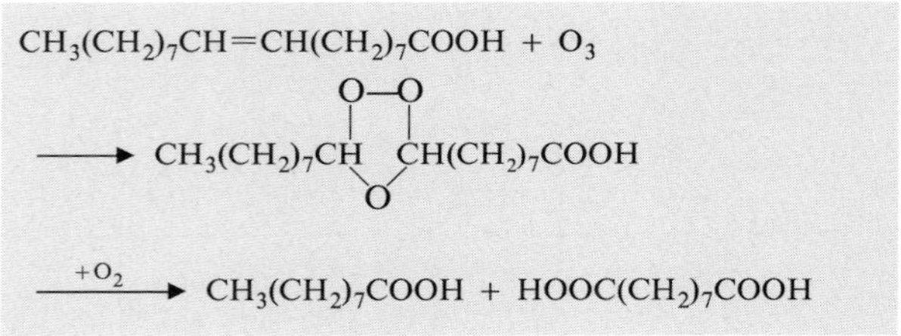

$$CH_3(CH_2)_7CH{=}CH(CH_2)_7COOH + O_3$$

$$\longrightarrow CH_3(CH_2)_7CH \overset{O-O}{\underset{O}{\vert \quad \vert}} CH(CH_2)_7COOH \qquad (13)$$

$$\overset{+O_2}{\longrightarrow} CH_3(CH_2)_7COOH + HOOC(CH_2)_7COOH$$

other uses for fatty acids:

esters, amides, Al-, Mg-, Zn-salts for emulsifying and thickening agents, amines for corrosion inhibition, fabric softeners, adhesives, and flotation agents

(world production of fatty amines currently ca. 300 000 tonnes per year)

About 30−40% of all fatty acid production is used for alcohol manufacture, and a similar amount is used as the sodium salt in the cleaning industry. Fatty acid derivatives such as esters, amides, and metal soaps (Al, Mg, and Zn soaps) are used as emulsifiers and thickeners. Fatty acid nitriles from the dehydration of ammonium salts (from fatty acid amides) are hydrogenated to primary fatty amines. Fatty amines, fatty diamines, and quaternary ammonium salts from the reaction of N,N-dialkylated fatty amines with alkyl halides (usually methyl chloride) are used as corrosion inhibitors, fabric softeners, adhesives, and flotation agents.

Dicarboxylic acids (*e. g.,* azelaic acid) are used for alkyd resins, polyamides, polyesters, plasticizers and lubricants. Henkel (Emery Group) is the only manufacturer of azelaic acid in the USA.

To 2:

hydroformylation of olefins with two constructive synthetic principles for the C skeleton:

The manufacture of alcohols is the most important industrial application of hydroformylation. As already discussed with in Section 6.1, there are two basic processes:

1. The olefins are converted into alcohols by hydroformylation followed by hydrogenation. These alcohols possess one more carbon atom than the olefin feedstock. Isooctanol[1] from isoheptenes, isononanol from diisobutene, isodecanol from tripropene and isotridecanol from tetrapropene are all members of this group.

2. By an aldol condensation followed by hydrogenation, an 'oxo' aldehyde is converted into an alcohol possessing double the number of carbon atoms. This process is used commercially to manufacture 2-ethylhexanol from *n*-butanal (*cf.* Section 6.1.4.3) and isooctadecanol from isononanal.

To 3 and 4:

These two manufacturing routes will be discussed in the following sections.

8.2.1. Oxidation of Paraffins to Alcohols

Linear secondary alcohols are manufactured from *n*-paraffins, which can be obtained from kerosene fractions (*cf.* Section 3.3.3). The development of paraffin oxidation began around 1930 in Germany. It was commercialized in about 1940 when hydrocarbons from the Fischer–Tropsch synthesis were oxidized with air to give a complex mixture of alcohols, ketones, esters, and acids.

A substantial increase in selectivity to alcohols was obtained with the use of boric acid. In Bashkirov's method for oxidizing *n*-paraffins ($C_{10}-C_{20}$) with air or oxygen in the presence of boric acid (mainly as metaboric acid, $HO-B=O$, under reaction conditions), secondary alcohols were obtained without chain cleavage:

$$R^1-CH_2-CH_2-R^2 + 0.5\,O_2 \xrightarrow{[H_3BO_3]} \begin{array}{c} R^1-CH-CH_2-R^2 \\ | \\ OH \\ + \\ R^1-CH_2-CH-R^2 \\ | \\ OH \end{array} \quad (14)$$

The secondary alkyl hydroperoxides formed initially react with boric acid to give thermally stable, non-oxidizing boric acid esters. The hydroperoxides are thus prevented from decomposing

Side notes

1. olefin C skeleton + C: used industrially for isoolefin mixtures → isoalcohol mixtures:

iso-C_7 olefin → iso-C_8 alcohol
iso-C_8 olefin → iso-C_9 alcohol
iso-C_9 olefin → iso-C_{10} alcohol
iso-C_{12} olefin → iso-C_{13} alcohol

2. aldol condensation of 'oxo' aldehydes followed by hydrogenation:

n-C_4 aldehyde → iso-C_8 alcohol
iso-C_9 aldehyde → iso-C_{18} alcohol

process principle of alkane oxidation:

air oxidation of alkanes catalyzed by metal ions gives *sec*-alcohols from radical reaction with intermediate alkane hydroperoxides

operation of paraffin oxidation:

'Bashkirov Oxidation' through boric acid ester from alkyl hydroperoxides with statistical distribution of O—O—H groups along paraffin chain

$$R^1-CH-CH_2-R^2 \xrightarrow{HO-B=O}$$
$$\quad\quad | $$
$$\quad\quad O-O-H$$

$$R^1-CH-CH_2-R^2$$
$$\quad\quad | $$
$$\quad O-O-B^{\ominus}-OH_2^{\oplus} \longrightarrow$$
$$\quad\quad\quad || $$
$$\quad\quad\quad O$$

$$R^1-CH-CH_2-R^2$$
$$\quad\quad | \quad\quad\quad + O + H_2O$$
$$\quad\quad O-B=O$$

characteristics of boric acid route:

stable boric acid esters from direct reaction of H_3BO_3 with hydroperoxides prevent further oxidation of alcohols

---

[1] The term 'isooctanol' is a commercial expression which has no relationship to 'iso' in systematic nomenclature.

H_3BO_3 is recovered after ester saponification and recycled, *i. e.*, H_3BO_3 cycle

to alcohols and being further oxidized. In commercial plants such as in Japan (Nippon Shokubai) and the CIS (UCC's plant in the USA was shut down in 1976), the paraffin is oxidized at atmospheric pressure with air at $140-190 \,°C$, generally in the presence of about 0.1 wt% $KMnO_4$ and $4-5$ wt% boric acid until a $15-25\%$ conversion is obtained. By using a cocatalyst such as NH_3 or amine in addition to boric acid, Nippon Shokubai obtains the same selectivity with a higher conversion. The feedstock paraffins have a much lower boiling point than the boric acid esters, and can be easily distilled from the oxidized mixture. The boric acid esters are later hydrolyzed.

Since the boric acid esters are thermally labile and, particularly in acid, can be cleaved to boric acid and olefins, it may be preferable to hydrolyze them directly while still in the oxidation medium.

industrial plants employing Bashkirov paraffin oxidation:

Japan, CIS, USA until 1977

Boric acid can be obtained from the aqueous phase and, after crystallization and dehydration, recycled to the process. With the direct hydrolysis, the organic phase is separated by distillation into unreacted paraffin and the oxidation medium which is then hydrogenated to convert the appreciable amounts of ketones present into the corresponding alcohols. The selectivities are about 70% to alcohols, 20% to ketones, and 10% to carboxylic acids.

oxidation takes place without chain degradation or structural isomerism

The oxidation yields almost exclusively (*i. e.*, up to 98%) secondary alcohols with the same chain length as the paraffin feedstock and a statistical distribution of OH groups.

uses of alcohols from paraffin oxidation:

in form of $ROSO_3H$ or $R(OCH_2CH_2)_nOSO_3H$ as alkali salts or directly as ethoxylates for surfactants and detergents

The bulk of these alcohols is processed to give surfactants and detergents either by direct esterification with H_2SO_4 or SO_3 to ether sulfates, or after previous ethoxylation (*cf.* Section 8.2.2). The secondary alcohol ethoxylates are already excellent materials for detergents before esterification.

Another important use of Bashkirov oxidation in alcohol manufacture is the commercial oxidation of cyclohexane (*cf.* Section 10.1.1), and the production of cyclododecanol as an intermediate in the synthesis of nylon-12 (*cf.* Section 10.1.2).

8.2.2. Alfol Synthesis

Ziegler synthetic alcohols – principles of process:

growth reaction with $H_2C{=}CH_2$ in presence of triethylaluminum leads to linear primary alcohols ('Alfols' or 'Ziegler alcohols')

K. Ziegler's organo-aluminum reactions provide a route to unbranched primary alcohols with an even number of carbons. The Alfol process, as it is known, allows the same fatty alcohols to be synthesized from ethylene as arise from natural products.

The Alfol process proceeds in four steps. First, the reactant triethylaluminum is produced. Then, a very finely divided Al powder and triethylaluminum are hydrogenated with hydrogen to diethylaluminum hydride at $110-140\,°C$ and $50-200$ bar:

four-step Alfol synthesis:

1. catalyst manufacture
2. building of C chain
3 oxidation
4. hydrolysis

$$Al + 2\,Al(C_2H_5)_3 + 1.5\,H_2 \longrightarrow 3\,AlH(C_2H_5)_2 \qquad (15)$$

Next, triethylaluminum is obtained from the reaction with ethylene at about $100\,°C$ and 25 bar:

to 1:

$Al(C_2H_5)_3$ theoretically increases around 50% (*cf.* eq $15-16$)

in practice less, due to:

$H_2 + Al(C_2H_5)_3 \rightarrow AlH(C_2H_5)_2 + C_2H_6$

$$AlH(C_2H_5)_2 + H_2C=CH_2 \longrightarrow Al(C_2H_5)_3 \qquad (16)$$

In the second step, the strongly exothermic chain-growth reaction takes place in a flow tube at about $120\,°C$ and an ethylene pressure of $100-140$ bar. This conversion is known as the insertion reaction (*cf.* Section 3.3.3.1):

to 2:

higher trialkyl-Al compounds of twofold importance:

1. for oxidation to alcohols
2. for thermolysis to α-olefins

$$Al(C_2H_5)_3 + 3\,n\,C_2H_4 \longrightarrow Al[(CH_2-CH_2)_n-C_2H_5]_3 \qquad (17)$$

($n = 1$ to ca. 11)

If necessary, this growth reaction can be coupled with a transalkylation step in which the trialkylaluminum compounds react with α-olefins in the $C_{12}-C_{16}$ range until equilibration is attained. In this way, a substantially narrower chain distribution can be obtained in the alcohol mixture (Epal process of Ethyl Corporation, now Albemarle).

typical C number distribution and comparison with transalkylation (in %):

C number	basic process	*trans* alkylation
6	10	1
8	17	3
10	21	8
12	19	35
14	15	26
16	10	17
18	5	9
20	3	1

At the same time, the higher trialkylaluminum compounds are also suitable feedstocks for the manufacture of straight-chain α-olefins (the so-called Ziegler olefins; *cf.* Section 3.3.3.1).

In the third step of the Alfol synthesis, the trialkylaluminum compounds are oxidized at $50-100\,°C$ in an exothermic reaction with extremely dry air to the corresponding alkoxides:

to 3:

auto-oxidation of the higher trialkyl-Al compounds to Al-alcoholates with hydroperoxide intermediate

$$Al[(CH_2-CH_2)_n-C_2H_5]_3 + 1.5\,O_2$$
$$\longrightarrow Al[O(CH_2-CH_2)_n-C_2H_5]_3 \qquad (18)$$

Small amounts of byproducts such as esters, ethers, acids, and aldehydes result from this reaction. In the final step, the Al-

to 4:

hydrolysis in presence of H_2SO_4 or alone with H_2O

workup of the Alfols:

due to possible pyrolysis, the higher Alfols are vacuum distilled

alkoxides are saponified – with dilute H_2SO_4 or water – to the alcohols and a pure $Al_2(SO_4)_3$ solution or to extremely pure aluminum hydroxide:

$$Al[O(CH_2—CH_2)_n—C_2H_5]_3 + 3H_2O$$
$$\longrightarrow 3CH_3(CH_2—CH_2)_nCH_2OH + Al(OH)_3 \quad (19)$$

The alcohol selectivity reaches $85-91\%$. After separation of the alcohol and water phases, the organic phase is fractionally distilled.

above C_{14}, soft, colorless solids

uses of Alfols:

biodegradable detergents (sulfates and ether sulfates, i.e., H_2SO_4-esters of ethoxylated products)

uses of byproducts:

extremely pure Al_2O_3 as catalyst support

The strictly linear Ziegler alcohols are particularly suitable for biodegradable sulfates and ether sulfates for use in detergents. Important manufacturers are Vista Chemical (formerly Conoco), Ethyl Corporation and Shell in the USA; RWE-DEA (formerly Condea), Shell, PCUK, and ICI in Western Europe; Mitsui and Kao-Ethyl in Japan; and the CIS.

8.3. Polyhydric Alcohols

commercially important polyols:

1,2-diols:
ethylene-, propylene glycol

1,3-diols:
propanediol, dimethylpropanediol (neopentyl glycol), butanediol

1,4-diols:
butyne-, butene-, butanediol

1,6-diols:
hexanediol and

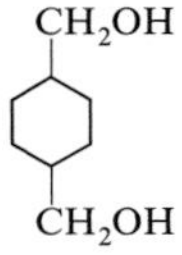

CH$_2$OH

CH$_2$OH

triols:
glycerol, trimethylolpropane

tetraols:
pentaerythritol

The most important group of polyhydric alcohols are the 1,2-diols, or glycols, which can be readily manufactured from epoxides, for example, ethylene glycol (cf. Section 7.2.1) and propylene glycol (cf. Section 11.1.2). The other di- and triols will be discussed elsewhere in conjunction with their precursors; e.g., 1,4-butanediol (cf. Sections 4.3; 5.3; 13.2.3.4), 1,3-butanediol (cf. Section 7.4.3), 1,6-hexanediol (cf. Section 10.1.1), 1,4-dimethylolcyclohexane (cf. Section 14.2.4), and glycerol (cf. Section 11.2.3). Another group of polyhydric alcohols results from a mixed aldol reaction between formaldehyde and acetaldehyde, isobutyraldehyde, or n-butyraldehyde. An aldehydic intermediate is formed initially, and must be reduced to the polyhydric alcohol. Pentaerythritol, neopentyl glycol, and trimethylolpropane are examples of these types of polyols.

8.3.1. Pentaerythritol

pentaerythritol is the simplest tetravalent alcohol (erythritol) with 5 C atoms:

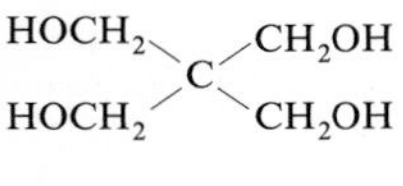

HOCH$_2$ — C — CH$_2$OH
HOCH$_2$ CH$_2$OH

Of the three formaldehyde secondary products mentioned, pentaerythritol or 2,2-bis(hydroxymethyl)-1,3-propanediol is produced in largest quantities. Production figures for pentaerythritol are given in the adjacent table. In 1991, production capacities for pentaerythritol in the USA, Japan, and Germany were 72000, 45000, and 35000 tonnes per year, respectively. Pentaerythritol is generally manufactured in a batch process but can also be continuously produced by reacting formaldehyde with acetaldehyde in an aqueous Ca(OH)$_2$ or NaOH solution at 15–45 °C. In the first step, a threefold mixed aldolization is followed by a Cannizzaro reaction, *i. e.*, reduction of trimethylol acetaldehyde with formaldehyde and formation of formic acid as a coproduct:

production of pentaerythritol (in 1000 tonnes):

	1990	1992	1994
USA	55	54	60

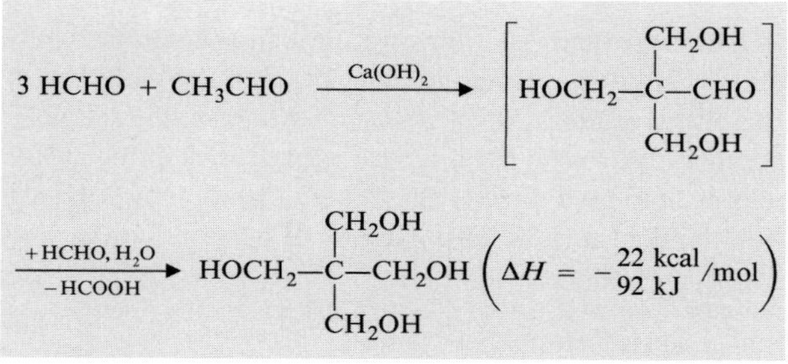

(20)

manufacture by classic Tollens method:

reaction of acetaldehyde with formaldehyde in presence of an alkaline catalyst (Ca(OH)$_2$ or NaOH) in a threefold aldolization with subsequent Cannizzaro reaction (intermolecular H transfer)

Theoretically, four moles formaldehyde are necessary per mole acetaldehyde. In practice, up to a fourfold excess of formaldehyde is used to limit the formation of dipentaerythritol, the monoether of pentaerythritol [(HOCH$_2$)$_3$C—CH$_2$]$_2$O.

avoiding HCOOH coproduct by catalytic hydrogenation of the aldol intermediate has not been successful

After neutralizing the excess Ca(OH)$_2$ with HCOOH the processing of the reaction mixture is done by stepwise concentration and fractional crystallization of pentaerythritol.

isolation of pentaerythritol:

high loss fractional crystallization of Ca-(OOCH)$_2$ and pentaerythritol

The analytically proven selectivity of about 91% (based on CH$_3$CHO) in the crude product is in practice seldom achieved because the isolation involves high losses.

Most pentaerythritol (about 90% in the USA) is used for the manufacture of alkyd resins, which are raw materials for the paint industry (*cf.* Section 11.2.1). The tetranitrate is used as an explosive, and the esters of the higher fatty acids are used as oil additives, plasticizers, and emulsifying agents.

use of pentaerythritol:

alkyd resins, explosives, plasticizers, additives, emulsifiers

8.3.2. Trimethylolpropane

Quantitatively, trimethylolpropane or 1,1,1-tris(hydroxymethyl)-propane ranks second amongst the polyols obtained from aldol hydrogenation. About 100000 tonnes are produced worldwide each year. Its manufacture is analogous to that of pentaerythritol; that is, mixed aldolization of an excess of formaldehyde with *n*-butyraldehyde in the presence of Ca(OH)$_2$, NaOH, or basic ion-exchangers followed by reduction:

manufacture of trimethylolpropane

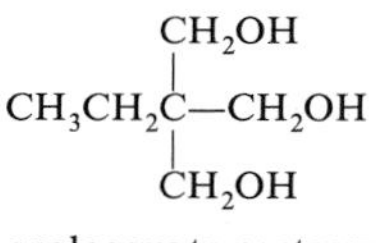

analogous to pentaerythritol by double aldolization of *n*-butyraldehyde and formaldehyde followed by Cannizzaro reaction

numerous attempts to find new routes without HCOOH coproduct have all been unsuccessful

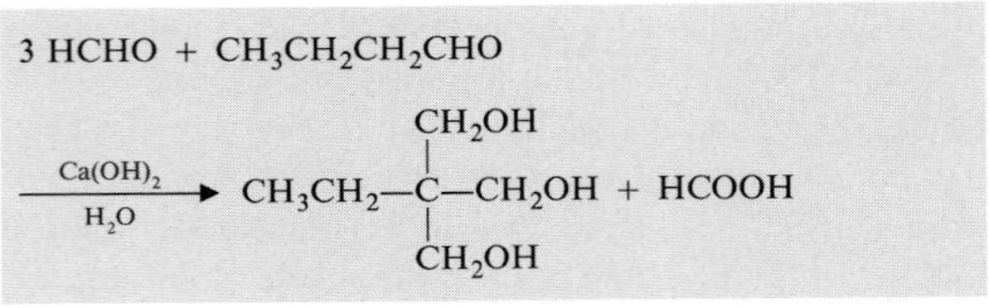

$$3\ HCHO\ +\ CH_3CH_2CH_2CHO$$

$$\xrightarrow[H_2O]{Ca(OH)_2}\ CH_3CH_2-\overset{\displaystyle CH_2OH}{\underset{\displaystyle CH_2OH}{\overset{|}{\underset{|}{C}}}}-CH_2OH\ +\ HCOOH \qquad (21)$$

isolation by extraction, purification by distillation

Trimethylolpropane is usually extracted from the reaction mixture with solvents (*e. g.*, acetic acid esters, cyclohexanol) and purified by distillation. The yield reaches about 90%.

application as glycerol substitute or after reaction with EO or PO as polyester or polyurethane component

This economical synthesis from two inexpensive, commercially available components has led in many cases to the replacement of glycerol (for example, in alkyd resins) by trimethylolpropane. Furthermore, it has the usual polyol applications and can, for example, be employed directly or after ethoxylation or propoxylation in the manufacture of polyesters or polyurethanes.

8.3.3. Neopentyl Glycol

manufacture of neopentyl glycol

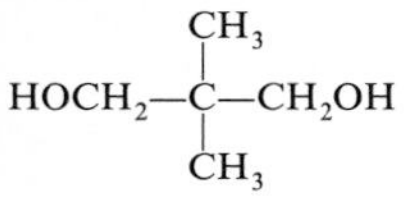

$$HOCH_2-\overset{\displaystyle CH_3}{\underset{\displaystyle CH_3}{\overset{|}{\underset{|}{C}}}}-CH_2OH$$

Neopentyl glycol, or 2,2-dimethyl-1,3-propanediol, can be readily manufactured at low cost by the mixed aldolization of formaldehyde and isobutyraldehyde. The world production capacity in 1991 was about 160 000 tonnes per year, with 90 000, 46 000, and 10 000 tonnes per year in Western Europe, the USA, and Japan, respectively. According to these figures, it is the lowest-volume product in this group of polyols, but the market is steadily increasing.

by mixed aldolization of isobutyraldehyde and formaldehyde with alkaline catalysts such as alkali hydroxides and acetates, amines, ion-exchangers

further hydrogenation to neopentyl glycol by Cannizzaro reaction or catalytically with H_2

With equimolar amounts of the two aldehydes, the aldol reaction leads – in the presence of basic substances – with very high selectivities to the isolable intermediate hydroxypivalaldehyde. This is converted either with an excess of formaldehyde, or by catalytic hydrogenation in the gas or liquid phase over Co, Cu, or nickel catalysts at temperatures over 80 °C, to give neopentyl glycol with more than 90% selectivity:

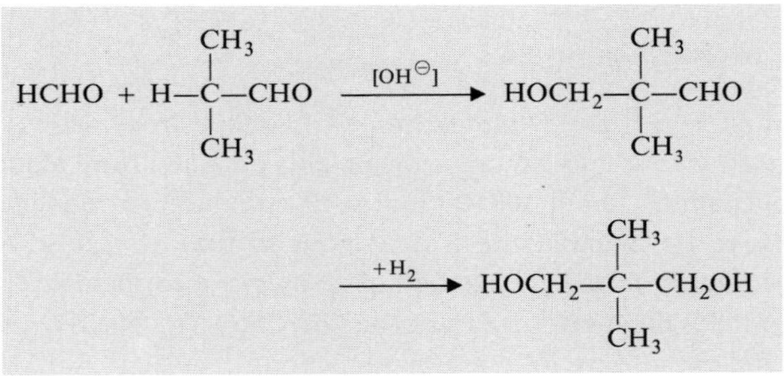

$$HCHO\ +\ H-\overset{\displaystyle CH_3}{\underset{\displaystyle CH_3}{\overset{|}{\underset{|}{C}}}}-CHO\ \xrightarrow{[OH^{\ominus}]}\ HOCH_2-\overset{\displaystyle CH_3}{\underset{\displaystyle CH_3}{\overset{|}{\underset{|}{C}}}}-CHO$$

$$\xrightarrow{+H_2}\ HOCH_2-\overset{\displaystyle CH_3}{\underset{\displaystyle CH_3}{\overset{|}{\underset{|}{C}}}}-CH_2OH \qquad (22)$$

Neopentyl glycol is used in polyesters (for plastics in airplane and ship construction), synthetic resin paints (principally for solvent-free powder coatings), synthetic lubricating oils, and plasticizers. The neopentyl structure of the reaction products leads to particular stability toward hydrolysis, heat, and light.

One neopentyl glycol derivative that is used industrially is 2,2-dimethyl-1,3-propanediol-monohydroxypivalic acid ester. It is obtained by disproportionation (Tishchenko reaction) of hydroxypivalaldehyde, the intermediate from neopentyl glycol manufacture:

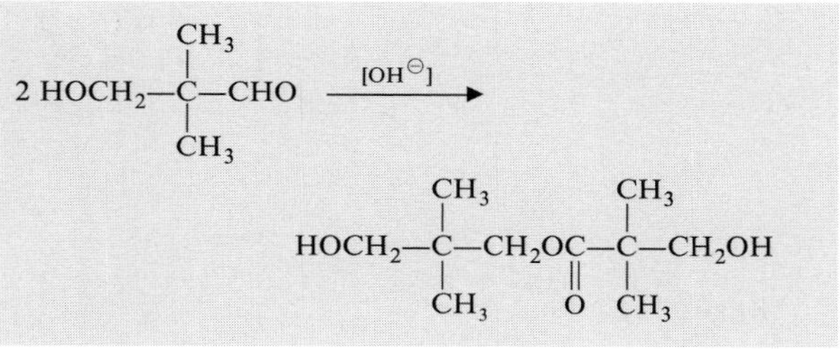

(23)

The reaction takes place in the presence of alkaline earth hydroxides at $80-130\,°C$. In 1976, BASF started up a 1200 tonne-per-year plant in Germany, which has since been expanded. UCC has discontinued production of 2,2-dimethyl-1,3-propanediol-monohydroxypivalic acid ester in the USA. The neopentyl glycol monoester of hydroxypivalic acid is increasingly used in the manufacture of high quality, extremely resistant polyester resins and polyurethanes.

use of neopentyl glycol:

as condensation product for synthesis of polyesters, synthetic resin paints, lubricants, and plasticizers

steric hindrance from neopentyl structure impedes ester formation, but makes esters, when formed, particularly stable

derivative of neopentyl glycol:

monoester with hydroxypivalic acid from alkaline-catalyzed Tishchenko reaction of hydroxypivalaldehyde (*cf.* acetaldehyde → ethyl acetate, Section 7.4.4)

9. Vinyl-Halogen and Vinyl-Oxygen Compounds

Due to their high reactivity, commercial availability, and versatile applicability, vinyl compounds have become key products in industrial chemistry.

The extensive group of vinyl compounds includes products such as styrene, acrylic acid, acrylonitrile, and methacrylic acid, which will be described elsewhere along with their characteristic precursors; and vinyl halides, vinyl esters and vinyl ethers, which will be dealt with in this chapter.

The vinyl halides also include multisubstituted ethylenes such as vinylidene chloride (VDC), vinylidene fluoride, trichloro- and tetrachloroethylene, and tetrafluoroethylene.

industrially important vinyl compounds:

$H_2C=CH-\langle\!\bigcirc\!\rangle$

$H_2C=CH-COOH(R)$

$H_2C=CH-CN$

$H_2C=C-COOH(R)$
$\qquad\quad |$
$\qquad\;\; CH_3$

$H_2C=CH-Cl(F)$

$H_2C=CH-OAc$

$H_2C=CH-OR$

$H_2C=CCl_2(F_2)$

$Cl_2C=CHCl$

$Cl_2C=CCl_2$

$F_2C=CF_2$

The highest volume products of the vinyl halides, -esters, and -ethers are vinyl chloride (VCM = vinyl chloride monomer) and vinyl acetate (VAM = vinyl acetate monomer). Their growth has essentially been determined by the new catalytic processes based on ethylene which have largely replaced the classical acetylene processes.

largest volume vinyl halides and esters:

$H_2C=CH-Cl$

$H_2C=CH-OCCH_3$
$\qquad\qquad\;\; \overset{\|}{O}$

9.1. Vinyl-Halogen Compounds

9.1.1. Vinyl Chloride

The supreme importance of vinyl chloride as a universal monomer for the manufacture of homo-, co-, and terpolymers is reflected in the dramatic expansion of its industrial production, which began in the 1930s and reached a worldwide production capacity of about 24×10^6 tonnes per year in 1995. In this year, the capacities in the USA, Western Europe, and Japan were about 6.3, 6.3, and 2.6×10^6 tonnes per year, respectively. Vinyl chloride, along with its precursor 1,2-dichloroethane, has thus become one of the greatest consumers of chlorine; for example, in 1995 about 37%, 46%, and 32% of the total chlorine

together with precursor ($ClCH_2CH_2Cl$ = EDC), VCM follows ethylene, propene, urea as the fourth largest volume organic product in the USA (1994)

EDC production (in 10^6 tonnes):

	1990	1992	1994
W. Europe	9.73	8.90	9.84
USA	6.28	7.23	7.60
Japan	2.66	2.71	2.79

production in the USA, Western Europe and Japan was employed in the manufacture of vinyl chloride. The pressure to optimize the process led to constant improvements and to an increase in the size of production units. The largest vinyl chloride production plant to date has a capacity of 635 000 tonnes per year, and started operation in the USA in 1991.

PVC recycling in a specialized pyrolysis process converts chlorine to available and reusable HCl

A new and attractive development has to do with the recovery of chlorine in the form of hydrogen chloride from polyvinyl chloride, the largest volume product made from vinyl chloride. Discarded PVC can be reycled by pyrolysis at 1200 °C in a rotary kiln and the resulting HCl can be isolated, purified, and reused for the production of vinyl chloride. Thus the fraction of total chlorine use for chlorine-containing end products, currently about 46% in Western Europe, can be lessened in the future by this new form of chlorine recycling.

Production figures for vinyl chloride in several countries are summarized in the adjacent table.

VCM production (in 10^6 tonnes):

	1990	1992	1994
W. Europe	5.92	5.31	5.40
USA	4.83	5.13	6.28
Japan	2.28	2.28	2.31

9.1.1.1. Vinyl Chloride from Acetylene

traditional VCM manufacture by HgCl$_2$-catalyzed HCl addition to acetylene (vinylation of hydrogen chloride):

older liquid-phase processes with dissolved or suspended catalyst, later generally gas phase with fixed- or fluidized-bed catalysts

Griesheim-Elektron developed the first industrial process for the manufacture of vinyl chloride. It was based on the addition of hydrogen chloride to acetylene, which was initially obtained solely from carbide. In the subsequent period, acetylene from petroleum was also employed in vinyl chloride manufacture. At first, the hydrogen chloride was made by the reaction of Cl$_2$ and H$_2$, but was gradually replaced by hydrogen chloride from chlorination. The vinylation of hydrogen chloride takes place according to the following equation:

$$HC\equiv CH \ + \ HCl \xrightarrow{\text{[cat.]}} H_2C=CHCl \ \left(\Delta H = -\frac{24 \ \text{kcal}}{99 \ \text{kJ}}/\text{mol} \right) \quad (1)$$

process principles:

heterogeneously catalyzed gas-phase reaction, very selective and with simple apparatus (tube-bundle reactor)

byproduct formation:

CH$_3$CHO from HC≡CH + H$_2$O
CH$_3$CHCl$_2$ from H$_2$C=CHCl + HCl

HgCl$_2$ on activated charcoal serves as a catalyst at 140–200 °C. The conversion of acetylene is 96–97%, with a selectivity to vinyl chloride of about 98% (based on C$_2$H$_2$). The isolation of the vinyl chloride is relatively simple as the only byproducts are small amounts of acetaldehyd (from residual moisture in the gas feed) and 1,1-dichloroethane (from HCl addition to vinyl chloride).

Despite low investment and operating costs, this route (based exclusively on acetylene) has been largely abandoned in favor of ethylene which represents a cheaper feedstock base. Hence, the last plant operating with this procedure was closed down in Japan in 1989. In 1991, there were two vinyl chloride plants based on acetylene still in operation in Western Europe, one in the USA, and four in the CIS. In 1993, the three plants in Western Europe and the USA, as well as a Hüls plant in Germany using this energy-intensive process, were shut down. In countries with inexpensive coal, such as South Africa, production of vinyl chloride from carbide-based acetylene will be economical for the longer term.

9.1.1.2. Vinyl Chloride from Ethylene

Today, vinyl chloride is almost exclusively manufactured by thermal cleavage (dehydrochlorination) of 1,2-dichloroethane (EDC). The feedstock for the thermolysis can be obtained from two routes:

1. By the older method, addition of chlorine to ethylene
2. By the more modern process, oxychlorination of ethylene with hydrogen chloride and O_2 or air.

To 1:

Ethylene chlorination generally takes place in the liquid phase in a bubble-column reactor using the reaction product EDC as reaction medium with dissolved $FeCl_3$, $CuCl_2$, or $SbCl_3$ as catalyst at 40–70 °C and 4–5 bar:

$$H_2C{=}CH_2 + Cl_2 \rightarrow ClCH_2{-}CH_2Cl \left(\Delta H = -\begin{array}{c}43 \text{ kcal}\\180 \text{ kJ}\end{array}/\text{mol}\right) \quad (2)$$

The addition of chlorine probably takes place according to an electrophilic ionic mechanism in which the catalyst causes polarization of the chlorine molecule, thereby facilitating an electrophilic attack.

Selectivity to 1,2-dichloroethane can reach 98% (based on C_2H_4) and 99% (based on Cl_2).

The chlorination of ethylene can also be carried out in the gas phase at 90–130 °C. Formation of 1,2-dichloroethane probably results from a radical chain mechanism, initiated by chlorine radicals formed at the reactor surface by homolysis of chlorine molecules.

VCM manufacture from ethylene through intermediate stage $ClCH_2CH_2Cl$ (EDC) with subsequent thermal dehydrochlorination

EDC manufacture from ethylene by

1. Cl_2 addition
2. oxychlorination

process principles of Cl_2 addition (liquid phase):

homogeneously catalyzed electrophilic reaction with carbenium ion $^{\oplus}CH_2{-}CH_2Cl$ as probable intermediate

process principles of Cl_2 addition (gas phase):

uncatalyzed radical chain reaction initiated by chlorine radicals from homolytic surface reaction

To 2:

process principles of oxychlorination:

Cu salt-catalyzed reaction of C_2H_4, HCl, and air (O_2), separable into a reaction step:

$C_2H_4 + 2\ CuCl_2 \rightarrow ClCH_2CH_2Cl + 2\ CuCl$

and a regeneration sequence:

$2\ CuCl + 0.5\ O_2 \rightarrow CuO \cdot CuCl_2$
$CuO \cdot CuCl_2 + 2\ HCl \rightarrow 2\ CuCl_2 + H_2O$

The oxychlorination of ethylene is generally conducted in the gas phase. When ethylene is reacted with anhydrous HCl and air or oxygen at $220-240\,°C$ and $2-4$ bar, it is converted into EDC and H_2O:

$$H_2C{=}CH_2 + 2\ HCl + 0.5\ O_2 \rightarrow ClCH_2{-}CH_2Cl + H_2O$$

$$\left(\Delta H = -\ \frac{57\ kcal}{239\ kJ}\Big/mol \right) \quad (3)$$

function of catalyst components:

1. $CuCl_2$ as chlorinating agent is volatile under reaction conditions
2. alkali chloride (KCl) lowers the volatility by formation of a salt melt
3. rare earth chlorides increase the O_2 absorption by the melt

Supported $CuCl_2$ serves as catalyst. The supports often contain activators and stabilizers such as chlorides of the rare earths and alkali metals. No free chlorine is formed under the reaction conditions; $CuCl_2$ is the chlorinating agent which is subsequently regenerated with air and HCl through the intermediate oxychloride. This is similar to the oxychlorination of benzene to chlorobenzene (*cf.* Section 13.2.1.1). Catalysts of this type, which are employed for example by Distillers and Shell, allow EDC selectivities of about 96% to be attained. The ethylene conversion is almost quantitative when a slight excess of HCl and air is used.

industrial operation of ethylene oxychlorination:

1. gas phase with fluidized-bed reactor or fixed-bed reactor
2. liquid phase

In the USA in 1964, Goodrich, Dow, and Monsanto were among the first firms to operate an industrial oxychlorination process. Today, numerous other firms have developed their own processes based on a similar principle.

characteristics of catalytic gas-phase oxychlorination:

compared to chlorination, by enthalpy of formation of H_2O (from $2\,HCl + 0.5\,O_2$) – more exothermic reaction
exact temperature control avoids:
superheating
excessive chlorination
total combustion to CO_2

There is a basic difference in the process operation which affects the heat removal from this strongly exothermic reaction. Ethyl Corp., Goodrich, Mitsui Toatsu Chemicals, Monsanto, Scientific Design (based on the Monsanto process), and Rhône-Poulenc use a fluidized-bed reactor, while the other manufacturers employ a fixed-bed reactor. Another method to limit the local development of heat is the use of catalysts diluted with inert materials.

characteristics of liquid-phase oxychlorination:

use of aqueous HCl facilitates heat removal but aggravates corrosion control problems

Besides the gas-phase processes for oxychlorination, Kellogg has developed a method involving an aqueous $CuCl_2$ solution acidified with hydrochloric acid. Ethylene can be converted into 1,2-dichloroethane by oxychlorination with $7-25\%$ conversion and 96% selectivity at $170-185\,°C$ and $12-18$ bar. The advantages of this modification lie with the use of aqueous hydrochloric acid and good heat removal by H_2O evaporation. However, considerable corrosion problems arise from the handling of the hot aqueous hydrochloric acid.

The further conversion of 1,2-dichloroethane to vinyl chloride used to be carried out in the liquid phase with alkali. Today, gas-phase dehydrochlorination is used exclusively:

$$ClCH_2CH_2Cl \rightarrow H_2C{=}CHCl + HCl \left(\Delta H = +\frac{17\ kcal}{71\ kJ}/mol \right) \quad (4)$$

The endothermic cleavage of EDC is conducted at 500–600 °C and 25–35 bar at high flow rates in tubes made of special steels (Ni, Cr) with high heat resistance. This is a thermal reaction that proceeds by a radical chain mechanism. In many cases, carbon tetrachloride is added in small amounts as an initiator. More important is the purity of the EDC used (> 99.5%), since impurities can easily inhibit the thermolysis (radical trapping). Catalytic cracking at 300–400 °C on pumice (SiO_2, Al_2O_3, alkalis) or on charcoal, impregnated with $BaCl_2$ or $ZnCl_2$, has not found more widespread application due to the limited life of the catalysts.

Conversion of EDC in thermal cracking amounts to 50–60% with a selectivity to vinyl chloride of greater than 98% (based on EDC). The reaction mixture is directly quenched with cold EDC, releasing gaseous hydrogen chloride. After separation of the vinyl chloride by distillation, the EDC is fed back to the dehydrochlorination step.

Modern industrial processes for the manufacture of vinyl chloride are characterized by an extensive and thus very economical integration of the above-mentioned partial steps of ethylene-chlorine addition, EDC thermolysis, and ethylene oxychlorination. In this integrated operation, chlorine is introduced to the process by addition to ethylene, and hydrogen chloride from the thermolysis is used in the oxychlorination.

As an alternative to the reuse of hydrogen chloride from thermolysis in the oxychlorination, processes have been developed for converting it into chlorine.

For example, aqueous hydrochloric acid in a particular concentration range (25 → 10 wt%) can be converted by electrolysis into Cl_2 and H_2 using processes from Hoechst-Uhde, De Nora, or Mobay Chemical.

The Shell-Deacon process, another method for recovering Cl_2, involves oxidizing gaseous HCl or aqueous hydrochloric acid to chlorine and water over a supported catalyst containing the

VCM manufacture by HCl elimination from EDC:

previously, dehydrochlorination in liquid phase with alkali

today, dehydrochlorination in gas phase by two processes:

1. thermal (favored method)
2. catalytic (infrequent; *e.g.*, SBA process, Wacker process)

principle of thermal HCl cleavage:

$ClCH_2CH_2Cl \rightarrow Cl{\cdot} + {\cdot}CH_2CH_2Cl$
$Cl{\cdot} + ClCH_2CH_2Cl \rightarrow HCl + ClCH_2\overset{\cdot}{C}HCl$
$ClCH_2\overset{\cdot}{C}HCl \rightarrow Cl{\cdot} + H_2C{=}CHCl$

process characteristics of thermolysis:

selectivity to VCM favored by high flow rate, exact temperature control, careful purification of EDC, limitation of conversion of EDC

modern VCM manufacture characterized by integration of three partial reactions:

1. liquid-phase C_2H_4 chlorination
2. gas-phase EDC dehydrochlorination
3. gas-phase C_2H_4 oxychlorination (fluidized- or fixed-bed)

integrated Cl_2 recovery from HCl in oxychlorination of C_2H_4 also possible externally by:

1. electrolysis, *e.g.*, Hoechst–Uhde, De Nora, Mobay Chemical
2. Deacon process (Shell)
3. KEL chlorine process (Kellogg)

common reaction principle of Shell–Deacon and KEL chlorine processes:

$4 HCl + O_2 \rightarrow 2 Cl_2 + 2 H_2O$

$$\left(\Delta H = - \frac{28 \text{ kcal}}{117 \text{ kJ}} \Big/ \text{mol} \right)$$

characteristic catalytic step:

Deacon process:

$2 CuCl_2 \rightarrow Cl_2 + Cu_2Cl_2$

KEL chlorine process:

$NO_2 + 2 HCl \rightarrow Cl_2 + NO + H_2O$

active components $CuCl_2$, KCl, and $DiCl_3$ (Di = didymium, *i.e.*, industrially resulting mixtures of rare earths) at 350–400 °C. This process has been used commercially in The Netherlands (41 000 tonnes per year Cl_2, no longer in operation) and in India (27 000 tonnes per year Cl_2). A modified Deacon process, the Mitsui-Toatsu (MT-chlorine) process, is catalyzed with an active chromium oxide/SiO_2 system. A plant with a production capacity of 30 000 tonnes per year is in operation in Japan.

A third route is the KEL chlorine process (Kellogg), in which the oxidation of HCl (anhydrous, or in aqueous solution) to Cl_2 at 4 bar and 260–320 °C is catalyzed by oxides of nitrogen, and H_2SO_4 (as nitrosylsulfuric acid) is present as a circulation "catalyst carrier". A 270 000 tonne-per-year plant using this process has been operated by Du Pont in the USA since 1974.

ethylene–acetylene combined processes were intermediate solution to problem of total Cl_2 utilization, but both ethylene and acetylene had to be available at the same time

During an interim period, before the modern processes for complete industrial utilization of chlorine based solely on ethylene had made their breakthrough, vinyl chloride was manufactured in a balanced ethylene–acetylene combined process. In analogy to the traditional VCM process, hydrogen chloride from the thermolysis of EDC was added to acetylene.

The combined use of ethylene and acetylene was particularly advantageous where thermal cracking of crude oil fractions supplied mixtures of C_2H_4 and C_2H_2.

Combined processes were operated by, for example, Goodrich, Monsanto, and UCC in the USA, by Hoechst/Knapsack in Germany, and by Kureha in Japan.

9.1.1.3. Potential Developments in Vinyl Chloride Manufacture

modern three-step cost-intensive VCM manufacturing processes show two possibilities for cost savings:

1. single-step, direct substitutional chlorination of C_2H_4

   $PtCl_2$–$CuCl_2$ supported catalysts in the gas phase (ICl)

   $PdCl_2$ in the liquid phase (Distillers)

   catalyst melts
   (Hoechst)

Considerations of improvement in the VCM manufacturing processes mainly concentrate on eliminating the multiple stages of combined processes (chlorination, oxychlorination, thermolysis), which entail considerable capital and process costs. A basic simplification of VCM manufacture could, for example, be achieved by direct chlorine substitution or oxychlorination of ethylene to vinyl chloride. The patent literature contains numerous examples with noble metal catalysts in the gas and liquid phase as well as in a melt of the catalytically active compounds. Until now, they have not led to an industrial process due to the high temperature necessary and to consequent side reactions.

The use of ethane rather than ethylene has also been suggested. In the Lummus–Armstrong Transcat process, ethane is converted to vinyl chloride by reacting directly with chlorine in a melt containing the chlorides of copper-based oxychlorination catalysts. In the first stage, the following chlorination, oxychlorination, and dehydrochlorination reactions occur simultaneously in a reactor containing a salt melt at 450–500 °C:

$$
\begin{aligned}
CH_3CH_3 + Cl_2 &\rightarrow CH_3CH_2Cl + HCl \\
CH_3CH_2Cl &\rightarrow H_2C{=}CH_2 + HCl \\
H_2C{=}CH_2 + Cl_2 &\rightarrow ClCH_2{-}CH_2Cl \\
ClCH_2{-}CH_2Cl &\rightarrow H_2C{=}CHCl + HCl
\end{aligned}
\tag{5}
$$

The melted copper salt participates in the chlorination process and, on transferring chlorine, is reduced from the divalent to the monovalent state:

$$
CuO \cdot CuCl_2 + 2\,HCl + H_2C{=}CH_2 \rightarrow 2\,CuCl \\
+ ClCH_2{-}CH_2Cl + H_2O
\tag{6}
$$

At an ethane conversion of about 30%, selectivity to vinyl chloride is less than 40%.

In order to regenerate the salt melt, it is fed to the oxidation reactor where, by means of HCl and air, it is brought to the oxychlorination level ($CuO{-}CuCl_2$) once again.

With ethane, however, the involved technology for circulating the catalyst melt has only been employed in a pilot plant (*cf.* Section 2.3.6.1).

In a newer gas-phase process from ICI, ethane undergoes complete conversion to ethylene and a mixture of chlorinated derivatives of ethane and ethylene in an oxychlorination reaction with chlorine and oxygen over zeolite catalysts containing silver metal and, *e. g.*, manganese salts at 350–400 °C. The selectivity to vinyl chloride can be as high as 51%.

9.1.1.4. Uses of Vinyl Chloride and 1,2-Dichloroethane

Most vinyl chloride (currently about 95% worldwide) is used as a monomer and, to a lesser extent, as a comonomer in polymerizations. Only a small amount, along with its precursor EDC, serves as a starting material for the numerous chlorine deriva-

2. use of ethane instead of ethylene for direct VCM manufacture:

2.1 Lummus–Armstrong 'Transcat process'

process characteristics:
two-step, heterogenously catalyzed conversion of ethane, Cl_2, O_2 to VCM in copper chloride/alkali chloride melt

1st step:

reactions take place simultaneously (chlorination, oxychlorination, dehydrochlorination)

2nd step:

regeneration of melt carried out separately

2.2. ICI oxychlorination process

process characteristics:

one-step, heterogeneously (Ag, Mn-, Co-, Ni-salt/zeolites) catalyzed conversion of C_2H_6, Cl_2, O_2 to VCM in the gas phase

uses of VCM and EDC:

1. VCM for PVC and numerous copolymers

main application areas for PVC:

building and electrical industries, packaging, automobile construction

1,1,1-trichloroethane production from:

1. EDC by chlorination to $CH_2ClCHCl_2$, HCl elimination to $H_2C=CCl_2$, and HCl addition
2. C_2H_6 by chlorination
3. VCM by HCl addition to CH_3CHCl_2 and chlorination

Route 3 is most important commercial route

tives of ethane and ethylene used as solvents and extractants which were once derived solely from acetylene.

The most important of these is 1,1,1-trichloroethane, which – apart from a three-step route from EDC or by chlorination of ethane – is chiefly produced from vinyl chloride. First, hydrogen chloride is added to vinyl chloride in the liquid phase in the presence of finely-divided $FeCl_3$ to form 1,1-dichloroethane, which also constitutes the reaction medium. Selectivities of $95-98\%$ are attained. The second step is a gas-phase chlorination; generally the Ethyl Corporation process in a fluidized bed at $370-400\,°C$ and $3-5$ bar over an SiO_2 catalyst is used:

$$H_2C=CHCl + HCl \rightarrow H_3C-CHCl_2 \xrightarrow[-HCl]{+Cl_2} H_3C-CCl_3 \quad (7)$$

The hydrogen chloride is recycled to the first reaction step. Selectivity to 1,1,1-trichloroethane is as high as 82%.

1,1,1-trichloroethane production (in 1000 tonnes):

	1989	1991	1993
USA	353	294	205
W. Europe	222	245	108
Japan	164	167	78

ethylenediamine manufacture from:

1. EDC by Cl/NH_2 substitution
2. $H_2NCH_2CH_2OH$ by OH/NH_2 substitution

both ammonolysis routes burdened by formation of byproducts

The main use of 1,1,1-trichloroethane, after addition of stabilizers, is as a cold cleaning agent for metals and textiles. The production capacity of 1,1,1-trichloroethane in 1995 was about 250000, 140000, and 15000 tonnes in the USA, Western Europe, and Japan, respectively. Production numbers for these countries are given in the adjacent table.

1,2-Dichloroethane is also used for the manufacture of ethylenediamine (1,2-diaminoethane); it is reacted with an excess of aqueous ammonia at about $180\,°C$ under pressure. Diethylenetriamine and triethylenetetramine are also formed in secondary reactions of the ethylenediamine with 1,2-dichloroethane. The mixture of amine hydrogen chlorides is converted with caustic soda to the free amines, which are then separated by rectification. The yield of ethylenediamine is about 60%. Ethylenediamine is an intermediate in the production of rubber chemicals, pharmaceuticals, and, above all, for the complexing agent ethylenediaminetetraacetic acid (EDTA).

Another route for the manufacture of ethylenediamine is based on the reaction of ethanolamine with NH_3 (cf. Section 7.2.3).

2. VCM and EDC as intermediates for manufacture of solvents and extractants, e. g.,

Cl_3C-CH_3 (trichloroethane)
$Cl_2C=CHCl$ (trichloroethylene)
$Cl_2C=CCl_2$ (tetrachloroethylene)
as well as
$H_2NCH_2CH_2NH_2$ (ethylenediamine)

1,2-Dichloroethane is also used as a solvent for resins, asphalt, bitumen, and rubber, and as an extractant for fats and oils. In combination with 1,2-dibromoethane, it is used with the antiknock agents tetraethyl- and tetramethyllead to convert almost nonvolatile lead compounds into readily volatile ones. However, because of the reduction of lead additives in gasoline, this use is declining sharply.

Vinyl chloride and 1,2-dichloroethane are also starting materials for the manufacture of vinylidene chloride (*cf.* Section 9.1.2) and tri- and tetrachloroethylene (*cf.* Section 9.1.4).

and the monomer $H_2C=CCl_2$ (vinylidene chloride, VDC)

9.1.2. Vinylidene Chloride

Vinylidene chloride (more properly named vinylidene dichloride, VDC) is manufactured by chlorinating vinyl chloride or 1,2-dichloroethane to 1,1,2-trichloroethane which is then dehydrochlorinated with aqueous $Ca(OH)_2$ or NaOH at about 100 °C to VDC with a selectivity of over 90%:

vinylidene chloride manufacture (two-step):

1. conversion of VCM or 1,2-dichloroethane into 1,1,2-trichloroethane
2. dehydrochlorination with aqueous alkali

$$
\begin{array}{l}
H_2C=CHCl \quad + \; Cl_2 \\
ClCH_2-CH_2Cl \;\; + \; Cl_2 \xrightarrow{-HCl}
\end{array}
\!\!\longrightarrow ClCH_2CHCl_2
$$

$$
\xrightarrow{-HCl} H_2C=CCl_2
$$

(8)

Gas-phase dehydrochlorination has been investigated many times; the selectivity to vinylidene chloride is, however, sharply lowered by the formation of 1,2-dichloroethylene.

Vinylidene chloride can be readily copolymerized with vinyl chloride and other monomers. The polymers are used to coat cellulose hydrate films and polypropene in order to reduce air and water permeability.

uses of vinylidene chloride:

as comonomer with VCM (85% VDC + 15% VCM, *e. g.*, 'Saran') for coating of films (cellulose hydrate and polypropene), with AN (30% VDC + 70% AN) for flame-resistant acrylic fibers

The current annual production of vinylidene chloride in the western world is about 150 000 – 200 000 tonnes.

9.1.3. Vinyl Fluoride and Vinylidene Fluoride

Vinyl fluoride (VFM, vinyl fluoride monomer) can be manufactured by two routes:

vinyl fluoride manufacture by two routes:

1. vinylation of HF with acetylene

1. By catalytic addition of HF to acetylene:

$$
HC\equiv CH + HF \xrightarrow{[cat.]} H_2C=CHF
$$

(9)

2. By catalytic substitution of chlorine with fluorine in vinyl chloride with 1-chloro-1-fluoroethane as intermediate, with subsequent dehydrochlorination:

2. HF addition to VCM followed by dehydrochlorination

$$
H_2C=CHCl + HF \rightarrow CH_3CHClF \xrightarrow{-HCl} H_2C=CHF
$$

(10)

characteristics of HF vinylation:

HF addition to HC≡CH requires catalyst, and generally stops at vinyl stage

To 1:

The vinylation of hydrogen fluoride is conducted with heterogeneous catalysts such as mercury compounds on supports at $40-150\,°C$ or with fluorides of Al, Sn, or Zn at higher temperatures ($250-400\,°C$). 1,1-Difluoroethane, which is formed as a byproduct, can be converted into additional VFM by HF cleavage.

With the Hg compounds, a VFM selectivity of 60 to 85% can be obtained. The disadvantage of the Hg catalysts is that they rapidly lose their activity.

Du Pont operates a vinyl fluoride plant based on a similar technology with a capacity of 2700 tonnes per year (1989).

characteristics of F exchange for Cl in VCM:

HF addition to VCM catalyst-free, mild conditions

HCl elimination under severe pyrolysis conditions, with or without catalyst

To 2:

The noncatalytic addition of hydrogen fluoride to vinyl chloride takes place relatively easily in an autoclave at slightly raised temperatures.

The elimination of HCl from the intermediate 1-chloro-1-fluoroethane can be conducted catalyst-free at $500-600\,°C$ in Cr-Ni steel reaction tubes or in the presence of copper powder or Cu-Ni alloys.

use of vinyl fluoride:

polymerization to poly(vinyl fluoride) (PVF)

$$n\,H_2C{=}CHF \rightarrow +CH_2{-}CH+_n$$
$$\overset{|}{F}$$

use of PVF:

for particularly weather resistant coatings in the form of films or dispersions

Vinyl fluoride is mainly used as a monomer for polymerization to polyvinyl fluoride (PVF). The polymerization takes place in aqueous suspension at $85\,°C$ and 300 bar, in the presence of peroxides such as dibenzoyl peroxide.

Polyvinyl fluoride is a high-grade polymer characterized by its exceptional durability. The relatively high price, however, limits its applications to a few special cases, *e.g.*, films for weather resistant coatings for various materials, as well as films for metallizing of piezoelectrical components. Tedlar®, Solef®, and Dalvor® are commercial products manufactured by Du Pont, Solvay, and Diamond Shamrock, respectively.

vinylidene fluoride manufacture by two routes:

1. HCl elimination from $CF_2Cl{-}CH_3$
2. Cl_2 elimination from $CF_2Cl{-}CH_2Cl$

Vinylidene fluoride can be obtained by alkaline or thermal elimination of hydrogen chloride from 1-chloro-1,1-difluoroethane:

$$CF_2Cl{-}CH_3 \xrightarrow{\;-HCl\;} F_2C{=}CH_2 \tag{11}$$

Another manufacturing route starts from 1,2-dichloro-1,1-difluoroethane which is converted into vinylidene fluoride by elimination of chlorine using zinc or nickel at $500\,°C$:

$$CF_2Cl—CH_2Cl + Zn \rightarrow F_2C=CH_2 + ZnCl_2 \qquad (12)$$

The copolymer of hexafluoropropene and vinylidene fluoride (Viton®) is commercially important as a vulcanizable elastomer.

Polyvinylidene fluoride is used in the construction of chemical apparatus and in the electrical industry. Due to its good physical properties it is formed into flexible films which, after special polarization, have excellent electrical properties and are used, for example, for headphone membranes.

use of vinylidene fluoride:

in copolymers for fluoro elastomers, *e.g.*, from $F_3C—CF=CF_2/F_2C=CH_2$

9.1.4. Trichloro- and Tetrachloroethylene

Trichloro- and tetrachloroethylene (also known as perchloroethylene) have developed into very important commercial products. Production figures for trichloro- and perchloroethylene in several countries are summarized in the adjacent table. The production capacity in 1995 in Western Europe, the USA, and Japan was 0.14, 0.15, and 0.10 × 10^6 tonnes, respectively, for trichloroethylene and 0.41, 0.22, and 0.10 × 10^6 tonnes, respectively, for perchloroethylene.

tri- and tetrachloroethylene production (in 1000 tonnes):

	$Cl_2C=CHCl$			$Cl_2=Cl_2$		
	1987	1990	1992	1987	1990	1993
USA	127	107	91	215	174	123
Japan	71	57	61	84	84	64
W. Europe	180	NA	125	305	NA	174

NA = not available

The older manufacturing processes are based on acetylene and tetrachloroethane, which is formed from chloroaddition in the presence of $FeCl_3$ at 70–85 °C. Trichloroethylene, obtained from dehydrochlorination of the tetrachloroethane, yields tetrachloroethylene after a second chlorination and dehydrochlorination.

manufacture of tri- and tetrachloroethylene:

1. older route by chlorination of $HC\equiv CH$ to $Cl_2CH—CHCl_2$ intermediate
2. newer route by catalytic gas-phase chlorination/dehydrochlorination reaction sequence, starting from $ClCH_2—CH_2Cl$

The HCl elimination is generally done in the gas phase at 250–300 °C over activated carbon that has been impregnated with $BaCl_2$. The overall yield is about 90%.

Particularly in the USA and Japan, 1,2-dichloroethane is the favored and, since the closing of the last acetylene-based plant in 1978, the only starting material for their manufacture. This can, in a combined chlorination and dehydrochlorination reaction, be converted into a mixture of both products:

$$ClCH_2—CH_2Cl + Cl_2 \xrightarrow{\text{[cat.]}}$$
$$Cl_2C=CHCl + Cl_2C=CCl_2 + HCl \qquad (13)$$

The conversion is carried out in the range 350–450 °C using fluidized-bed catalysts (*e.g.*, Diamond Shamrock) or fixed-bed catalysts (Donan Chemie) which usually contain $CuCl_2$, $FeCl_3$, or $AlCl_3$.

manufacturing variants:

chlorination/dehydrochlorination of EDC replaced by combination of oxychlorination/dehydrochlorination with advantage of simultaneous use of HCl

In another process modification, the accumulation of large amounts of hydrogen chloride is avoided by a concurrent HCl oxidation. The process, with simultaneous oxychlorination/dehydrochlorination with chlorine or anhydrous hydrogen chloride as the chlorine source, was developed by PPG:

$$ClCH_2{-}CH_2Cl + Cl_2 + O_2 \xrightarrow{\text{[cat.]}}$$
$$Cl_2C{=}CHCl + Cl_2C{=}CCl_2 + H_2O \qquad (14)$$

commercial application of oxychlorination modifications:

PPG process

The $CuCl_2$-containing catalyst is arranged in a vertical bundle of reactors, each of which is a separate fluidized-bed unit. At 420 to 450 °C and slightly raised pressure, a mixture of tri- and tetrachloroethylene is obtained with a total selectivity of 85 – 90%. The composition of the product mixture is determined by the 1,2-dichloroethane/Cl_2 ratio. The total world capacity of trichloro-/tetrachloroethylene plants using the PPG process was about 455 000 tonnes in 1991.

Tetrachloroethylene (perchloroethylene) is also formed, along with carbon tetrachloride, during the chlorolysis of propene or chlorine-containing residues (*cf.* Section 2.3.6.1).

uses of tri- and tetrachloroethylene:

solvent, extractant, and degreasing agent

Trichloroethylene is chiefly used as a solvent and extractant for fats, oils, waxes, and resins. A smaller amount is used in the manufacture of monochloroacetic acid (*cf.* Section 7.4.1.5) by the hydrolysis of trichloroethylene with about 75% H_2SO_4 at 130 – 140 °C:

$$CHCl{=}CCl_2 + 2\,H_2O \rightarrow ClCH_2COOH + 2\,HCl \qquad (15)$$

This so-called trichloroethylene process is practiced by Rhône-Poulenc.

increasing use of trichloroethylene as environmentally friendly replacement for HFCs (type 134 a) as refrigerants

An increasing demand for trichloroethylene as an alternative refrigerant to fluoroalkanes is anticipated. Currently, ICI is the world's largest manufacturer of trichloroethylene.

intermediate for Cl, F-alkanes, *e.g.*, $CF_2Cl{-}CFCl_2$ (type 113) with special application – cleaning of working parts made of mixed materials (metal, plastic)

Perchloroethylene is used primarily as a dry cleaning solvent, as an extractant and solvent for animal and vegetable fats and oils, and as a degreasing agent in metal and textile processing. It is also a precursor for fluorine compounds such as 1,2,2-trichloro-1,1,2- trifluoroethane, which is employed as a refrigerant and as a solvent for metal cleaning and metal–plastic cleaning. Since working with tri- and perchloroethylene poses health risks, they are being increasingly replaced by fluorine compounds.

9.1.5. Tetrafluoroethylene

Unlike tetrachloroethylene, tetrafluoroethylene is gaseous, and, in contrast to many other fluoroalkanes, it is poisonous. It is used mainly in the manufacture of polytetrafluoroethylene (PTFE), and to a limited extent for the manufacture of copolymers and telomers. The worldwide capacity for PTFE in 1989 was about 43 000 tonnes per year, of which the USA, Western Europe, and Japan accounted for ca. 11 000, 14 000, and 5000 tonnes per year, respectively.

$F_2C{=}CF_2$, gaseous, poisonous, mainly used as starting material for PTFE, the world's most important fluoroplastic, which accounts for 60% of the total fluoropolymer production

The manufacture of tetrafluoroethylene is accomplished by thermolysis of chlorodifluoromethane (*cf.* Section 2.3.6.2) with elimination of hydrogen chloride:

manufacture of $F_2C{=}CF_2$:

thermal dehydrochlorination with simultaneous dimerization (difluorocarbene as possible intermediate)

$$2\ CHClF_2 \xrightarrow{\Delta} F_2C{=}CF_2 + 2\ HCl \qquad (16)$$

In the commercial process, chlorodifluoromethane is fed through a flow tube made of corrosion-resistant material (*e. g.*, platinum) at 600–800 °C. Adequate selectivity to tetrafluoroethylene can be attained either by limiting the conversion to about 25%, or by adding steam or CO_2 and limiting conversion to about 65%.

process characteristics:

$F_2C{=}CF_2$ selectivity increase with limited conversion or lowering of partial pressure (addition of H_2O or CO_2)

The tetrafluoroethylene selectivity then reaches roughly 90%. Byproducts are hexafluoropropene, and linear and cyclic perfluoro and chlorofluoro compounds.

byproducts:

$F_3C{-}CF{=}CF_2$

$H{+}CF_2{\rightarrow}_n Cl$

In the presence of traces of oxygen, tetrafluoroethylene tends to polymerize spontaneously through an intermediate peroxide step. For this reason, it must be stored at low temperature with exclusion of air and with addition of stabilizers (dipentenes or tri-*n*-butylamine).

Tetrafluoroethylene in its very pure form is a feedstock for polymers with high temperature stability (short-term to about 300 °C) and chemical resistance. It is also copolymerized with ethylene in the manufacture of thermoplastic materials for many applications.

$F_2C{=}CF_2$ must be 99.9999% pure for polymerization to ${+}CF_2CF_2{\rightarrow}_n$; a few ppm impurities can make polymers unsuitable for use

9.2. Vinyl Esters and Ethers

vinyl-oxygen compounds:

most important product groups are vinyl esters and vinyl ethers as acyl- or alkyl derivatives of the unstable vinyl alcohol

$$H_2C{=}CHOCR \atop \overset{\|}{O} \Bigg\} \rightarrow [H_2C{=}CHOH] \atop H_2C{=}CHOR \qquad \overset{\downarrow}{CH_3CHO}$$

commercial importance of vinyl-oxygen compounds:

vinyl acetate is dominant vinyl ester, limited vinyl ether application

Vinyl esters and ethers can be regarded as derivatives of the hypothetical vinyl alcohol, *i.e.*, the enol form of acetaldehyde. As such, they are readily saponified in weakly acidic solution to acetaldehyde and the corresponding acid or alcohol.

Their industrial use as an intermediate is based solely on the readily polymerizable vinyl double bond.

Vinyl acetate (VAM, vinyl acetate monomer) is the most important of the esters of vinyl alcohol, mainly due to its use in the manufacture of homo- and copolymer dispersions. Vinyl ethers, which are versatile monomers with wide application, have a significantly smaller production volume than the vinyl esters.

9.2.1. Vinyl Acetate

vinyl acetate production (in 1000 tonnes):

	1990	1992	1994
USA	1157	1207	1377
W. Europe	553	570	582
Japan	500	540	537

In many countries, vinyl acetate production has stabilized following a dramatic growth period. Production figures for several countries are summarized in the adjacent table. The world capacity for vinyl acetate in 1995 was about 3.8×10^6 tonnes per year, with about 1.5, 0.59, and 0.70×10^6 tonnes per year in the USA, Japan, and Western Europe.

The largest producer of vinyl acetate is Hoechst, mainly in the USA and Europe, with a capacity of roughly 1.0×10^6 tonnes per year (1995).

vinyl acetate manufacturing processes:

until a few years ago, three different routes:

1. $HC{\equiv}CH + AcOH$
2. $CH_3CHO + Ac_2O$
3. $H_2C{=}CH_2 + AcOH + O_2$

today new vinyl acetate plants based solely on ethylene

In 1970, vinyl acetate was still being produced by three processes. Since then, the two older processes — the addition of acetic acid to acetylene and the two-step reaction of acetaldehyde with acetic anhydride through the intermediate ethylidene diacetate — have been increasingly replaced by the modern acetoxylation of ethylene process. In new plants, this process is used almost exclusively. In 1996, about 88% of the vinyl acetate capacity worldwide was based on ethylene acetoxylation.

9.2.1.1. Vinyl Acetate Based on Acetylene or Acetaldehyde

vinyl acetate based on acetylene:

single-step, heterogeneously catalyzed, very selective gas-phase addition of AcOH to C_2H_2 in fixed- or fluidized-bed reactor

The first industrial manufacture of vinyl acetate involved addition of acetic acid to acetylene in the gas phase over $Zn(OAc)_2$/charcoal at 170–250 °C:

$$HC{\equiv}CH + CH_3COOH \xrightarrow{\text{[cat.]}} H_2C{=}CHOCCH_3 \atop \overset{\|}{O} \qquad (17)$$

$$\left(\Delta H = -\,{28\ \text{kcal} \atop 118\ \text{kJ}}\big/\text{mol} \right)$$

At a 60–70% acetylene conversion, the selectivity to vinyl acetate reaches 93% (based on C_2H_2) and 99% (based on AcOH). For a time, $Hg(OAc)_2$ and Zn silicate were also used as catalysts. Although these processes are distinguished by high selectivities, simple reaction control, and insignificant catalyst costs, the lower ethylene cost coupled with its storage and transport capability has plagued the acetylene route. However, several vinyl acetate plants based on acetylene in Western and Eastern Europe have been able to maintain their processes. In 1994, 9% of the vinyl acetate production in Western Europe, Eastern Europe, and Asia was still based on acetylene. In the USA, all acetylene-based plants have been shut down.

commercial use of AcOH vinylation with C_2H_2 in only a few countries today, mainly in the former Eastern Bloc

Vinyl propionate can be manufactured from acetylene and propionic acid in a similar process. One such producer is BASF in Germany.

The second traditional process was practiced by Celanese until 1970 in a 25000 tonne-per-year plant in the USA. Plants in India and China are still in operation. In the acetaldehyde–acetic anhydride process both components react in the first stage at 120–140 °C in the liquid phase with an $FeCl_3$ catalyst to form ethylidene diacetate:

vinyl acetate manufacture based on acetaldehyde in two steps:

1. reaction of CH_3CHO with Ac_2O to ethylidene diacetate (intermediate)
2. proton-catalyzed (*e.g.*, *p*-toluenesulfonic acid) AcOH elimination

$$CH_3CHO + (CH_3C)_2O \xrightarrow{\text{[cat.]}} CH_3CH(OCCH_3)_2 \qquad (18)$$

The ethylidene diacetate is then cleaved to vinyl acetate and acetic acid at 120 °C using acidic catalysts:

today acetaldehyde route to vinyl acetate seldom used commercially

$$CH_3CH(OCCH_3)_2 \xrightarrow{\text{[H}^{\oplus}\text{]}} H_2C=CHOCCH_3 + CH_3COOH \qquad (19)$$

9.2.1.2. Vinyl Acetate Based on Ethylene

The catalytic modern manufacturing process for vinyl acetate from ethylene and acetic acid is based on an observation made by J. J. Moiseev and co-workers. They found that the palladium-catalyzed oxidation of ethylene to acetaldehyde becomes an

vinyl acetate from ethylene:

single step, *i.e.*, direct Pd-catalyzed acetoxylation of C_2H_4 with nucleophilic attack of $OAc^{\ominus}$ instead of $OH^{\ominus}$ as with CH_3CHO

$$\begin{bmatrix} CH_2 & \overset{Cl}{\underset{|}{}} \\ \| & \to Pd{-}OAc \\ CH_2 & \underset{|}{} \\ & Cl \end{bmatrix}^{\ominus} \to [AcOCH_2CH_2PdCl_2]^{\ominus}$$

$$\begin{bmatrix} CHOAc \\ \| & \to PdHCl_2 \\ CH_2 \end{bmatrix}^{\ominus}$$

$$\to VAM + Pd + 2\,Cl^{\ominus} + H^{\oplus}$$

stoichiometric acetoxylation in liquid phase becomes catalytic process with redox system

$Pd^0 + 2CuCl_2 \to PdCl_2 + 2CuCl$
$2CuCl + 2HCl + 0.5\,O_2 \to 2CuCl_2 + H_2O$

acetoxylation reaction when conducted in a solution of acetic acid and in the presence of sodium acetate. The divalent palladium is thereby reduced to the metallic state:

$$H_2C{=}CH_2 + PdCl_2 + 2\,CH_3COONa$$
$$\xrightarrow{} H_2C{=}CHO\underset{\underset{O}{\|}}{C}CH_3 + 2\,NaCl + Pd + CH_3COOH \quad (20)$$

This stoichiometric reaction can be converted into a catalytic process if a redox system is also present which regenerates the $Pd^{2\oplus}$ from Pd^0. Cu(II)-salts are generally used for this purpose. In analogy to the oxidation of ethylene to acetaldehyde (*cf.* Section 7.3.1.1), $Cu^{2\oplus}$ can readily oxidize the zerovalent palladium to $Pd^{2\oplus}$. The copper cation, which is reduced to the monovalent state, is regenerated with air to $Cu^{2\oplus}$. The multi-step reaction can be summarized in the net equation:

$$H_2C{=}CH_2 + CH_3COOH + 0.5\,O_2$$
$$\xrightarrow{\text{[cat.]}} H_2C{=}CHO\underset{\underset{O}{\|}}{C}CH_3 + H_2O \left(\Delta H = -\,{\small\begin{matrix}42\ \text{kcal}\\176\ \text{kJ}\end{matrix}}\Big/\text{mol}\right) \quad (21)$$

C_2H_4 acetoxylation first operated in liquid phase (ICI, Celanese)

Based on this catalyst principle, ICI and Celanese (with a license from ICI) developed industrial liquid-phase processes which led to the construction of large-scale plants. Hoechst independently developled a liquid-phase process to a semicommercial state.

reaction principle of liquid-phase acetoxylation:

analogous to Wacker–Hoechst CH_3CHO process

$Pd^{\oplus}$-, $Cu^{\oplus}$-catalyzed acetoxylation, where rate is determined by $Cl^{\ominus}/OAc^{\ominus}$ ratio

special characteristics of liquid-phase process:

1. secondary reaction $H_2O + C_2H_4 \to$ acetaldehyde can, after further oxidation to AcOH, be used for own requirement
2. highly corrosive catalyst solution necessitates resistant materials

The liquid-phase processes resembled Wacker–Hoechst's acetaldehyde process, *i.e.*, acetic acid solutions of $PdCl_2$ and $CuCl_2$ are used as catalysts. The water of reaction from the oxidation of CuCl to $CuCl_2$ forms acetaldehyde in a secondary reaction with ethylene. The ratio of acetaldehyde to vinyl acetate can be regulated by the operating conditions. Thus, in principle, this process can supply its own requirements of acetic acid through oxidation of acetaldehyde. The reaction takes place at $110-130\,°C$ and $30-40$ bar. The vinyl acetate selectivity reaches 93% (based on CH_3COOH). The net selectivity to acetaldehyde and vinyl acetate is about 83% (based on C_2H_4), the by-products being CO_2, formic acid, oxalic acid, butene, and chlorinated compounds.

industrial operation of liquid-phase acetoxylation was discontinued in favor of gas-phase process

The presence of chloride ions and formic acid makes this reaction solution very corrosive, so that titanium or diabon must be used for many plant components. After only a few years operation, ICI shut down their plant in 1969 due to corrosion problems.

In 1970, Celanese, and likewise the other companies, also stopped operation of the liquid-phase process for economic reasons.

Parallel to the liquid-phase process, ethylene acetoxylation was also developed as a gas-phase process. There is a distinctive difference between the two processes: in the liquid phase, in the presence of palladium salts and redox systems, both vinyl acetate and acetaldehyde are formed, while in the gas-phase process using palladium or chlorine-free palladium salts, vinyl acetate is formed almost exclusively. Furthermore, there are no noticeable corrosion problems in the gas-phase process. The most suitable construction material is stainless steel.

variant of vinyl acetate manufacture:

gas-phase acetoxylation of ethylene with

1. Pd metal catalyst (Bayer)
2. Pd salt catalyst (Hoechst or USI (now Quantum))

characteristic differences in gas-phase/liquid-phase acetoxylation:

lesser amount of CH_3CHO and relatively little corrosion

In the industrial operation of the Bayer process, developed together with Hoechst, the palladium metal catalyst is manufactured by reducing a palladium salt impregnated on a support (generally SiO_2).

Hoechst and USI (now Quantum) developed, independently of one another, a supported palladium acetate catalyst. The metal and salt catalysts contain alkali acetates and other components (*e. g.*, Cd, Au, Pt, Rh) which serve to increase activity and selectivity. During the course of the reaction, the catalyst changes, especially in its alkali acetate content. The alkali acetates migrate from the catalyst under the reaction conditions and must be constantly renewed.

similarities of both gas-phase catalyst types:

alkali acetate addition to increase activity and selectivity with the disadvantage of migration from catalyst bed to product stream

The operating conditions for both processes are similar. The strongly exothermic reaction is conducted in a tubular reactor with a fixed-bed catalyst at $175-200\,°C$ and $5-10$ bar. The explosion limit restricts the O_2 content in the feed mixture, so that the ethylene conversion (ca. 10%) is also relatively small. The acetic acid conversion is $20-35\%$, with selectivities to vinyl acetate of up to 94% (based on C_2H_4) and about $98-99\%$ (based on AcOH). The main byproduct, CO_2 from the total oxidation of ethylene, is removed with a carbonate wash.

process characteristics:

exothermic, heterogeneously catalyzed acetoxylation in which the conversion of the reaction components must be constrained due to explosion limits, heat evolution, and selectivity requirements

byproducts:

CO_2, CH_3CHO, CH_3COOCH_3, $CH_3COOC_2H_5$

In 1991, the capacity of plants utilizing the Bayer/Hoechst process was more than 1.8×10^6 tonnes per year, and about 0.36×10^6 tonnes per year for those using the USI process.

After a multistep distillation, the vinyl acetate purity is 99.9 wt%, with traces of methyl acetate and ethyl acetate, which do not affect the polymerization.

vinyl acetate isolation and purification by multistep distillation due to special purity requirements for polymerization uses of vinyl acetate:

main uses vary from country to country:

USA
Western Europe $\Big\}$ 50–60% PVA

Japan: over 70% for production of poly(vinyl alcohol) for fibres (Vinylon)

poly(vinyl alcohol) production (in 1000 tonnes):

	1987	1989	1991
Japan	147	193	207
USA	68	77	77
W. Europe	56	64	68

Most vinyl acetate is converted into poly(vinyl acetate) (PVA) which is used in the manufacture of dispersions for paints and binders, and as a raw material for paints. It is also copolymerized with vinyl chloride and ethylene and, to a lesser extent, with acrylic esters. A considerable portion of the vinyl acetate is converted into poly(vinyl alcohol) by saponification or transesterification of poly(vinyl acetate). The world production capacity for poly(vinyl alcohol) in 1991 was over 0.68×10^6 tonnes per year, of which Japan, the USA, and Western Europe accounted for 250000, 120000, and 80000 tonnes per year, respectively. The production figures for these countries are listed in the adjacent table. Its use varies from country to country: in the USA and Western Europe its main application is as a raw material for adhesives, while its use for fibers dominates in Japan. It is also employed in textile finishing and paper gluing, and as a dispersion agent (protective colloid).

modifications of poly(vinyl alcohol):

1. varying degree of hydrolysis of PVA determines properties of polyol

2. acetal formation with aldehydes

$$\left[\begin{array}{c} CH_2 \quad CH_2 \\ -CH \quad CH- \\ OH \quad OH \end{array} \right]_n + RCHO \xrightarrow{[H^\oplus]}$$

$$\left[\begin{array}{c} CH_2 \quad CH_2 \\ -CH \quad CH- \\ O \quad O \\ CH \\ R \end{array} \right]_n \quad \text{R preferably } C_3H_7$$

Acetals are formed by reacting the free OH groups in poly(vinyl alcohol) with aldehydes. Industrially significant products are manufactured using formaldehyde, acetaldehyde, and butyraldehyde. Polyvinylbutyral is, for example, used as an intermediate layer in safety glasses. The global demand for polyvinylbutyral is currently estimated at 60–70000 tonnes per year.

The percentage of vinyl acetate used for polymers in the world, the USA, Japan, and Western Europe is compared in the following table:

Table 9-1. Use of vinyl acetate (in %).

	World		USA		Japan		Western Europe	
	1984	1991	1984	1994	1984	1991	1984	1992
Poly(vinyl acetate) (homo- and copolymers)	47	48	55	61	15	15	57	$\Big\}$ 75
Poly(vinyl alcohol)	29	32	21	23	73	69	16	
Vinyl chloride/vinyl acetate copolymers	5	3	7	1	1	1	3	4
Ethylene/vinyl acetate resins	6	7	6	7	4	13	5	7
Miscellaneous uses*)	13	10	11	8	7	2	19	14
Total consumption (in 10^6 tonnes)	2.11	2.94	0.71	1.04	0.46	0.55	0.50	0.74

*) *e.g.*, polyvinylbutyral

9.2.1.3. Possibilities for Development of Vinyl Acetate Manufacture

In evaluating the potential development of the vinyl acetate processes, it should be noted that the cost of acetic acid − which makes up 70% by weight of the vinyl acetate − basically determines the cost of vinyl acetate. In the future, inexpensive acetic acid will probably have an even greater effect on the economics of the otherwise very attractive gas-phase ethylene acetoxylation process.

A new manufacturing route for vinyl acetate may become of interest from another standpoint. Based on its manufacture of glycol by the catalytic oxidation of ethylene with acetic acid (*cf.* Section 7.2.1.1), Halcon has developed the thermolysis of the intermediate product (1,2-diacetoxyethane) to vinyl acetate:

$$CH_3COCH_2CH_2OCCH_3 \xrightarrow{\Delta} H_2C{=}CHOCCH_3 \quad (22)$$
$$\underset{O}{\|} \quad\quad \underset{O}{\|} \quad\quad\quad\quad \underset{O}{\|}$$
$$+\ CH_3COOH$$

With a 20% conversion to 1,2-diacetoxyethane at 535 °C, a selectivity to vinyl acetate of 87% is obtained.

By integrating the vinyl acetate and the glycol manufacture, an advantageous economic size for the first stage − acetoxylation and AcOH addition to ethylene − can be realized.

In another Halcon process for vinyl acetate, the intermediate 1,1-diacetoxyethane is formed by the stepwise carbonylation and hydrogenation of methyl acetate in the presence of $RhCl_3/CH_3I$ and 3-picoline at 150 °C and 70 bar with a selectivity of almost 80%. The acid-catalyzed elimination of acetic acid from 1,1-diacetoxyethane takes place in the liquid phase at about 170 °C:

$$CH_3COOCH_3 \xrightarrow[\substack{1.\ +CO \\ 2.\ +H_2}]{[cat.]} CH_3CH(OCCH_3)_2$$
$$\underset{O}{\|}$$

$$\quad (23)$$

$$\xrightarrow[-HOAc]{[H^{\oplus}]} H_2C{=}CHOCCH_3$$
$$\underset{O}{\|}$$

A selectivity to vinyl acetate of 87% has been reported at a conversion of 70%. It would thus be possible to develop a process

future vinyl acetate developments will be basically influenced by cost of acetic acid

economically interesting alternative vinyl acetate manufacture in new Halcon process:

1. integration of vinyl acetate and ethylene glycol manufacture through common intermediate $AcOCH_2CH_2OAc$

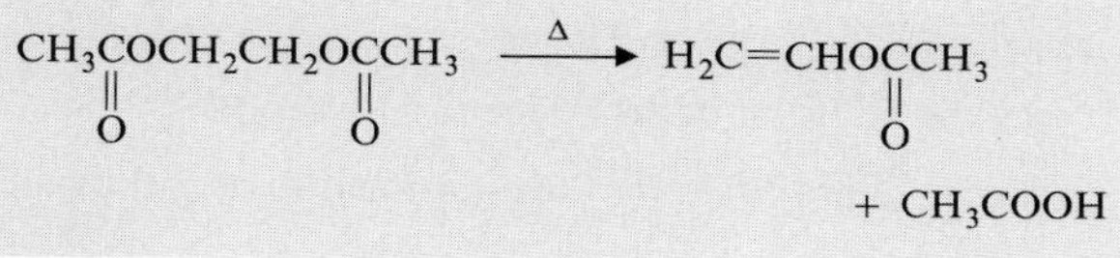

2. hydrogenative carbonylation of methyl acetate and AcOH elimination from the intermediate 1,1-diacetoxyethane

synthesis gas possible as sole feedstock

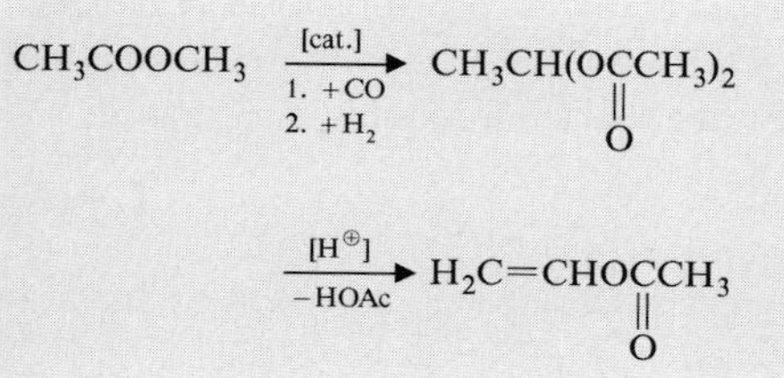

for vinyl acetate based solely on synthesis gas. This has not yet been achieved commercially.

9.2.2. Vinyl Esters of Higher Carboxylic Acids

acid components in commercially important higher vinyl esters:

1. fatty acids
2. Koch acids
3. 'oxo' carboxylic acids

There are three main types of acid components in the industrially important vinyl esters of higher carboxylic acid: fatty acids, *e. g.*, stearic or oleic acids; both groups of Koch acids, *e. g.*, the 'Neo acids' (Exxon) or the 'Versatic acids' (Shell; *cf.* Section 6.3); and carboxylic acids obtained by conversion of 'oxo' synthesis products, *e. g.*, 2-ethylhexanoic or isononanoic acid (*cf.* Section 6.1.4.2), which are synthesized economically from low-cost olefins.

propionic acid can be included with the higher carboxylic acids due to analogous manufacture of its vinyl ester from $C_2H_5COOH + C_2H_2$

Propionic acid, as a lower carboxylic acid, is included in the first group as, in contrast to acetic acid, it is converted to the vinyl ester only by the addition of acetylene (*cf.* Method 1 below).

manufacturing processes for higher vinyl esters:

1. 'Reppe vinylation' with C_2H_2, Zn- or Hg-salt

2. transvinylation with *e. g.*, vinyl acetate, VCM, or vinyl ether with catalyst system based on Pd or Hg

There are two main processes for the esterification of carboxylic acids:

1. By vinylation of the acids with acetylene in the presence of their zinc or mercury salt.

2. By catalytic transvinylation, *i. e.*, transfer of the vinyl group *e. g.*, from vinyl acetate, chloride, or ether, to the carboxylic acid.

To 1:

process characteristics of vinylation:

higher carboxylic acids in liquid phase; lower, such as propionic and butanoic acids, still in gas phase

branched Koch acids more rapidly vinylated (*e. g.*, three times faster than lauric acid)

Vinylation of higher carboxylic acids is usually done in the liquid phase. For example, stearic acid is reacted with dilute acetylene at $10-15$ bar and $165\,^\circ\text{C}$ in the presence of $5-10$ mol% zinc stearate, although HgO/H_2SO_4 can also be employed as catalyst. Inert gases such as N_2 or propane serve as diluents. Branched acids from the Koch reaction can usually be vinylated at a higher conversion rate than, for example, stearic acid:

$$C_{18}H_{37}COOH + HC\equiv CH \xrightarrow{\text{[cat.]}} C_{18}H_{37}COOCH=CH_2 \quad (24)$$

With an acid conversion of $95-97\%$, the selectivity to vinyl ester is up to 97% (based on stearic acid).

To 2:

The second process, transvinylation, is generally run with an excess of vinyl acetate in order to displace the equilibrium. The reaction takes place in the presence of $PdCl_2$ and alkali chloride or Hg(II) salts and H_2SO_4, and usually under reflux conditions:

process principles of the transvinylation:

homogeneously catalyzed transvinylation requires, as equilibrium reaction, large excess of lowest boiling vinyl component

$$
\begin{aligned}
&C_{18}H_{37}COOH \ + \ H_2C{=}CHOCCH_3 \\
&\qquad\qquad\qquad\qquad\quad \underset{O}{\|} \\
&\xrightarrow[\text{[cat.]}]{} \ C_{18}H_{37}COOCH{=}CH_2 \ + \ CH_3COOH
\end{aligned}
\tag{25}
$$

Both catalyst systems effect a transfer of the vinyl group and no transesterification, *i. e.*, cleavage of the vinyl acetate, takes place between the vinyl C atom and the O atom of the carboxylate group.

catalysts for transvinylation:

1. $PdCl_2$ + LiCl
2. $Hg(OAc)_2$ + H_2SO_4

Vinyl chloride is also suitable for transvinylation. For example, pivalic acid can be reacted with vinyl chloride in the presence of sodium pivalate and $PdCl_2$ to give vinyl pivalate in 97% yield:

$$
(CH_3)_3CCOOH \ + \ H_2C{=}CHCl \ \xrightarrow{\text{[cat.]}}
$$
$$
(CH_3)_3CCOOCH{=}CH_2 \ + \ HCl
\tag{26}
$$

Vinyl esters of the higher carboxylic acids are used mainly as comonomers with vinyl chloride and acetate, acrylonitrile, acrylic acid and its esters, as well as styrene, maleic anhydride, and maleic acid esters for the modification of homopolymers in the paint industry. The homopolymers are used as additives for improving the viscosity index of lubricants, for example.

uses of higher vinyl esters:

in numerous copolymers usually with effect of internal plasticizing

as homopolymer, *e. g.*, for oil additives

Shell operates a plant in the Netherlands for the vinylation of 'Versatic acids'. Wacker manufactures vinyl esters of isononanoic acid and lauric acid.

9.2.3. Vinyl Ethers

The industrial processes for the manufacture of vinyl ethers are based on the reaction of alcohols with acetylene. With this method, discovered by Reppe, alkali hydroxides or alcoholates are used as catalysts:

vinyl ether manufacture:

favored Reppe vinylation of alcohols with acetylene and K-alcoholate as catalyst

$$ROH + HC\equiv CH \xrightarrow{\text{[ROK]}} ROCH=CH_2 \qquad (27)$$

$$\left(\Delta H \approx -\frac{30\ \text{kcal}}{125\ \text{kJ}}\big/\text{mol}\right)$$

reaction pressure dependent on boiling point of alcohol; *e. g.,* CH_3OH at 20 bar, *i*-C_4H_9OH at $4-5$ bar

The reaction is usually conducted in the liquid phase and, depending on the boiling point of the alcohol, either at normal pressure or 3 to 20 bar at 120 to 180 °C.

process characteristics:

countercurrent vinylation, relatively low expenditure for apparatus, low risk (due to N_2 dilution of C_2H_2) and good selectivity

commerical vinyl ethers

$R = CH_3, C_2H_5, \text{iso-}C_3H_7,$

$\quad n, \text{iso-}C_4H_9, CH_3(CH_2)_3\underset{\underset{C_2H_5}{|}}{CH}CH_2$

Acetylene, diluted with N_2, is introduced from below, counter-current to the alcohol trickling down the reaction column. The vinyl ether passes overhead with unreacted acetylene and N_2. Selectivities can be as high as 95%. After separation, the ether can be purified by distillation. Methyl-, ethyl-, isopropyl-, *n*-butyl-, isobutyl-, and 2-ethylhexyl vinyl ethers are commercially important.

Acetylene-independent processes are based on an oxidative addition of alcohols to ethylene in the presence of $PdCl_2/CuCl_2$; they have not yet been practiced commercially.

Producers include GAF and UCC in the USA and BASF in Germany.

uses of vinyl ethers:

in homo- and copolymers as additives in paint, adhesive, textile, and leather industries

in synthesis as intermediate, *e. g.,*

$H_2C=CH-OR + R'OH \rightarrow H_3C-CH\underset{\diagdown OR'}{\overset{\diagup OR}{}}$

Vinyl ethers are used to a limited extent in the manufacture of special homo- and copolymers. These are utilized in paint and adhesive manufacture, and as auxiliary agents in the textile and leather industry. Vinyl ethers are also used as intermediates in organic syntheses, *e. g.,* for manufacture of simple or mixed acetals.

10. Components for Polyamides

Polyamides are produced from the polycondensation of diamines with dicarboxylic acids, or of aminocarboxylic acids alone. Another possibility for their formation is the ring-opening polymerization of lactams.

In these linear macromolecules, the amide groups alternate with CH_2 chains of a given length. Carboxyl or amino groups are at the terminal positions.

The polyamides are designated using a code, according to which *one* figure represents synthesis from a single bifunctional compound (aminocarboxylic acid or lactam), and *two* figures signify synthesis from two components (diamine and dicarboxylic acid). The digits themselves represent the number of carbon atoms in the components.

Table 10-1 summarizes the designation and starting materials for the most important nylon types:

polyamide manufacturing by:
1. polycondensation of:
    1.1. diamines and dicarboxylic acids
    1.2. aminocarboxylic acids
2. polymerization of lactams

polyamide structure:

CONH units in alternating sequence with CH_2 chains

coded description of polyamides based on designation of starting material:

no. of figures = no. of components

value of digits = no. of C atoms in components

Table 10-1. Codes and starting materials for commercial polyamides.

Nylon code	Starting materials	Formula
6	ε-Caprolactam	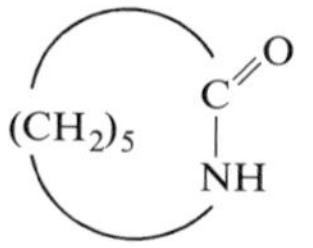 $(CH_2)_5$ ring with $C{=}O$ and NH
6,6	HMDA*/adipic acid	$H_2N(CH_2)_6NH_2$ /$HOOC(CH_2)_4COOH$
6,10	HMDA/sebacic acid	/$HOOC(CH_2)_8COOH$
6,12	HMDA/1,12-dodecanedioic acid	/$HOOC(CH_2)_{10}COOH$
11	ε-Aminoundecanoic acid	$H_2N(CH_2)_{10}COOH$
12	Lauryl lactam or	$(CH_2)_{11}$ ring with $C{=}O$ and NH
	ε-aminododecanoic acid	$H_2N(CH_2)_{11}COOH$

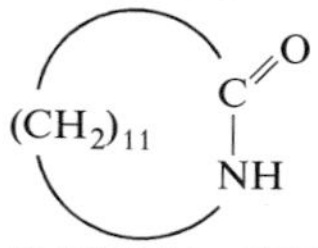

* HMDA = hexamethylenediamine.

Nylon 6,6 (the polycondensate of adipic acid and hexamethylenediamine) is the parent substance of the polyamides. It was first synthesized by W. H. Carothers in 1935 and produced by Du

nylon 6,6, the first polyamide to be synthesized, and nylon 6 are largest volume polyamides today

production of polyamide fibers
(in 1000 tonnes):

	1991	1993	1995
USA	1150	1206	1223
W. Europe	545	650	744
Japan	265	240	215

worldwide production of synthetic fibers
(in 10^6 tonnes):

	1992	1994	1995
polyester	9.63	11.05	11.92
polyamides	3.78	3.86	4.04
polyacrylic	2.41	2.45	2.42
others	1.38	3.61	1.82
total	17.20	20.97	20.20

synthetic fibers: made from synthetic polymers

man-made fibers: synthetic fibers and chemically modified natural fibers *e. g.,* cellulosic fibers

aromatic polyamides, *i. e.,* amide groups alternating with aromatic rings

examples (Du Pont aramides):

Kevlar:

Nomex:

uses of aramide:

flame resistant fibers, electrical insulation material, tire cord

important commercially available dicarboxylic acids:

$HOOC-(CH_2)_4-COOH$
$HOOC-(CH_2)_{10}-COOH$
little used:
$HOOC-(CH_2)_6-COOH$

Pont on an industrial scale soon after (1939). In 1938, P. Schlack succeded in producing nylon 6 from ε-caprolactam. In 1991, world polyamide production amounted to 3.67×10^6 tonnes; nylons 6 and 6,6 accounted for more than 95%. In 1991, the production capacities for nylon 6 and nylon 6,6 were about 2.75 and 1.92×10^6 tonnes per year, respectively.

Other polyamides are only important in specialized applications.

Production figures for polyamide fibers in the USA, Western Europe, and Japan are given in the adjacent table.

Polyamides are used mainly for the manufacture of synthetic fibers, particularly for clothing and carpets. After polyesters, which in 1995 accounted for over 59% of all synthetic fibers, polyamides constituted 20%, down insignificantly from 1992. Polyamides are also used as thermoplastics for injection molding, especially in mechanical engineering and electronics, and extrusion, accounting for 10–25% of polyamide production, depending on the country.

In the last few years, aromatic polyamides (aramides) have become increasingly important. These are made from terephthalic or isophthalic acid, and *m*- or *p*-phenylenediamine. Du Pont developed two aramide fibers, Kevlar® and Nomex®, and began production in the late 1960s. Since then, other production facilities have been started by other firms in the USA, Western Europe, Japan, and the CIS. Aramides are mainly used in the rubber industry. The properties of the aramides are so different from those of the aliphatic polyamides, *e. g.,* heat resistance and low flammability, that they are not classified as nylons.

10.1. Dicarboxylic Acids

Adipic acid is the most important commercial dicarboxylic acid. The second most important dicarboxylic acid is 1,12-dodecanedioic acid, which can be readily manufactured from cyclododecane *via* the cyclododecanone/ol mixture. Suberic acid from the oxidation of cyclooctane could well become increasingly important due to the convenient manufacture of its precursor from cyclooctadiene, the cyclodimer of butadiene. Up to now, however, it has only been used to a limited extent for the manufacture of nylon 6,8.

Sebacic acid, on the other hand, is – despite its industrial use in nylon 6,10 and softeners – usually obtained from natural products such as castor oil (*cf.* Section 10.2.2) by alkaline cleavage. However, high price and limited availability have restricted wider application. Thus, in 1973, Du Pont stopped producing nylon 6,10 in favor of nylon 6,12. Currently there is only a single producer of sebacic acid in the United States and one in Japan. The electrosynthesis of dimethyl sebacate in the Kolbe synthesis, *i.e.*, by decarboxylative dimerization of monomethyl adipate (*e.g.*, in a newer Asahi Chemical process), has still not been used commercially.

dicarboxylic acid from natural products:

$$HOOC-(CH_2)_8-COOH$$

electrosynthesis of sebacic acid from mono-methyl adipate still without application

$$2\ CH_3OOC(CH_2)_4COO^{\ominus} \xrightarrow{-2e}$$
$$CH_3OOC(CH_2)_8COOCH_3 + 2\ CO_2$$
$$2\ H^{\oplus} \xrightarrow{+2e} H_2$$

Another route for the manufacture of sebacic acid has been developed by Sumitomo. *cis*-Decalin is obtained from a two-step hydrogenation of naphthalene, and then oxidized to decalin hydroperoxide. This is rearranged to 6-hydroxycyclodecanone, and – after elimination of water and hydrogenation – a mixture of cyclodecanone and cyclodecanol is obtained. This mixture is a good raw material for selective oxidation with HNO_3 to sebacic acid:

new multistep sebacic acid synthesis from Sumitomo with *cis*-favored hydrogenation of naphthalene for the further selective oxidation (most important reaction step); still not used commercially

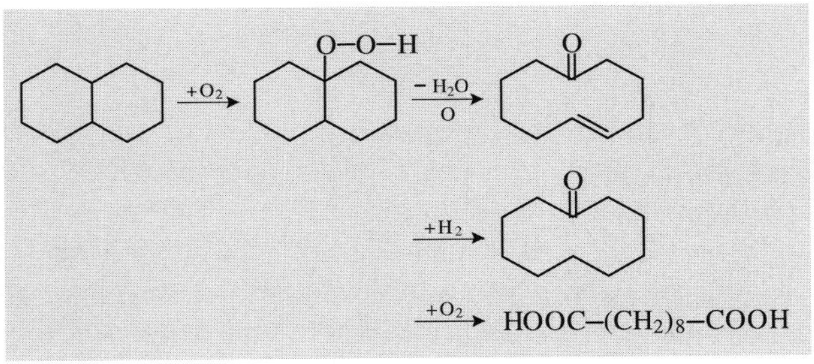

(1)

Despite the multistep, specifically catalyzed reaction sequence, an overall yield of 15–30% (based on naphthalene) is obtained. This process has not been used commercially.

10.1.1. Adipic Acid

In 1993, the worldwide production capacity for adipic acid was about 2.2×10^6 tonnes per year, with 810000, 950000, and 120000 tonnes per year in the USA, Western Europe, and Japan, respectively. Production figures for adipic acid in these countries are summarized in the adjacent table. The largest producer is Du Pont in the USA, with a capacity of about 450000 tonnes per year (1991).

adipic acid production (in 1000 tonnes):

	1988	1990	1993
USA	726	744	765
W. Europe	668	600	548
Japan	69	90	84

economical manufacturing route to adipic acid:

oxidative C_6 ring cleavage of cyclohexanol/-one mixture (intermediate) in one of two process variants:

1. direct two-step oxidation
2. oxidation with boric acid ester as further intermediate step

process principles of the first oxidation step:

homogeneously catalyzed liquid-phase oxidation of cyclohexane with intermediate cyclohexyl hydroperoxide

catalytic principle:

1. initiation of radical chain:

$$\underset{H}{\overset{H}{>}}C< + M^{3\oplus} \rightarrow \underset{\bullet}{\overset{H}{>}}C< + H^{\oplus} + M^{2\oplus}$$

2. cleavage of hydroperoxide:

$$\underset{O-O-H}{\overset{H}{>}}C< + M^{2\oplus} \rightarrow$$

$$\underset{O\bullet}{\overset{H}{>}}C< + OH^{\ominus} + M^{3\oplus}$$

$$\underset{O\bullet}{\overset{H}{>}}C< + \underset{H}{\overset{H}{>}}C< \rightarrow \underset{OH}{\overset{H}{>}}C< + \underset{\bullet}{\overset{H}{>}}C<$$

byproduct formation by oxidative degradation to:

$HOOC-(CH_2)_n-COOH$ $n = 3; 2$

and esterification with cyclohexanol

process characteristics:

low cyclohexane conversion to avoid oxidative side reactions leads to high cyclohexane recycle

The commercially favored route to adipic acid is the oxidative cleavage of cyclohexane. Direct oxidation processes to the acid stage have not been accepted due to unsatisfactory selectivities. Adipic acid is therefore manufactured in a two-stage process with the intermediates cyclohexanol/cyclohexanone (KA = ketone/alcohol). There are two possible manufacturing routes; the addition of boric acid in the second results in a higher selectivity.

In the first step of the oxidation, cyclohexane is oxidized to a KA mixture at $125-165\,°C$ and $8-15$ bar. The reaction is conducted in the liquid phase with air and Mn- or Co-salts, e. g., the acetate or the naphthenate, as catalysts:

$$\left(\Delta H = -\begin{matrix}70\ \text{kcal}\\294\ \text{kJ}\end{matrix}\Big/\text{mol}\right) \quad (2)$$

Cyclohexyl hydroperoxide, the primary product of this radical reaction, reacts further by means of a catalyst to the alcohol and ketone. A small quantity of adipic acid is also formed at this stage, along with glutaric and succinic acid from stepwise oxidative degradation. They are usually present as the cyclohexyl esters. Cyclohexane conversion is therefore limited to $10-12\%$ in order to increase the alcohol/ketone selectivity to $80-85\%$. The unreacted cyclohexane is distilled off and recycled to the oxidation. The acids are extracted with aqueous alkali, the esters being simultaneously hydrolyzed. Cyclohexanone and cyclohexanol (ratio $1:1$) are obtained in 99.5% purity by distillation.

Industrial processes are operated, for example, by BASF, Bayer, Du Pont, ICI, Inventa, Scientific Design, and Vickers-Zimmer.

process principles of 'boric acid oxidation route':

in the presence of H_3BO_3, cyclohexyl hydroperoxide reacts directly to the cyclohexanol boric acid ester (stable towards oxidation), e. g., $(C_6H_{11}O)_3B$

feedstocks for 2nd step of adipic acid manufacture:

In analogy to the oxidation of paraffins to secondary alcohols (*cf.* Section 8.2.1), the selectivity of the reaction can be increased by the presence of boric acid. In this case, it increases to over 90% with a mole ratio of alcohol to ketone of about $9:1$. The cyclohexane conversion remains almost unchanged at $12-13\%$.

In the second stage of the KA oxidation to adipic acid, products from other processes can also be employed, e. g., pure cyclohexanol as obtained from the liquid-phase hydrogenation of phenol with a Pd catalyst at 150 °C and 10 bar. In 1991 the production capacity for adipic acid from phenol was 15 000 tonnes, or about 2% of total adipic acid capacity, in the USA. In Western Europe, only 6% of the adipic acid was still produced from phenol in 1991; in Japan, this route is no longer used.

The primary oxidized products are generally oxidized further without separation of the KA mixture:

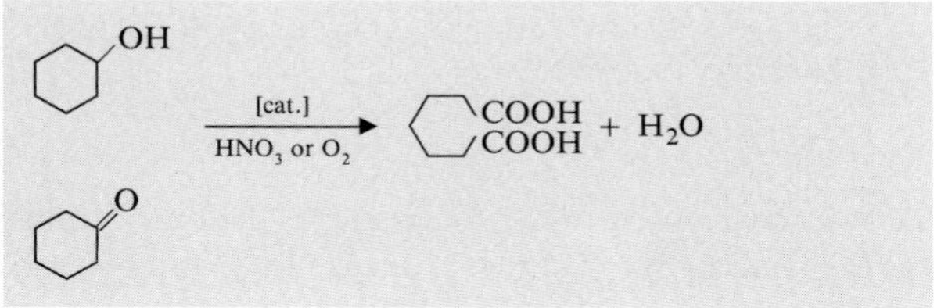

$$\text{(3)}$$

Two processes can be used for this oxidation:

1. with HNO_3 and NH_4-metavanadate/Cu-nitrate, or
2. with air and Cu-Mn-acetate

In the first process, the KA is oxidized with 60% HNO_3 at 50–80 °C and atmospheric pressure in the presence of the catalysts cited. The selectivity to adipic acid is up to 96%. In the course of the reaction, the cyclohexanol is first oxidized to cyclohexanone, which then – after α-nitrosation to the intermediate 2-nitrosocyclohexanone – is further oxidized. A mixture of nitrogen oxides which can be reoxidized to HNO_3 is emitted from the HNO_3.

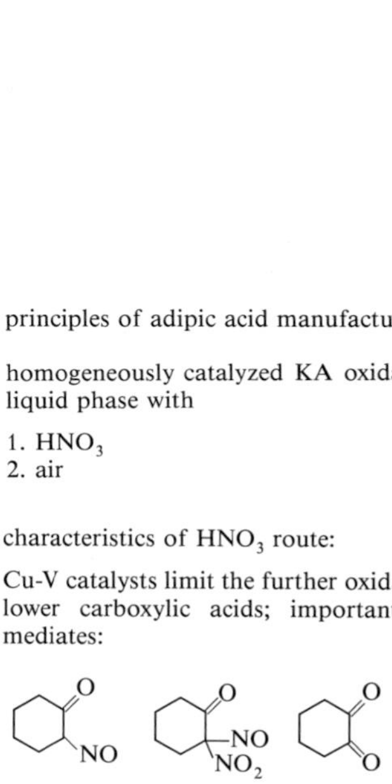

In the second process, reaction mixtures rich in cyclohexanone are preferred. The air oxidation is conducted in the liquid phase, usually in acetic acid as solvent, at 80–85 °C and 6 bar in the presence of Cu- and Mn-acetate. Adipic acid crystallizes from the product solution as it cools. The crude product is purified by recrystallization. The selectivity to adipic acid is said to approach that of the HNO_3 process. The advantage of the air oxidation lies mainly with the absence of corrosive HNO_3. Scientific Design has developed an up-to-date procedure for the process involving both stages, starting from cyclohexane and going through the KA mixture to adipic acid. A 10 000 tonne-per-year plant was operated by Rohm & Haas in the USA for several years, but was abandoned due to poor product quality. Other processes using boric acid were introduced by Stamicarbon and IFP.

1. KA (without separation) from cyclohexane oxidation
2. cyclohexanol from ring hydrogenation of phenol

principles of adipic acid manufacture:

homogeneously catalyzed KA oxidation in liquid phase with

1. HNO_3
2. air

characteristics of HNO_3 route:

Cu-V catalysts limit the further oxidation to lower carboxylic acids; important intermediates:

characteristics of air oxidation:

Cu-, Mn-catalyzed liquid-phase oxidation leads to same selectivity as HNO_3 route, moreover noncorrosive

process example:

Scientific Design two-step process for adipic acid:

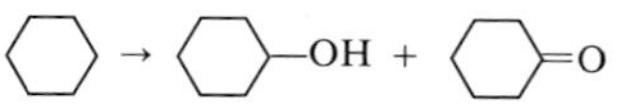

$$\rightarrow HOOC(CH_2)_4COOH$$

new C$_4$-based process for adipic acid:

Monsanto route by PdCl$_2$-catalyzed carbonylation of CH$_3$OCH$_2$CH=CHCH$_2$OCH$_3$

BASF route by Co/CO/N-ligand-catalyzed hydrocarbonylation of butadiene through methyl ester intermediate

Newer developments involve synthesis of adipic acid from butadiene; for example, by carbonylation of the intermediate 1,4-dimethoxy-2-butene (Monsanto), or two-step carbonylation of butadiene in the presence of methanol (BASF):

$$H_2C=CH-CH=CH_2 + CO + CH_3OH$$

$$\xrightarrow{\text{[cat.]}} CH_3CH=CHCH_2COOCH_3$$

$$\xrightarrow[+CO,\ +CH_3OH]{\text{[cat.]}} CH_3OOC(CH_2)_4COOCH_3$$

$$\xrightarrow[-CH_3OH]{+H_2O} HOOC(CH_2)_4COOH \qquad (4)$$

A cobalt carbonyl system with, for example, pyridine ligands serves as the catalyst. The first step is run at 130 °C and 600 bar; the second takes place at 170 °C and 150 bar. The yield from both steps is 72%, based on butadiene. BASF has developed this process in a pilot plant stage, but it has not yet been used commercially.

uses of adipic acid:

1. together with hexamethylenediamine in form of 1 : 1 salt

   [$^\ominus$O$_2$C(CH$_2$)$_4$CO$^\ominus$$^\oplus$H$_3$N(CH$_2$)$_6NH_3$$^\oplus$]

   for polycondensation
2. starting material for hexamethylenediamine

Adipic acid is most important for the manufacture of nylon 6,6; that is, up to about 70% of the total production (*e. g.*, in 1995 in the USA, Western Europe, and Japan, 68%, 46%, and 33%, respectively) can go into nylon 6,6 mainly for fibers. Adipic acid is used not only as the acid component, but also to a large extent as a precursor for the diamine component (*cf.* Section 10.2.1).

In the polycondensation, adipic acid is introduced as the 1 : 1 salt, hexamethylenediammonium adipate. The neutral salt crystallizes from methanol after adding equal quantities of acid and amine. Impurities and the excess reactants remain in solution. Today, the salt formation is usually conducted in aqueous solution.

3. component for polyesters
4. in esters, *e. g.*, with 2-ethylhexanol, as plasticizer and lubricant additive
5. after hydrogenation to HOCH$_2$(CH$_2$)$_4$CH$_2$OH for polyesters, polyurethanes, and for HMDA

In addition to this main application, the esters of adipic acid with polyalcohols are used for alkyd resins and plastic materials. Other esters serve as plasticizers or as additives for lubricants. The hydrogenation of adipic acid at 170−240 °C and 150−300 bar over Cu-, Co-, or Mn-catalysts, or of its ester, leads to the formation of 1,6-hexanediol, most of which is used for polyesters and polyurethanes. Another use involves the amination of 1,6-hexanediol to hexamethylenediamine (*cf.* Section 10.2.1).

The worldwide production capacity for 1,6-hexanediol is currently about 22000 tonnes per year. Manufacturers include BASF and Ube Ind.

10.1.2. 1,12-Dodecanedioic Acid

Cyclododecatriene is the starting material for the manufacture of 1,12-dodecanedioic acid (also known as 1,10-decanedicarboxylic acid).

precursor for C_{12} dicarboxylic acid:

cyclododeca-1,5,9-triene (CDT)

In 1955, G. Wilke and co-workers found a simple method to synthesize cyclododeca-1,5,9-triene (CDT) by the trimerization of butadiene with a Ziegler catalyst consisting of $TiCl_4$ and $(C_2H_5)_2AlCl$. It is necessary to increase the Ti : Al atomic ratio to $1:4-1:5$ as otherwise this catalyst (*e. g.*, with a $1:1$ ratio) would preferentially transform butadiene into 1,4-*trans*-polybutadiene.

CDT manufacture (3 isomers) by butadiene trimerization with help of Ti-, Ni-, Cr-catalysts, which are generally reduced with organo-Al compounds

Later it was also found that Ni^0-complexes, *e. g.*, bis (π-allyl)-nickel, or chromium complexes could catalyze the cyclotrimerization of butadiene.

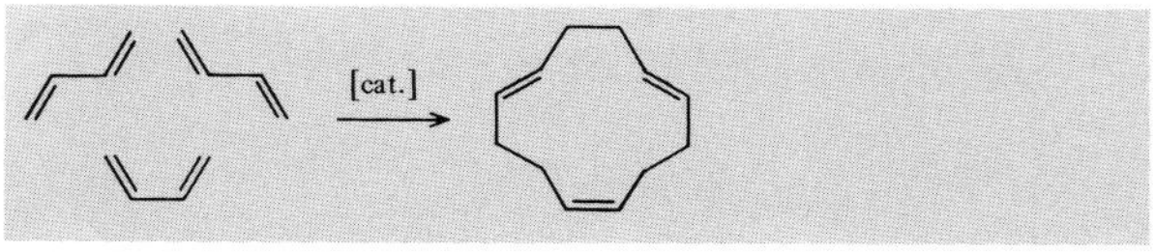

(5)

By this means, an industrial pathway to the alicyclic and open chained C_{12} compounds became available.

The first industrial plant was constructed by Hüls in 1975 with a capacity of 12 000 tonnes per year. Since then, the capacity has been increased to 18 000 tonnes per year. The net selectivity (based on C_4H_6) to the *trans-trans-cis* isomer is 90% with the catalyst system $TiCl_4-Al(C_2H_5)Cl_2-Al(C_2H_5)_2Cl$. Byproducts are cycloocta-1,5-diene, vinylcyclohexene, and butadiene oligomers.

in industrial process, *e. g.*, Hüls, Ti-catalysis mainly leads to *trans-trans-cis*-CDT with little *trans-trans-trans*-CDT

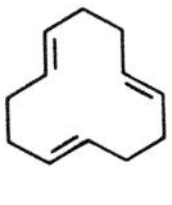

(all *trans*)

byproducts:

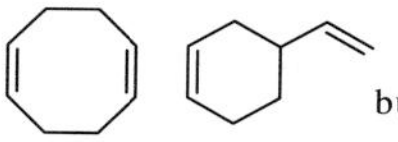

Another West European plant is operated by Shell in France. It has a capacity of 10 000 tonnes per year cyclododecatriene and cyclooctadiene, which can be produced alternatively as required. Other plants are in operation in the USA (Du Pont) and Japan.

butadiene oligomers

Cyclododecatriene is converted into dodecanedioic acid in a three-step reaction sequence. Initially, cyclododecane is manufactured nearly quantitatively in a liquid-phase hydrogenation at 200 °C and $10-15$ bar in the presence of Ni catalysts. Cyclododecane is then oxidized to a cyclododecanol/cyclododecanone mixture at $150-160$ °C and atmospheric pressure in the presence of boric acid in a manner analogous to the route involving cyclohexane and oxygen or air:

manufacture of C_{12} dicarboxylic acid in three steps:

1. hydrogenation to cycloalkane
2. mild oxidation to C_{12} alcohol/ketone
3. oxidative ring cleavage to dicarboxylic acid

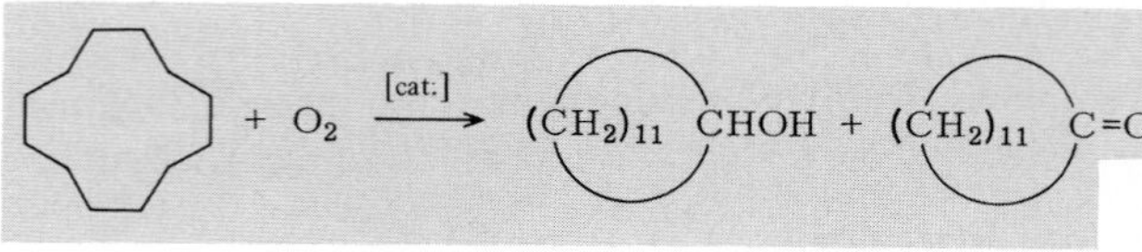

$$(\text{CH}_2)_{11} \quad + \quad O_2 \quad \xrightarrow{\text{[cat.]}} \quad (\text{CH}_2)_{11} \quad \text{CHOH} \quad + \quad (\text{CH}_2)_{11} \quad \text{C=O} \tag{6}$$

With a 25−30% conversion, the selectivity to alcohol/ketone mixture reaches an optimum of 80−82% and a ratio of 8 to 10:1.

double use of intermediate cyclododecanol/cyclododecanone:

1. HOOC(CH$_2$)$_{10}$COOH
2. 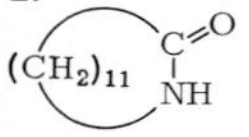
$(\text{CH}_2)_{11}$ $\overset{\text{C}\nearrow\text{O}}{\underset{\text{NH}}{|}}$

Cyclododecanol and cyclododecanone are produced by Hüls (Germany) and Du Pont (USA).

Cyclododecanol/cyclododecanone are also precursors for lauryl lactam (*cf.* Section 10.3.2).

The next stage, involving oxidative ring cleavage of the alcohol/ketone mixture, is done commercially with HNO$_3$:

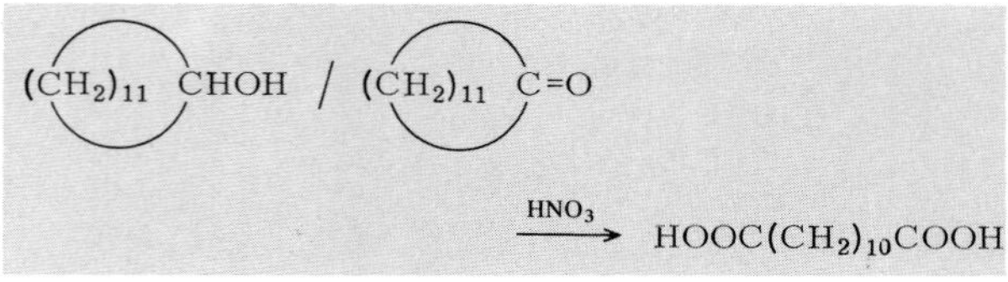

$$(\text{CH}_2)_{11} \quad \text{CHOH} \quad / \quad (\text{CH}_2)_{11} \quad \text{C=O}$$

$$\xrightarrow{\text{HNO}_3} \quad \text{HOOC(CH}_2)_{10}\text{COOH} \tag{7}$$

uses of C$_{12}$ dicarboxylic acid:

1. polyamide
2. polyester
3. ester for lubricants

In addition to its main use in polyamides and polyesters, 1,12-dodecanedioic acid is also used, as its diester, in lubricants.

10.2. Diamines and Aminocarboxylic Acids

10.2.1. Hexamethylenediamine

manufacture of α,ω-alkylenediamines generally from α,ω-alkylenedinitriles, *e. g.*, hexamethylenediamine (HMDA), or 1,6-diaminohexane, by hydrogenation of adiponitrile (ADN)

Only a few of the many synthetic pathways to amines and diamines are suitable for industrial operation. The α, ω-alkylenediamines, such as hexamethylenediamine, are almost exclusively obtained by hydrogenation of the corresponding dinitriles. Thus, all commercial scale manufacturing routes to 1,6-diaminohexane (hexamethylenediamine, HMDA) are only variations of the adiponitrile manufacture.

amination of 1,6-hexanediol to HMDA (Celanese process) still of minor importance until 1981

manufacture of 1,6-hexanediol by hydrogenation of:

1. caprolactone
2. ω-hydroxycaproic acid
3. adipic acid

There was, however, an alternative technology practiced by Celanese (USA) until 1981 in a plant with a capacity of almost 30000 tonnes per year. In this process, cyclohexanone was first oxidized to caprolactone (*cf.* Section 10.3.1.2), which was then hydrogenated to 1,6-hexanediol in almost quantitative yield at about 250 °C and 280 bar in the presence of, for example, Raney copper or copper chromite. ω-Hydroxycaproic acid or adipic acid can also be used as starting materials for 1,6-hexanediol manufacture (cf. Section 10.1.1). Finally, 1,6-hexanediol was aminated with ammonia at 200 °C and 230 bar in the presence of Raney nickel to hexamethylenediamine. The yield approached 90% and the main by-products were hexamethyleneimine and 1,6-aminohexanol:

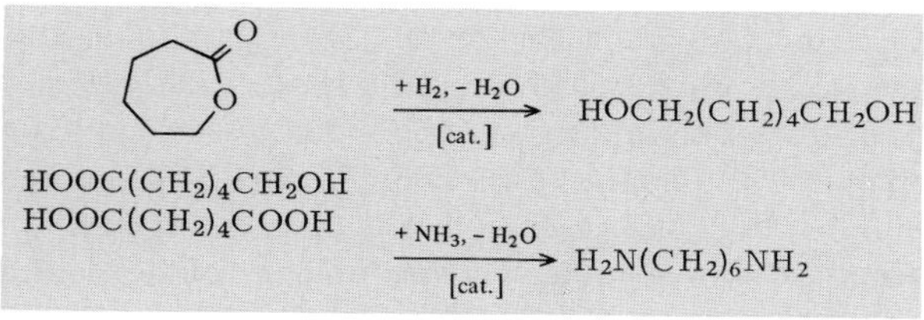

$$
\begin{array}{c}
\text{[lactone]} \xrightarrow[\text{[cat.]}]{+\,H_2,\,-\,H_2O} HOCH_2(CH_2)_4CH_2OH \\[2mm]
HOOC(CH_2)_4CH_2OH \\
HOOC(CH_2)_4COOH \quad \xrightarrow[\text{[cat.]}]{+\,NH_3,\,-\,H_2O} H_2N(CH_2)_6NH_2
\end{array}
\tag{8}
$$

10.2.1.1. Manufacture of Adiponitrile

The worldwide production capacity for adiponitrile in 1990 was about 1.0×10^6 tonnes per year, with 0.61, 0.31, and 0.03×10^6 tonnes per year in the USA, Western Europe, and Japan, respectively.

Four basically different routes have been developed for the commercial scale manufacture of adiponitrile (adipic acid dinitrile, ADN):

four ADN manufacturing routes with synthetic scheme for C_6–N_2 chain:

1. Dehydrative amination of adipic acid with NH_3 in the liquid or gas phase going through diamide intermediate

   1. dehydrative amination of adipic acid with NH_3 (C_6 + 2 N)

2. Indirect hydrocyanation of butadiene with 1,4-dichlorobutene intermediate

   2. indirect hydrocyanation of butadiene with HCN (C_4 + 2 CN)

3. Direct hydrocyanation of butadiene with HCN

   3. direct hydrocyanation of butadiene (C_4 + 2 CN)

4. Hydrodimerization of acrylonitrile in an electrochemical process

   4. hydrodimerization of acrylonitrile (C_3N + C_3N)

To 1:

In the first method, a recently developed process, adipic acid is converted with NH_3 through the intermediates diammonium adipate and adipic acid diamide into the dinitrile. The reaction is conducted in a melt at 200–300 °C, or with adiponitrile and the intermediates as solvent, in the presence of a catalyst such as H_3PO_4 which is soluble in the reaction medium:

adipic acid route:

proton-catalyzed dehydration of NH_4 salt with amide intermediate in the melt or in solution

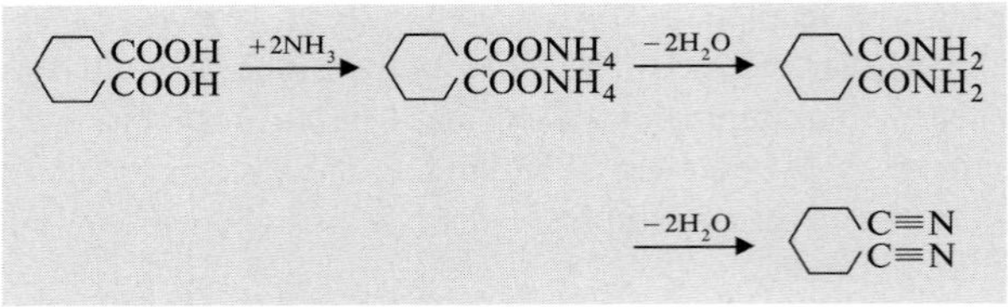

$$\text{(9)}$$

variation of adipic acid route:

heterogeneously catalyzed gas-phase dehydration in presence of NH_3 first in fixed bed, later in fluidized bed

The older process is conducted in the gas phase at 300–350 °C using, *e. g.*, boron phosphate catalysts with a great excess of NH_3. The disadvantage is the decomposition of adipic acid on evaporation, which limits the selectivity to only 80% and also necessitates a catalyst regeneration. The main improvement in this process has been the conversion from a fixed bed to a fluidized bed with an H_3PO_4/SiO_2 catalyst and continuous regeneration. The selectivity to adiponitrile can reach 90%. However, many plants have been taken out of operation in recent years for economic reasons.

To 2:

indirect butadiene hydrocyanation route (in three steps):

1. chlorination of butadiene in gas phase to product mixture with typical distribution:

   36% 3,4-dichloro-1-butene
   17% *cis*-1,4-dichloro-2-butene
   43% *trans*-1,4-dichloro-2-butene
   4% mono- and trichlorobutenes

The butadiene hydrocyanation can, as in the process developed by Du Pont in 1951, take place indirectly by chlorination of butadiene. Initially, the chlorination was carried out in the liquid phase; today it is usually conducted in the gas phase at 200–300 °C without catalyst. A mixture of 3,4-dichloro-1-butene and *cis*- and *trans*-1,4-dichloro-2-butene is obtained with a selectivity of about 96%:

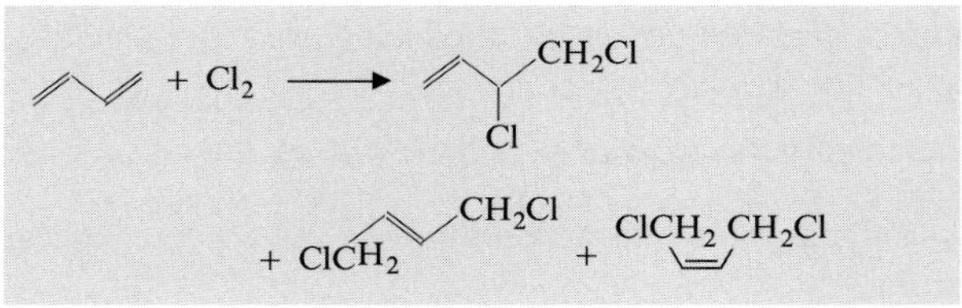

$$\text{(10)}$$

2. Cl substitution by $CN^{\ominus}$ with simultaneous isomerization (allyl rearrangement) in liquid phase to 1,4-dicyano-2-butenes (*cis* and *trans*)

The dichlorobutenes are then reacted with HCN or an alkali cyanide in the liquid phase at 80 °C to butenedinitriles. The formation of 3,4-dicyano-1-butene is not disadvantageous since, in the presence of the copper–cyano complex, an allyl rearrangement takes place under the hydrocyanation conditions. A mixture of the *cis/trans* isomers of 1,4-dicyano-2-butene is obtained with about 95% selectivity:

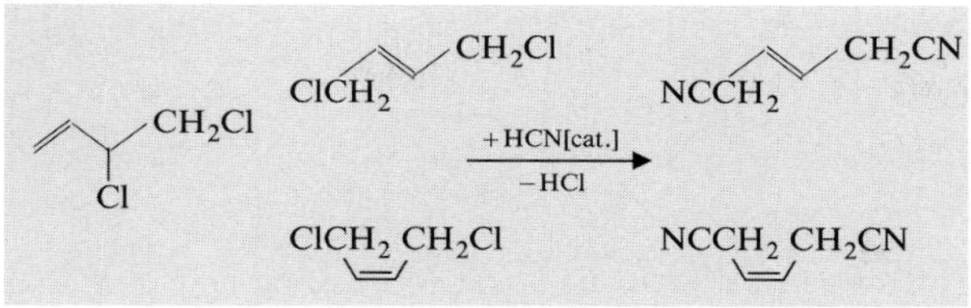

(11)

The mixture can then be hydrogenated with 95–97% selectivity to adiponitrile at 300 °C in the gas phase using a Pd catalyst. Du Pont used this technology in two adiponitrile plants until 1983.

3. double bond hydrogenation in gas phase to ADN

To 3:

The third method – the direct hydrocyanation – was also developed by Du Pont. The first step, HCN addition to butadiene, gives a mixture of isomers of pentene nitriles and methylbutene nitriles, which is then isomerized predominantly to 3- and 4-pentene nitriles. In the second step, adiponitrile is formed from the anti-Markovnikov addition of HCN:

butadiene direct hydrocyanation route:

two-step liquid-phase addition of HCN in presence of Ni^0 phosphine or phosphite complexes, *e. g.*, $[(C_6H_5)_3PO]_4 Ni^0$, with Al-, Zn-promoters involves intermediate structural and double-bond isomerization

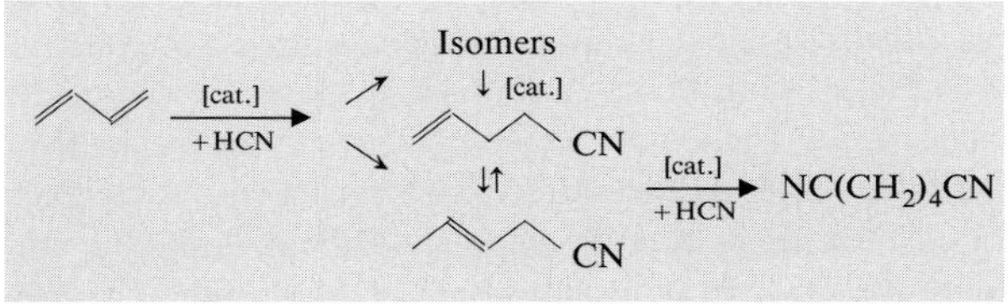

(12)

The reaction takes place at atmospheric pressure and 30–150 °C in the liquid phase using a solvent such as tetrahydrofuran. Ni^0-complexes with phosphine or phosphite ligands and metal salt promoters, *e. g.*, zinc or aluminum chlorides, are suitable catalysts. The net reaction is reported to lead to adiponitrile in high selectivity.

Du Pont produces adiponitrile in the USA mainly by this process, and, together with Rhône-Poulenc (Butachimie), has operated a plant in France for direct hydrocyanation with a capacity of 100000 tonnes per year since 1977.

To 4:

The fourth method is known as the Monsanto EHD process (**e**lectro-**h**ydro**d**imerization). It is based on the hydrogenative dimerization of acrylonitrile (AN) to adiponitrile (ADN):

AN electro-hydrodimerization route to ADN (Monsanto EHD process):

possible course of net cathodic reaction:

1. dianion formation $H_2C^{\ominus}—CH^{\ominus}—CN$ by transfer of two electrons
2. dimerization by coupling to second AN molecule
3. charge stabilization by proton addition from H_2O

$$\begin{array}{l} H_2C{=}CHCN \xrightarrow{+2e^{\ominus}} H_2\overset{\ominus}{C}{-}\overset{\ominus}{C}HCN \\ \xrightarrow{+AN} NC\overset{\ominus}{C}HCH_2{-}CH_2\overset{\ominus}{C}HCN \xrightarrow{+2H^{\oplus}} NC(CH_2)_4CN \end{array} \quad (13)$$

The net cathodic reaction can be represented as follows:

$$2\,H_2C{=}CHCN\ +\ 2e^{\ominus}\ +\ 2\,H^{\oplus} \rightarrow NC(CH_2)_4CN \qquad (14)$$

This reductive or cathodic dimerization was first practiced on a commercial scale in the USA in 1965. Other plants have been built in the USA and, since 1978, also in Western Europe. This electrochemical approach was also developed to a commercial process by Phillips Petroleum, and in Europe by ICI, Rhône-Poulenc, and UCB. In Japan, Asahi Chemical has practiced its own technology in a 26 000 tonne-per-year plant since 1971.

characteristics of industrial operation:

turbulent circulation produces emulsion from aqueous solution of conducting salt and organic AN phase

emulsified phases with large interfaces facilitate mass transfer and make separation of cathode–anode cells with magnetite or Fe-anodes unnecessary

Ion-exchange membranes were initially used to separate the anode and cathode regions. In the latest advances in electrolysis cells, developed by Asahi, BASF, and UCB as well as by Monsanto, no such mechanical separation is necessary. Instead, a finely divided two-phase emulsion is rapidly pumped through the cathode–anode system. The aqueous phase contains the conducting salt and a small amount of acrylonitrile (determined by its solubility), while the organic phase consists of acrylonitrile and adiponitrile. The loss of acrylonitrile in the aqueous phase by reaction is offset by the more facile transfer of acrylonitrile from the emulsified organic phase. Although graphite and magnetite (Fe_3O_4) can be used as cathode and anode, respectively, the most recent patents refer to an advantageous membrane-free procedure with a Cd cathode and Fe anode.

conducting salt (McKee salt), *e. g.*, tetraalkylammonium tosylate has two functions:

1. increase the conductivity
2. displace H_2O from cathode (depending on alkyl size) to minimize hydrogenation to propionitrile

The conducting salt – a tetraalkylammonium salt – screens the cathode so completely with its hydrophobic alkyl groups that no water electrolysis – with hydrogen formation – can occur. Thus the hydrogenation of acrylonitrile to propionitrile is almost entirely suppressed, and only the organophilic acrylonitrile can be dimerized at the cathode.

workup of electrolysis product:

organic phase separated by distillation, aqueous phase with conducting salt recycled

byproduct formation:

CH_3CH_2CN from AN hydrogenation $(NCCH_2CH_2)_2O$ as bis-adduct of AN with H_2O

alternative AN hydrodimerizations:

1. by 'chemical' reduction of AN with amalgams

After passing through the electrolysis cell, part of the organic phase – the unreacted acrylonitrile and adiponitrile product – is separated and distilled. The selectivity to adiponitrile is about 90%, with byproducts propionitrile and biscyanoethyl ether.

A chemical method for reductive dimerization uses alkali or alkaline earth amalgams as reducing agents. Since the amalgams form the corresponding salts in a mercury cell during the electrolysis, this route can be regarded as an indirect electrochemical method.

Other process developments include the hydrodimerization of acrylonitrile in the gas phase at $200-350\,°C$ and $1-3$ bar using ruthenium catalysts in the presence of hydrogen (*e. g.*, Kuraray) and the dimerization of acrylonitrile over organophosphorus complexes followed by hydrogenation (ICI).

2. by gas-phase hydrodimerization of $AN + H_2$ with Ru catalysts

3. by two-step dimerization/hydrogenation of AN over organophosphorus complexes

10.2.1.2. Hydrogenation of Adiponitrile

Adiponitrile can be hydrogenated with hydrogen to hexamethylenediamine (HMDA) using high pressure conditions of $600-650$ bar at $100-135\,°C$ with Co–Cu catalysts or $300-350$ bar at $100-180\,°C$ with Fe catalysts:

hydrogenation of ADN to HMDA with:

1. high pressure conditions using Co–Cu or Fe catalysts in trickle phase

$$NC(CH_2)_4CN + 4\,H_2 \xrightarrow{\text{[cat.]}} H_2N-(CH_2)_6-NH_2 \qquad (15)$$

$$\left(\Delta H = -\frac{75 \text{ kcal}}{314 \text{ kJ}}\,/\text{mol}\right)$$

The formation of polyamines and hexamethyleneimine is suppressed in the presence of NH_3. These can form from both nitrile groups by hydrogenation to the aldimine, reaction of the aldimine with amine to split off NH_3 and give the azomethine, followed by hydrogenation to the secondary amine. The secondary amine is also capable of adding to the aldimine intermediates. The selectivity to hexamethylenediamine is about 90 to 95%.

byproducts of ADN hydrogenation according to reaction principle:

$$RCN \xrightarrow{+H_2} RCH = NH \text{ (aldimine)}$$
$$RCH = NH + H_2NCH_2R \xrightarrow{-NH_3}$$
$$RCH = N-CH_2R \text{ (azomethine)}$$
$$RCH = N-CH_2R \xrightarrow{+H_2} RCH_2NHCH_2R$$

Other hydrogenation processes use Ni catalysts or Ni catalysts modified with Fe or Cr in the liquid phase, *e. g.*, aqueous caustic soda at lower H_2 pressures up to about 30 bar and at about 75 °C. The selectivity to hexamethylenediamine can be as high as 99%.

2. low pressure conditions with Ni- or Fe/Ni- or Cr/Ni-catalysts in liquid phase (industrially used catalysts *e. g.*, from Rhône-Poulenc or Rhodiatoce)

Hexamethylenediamine, as well as being of supreme importance for the manufacture of nylon 6,6 (*e. g.*, in 1992 in the USA 87%, in Western Europe 93% and in Japan 91% of the production of HMDA), is, after reaction with phosgene to form the diisocyanate, playing an increasing role as a component of foams and resins (*cf.* Section 13.3.3).

uses of HMDA:

1. primarily for nylon 6,6
2. after reaction with phosgene as diisocyanate for polyadditions
3. small amounts for nylon 6,10; nylon 6,12; and nylon 6,9

Lesser quantities are consumed in the manufacture of nylon 6,10 (HMDA and sebacic acid), nylon 6,12 (HMDA and 1,12-dodecanedioic acid), and nylon 6,9 (HMDA and azelaic acid). In 1993, world capacity for HMDA production was 1.20×10^6 tonnes per year, with the USA, Western Europe, and Japan having capacities of 600000, 410000, and 50000 tonnes per year, respectively.

production of HMDA (in 1000 tonnes):

	1987	1989	1992
USA	526	546	541
W. Europe	289	301	330
Japan	39	52	48

Production figures in these countries can be found in the adjacent table.

10.2.1.3. Potential Developments in Adiponitrile Manufacture

adiponitrile and hexamethylenediamine possess versatile manufacturing base:

1. olefin (propene-acrylonitrile)
2. diene (butadiene)
3. aromatics (benzene-cyclohexane-cyclo-hexanol-cyclohexanone-adipic acid)

characteristics of olefin-based process:

favorable materials costs from economically manufactured mass product AN

possible areas for improvements:

high process and capital costs of Monsanto route possibly reduced by:

extension of potential region, *e.g.*, (UCB) nonaqueous solvent

improved electrodes, *e.g.*, (Monsanto) Cd cathode, Fe anode, or change to single-chamber process

Since propene, butadiene, and benzene are widely differing feedstocks for hexamethylenediamine or its precursor adiponitrile, the manufacturing process is very versatile and can readily adapt to a changing market situation. The position of the individual feedstocks is as follows:

The commodity chemical acrylonitrile, manufactured from propene in the Sohio process, is an economical feedstock for electrochemical hydrodimerization. Since electricity accounts for only a small part of the manufacturing costs, the cost of adiponitrile is determined by process and capital costs, and this is where any cost improvement should begin. For example, a UCB development using nonaqueous media can eliminate H_2O decomposition and extend the potential region. New proposals from Monsanto concern improved anodic and cathodic materials and the shift to diaphragm-free — *i.e.*, single-chamber — electrolysis processes. As electrochemical processes are very selective through adjustment of the working potential, there is minimal byproduct formation. The Monsanto process is thus particularly nonpolluting which is an important factor in its favor.

Du Pont, which was the largest producer of adiponitrile based on 1,4-dichlorobutene, independently developed the direct hydrocyanation of butadiene, thereby avoiding an alkali chloride electrolysis and the pollution problems inherent with NaCl.

characteristics of the diene-based process:

single-step HCN addition avoids NaCl electrolysis and NaCl removal and is thus a more economical hydrocyanation route and, on a cost basis, most economical HMDA route

characteristics of aromatic-based process:

shortage of aromatics (due to use for octane improvement in low-lead gasolines) will have unfavorable effect on raw material supply and economics of process

For the near future at least, benzene, the third type of feedstock, has the bleakest prospects. The reduction of the lead content in gasoline, already under way in many countries, makes an addition of aromatics essential to improve the octane number. This could lead to a shortage of aromatics.

10.2.2. ω-Aminoundecanoic Acid

manufacture of ω-aminoundecanoic acid in five-step synthesis:

1. transesterification of castor oil to methyl ester of ricinoleic acid
2. pyrolysis to undecenoic acid (*cf.* Section 10.1) and heptanal

ω-Aminoundecanoic acid is the precursor of nylon 11. The starting material is either the natural product castor oil — the glycerol ester of ricinoleic acid — or ricinoleic acid itself. Glycerol triricinolate is first transesterified with methanol to the methyl ester and then pyrolytically cleaved at about 300 °C into a C_7 aldehyde and the methyl ester of undecenoic acid:

$$CH_3(CH_2)_5CH \!-\! CH_2CH \!=\! CH(CH_2)_7COOCH_3$$
$$|$$
$$OH$$

$$\downarrow \Delta$$

$$CH_3(CH_2)_5CH + H_2C \!=\! CHCH_2(CH_2)_7COOCH_3$$
$$\|$$
$$O$$

(16)

The methyl ester is saponified and converted into 11-bromo-undecanoic acid by peroxide-catalyzed HBr addition. This important step can easily be carried out on a commercial scale. A solution of the unsaturated acid in toluene–benzene flows down the column-shaped reactor countercurrent to a HBr and air stream from below. Approximately 96% of the resulting bromide has the bromine in a terminal position. The product is reacted with NH_3 to form the ammonium salt of ω-aminoundecanoic acid, which is released by acidification:

3. anti-Markovnikov HBr addition (peroxide or UV light) to undecenoic acid
4. ammonolysis to ω-aminoundecanoic acid
5. conversion of NH_4 salt into free acid

$$H_2C \!=\! CH(CH_2)_8COOH + HBr \xrightarrow{\text{[peroxide]}}$$

$$Br \!-\! CH_2CH_2(CH_2)_8COOH \xrightarrow[\text{2. } + H_2O]{\text{1. } +NH_3} H_2N \!-\! (CH_2)_{10}COOH$$

(17)

The overall selectivity to ω-aminoundecanoic acid is about 67%. Nylon 11 resulting from the polycondensation is mainly produced in France by ATO Plastiques (Rilsan®), in the USA, and in Brazil. A sufficient and inexpensive source of castor oil is necessary to produce it economically. This dependence on the main producing countries — North Africa, Brazil and India (the largest producer) — introduces uncertainties for future growth. World castor oil production figures are summarized in the adjacent table.

characteristics of ω-aminoundecanoic acid manufacture and thus nylon 11 manufacture:

dependence of raw material on agricultural products means variable supply and growth limitation

castor oil production (in 10^6 tonnes)

	1980	1992	1993
World	0.36	1.11	1.06

10.3. Lactams

10.3.1. ε-Caprolactam

ε-Caprolactam, the cyclic amide of ε-aminocaproic acid, is the most industrially important lactam. Its primary use is in the manufacture of nylon 6 (Perlon®).

The worldwide manufacturing capacity for ε-caprolactam in 1995 was about 3.7×10^6 tonnes per year, of which 0.95, 0.70, 0.56, and 0.56×10^6 tonnes per year were in Western Europe, the USA, the CIS, and Japan, respectively. Production figures are given in the adjacent table.

ε-caprolactam has become a commodity chemical with basically a single outlet — nylon 6

ε-caprolactam production (in 1000 tonnes):

	1989	1991	1994
W. Europe	775	841	843
USA	595	582	685
Japan	469	531	519

extensive worldwide research led to numerous but still not ideal manufacturing routes

$(NH_4)_2SO_4$-free route not yet used commercially

manufacturing processes for ε-caprolactam can be divided into two groups:

1. processes with cyclohexanone oxime as the intermediate product (dominates worldwide)

2. process type with different intermediates (minor importance)

classical cyclohexanone oxime route with cyclohexanone feedstock and Beckmann rearrangement as characteristic reaction

first step in classical ε-caprolactam process:

four routes for manufacture of cyclohexanone:

1. cyclohexane oxidation to cyclohexanone/cyclohexanol followed by dehydrogenation of cyclohexanol
2. ring hydrogenation of phenol and dehydrogenation of cyclohexanol, older method – two steps; newer process – single step
3. cyclohexylamine dehydrogenation and cyclohexylimine hydrolysis
4. partial hydrogenation of benzene to cyclohexene followed by hydration (planned for the future)

Despite the numerous manufacturing routes to ε-caprolactam, each of the processes still has room for improvement. All are multistep, with the unavoidable formation of ammonium sulfate or other byproducts.

The many – and for the most part industrially operated – ε-caprolactam processes can be divided into two groups. One is characterized by the intermediate cyclohexanone oxime, which is synthesized from cyclohexanone or from other precursors. This is by far the most important manufacturing route, currently accounting for more than 95% of worldwide ε-caprolactam production.

The other group encompasses caprolactam processes which involve other intermediates and other process steps.

10.3.1.1. ε-Caprolactam from the Cyclohexanone Oxime Route

A description of the classical manufacture of ε-caprolactam can be used to present not only the problems of this route but also its modern variations. The synthesis consists of three 'organic' and one 'inorganic' steps:

1. Manufacture of cyclohexanone
2. Oxime formation from cyclohexanone with hydroxylamine
3. Beckmann rearrangement of cyclohexanone oxime to ε-caprolactam
4. Manufacture of hydroxylamine

To 1:

Most cyclohexanone is made from cyclohexane. A second route starts with phenol. Another route used only in the CIS (capacity 20000 tonnes per year) uses cyclohexylamine, which is catalytically dehydrogenated and then hydrolyzed with steam to cyclohexanone. In 1995 about 57% and 55% of the e-caprolactam production in the USA and Western Europe, respectively, was based on cyclohexane oxidation and most of the remainder (i.e., 43 and 45%, respectively) came from phenol hydrogenation.

Typical ε-caprolactam processes with a phenol feedstock were developed by Allied Chemical, Montedipe, and Leuna-Werke.

A new process for the production of cyclohexanol from Asahi Chemical is at the pilot-plant stage. Here, benzene is partially hydrogenated on a ruthenium catalyst to cyclohexene, which is then hydrated to cyclohexanol under acid catalysis. Process conditions for the proposed 60000 tonne-per-year plant have not yet been disclosed.

In the first-mentioned cyclohexanone oxidation route (*cf.* Section 10.1.1), cyclohexanone is distilled from the cyclohexanone/cyclohexanol mixture and the cyclohexanol portion is catalytically dehydrogenated at $400-450\,°C$ and atmospheric pressure over Zn or Cu catalysts:

$$\Delta H = -\frac{15\ kcal}{65\ kJ}\ /mol \qquad (18)$$

The cyclohexanol conversion is about 90%, with a selectivity to cyclohexanone of 95%.

Earlier, phenol could only be converted into cyclohexanone in a two-step process: after ring hydrogenation with nickel catalysts at $140-160\,°C$ and 15 bar, the dehydrogenation was conducted analogous to equation 18. Simplification of this route was made possible by selective hydrogenation with Pd catalysts:

$$(19)$$

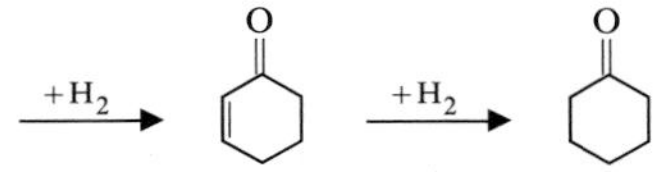

Phenol is completely converted in the gas phase at $140-170\,°C$ and $1-2$ bar using a supported Pd catalyst containing alkaline earth oxides (*e.g.*, Pd—CaO/Al$_2$O$_3$). The selectivity to cyclohexanone is greater than 95%.

To 2:

Oxime formation with cyclohexanone is done with a hydroxylamine salt, usually the sulfate, at $85\,°C$:

$$\Delta H = -\frac{10\ kcal}{42\ kJ}\,/mol \qquad (20)$$

In order to displace the equilibrium, NH$_3$ must be continuously introduced to maintain a pH of 7. The first ammonium sulfate formation takes place at this stage in the process.

characteristics of cyclohexanol dehydrogenation process:

heterogeneously catalyzed gas-phase dehydrogenation at atmospheric pressure in heated tubular oven

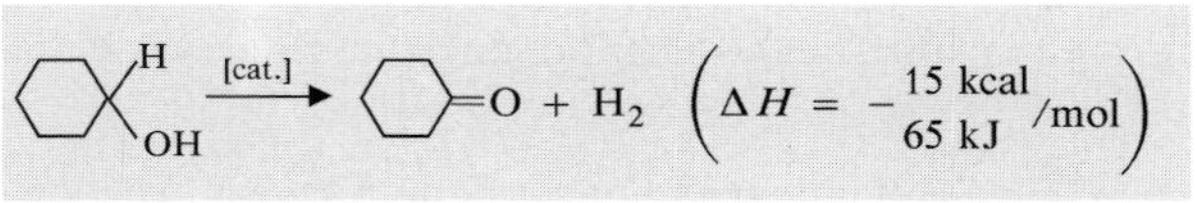

characteristics of phenol hydrogenation process:

selective Pd-metal-catalyzed gas-phase single-step hydrogenation to cyclohexanone

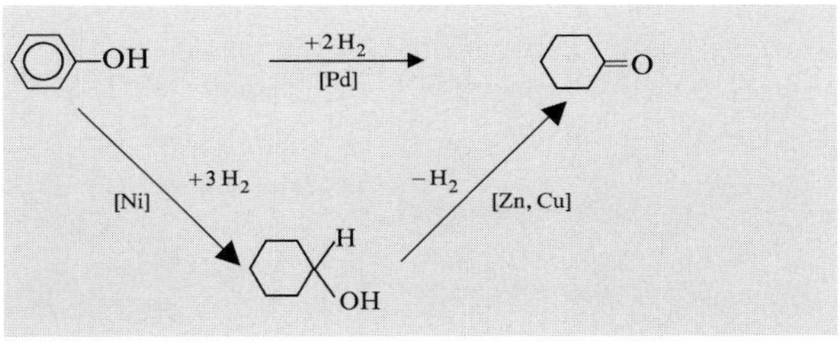

mechanism of the selective phenol hydrogenation:

1. phenol activation by adsorption in keto form:

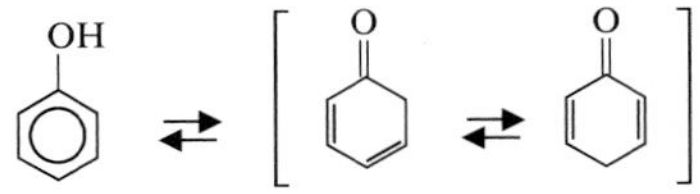

2. hydrogenation in two steps:

second step in classical ε-caprolactam process:

process principles of oxime formation from cyclohexanone:

reaction with hydroxylamine sulfate also forms H$_2$SO$_4$, which is removed by

1. NH$_3$ or
2. with regenerable phosphate buffer (HPO process)

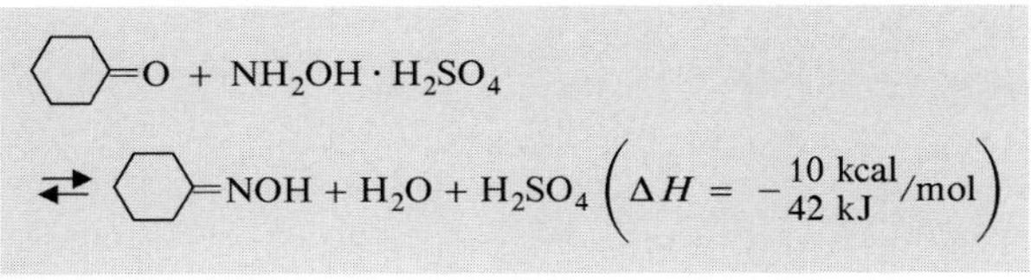

The same constant pH value can be attained without salt formation in the DSM/Stamicarbon HPO process (**h**ydroxylamine-**p**hosphate-**o**xime) with hydroxylamine in a buffer solution containing H_3PO_4. The buffer solution – "liberated" during oxime formation – is recycled to the hydroxylamine production.

two alternatives for manufacturing cyclo-hexanone oxime:

Toray and Du Pont practice two other methods. Both involve cyclohexanone oxime, but avoid its manufacture from cyclohexanone.

1. photonitrosation of cyclohexane in Toray PNC process

Toray developed a commercial process from a known reaction, the **photo**nitrosation of **c**yclohexane directly to cyclohexanone oxime (PNC process):

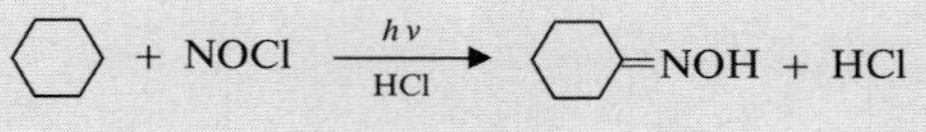

$$\bigcirc + NOCl \xrightarrow[HCl]{h\nu} \bigcirc=NOH + HCl \qquad (21)$$

process characteristics:

UV light induces decomposition of NOCl to NO and Cl radicals with subsequent formation of nitrosocyclohexane

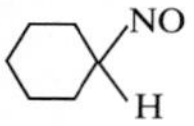

which rearranges to the oxime

A gas mixture consisting of HCl and nitrosyl chloride (NOCl) is fed into cyclohexane at a temperature below $20\,°C$. The reaction, which is initiated by Hg light, results in the formation of cyclohexanone oxime with 86% selectivity (based on C_6H_{12}). Toray uses this process in two plants (startups in 1963 and 1971) with a total production capacity of 174000 tonnes (1995). NOCl is obtained from the reaction of HCl with nitrosyl sulfuric acid:

$$2\,H_2SO_4 + NO + NO_2 \rightarrow 2\,NOHSO_4 + H_2O \qquad (22)$$

$$NOHSO_4 + HCl \rightarrow NOCl + H_2SO_4 \qquad (23)$$

H_2SO_4 and HCl go through the process without either salt formation or great losses.

2. cyclohexane nitration and partial hydrogenation to oxime used industrially (Du Pont, 1963–1967)

The Du Pont process (Nixan process, **Ni**trocyclohe**xan**e process) was practiced for a time in a 25000 tonne-per-year oxime manufacturing plant. Nitrogen was introduced into cyclohexane by nitration with HNO_3 in the liquid phase, or NO_2 in the gas phase. The nitrocyclohexane was then catalytically hydrogenated to the oxime:

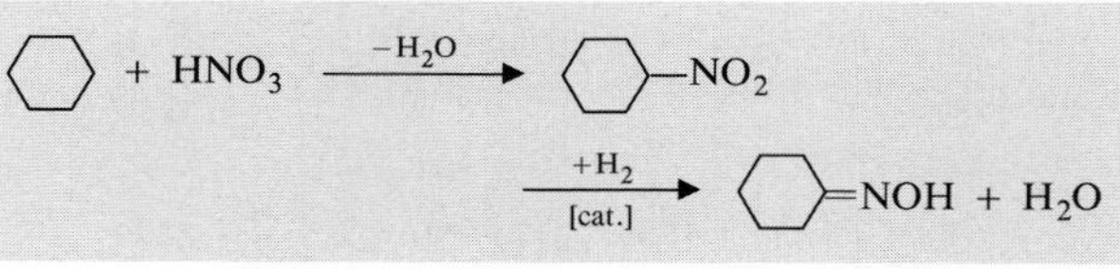

$$\bigcirc + HNO_3 \xrightarrow{-H_2O} \bigcirc-NO_2 \qquad (24)$$

$$\xrightarrow[\text{[cat.]}]{+H_2} \bigcirc=NOH + H_2O$$

third step in classical ε-caprolactam process:

Beckmann rearrangement of cyclohexanone oxime to amide of ε-aminocaproic acid

To 3:

All processes described so far proceed either directly or *via* intermediate stages to cyclohexanone oxime, which is then converted into ε-caprolactam by a rearrangement with H_2SO_4 or oleum discovered by E. Beckmann in 1886:

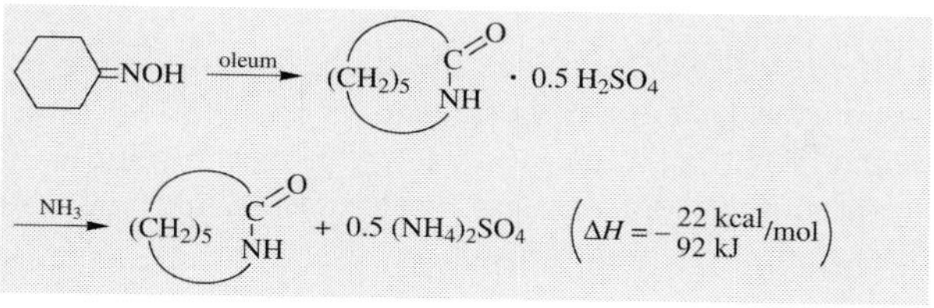

$$(25)$$

Commercial development of this process was done by BASF. In the continuous process the oxime solution, acidified with sulfuric acid, is passed through the reaction zone which is kept at the rearrangement temperature (90–120 °C). The rearrangement is complete within a few minutes, and the resulting lactam sulfate solution is converted into the free lactam with NH_3 in a neutralization vessel. It separates from the saturated ammonium sulfate solution as an oily layer, which, after extraction with benzene, toluene, or chlorinated hydrocarbons and stripping with water, is further purified and then distilled. The selectivities amount to almost 98%.

To 4:

Hydroxylamine sulfate for the conversion of cyclohexanone into the oxime is usually manufactured in a modified four-step Raschig process. Essentially, it consists of the reduction of ammonium nitrite with SO_2 at about 5 °C to the disulfonate, which is then hydrolyzed at 100 °C to hydroxylamine sulfate:

$$NH_4NO_2 + NH_3 + 2\, SO_2 + H_2O \rightarrow HON(SO_3NH_4)_2 \qquad (26)$$

$$HON(SO_3NH_4)_2 + 2\, H_2O \rightarrow NH_2OH \cdot H_2SO_4 + (NH_4)_2SO_4 \quad (27)$$

One mole $(NH_4)_2SO_4$ is formed for each mole of hydroxylamine sulfate. A further mole of ammonium sulfate is obtained from the oxime formation with cyclohexanone, where it is introduced to the reaction by hydroxylamine sulfate and NH_3 (to neutralize the oxime).

A basic improvement in hydroxylamine manufacture was achieved with the DSM HPO process (*cf.* 2nd process stage). Nitrate ions are reduced with hydrogen to hydroxylamine using a palladium-on-charcoal catalyst or Al_2O_3 with, *e.g.*, promoters such as germanium compounds, suspended in a phosphate buffer solution:

process principles:

H_2SO_4-catalyzed, liquid-phase ring expansion to basic lactam, which is released with NH_3 from the sulfate formed initially

SO_3 added to conc. H_2SO_4 increases rate of rearrangement

fourth step in classical ε-caprolactam process:

hydroxylamine manufacture by four-stage Raschig process:

1. $(NH_4)_2CO_3$ manufacture from NH_3, CO_2, H_2O
2. conversion into NH_4NO_2 with $NO \cdot NO_2$
3. reduction with SO_2 to disulfonate
4. hydrolysis to NH_2OH

alternatives to hydroxylamine manufacture:

1. Pd-catalyzed $NO_3^{\ominus}$ reduction to NH_2OH in phosphate buffer (H_3PO_4, NH_4NO_3, H_2O)

$$NO_3^{\ominus} + 2\,H^{\oplus} + 3\,H_2 \xrightarrow[PO_4^{3\ominus}]{[cat.]} NH_3OH^{\oplus} + 2\,H_2O \qquad (28)$$

principle of DSM HPO process:

combination of hydroxylamine and cyclo-hexanone oxime manufacture in phosphate buffer, run in cycle

The buffered hydroxylamine solution is then used, during the oximation of cyclohexanone in toluene, to dissolve and extract the oxime. After addition of HNO_3, the spent solution is recycled to nitrate hydrogenation. By this means, much less salt is formed than in the Raschig process.

Since $NO_3^{\ominus}$ is manufactured by oxidation of NH_3, the whole process can be regarded as an oxidation of NH_3 to NH_2OH. Overall selectivity to hydroxylamine from ammonia is 58%.

2. Pt/Pd-catalyzed NO reduction in BASF, Bayer, and Inventa processes, second in importance

Other processes for the manufacture of hydroxylamine, such as the platinum- or palladium-catalyzed reduction of NO with hydrogen in dilute mineral acid solution developed by BASF, Bayer, and Inventa, do in fact result in less salt. However, the starting materials must be very pure and an involved catalyst recovery is necessary. Commercial use of these processes is increasing.

10.3.1.2. Alternative Manufacturing Processes for ε-Caprolactam

process types with intermediates other than cyclohexanone oxime:

1. 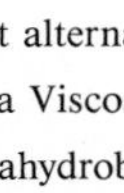 COOH   2. 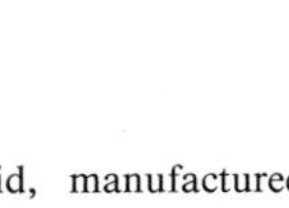 $(CH_2)_5$ ... C=O / O

3. 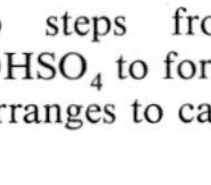 O ... NO₂

Three other commercial ε-caprolactam processes avoid cyclohexanone oxime and therefore the Beckmann rearrangement. They are:

1. The Snia Viscosa cyclohexanecarboxylic acid process

2. The UCC caprolactone process

3. The Techni-Chem nitrocyclohexanone process

first alternate route:

Snia Viscosa process:

hexahydrobenzoic acid, manufactured in two steps from toluene, reacts with $NOHSO_4$ to form a mixed anhydride which rearranges to caprolactam

To 1:

The first step in the Snia Viscosa process is an air oxidation of toluene to benzoic acid at 160–170 °C and 8–10 bar over a Co catalyst with a yield of ca. 30% (*cf.* Section 13.2.1.1). The acid is then hydrogenated almost quantitatively to cyclohexanecarboxylic acid in the liquid phase over a Pd/C catalyst at 170 °C and 10–17 bar.

nitrosylsulfuric acid production by absorption of N_2O_3 in oleum:

$$N_2O_3 + H_2SO_4 + SO_3 \rightarrow 2\,O{=}N{-}OSO_3H$$

The cyclohexanecarboxylic acid is then reacted with nitrosylsulfuric acid in oleum at temperatures up to 80 °C. The reaction apparently goes through formation of a mixed anhydride and several other intermediate steps to eliminate CO_2 and give ε-caprolactam as the sulfate:

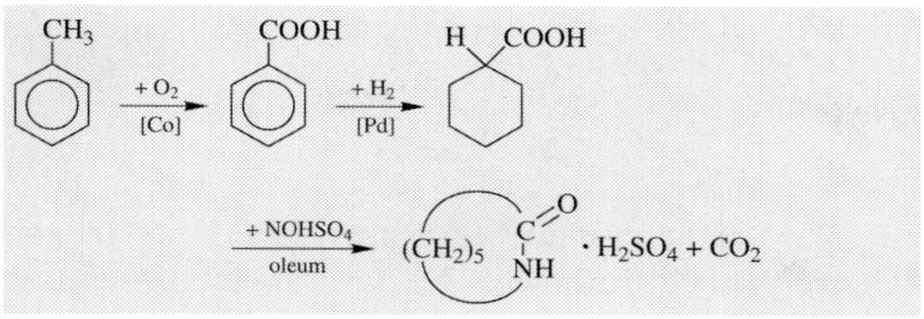

$$(29)$$

Processing of the acidic caprolactam solution takes place in the usual manner with NH_3. In a newer version, ε-caprolactam is extracted from the sulfuric acid solution with alkylphenols, and then stripped with H_2O.

Operated this way, the Snia Viscosa route can be a salt-free process. With a 50% conversion of cyclohexanecarboxylic acid, the selectivity to ε-caprolactam is 90%.

Two plants in Italy with a total capacity of 100000 tonnes per year and one 80000 tonne-per-year plant in the CIS use the Snia Viscosa process.

industrial practice of Snia Viscosa process: plants in Italy and the CIS

To 2:

In the UCC process, cyclohexanone is first oxidized to ε-caprolactone with peracetic acid at 50 °C and atmospheric pressure. The selectivities are 90% (based on cyclohexanone) and 85 – 90% (based on peracetic acid). The ε-caprolactone is then reacted with NH_3 at 170 bar and 300 – 400 °C to ε-caprolactam:

second alternate route:

UCC process:

analogous to the Baeyer-Villiger oxidation (ketone → ester) cyclohexanone is oxidized with peracids or hydroperoxides to ε-caprolactone and converted into the lactam with NH_3

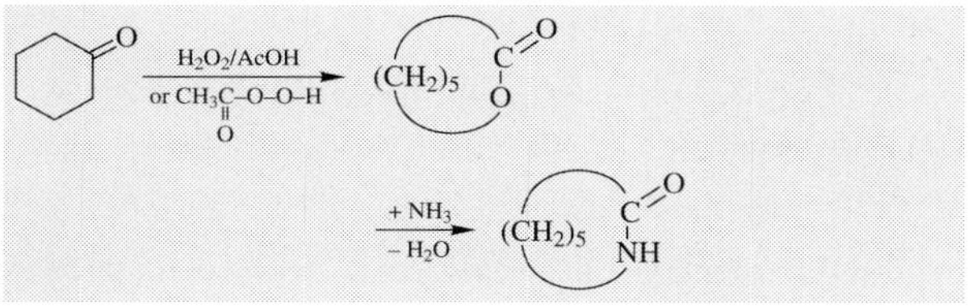

$$(30)$$

The selectivity for both steps is about 70%. The UCC process has been operating in a 25000 tonne-per-year plant since 1967. In 1972 the ammonolysis was discontinued, but the plant is still being used to manufacture ε-caprolactone.

Many firms have been working on similar processes, *i. e.*, synthesis of ε-hydroxycaproic acid or ε-caprolactone intermediates. Interox in England, for example, has been producing caprolactone since 1975, increasing their capacity to 10000 tonnes per year. Daicel in Japan also manufactures caprolactone.

caprolactone ammonolysis discontinued by UCC, instead other applications for caprolactone, *e. g.*, polyester manufacture

Caprolactone, as a bifunctional compound, is also suitable for the manufacture of polyesters for casting resins.

To 3:

third alternate route:

Techni-Chem process:

nitration of cyclohexanone (through enol-acetate intermediate which increases selectivity), hydrolytic ring opening, and reduction to aminocaproic acid (lactam precursor)

The advantage of the Techni-Chem process (USA) over all other processes is that no byproducts are formed. In this process, cyclohexanone is acetylated with ketene/acetic anhydride to cyclohexenyl acetate, which is then nitrated to 2-nitrocyclohexanone with elimination of acetic acid. The hydrolytic ring opening leads to nitrocaproic acid which is then hydrogenated to ε-aminocaproic acid. This can then be converted into ε-caprolactam at 300 °C and 100 bar:

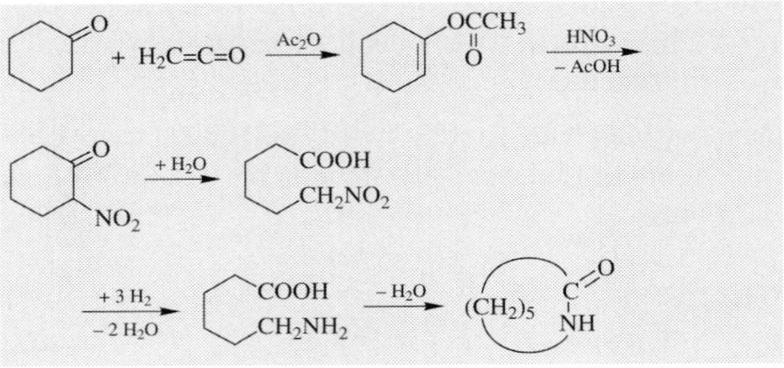

(31)

(32)

process characteristics:

sole ε-caprolactam process without unavoidable byproducts

One disadvantage of this process is the use of expensive ketene to manufacture an intermediate which can be more selectively nitrated. The resulting acetic acid can, of course, be recycled to the ketene manufacture.

This process has been practiced only at the pilot-plant scale.

10.3.1.3. Possibilities for Development in ε-Caprolactam Manufacture

classical ε-caprolactam processes are encumbered by $(NH_4)_2SO_4$ which is formed at two points:

1. during cyclohexanone oximation
2. during neutralization of H_2SO_4 from Beckmann rearrangement

The greatest disadvantage of the classical ε-caprolactam process is the coproduction of 5 kg ammonium sulfate per kilogram caprolactam. Since ammonium sulfate is a low-grade fertilizer (the H_2SO_4 formed on decomposition acidifies the soil) only limited credits are obtained, which contribute little to the economics of the process.

Most process improvements are therefore focused on saving H_2SO_4.

Compared to other oxime processes, DSM's HPO route involves the greatest reduction in consumption of H_2SO_4 and NH_3; only 1.8 kg ammonium sulfate is formed per kilogram ε-caprolactam.

Interesting proposals for reducing the amount of $(NH_4)_2SO_4$ have been made by many firms. The catalytic oxime–lactam rearrangement, for example, over a B_2O_3/Al_2O_3 catalyst at 340–360 °C (BASF), or a B_2O_3 catalyst in a fluidized bed at ca. 330 °C (Bayer), is reported to be more selective. Though this reaction has been studied for a long time, no economical technology has been developed.

Similarly, DSM/Stamicarbon have reported a new type of Beckmann rearrangement of cyclohexanone oxime in DMSO at 100 °C using a strongly acidic ion-exchange resin such as Amberlyst 15 or Amberlite 200.

According to another BASF proposal, neutralization of caprolactam sulfate with NH_3 can be replaced by electrodialysis. Polypropylene membranes separate the cathode and anode regions, where caprolactam and H_2SO_4 concentrate, respectively. The caprolactam can be obtained from the cathodic liquid in 98.5% yield by extraction with benzene.

Another solution to the ammonium sulfate problem is thermal decomposition of $(NH_4)_2SO_4$ to NH_3 and SO_2 (Inventa), or to N_2 and SO_2 at 1000 °C (DSM). The SO_2 can be used to manufacture hydroxylamine *via* the Raschig process.

A surprisingly simple production of cyclohexanone oxime/ ε-caprolactam has been suggested by Allied in which cyclohexanone is reacted with NH_3 and O_2 over either high-surface-area silica gel or Ga_2O_3 at 200 °C and 10 bar. At a conversion of about 50%, selectivity to a mixture of oxime and lactam approaches 68%:

(33)

A final acid-catalyzed Beckmann rearrangement of the oxime portion would complete a route distinguished by a small number of reaction steps and no salt formation. This method has not yet been proved practicable.

The oxidative cyclohexanone/NH_3 reaction to give cyclohexanone oxime has been developed by Montedison. The oxidation is carried out with H_2O_2 in the liquid phase over a TiO_2/SiO_2 catalyst at 40–90 °C and atmospheric pressure. A yield and selectivity of greater than 90% (based on cyclohexanone, H_2O_2)

process improvements aim at reducing H_2SO_4 consumption, *e. g.*, by:

Beckmann rearrangement with acidic catalysts in the gas phase (BASF, Bayer)

Beckmann rearrangement with acidic ion-exchangers in the liquid phase (DSM/Stamicarbon)

electrodialysis of caprolactam sulfate for separation into lactam and H_2SO_4 (BASF)

recycle of H_2SO_4 as SO_2 from $(NH_4)_2SO_4$ pyrolysis (Inventa, DSM)

single-step manufacture of cyclohexanone oxime/ε-caprolactam from cyclohexanone (Allied)

oxidative oximation of cyclohexanone with H_2O_2/NH_3 (Montedison); first commercial use by EniChem (1994)

are achieved. This process is being developed (1994) by EniChem in a 15 000 tonne-per-year pilot plant.

cost comparison of ε-caprolactam processes indicates:

DSM/Stamicarbon HPO process leads to lowest manufacturing costs

Toray photonitrosation has low material but high process costs

A comparison of manufacturing costs of caprolactam in the two most important modern processes indicates that the DSM/Stamicarbon process so far has lower production costs than the classical route (*e. g.*, BASF). In 1992, nearly a third of the worldwide caprolactam capacity was based on the DSM technology. Although the Toray route has the lowest material costs, process costs are considerably higher than rival processes due to the frequent change of Hg lamps and the high consumption of electricity.

10.3.1.4. Uses of ε-Caprolactam

uses of caprolactam:

1. nylon 6 manufacture

Most ε-caprolactam is used for the manufacture of nylon 6, which is the starting material for fibers with many uses in textile manufacture and in the industrial sector.

2. starting product for N-methyl-ε-capro-lactam, aprotic solvent for material separation processes used commercially

N-methyl-ε-caprolactam is a thermally and chemically stable derivative of ε-caprolactam which has been developed by Leuna as an aromatic extractant (*cf.* Section 12.2.2.2). It is obtained by the gas-phase methylation of ε-caprolactam with methanol over an Al_2O_3 catalyst at about 350 °C and atmospheric pressure:

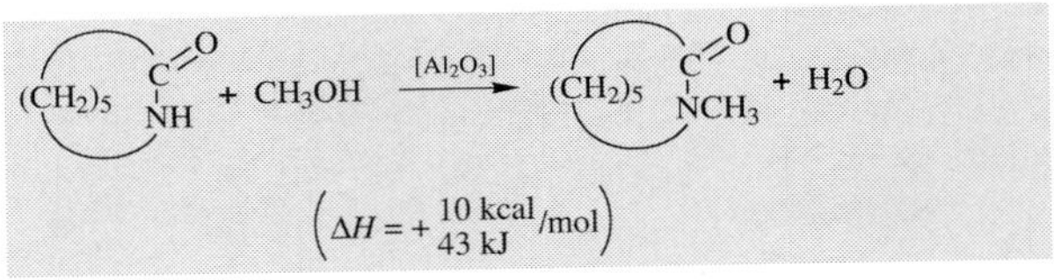

$$(CH_2)_5 \begin{array}{c} C{=}O \\ NH \end{array} + CH_3OH \xrightarrow{[Al_2O_3]} (CH_2)_5 \begin{array}{c} C{=}O \\ NCH_3 \end{array} + H_2O \qquad (34)$$

$$\left(\Delta H = + \begin{array}{c} 10\ kcal \\ 43\ kJ \end{array} /mol \right)$$

Commercial production began in 1977.

3. starting product for hexamethylenimine, commercial intermediate

ε-Caprolactam is also used in a process from Mitsubishi Chemical for the manufacture of hexamethylenimine, where it is selectively hydrogenated in the presence of a Co catalyst:

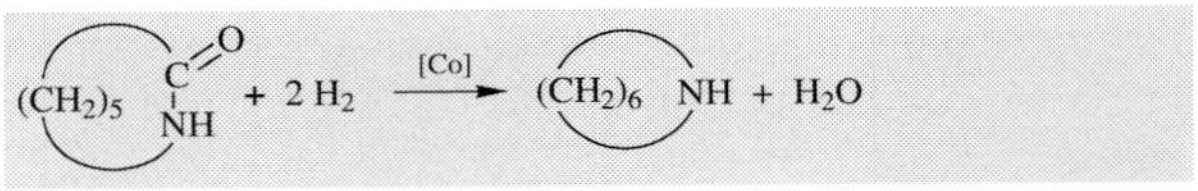

$$(CH_2)_5 \begin{array}{c} C{=}O \\ NH \end{array} + 2\,H_2 \xrightarrow{[Co]} (CH_2)_6\ NH + H_2O \qquad (35)$$

Exact process conditions and yields have not yet been disclosed.

A commercial plant began operation in 1982. Hexamethylenimine is an intermediate for such things as pharmaceuticals and pesticides.

For a time, ε-caprolactam was also used as an intermediate in the manufacture of L-lysine, an essential amino acid which cannot be synthesized by humans or animals.

Lysine, or α, ε-diaminocaproic acid, is formed by introducing an NH_2 group into the α position of the ε-aminocaproic acid or its cyclic amide, ε-caprolactam:

$$(CH_2)_5 \begin{matrix} COOH \\ NH_2 \end{matrix} \xrightarrow{\text{multistep}} H_2N-(CH_2)_4\underset{\underset{NH_2}{|}}{C}HCOOH \tag{36}$$

This takes place, after protecting the NH_2 group, by halogenation and then halogen exchange with NH_3.

This traditional pathway has been varied in numerous ways, *e. g.*, attempting the introduction of the α-amino group as a nitro or azido group. None of these processes is used commercially today.

Another synthetic route to lysine, based on a multistep process from Stamicarbon with the addition of acetaldehyde to acrylonitrile as the first step, is not used industrially because of its cost.

The DL mixture, which results in all syntheses, must be separated into its optical isomers, since only L-lysine is identical with the naturally occurring amino acid and therefore physiologically effective. Because of this, and the currently expensive chemical process for lysine racemate, the enzymatic process has proved to be superior.

The Toray route, used in a 4000 tonne-per-year plant in Japan, can be seen as a transition to this. DL-α-Amino-ε-caprolactam from the multistep reaction of cyclohexene with nitrosyl chloride is used as the raw material for hydrolysis on an immobilized L-hydrolase. The unhydrolyzed D-α-amino-ε-caprolactam is simultaneously racemized on an immobilized racemization catalyst, and is thus available for enzyme hydrolysis.

Currently, most L-lysine is produced by fermentation processes using, for example, molasses. The main producers are Ajinomoto and Kyowa Hakko in Japan, and, for the last few years, in France and the USA. The worldwide production capacity for L-lysine was about 338 000 tonnes per year in 1995, of which 155 000, 55 000, and 37 000 tonnes per year was in the USA, Western Europe, and Japan.

4. starting product for L-lysine, once used commercially by Du Pont

lysine manufacture in several steps (*e. g.*, from ε-aminocaproic acid):

1. protection of NH_2 group by benzoylation
2. α-bromination
3. substitution of Br by NH_2
4. hydrolysis

lysine manufacture by DSM process, with cyanoethylation of CH_3CHO as first step, not used commercially

L-lysine separation from racemate, *e. g.*, by salt formation with L-pyrrolidonecarboxylic acid

D-form is subsequently racemized

Toray process for L-lysine by synthesis of lysine precursor

followed by combined enantiomer-selective, enzyme-catalyzed hydrolysis and racemization

commercial L-lysine production by fermentation, especially in Japan and Southeast Asia

use of L-lysine:

as essential amino acid as supplement for certain foodstuffs and animal feed

Synthetic L-lysine is used to supplement foods such as corn, rice and several types of grain which have a lysine deficiency and thus possess limited biological value; and in animal feed, particularly as a supplement to soy and fish meal. In 1995, the world demand for lysine was about 260 000 tonnes per year.

10.3.2. Lauryl Lactam

lauryl lactam manufacture:

1. cyclododecanone oxime route analogous to caprolactam
2. nitrosyl chloride route not analogous to caprolactam
3. nitrosyl sulfuric acid route analogous to Snia Viscosa process

Lauryl lactam, the nylon 12 monomer, can be manufactured by an oxime rearrangement analogous to that used for caprolactam. Initially the cyclododecanol/cyclodecanone mixture, as in the manufacture of 1,12-dodecanedioic acid (*cf.* Section 10.1.2), is dehydrogenated in the liquid phase at $230-245\,°C$ and atmospheric pressure in the presence of Cu/Al_2O_3 or Cu/Cr catalysts to give an almost quantitative yield of cyclododecanone:

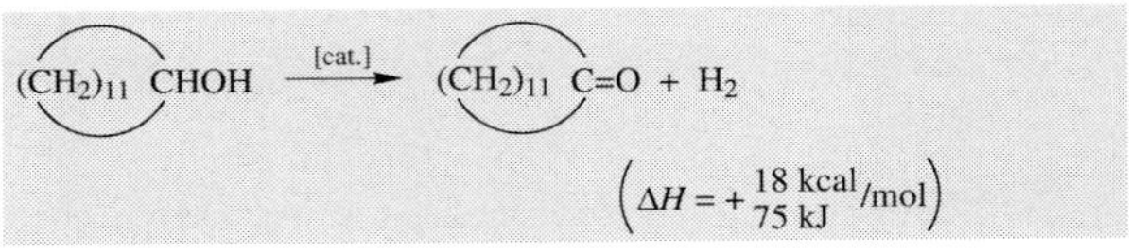

$$(CH_2)_{11}\ CHOH \xrightarrow{\text{[cat.]}} (CH_2)_{11}\ C{=}O + H_2 \tag{37}$$

$$\left(\Delta H = +\ \frac{18\ \text{kcal}}{75\ \text{kJ}}/\text{mol}\right)$$

process principles of oxime route:

cyclododecanone/NH_2OH reaction with oxime formation, acid-catalyzed Beckmann rearrangement

Cyclododecanone is then oximated and the rearrangement is carried out in oleum at 100 to 130 °C:

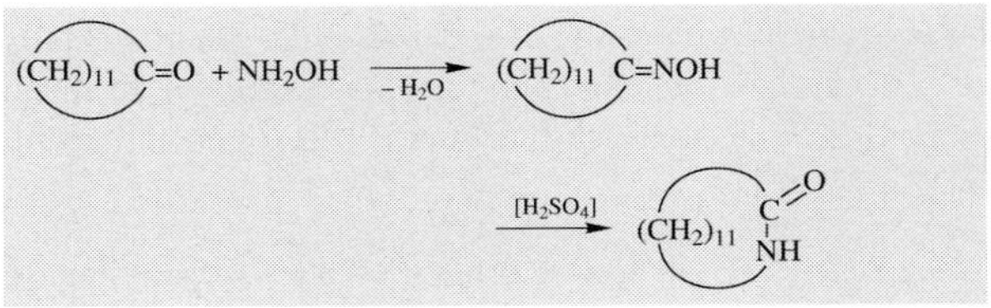

$$(CH_2)_{11}\ C{=}O + NH_2OH \xrightarrow{-H_2O} (CH_2)_{11}\ C{=}NOH \tag{38}$$

$$\xrightarrow{[H_2SO_4]} (CH_2)_{11}\ \underset{NH}{\overset{C\diagup O}{\mid}}$$

technological advantages compared to cyclohexanone oxime rearrangement:

lauryl lactam insoluble in H_2O, therefore easily isolated by crystal separation

Since lauryl lactam is insoluble in water, the neutralization step is unnecessary; the lactam can be separated by dilution with H_2O.

All three process steps have conversions of $95-100\%$, with selectivities of over 98%. Commercial processes are operated by Hüls and Nikon Rilsan.

process principles of NOCl route:

1. hydrogenation of cyclododeca-1,5,9-triene
2. photonitrosation (photooximation) of cyclododecane
3. Beckmann rearrangement

A second pathway to lauryl lactam or its oxime precursor involves the photochemically initiated reaction of NOCl (*cf.* Section 10.3.1.1) with cyclododecane at 70 °C. Cyclododecane is easily obtained commercially by the trimerization of butadiene to cyclododecatriene, which is then hydrogenated:

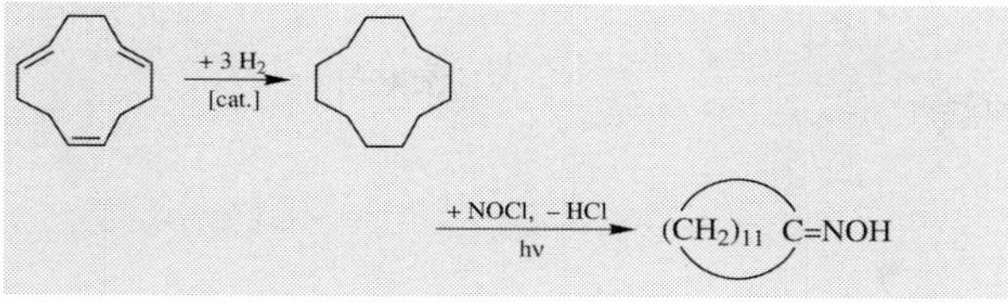

$$(39)$$

The conversion is almost 70%, with a selectivity of 88%.

This process is practiced in France by Aquitaine Organico (ATO) in an 8000 tonne-per-year plant.

Snia Viscosa's caprolactam manufacturing process can also be used for the reaction of cyclododecanecarboxylic acid with nitrosylsulfuric acid. Lauryl lactam is obtained directly:

process principles of $NOHSO_4$ route:

single-step lactam formation from cyclododecanecarboxylic acid + $NOHSO_4$

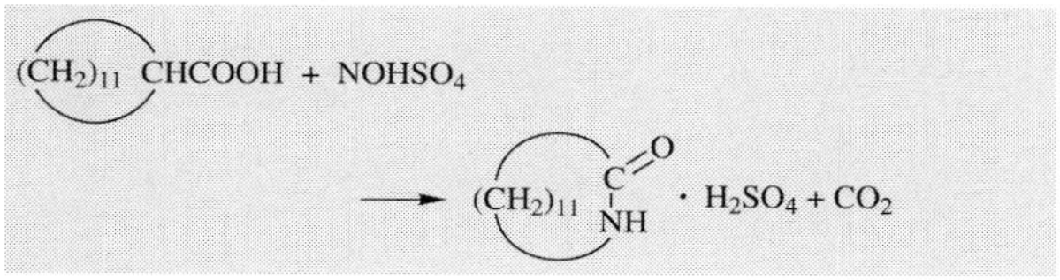

$$(40)$$

Cyclododecanecarboxylic acid can be obtained from an industrially developed carbonylation of cyclododeca-1,5,9-triene in the presence of palladium catalysts modified with phosphine, through the intermediate diene carboxylic acid. It can also be obtained by hydroformylation of cyclododecene, which goes through the formyl derivative.

cyclododecanecarboxylic acid feedstock from two processes:

1. carbonylation *via*

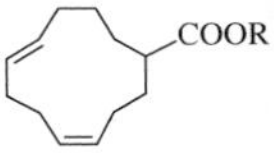

2. hydroformylation *via*

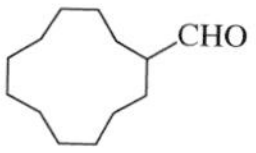

A newer process is based on initial research by BP, with further development by Ube Industries, in which cyclohexanone in the presence of NH_3 is converted to 1,1'-peroxydicyclohexylamine in a catalytic peroxidation:

Ube process for manufacture of ω-aminododecanoic acid (alternative monomer for nylon 12) from cyclohexanone

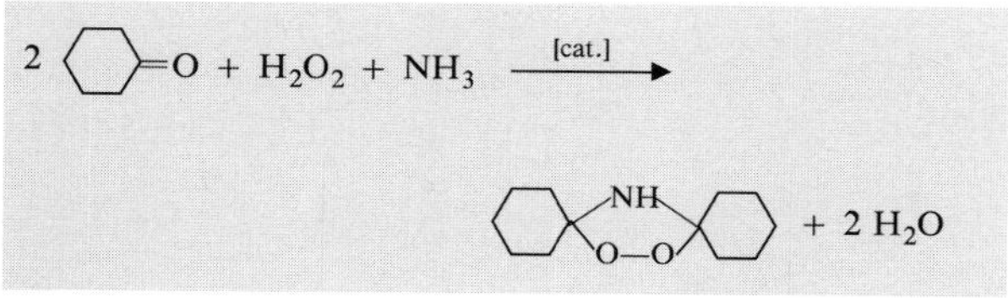

$$(41)$$

The reaction product is separated as an oily phase from the two-phase mixture, washed, and dissociated to ω-cyanoundecanoic acid in a pyrolysis reactor in the presence of steam. Byproducts are ε-caprolactam and cyclohexanone. The purified crystalline ω-cyanoundecanoic acid is dissolved in a solvent and catalytically hydrogenated to ω-aminododecanoic acid, which is then purified by crystallization:

process principles of the Ube route:

1. catalytic aminative peroxidation of cyclohexanone
2. thermolysis to 11-cyanoundecanoic acid
3. catalytic hydrogenation to 12-aminododecanoic acid

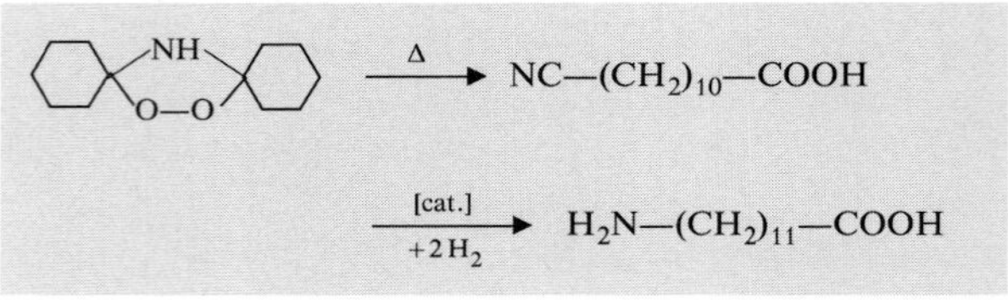

$$\text{(42)}$$

Further details of this process have not been disclosed.

Ube currently operates a plant whose capacity has been expanded to 3000 tonnes per year.

Nylon 12 is distinguished from nylon 6 or nylon 6,6 by its mechanical properties: lower density, lower softening point, and greater shape stability. As a thermoplastic, it is used in injection molding and extrusion.

11. Propene Conversion Products

After ethylene, propene is the most important raw material used in the production of organic chemicals. Approximately equal amounts of propene are used in Western Europe and the USA (cf. Table 11 − 1). The pattern of propene usage is also similar if the emphasis on hydroformylation (oxo products) in Germany and polypropylene in Japan are ignored.

In addition, propylene is also used for nonchemical purposes; for example, in alkylate and polymer gasolines, and in industrial gases.

main emphasis on use of propene for polypropylene, acrylonitrile for polyacrylonitrile

greatest difference in consumption structure: high share of hydroformylation products in Western Europe, especially in Germany

Table 11-1. Propene use (in %).

Product	World		USA		Western Europe		Japan		Germany	
	1988	1991	1986	1995	1986	1995	1985	1995	1986	1992
Polypropylene	41	47	35	42	36	49	45	50	17	26
Acrylonitrile	16	13	16	14	17	10	20	14	18	15
Propylene oxide	9	8	11	11	11	10	7	6	18	16
Isopropanol	6	4	6	4	6	4	3	2	6	5
Cumene	8	8	9	9	8	6	5	8	7	7
Oxo products	11	9	7	7	12	9	11	8	28	23
Oligomers	6	5	8	6	5	5	}9	2	3	5
Miscellaneous	3	6	8	7	5	7		10	3	3
Total use (in 10^6 tonnes)	28.0	32.0	7.4	11.1	7.5	11.8	2.9	4.4	2.4	2.8

Oxo products include *n*-butyraldehyde, isobutyraldehyde, *n*-butanol, isobutanol, and 2-ethylhexanol. The major oligomers are nonene, dodecene, and heptene.

Numerous minor products are included under 'miscellaneous', for example, acrylic acid and its esters, acrolein, allyl chloride, epichlorohydrin, allyl alcohol, and glycerol.

The pattern of propene use has changed, particularly with respect to polypropylene. The fraction of propene used for polypropylene has increased steadily over the last decade and accounts for nearly 50% of propene use in Japan.

polypropylene shows strongest growth in all countries

11.1. Oxidation Products of Propene

11.1.1. Propylene Oxide

propylene oxide (PO) is, like ethylene oxide, an important key product

Although propylene oxide is not as reactive as ethylene oxide, it is also the basis of a versatile and rapidly expanding chemistry.

additionally, number of PO secondary products is increased; asymmetric epoxides such as PO yield isomers on ring opening, unlike EO

The production figures for propylene oxide in USA, Western Europe, and Japan are presented in the adjacent table.

propylene oxide production (in 1000 tonnes):

	1990	1992	1994
USA	1492	1225	1678
W. Europe	1147	1221	1389
Japan	336	323	315

In 1993, the worldwide production capacity for propylene oxide was about 4.0×10^6 tonnes per year, of which 1.7, 1.4, and 0.36×10^6 tonnes per year were in the USA, Western Europe, and Japan, respectively. The largest manufacturers are Arco and Dow.

propylene oxide manufacturing routes:

as propene can be only unselectively oxidized to PO at present, only the following processes are available:

1. chlorohydrin process
2. indirect oxidation process (coupled oxidation)

All attempts to manufacture propylene oxide commercially by direct oxidation of propene (analogous to ethylene oxide production) have been unsuccessful. Since the hydrogen of the allyl methyl group is oxidized preferentially, the main product is acrolein; when the propene conversion is economical, the epoxide selectivity is low. Also, the large number of oxygen-containing C_1-, C_2-, and C_3-byproducts complicates processing of the reaction mixture and purification of the propylene oxide.

importance of chlorohydrin process:

PO capacity (in %):

	1976	1980	1995
USA	60	56	40
Japan	61	66	64
W. Europe	82	69	53

Thus the chlorohydrin process, no longer economical for ethylene oxide, is still important in propylene oxide manufacture, though it is in competition with other indirect oxidation processes using hydroperoxides. In 1996, the share of propylene oxide capacity based on the chlorohydrin process was 49% worldwide; the rest was based on indirect oxidation processes with hydroperoxides. A breakdown by countries is given in the adjacent table. The world's largest producer of propylene oxide using the chlorohydrin process is Dow.

11.1.1.1. Propylene Oxide from the Chlorohydrin Process

chlorohydrin process basically two steps:

1. HOCl addition
2. HCl elimination

The marked increase in propylene oxide production was promoted by the availability of ethylene chlorohydrin plants, which became unprofitable in the 1960s. After only minor modifications, propylene oxide could be manufactured in these units.

characteristics of chlorohydrin formation:

$Cl^\oplus$ from HOCl adds preferentially to terminal C of double bond

attainment of equilibrium
$Cl_2 + H_2O \rightleftharpoons HOCl + HCl$ necessary, as otherwise Cl_2 adds to propene to form $CH_3CHCl—CH_2Cl$

Analogous to ethylene oxide, propene is reacted at $35-50\,°C$ at $2-3$ bar in reaction columns with an aqueous chlorine solution in which HCl and HOCl are in equilibrium. The resulting $4-6\%$ mixture of α- and β-chlorohydrin (ratio 9:1) is dehydrochlorinated — without intermediate separation — at $25\,°C$ with an excess of alkali, e. g., 10% lime water or dilute sodium hydroxide solution from NaCl electrolysis:

$$2\ CH_3CH{=}CH_2 + 2\ HOCl$$

$$\longrightarrow\ CH_3\underset{\underset{\textstyle OH}{|}}{CH}{-}CH_2Cl + CH_3\underset{\underset{\textstyle Cl}{|}}{CH}{-}CH_2OH \qquad (1)$$

$$\xrightarrow{+\,Ca(OH)_2}\ 2\ H_3C{-}\underset{\underset{\textstyle O}{\diagdown\!\diagup}}{CH}{-}CH_2 + CaCl_2 + 2\ H_2O$$

addition of $Fe^{3\oplus}$ or $Cu^{2\oplus}$ suppresses Cl_2 addition

characteristics of epoxide formation:

$Ca(OH)_2$ has two functions:

1. neutralization of HCl
2. dehydrochlorination

The propylene oxide is then rapidly driven out of the reaction mixture (to avoid hydration) by directly applying steam and then purified by distillation. The selectivity is $87-90\%$ (based on C_3H_6). Main byproducts are 1,2-dichloropropane (selectivity $6-9\%$) and bischlorodiisopropyl ether (selectivity $1-3\%$).

As the total chlorine losses are in the form of economically nonutilizable $CaCl_2$ or NaCl solutions, work has been conducted from the beginning to find chlorine-recycle or chlorine-free oxidation systems. Instead of the inorganic oxidizing agent HOCl, organic compounds have been selected to transfer the O to propene (*cf.* Section 11.1.1.3).

total Cl_2 loss as $CaCl_2$ as well as byproducts $CH_3CHClCH_2Cl$ and

$$\underset{ClH_2C}{\overset{H_3C}{>}}CH{-}O{-}\underset{CH_2Cl}{\overset{CH_3}{<}}CH$$

ca. 40 tonnes aqueous salt solution with $5-6$ wt% $CaCl_2$ and 0.1 wt% $Ca(OH)_2$ result per tonne PO

11.1.1.2. Indirect Oxidation Routes to Propylene Oxide

The indirect epoxidation of propene is based on the observation that organic peroxides such as hydroperoxides or peroxycarboxylic acids can, in the liquid phase, transfer their peroxide oxygen selectively to olefins to form epoxides. The hydroperoxides are thereby converted into alcohols and the peroxyacids into acids (*cf.* Table $11-2$):

principle of indirect propene epoxidation:

radical oxidation of component supplying organic peroxide, which can transfer peroxide oxygen to propene in homogeneously catalyzed liquid-phase reaction

$$CH_3CH{=}CH_2 + \begin{matrix} R{-}O{-}O{-}H \\[4pt] R'{-}\underset{\underset{\textstyle O}{\|}}{C}{-}O{-}O{-}H \end{matrix} \xrightarrow{\ [cat.]\ }$$

$$H_3C{-}\underset{\underset{\textstyle O}{\diagdown\!\diagup}}{CH}{-}CH_2 + \begin{matrix} ROH \\[4pt] R'\underset{\underset{\textstyle O}{\|}}{C}OH \end{matrix} \qquad (2)$$

Basically, epoxidation can be done in two different ways. In one, a higher concentration of a hydroperoxide or a peroxycarboxylic acid is produced in a preliminary reaction – generally by autoxidation with air or O_2 – which then transfers its oxygen in a second step. In the second, cooxidation of propene with a peroxide generator is used to form a hydroperoxide or a peroxy-

process variations of indirect epoxidation:

1. two-step reaction, *i.e.*, autoxidation to peroxide and subsequent epoxidation to epoxide
2. single-step reaction, *i.e.*, cooxidation with intermediate peroxide formation and simultaneous epoxidation

carboxylic acid in situ, which then epoxidizes propene. The first method has established a dominant position in the industry.

Hydroperoxide- or peroxycarboxylic acid-supplying substances which are of interest in the economic exploitation of the co-oxidized substances are listed in Table 11−2, along with their oxidation products and potential secondary products.

Table 11-2. Subsidiary components for hydroperoxide formation.

Starting material	Peroxidizing agent	Co-oxidized substance(s)[1]	Secondary product(s)[1]
Acetaldehyde	AcOOH or $CH_3-C{\overset{OAc}{\underset{O-O-H}{\diagdown}}}$	Acetic acid or anhydride	−
Isobutane	$(CH_3)_3C-O-O-H$	*tert*-Butanol	Isobutene, MIBK, or methacrylic acid
Isopentane	$C_5H_{11}-O-O-H$	Isopentanol	Isopentene and isoprene
Cyclohexane	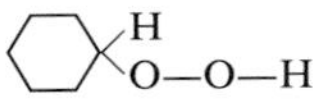 $\diagup$ H $\diagdown$ O−O−H	Cyclohexanone/ol	−
Ethylbenzene	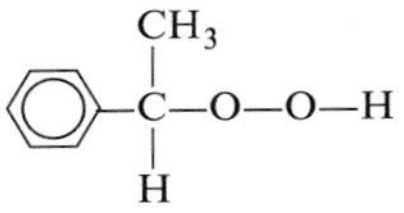 C with CH$_3$ and H, −O−O−H	Methylphenylcarbinol	Styrene
Cumene	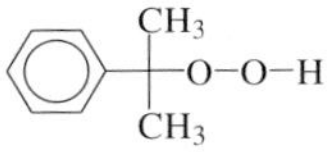 C with CH$_3$ and CH$_3$, −O−O−H	Dimethylphenylcarbinol	α-Methylstyrene

[1] for recycling or commercial use.

industrially most important example of a two-step epoxidation with hydroperoxide:

oxirane process of Atlantic Richfield/Halcon (now Arco) with variable subsidiary substances:

1. isobutane
2. ethylbenzene

Halcon and Atlantic Richfield, after independent development work, cooperated to establish the oxirane process. Currently about 40% of worldwide propylene oxide capacity is based on this technology. In the USA the share from the oxirane route rose dramatically from 9% in 1970 to more than 47% at the end of 1991. Another process which uses isobutane as the ancillary substance has been developed by Texaco.

The peroxidation can be conducted with either isobutane or ethylbenzene. However, isobutane and ethylbenzene are not readily interchangeable in the same plant due to differing reaction conditions and byproducts. Currently about 56% of the worldwide oxirane plant capacity for propylene oxide is based on isobutane, and 44% on ethylbenzene.

The hydrocarbon is oxidized in a peroxidation reactor in the liquid phase with air or O_2 and an initiator under the mildest possible conditions (*e. g.*, $120-140\,°C$ and about 35 bar) such that the ratio of hydroperoxide to alcohol from spontaneous decomposition of hydroperoxide is maximized. With isobutane as feedstock, the conversion is limited to about $35-40\%$ (Arco) or $45-50\%$ (Texaco); the selectivity to *tert*-butyl hydroperoxide is about $50-60\%$. Byproducts are *tert*-butanol and acetone. With ethylbenzene, the selectivity to hydroperoxide is about 87% at an ethylbenzene conversion of $15-17\%$. In this case, the byproducts are 1-phenylethanol and acetophenone.

characteristics of two-step process with hydroperoxide:

1. mild preoxidation in peroxidation bubble tray reactor, in order to achieve high ratio of hydroperoxide/alcohol

This crude product then epoxidizes the propene to propylene oxide either directly or, alternately, by enrichment of the hydroperoxide by distilling the unreacted hydrocarbon into a second reactor. The reaction is done in the liquid phase, *e. g.*, using *tert*-butanol as solvent, and in the presence of a generally Mo-based catalyst at $90-130\,°C$ and $15-65$ bar:

2. catalyzed epoxidation *via* metal peroxy compounds in stirred reactor

$$
(CH_3)_3CH \xrightarrow{+O_2} (CH_3)_3C{-}O{-}O{-}H
$$

$$
\xrightarrow[\text{[cat.]}]{+C_3H_6} H_3C{-}HC{-}CH_2 + (CH_3)_3COH \qquad (3)
$$

With *tert*-butyl hydroperoxide, the selectivity to propylene oxide is about 90% at a 10% propene conversion.

The usual epoxidation catalysts are Mo, V, Ti, and other heavy metal compounds or complexes soluble in hydrocarbons (*e. g.*, naphthenates).

The reaction mixture is worked up by distillation. *tert*-Butanol, which is formed along with propylene oxide, can be added directly to motor fuel for deicing or to raise the octane number. It can also be dehydrated to isobutene, which can be used to manufacture alkylate gasoline or diisobutene, for example, or hydrogenated to isobutane and recycled. In a recent development of the Oxirane or Mitsubishi Rayon process, *tert*-butanol can also be converted to methacrylic acid (*cf.* Section 11.1.4.2).

uses of (usually more than stoichiometric amount of) co-oxidized substance formed:

$(CH_3)_3COH$:

1. direct use

2. $\xrightarrow{-H_2O}$ isobutene

3. $\xrightarrow[+H_2]{-H_2O}$ isobutane

4. $\xrightarrow[+O_2]{-H_2O}$ methacrylic acid

With ethylbenzene, the methyl phenyl carbinol which forms is converted into styrene (*cf.* Section 13.1.2). The alcohol is dehydrated at atmospheric pressure in the gas phase at $180-280\,°C$ with 85% conversion and 95% selectivity using TiO_2/Al_2O_3 catalysts:

$C_6H_5\underset{\underset{OH}{|}}{C}HCH_3$:

$\xrightarrow{-H_2O}$ styrene

$$\text{C}_6\text{H}_5-\underset{\underset{\text{OH}}{|}}{\text{CH}}-\text{CH}_3 \xrightarrow{[\text{TiO}_2]} \text{C}_6\text{H}_5-\text{CH}=\text{CH}_2 + \text{H}_2\text{O} \qquad (4)$$

process variation of two-step epoxidation with hydroperoxide:

Shell process with heterogeneous carrier catalyst for epoxidation step

A new Shell process for propylene oxide (SMPO = **S**tyrene **M**onomer **P**ropylene **O**xide process) is very similar to the oxirane process. The only difference is in the epoxidation step, where a heterogeneous system based on V, W, Mo, or Ti compounds on an SiO_2 support is used instead of a homogeneous catalyst. A commercial unit for the production of 125 000 tonnes per year propylene oxide and 330 000 tonnes per year styrene began operation in Holland in 1978; the capacity of this unit has since been increased several times.

commercially used peroxycarboxylic acids:

peracetic acid, either two-step or one-step route with special significance for propene epoxidation

commercially tested peroxycarboxylic acids:

$$\underset{\underset{\displaystyle \text{R}=\text{C}_2\text{H}_5}{}}{\overset{\overset{\displaystyle \text{O}}{\|}}{\text{R}-\text{C}-\text{O}-\text{O}-\text{H}}}$$

The second group of organic peroxides used for the epoxidation of propene comprises the peroxycarboxylic acids, of which peracetic acid once had commercial significance. Several other peroxycarboxylic acids have been tested for the same purpose, for example perpropionic acid (*cf.* Section 11.1.1.3), perbenzoic acid (Metallgesellschaft), and *p*-methylperbenzoic acid (Mitsubishi).

The methods for peracetic acid manufacture — either in a separate preliminary stage or in situ — illustrate the difference between the more frequently used two-step process and the single-step co-oxidation method for the manufacture of propylene oxide.

manufacture of peracetic acid for two-step epoxidation:

1. autoxidation of CH_3CHO
2. reaction of AcOH with H_2O_2

process principles of 1:

uncatalyzed, mildly conducted radical autoxidation *via* acetaldehyde monoperacetate

$$\text{CH}_3\text{C}\overset{\displaystyle \text{O}}{\underset{\displaystyle \text{O}-\text{O}}{\diagdown}}\overset{\displaystyle \text{HO}}{\underset{\displaystyle}{\diagup}}\text{CH}-\text{CH}_3$$

In the two-step process, peracetic acid can be manufactured either by oxidation of acetaldehyde or by the reaction of acetic acid with H_2O_2.

The autoxidation of acetaldehyde with oxygen to peracetic acid has already been described in Section 7.4.1.1. The manufacture of peracetic acid from acetic acid and H_2O_2 takes place according to the equation:

$$\underset{\displaystyle \text{CH}_3\text{COH}}{\overset{\overset{\displaystyle \text{O}}{\|}}{}} + \text{H}_2\text{O}_2 \underset{\longleftarrow}{\overset{[\text{H}^\oplus]}{\longrightarrow}} \underset{\displaystyle \text{CH}_3\text{C}-\text{O}-\text{O}-\text{H}}{\overset{\overset{\displaystyle \text{O}}{\|}}{}} + \text{H}_2\text{O} \qquad (5)$$

process principles of 2:

acid-catalyzed acylation of H_2O_2 with AcOH to monoacyl product, *i.e.*, peracetic acid

Equilibrium can be reached in the liquid phase using a solvent at a temperature up to about 40 °C with catalytic amounts of H_2SO_4. In a subsequent vacuum distillation, the solvent — for example, methyl or ethyl acetate — functions as an entraining agent to remove water, and thus shift the equilibrium. Yields of 95% (based on AcOH) and 83% (based on H_2O_2) can be obtained. Other processes operate without an entraining agent and continuously distill an aqueous azeotrope containing 61% peracetic acid from the reaction mixture.

The propene epoxidation takes place in three connected bubble column reactors, usually with the solvent-containing peracid solution in the liquid phase at 9–12 bar and 50–80 °C. The propylene oxide selectivity is over 90%. Acetic acid is formed as coproduct; a use must be found for it.

The propylene oxide manufacture using peracetic acid from acetaldehyde was practiced by Daicel in a 12500 tonne-per-year plant in Japan from 1969 until 1980.

process principles of epoxidation step with peroxycarboxylic acids:

catalyzed or uncatalyzed radical oxygen transfer to propene to form PO and the co-product carboxylic acid (Prilezhaev reaction)

11.1.1.3. Possibilities for Development in the Manufacture of Propylene Oxide

The current situation in propylene oxide manufacture is marked in the chlorohydrin route by wastewater and byproduct problems, and, in all indirect epoxidation processes with hydroperoxides or peroxycarboxylic acids, by the production of cooxidates in larger amounts than propylene oxide itself.

However, plants based on the chlorohydrin process which have been built or expanded during the last few years have illustrated the process can be profitable with extensive integration of the generation, use, and recovery of chlorine. At the same time, it is also clear that none of the many new process developments has reached an economically convincing status.

Research toward improved propylene oxide production is directed toward one of the following objectives:

1. More economical variations of the chlorohydrin process
2. Different technology for addition of peroxide oxygen
3. Direct epoxidation routes using oxygen from air
4. Biochemical routes to epoxidation

problems of the preferred PO manufacturing routes:

in the chlorohydrin route, 1.4–1.5 kg chlorine lost as dilute $CaCl_2$ solution and 0.1–0.15 kg lost as dichloropropane per kg PO

in the epoxidation route, coproducts formed per kg PO:

2.5 kg styrene, or
2.1 kg *tert*-butanol, or
1.6 kg isobutene

PO manufacture with Cl or hydroperoxide auxiliary system cannot yet be replaced by direct oxidation

To 1:

Electrochemical variants of the chlorohydrin process (*e.g.*, Bayer, Kellogg, and BASF) which in principle do not consume chlorine have not yet been used commercially. HOCl, which is formed in situ from the chlorine released at the anode in the modified alkali metal chloride electrolysis, reacts with propene to form propylene chlorohydrin. This is then converted into propylene oxide near the cathode where the hydroxide ion concentration is high. NaCl, which also results, is recycled to the process. The net equation is as follows:

improvement by internal Cl_2 recycle with help of electrochemical chlorohydrin process, not yet used commercially

$$CH_3CH=CH_2 + H_2O \xrightarrow[\text{energy}]{\text{electrical}} H_3C-HC-CH_2 + H_2 \quad (6)$$

disadvantages of indirect electrochemical routes arise from high costs for:

investment
electricity
PO isolation

variation of electrochemical chlorohydrin process by addition of $(CH_3)_3COH$ as HOCl carrier with closed NaCl cycle (Lummus)

The disadvantages of this process include high capital costs, consumption of electricity, and energy requirements for the distillation of low concentrations of propylene oxide from the cathodic liquid.

Lummus Corporation has recently developed a new variation of the electrochemical chlorohydrin process for propene which uses *tert*-butyl hypochlorite as the HOCl carrier. In the first step, the electrolytic solution of the chloralkali electrolysis reacts with *tert*-butanol to form *tert*-butyl hypochlorite, which is easily separated from the concentrated NaCl solution:

$$Cl_2 + NaOH + (CH_3)_3COH \rightarrow (CH_3)_3COCl + NaCl + H_2O \quad (7)$$

After purification, the NaCl solution is recycled to the electrolytic cell, and *tert*-butyl hypochlorite reacts with propene and H_2O to form propylene chlorohydrin and *tert*-butanol. This is distilled overhead and recirculated to the hypochlorite generator:

$$(CH_3)_3COCl + H_2O + CH_3CH=CH_2 \rightarrow$$
$$CH_3CH-CH_2Cl + (CH_3)_3COH \quad (8)$$
$$\overset{|}{OH}$$

Propylene chlorohydrin is then dehydrochlorinated with the catholyte (NaOH), NaCl) and the purified NaCl solution is recycled to the chloralkali electrolysis. An overall yield of 90% propylene oxide has been reported. The only coproduct is hydrogen; small amounts of byproducts are formed in the propene chlorohydrin reaction. This process has not been used commercially.

To 2:

The large amount of coproducts formed in the indirect oxidation process can be avoided by using percarboxylic acids produced with H_2O_2.

process principle of propene epoxidation with perpropionic acid:

low-loss propionic acid cycle for selective O transfer from H_2O_2 to propene (Bayer/Degussa, Solvay/Laporte/Carbochimique)

In the Bayer/Degussa process, perpropionic acid from the reaction of propionic acid with aqueous H_2O_2 at $25-45\,°C$ is used for the oxidation of propene at $60-80\,°C$ and $5-14$ bar. With an almost complete reaction of perpropionic acid, the selectivities to propylene oxide are more than 94% (based on H_2O_2)

and 95 – 97% (based on propene). An important hallmark of this process is the extraction of perpropionic acid from the aqueous H_2O_2 solution with benzene or, in the Interox process (Solvay/Laporte/Carbochimique), with 1,2-dichloropropane.

Despite the efficient propionic acid cycle and the very good selectivity, the relatively high price of H_2O_2 remains an obstacle to commercialization of this process.

Processes based on a direct epoxidation of propene with H_2O_2, developed by many firms, suffer from similar cost disadvantages.

process principle of propene epoxidation with H_2O_2:

direct, selectively catalyzed, technically simple epoxidation with coproduct H_2O (*e. g.*, Ugine Kuhlmann, UCC)

Compounds of arsenic, boron, or molybdenum which give selectivities to propylene oxide of up to 95% are used for the catalytic epoxidation in solvents. Ugine Kuhlmann has developed a process using metaboric acid; the orthoboric acid obtained from the epoxidation is dehydrated to metaboric acid and recycled. UCC uses arsenic trioxide in its recycle process.

To 3:

The most urgent task is to develop a direct oxidation of propene by oxygen; much of this work, especially in the liquid phase, has been done in the CIS. A suspension of, *e. g.*, the rare earth oxides on silica gel in acetone is used as the catalyst. The selectivity to propylene oxide has been reported as 89% at a propene conversion of 15%.

possibilities for development of propene direct oxidation to PO:

rare earth oxide/SiO_2 catalyst suspended in acetone (CIS)

BP describes a gas-phase process with a Cr-modified Ag catalyst in which a 50% selectivity to PO can be attained at a 15% propene conversion.

Cr-modified Ag catalyst (BP)

Rhône-Progil reports an uncatalyzed liquid-phase oxidation with simultaneous removal of concomitantly formed lower carboxylic acids which would otherwise lead to a decrease in selectivity of propylene oxide.

noncatalytic liquid-phase oxidation (Rhône-Progil)

A homogeneous catalyst system, *e. g.*, Mo compounds for the liquid-phase oxidation in chlorobenzene, is described by Jeffersen Chemical. Selectivity to PO of about 59% at a 12% propene conversion has been reported.

homogeneously catalyzed liquid-phase oxidation (Jeffersen Chemical)

A new route is currently being piloted by Olin. Here the direct oxidation of propene is catalyzed with a lithium nitrate melt. Propylene oxide is produced at 180–250°C with a selectivity of up to 57% at a conversion of 24%.

$LiNO_3$ melt $(+KNO_3 + NaNO_3)$ as catalytic system (Olin Corp.)

To 4:

process principle of biosynthetic PO manufacture:

enzymatic propene epoxidation with biocatalytically produced dilute H_2O_2 with simultaneous increase in value as D-glucose is converted into the sweeter D-fructose

disadvantages:

multistep, dilute PO solution with KBr recycle, 3 kg fructose per kg PO

Biosynthetic manufacture of propylene oxide has attracted a lot of interest. The most well-known process is the Cetus process, which comprises four separate chemical and enzyme-catalyzed steps. The first is the enzymatic oxidative dehydration of D-glucose with the formation of H_2O_2. In the next two steps, propene in the presence of KBr is oxidized to propylene oxide on an immobilized enzyme system. The coproduct D-fructose is obtained by hydrogenation of the intermediate from D-glucose, so that the net equation is as follows:

$$\text{D-Glucose} + O_2 + CH_3CH{=}CH_2 \longrightarrow$$
$$\text{D-Fructose} + CH_3{-}\underset{\underset{O}{\diagdown\diagup}}{CH}{-}CH_2 + H_2O \quad (9)$$

Currently this process is not economical, although biosynthetic production of H_2O_2 has pricing advantages over the usual processes.

In contrast to the commercial use of ethylene oxide, only approximately one-third of the propylene oxide is hydrated to propylene glycol and dipropylene glycol (*cf.* Section 11.1.2). In spite of this, direct processes for manufacture of the glycol could spare some of the expensive propylene oxide.

alternate, direct propylene glycol manufacture from propene circumventing PO, *e. g.*,

two-step route with 1,2-diacetoxypropane as intermediate which is subsequently saponified (Halcon)

According to research conducted by Halcon, the diacetate of propylene glycol can be obtained with 93% selectivity by reacting propene with acetic acid and O_2 in the presence of TeO_2 and bromide ions at 150 °C. Hydrolysis at 110 °C and slightly raised pressure gives propylene glycol:

$$CH_3CH{=}CH_2 + 2\,AcOH + 0.5\,O_2 \xrightarrow[\text{[cat.]}]{-H_2O} \underset{\underset{OAc\ OAc}{|\ \ \ |}}{CH_3CH{-}CH_2}$$

$$\xrightarrow{+2\,H_2O} \underset{\underset{OH\ OH}{|\ \ \ |}}{CH_3CH{-}CH_2} + 2\,AcOH \quad (10)$$

As with ethylene glycol (*cf.* Section 7.2.1.1), the disadvantages of the two-step process resulting from the acetate saponification must be taken into account.

two-step route through propylene glycol monoacetate with subsequent hydrolysis, but preferably AcOH elimination to PO (Chem Systems)

Chem Systems has developed a comparable propene acetoxylation route using a TeO_2/I_2 catalyst system to obtain a mixture of propylene glycol and its monoacetate at the pilot plant scale.

Besides the total hydrolysis, propylene glycol monoacetate can also be converted to propylene oxide with a selectivity of 80% (based on propene) in a subsequent gas-phase elimination of acetic acid at 400 °C. Byproducts are propionaldehyde and acetone.

11.1.2. Secondary Products of Propylene Oxide

Propylene oxide is an important intermediate in the manufacture of 1,2-propylene glycol (usually referred to as simply propylene glycol), dipropylene glycol, various propoxylated products and several minor product groups, *e. g.*, the propylene glycol ethers and the isopropanolamines. In contrast to ethylene oxide, where ethylene glycol is the most important secondary product, propylene glycol ranks behind polypropylene glycols and propoxylated products for polyurethane manufacture. The use of propylene oxide worldwide and in the USA, Western Europe, and Japan can be broken down as follows:

most important products from propylene oxide:

1. propoxylated products by multiple addition of PO to propylene glycol or to polyhydric alcohols
2. propylene glycol

Table 11-3. Use of propylene oxide (in %).

Product	USA 1986	1993	Western Europe 1985	1994	Japan 1983	1993	World 1986	1991
Polypropylene glycol and propoxylated products for polyurethanes	61	60	66	61	74	71	64	65
Propylene glycol	28	25	22	23	16	16	21	21
Miscellaneous (glycol ethers, glycerine, isopropanolamine, etc.)	11	15	12	16	10	13	15	14
Total use (in 10^6 tonnes)	0.91	1.31	0.93	1.38	0.21	0.34	2.59	3.32

In 1993, the world production capacity for propylene glycol was about 0.97×10^6 tonnes per year, of which 0.47, 0.42, and 0.09 $\times 10^6$ tonnes per year were in the USA, Western Europe, and Japan, respectively.

Production figures for propylene glycol in these countries are listed in the adjacent table.

Proylene glycol is obtained from the oxide by hydration with a large excess of water in the presence of acidic or basic catalysts.

Propylene glycol is used as an antifreeze and brake fluid, and in the manufacture of alkyd and polyester resins. In a newer commercialized process from Rhône-Poulenc, propylene oxide

propylene glycol production (in 1000 tonnes):

	1989	1991	1994
USA	382	302	423
W. Europe	325	300	396
Japan	61	65	66

manufacture of propylene glycol:

catalyzed PO hydration, analogous to EO hydration

use of propylene glycol:

antifreeze
brake fluid
alkyd polyester resins
cosmetics, pharmaceuticals

can be used directly for the reaction with, for example, phthalic anhydride or maleic anhydride. Avoiding the hydration step has definite cost advantages. Like propylene glycol, dipropylene glycol is also used for resins. However, unlike ethylene glycol which is toxic to a certain extent (it is oxidized to oxalic acid in the body), propylene glycol can be used in cosmetic and pharmaceutical products.

manufacture and use of propoxylated products from addition of PO to:

1. propylene glycol, then with EO to give block polymers for detergents, emulsifying agents
2. polyhydric alcohols (*e.g.*, glycerol, trimethylolpropane, pentaerythritol) for polyurethanes

Polypropylene glycols can be manufactured by reacting propylene oxide with propylene glycol. Polypropylene glycols with a molecular weight of 1000 and above are almost insoluble in water, unlike the polyethylene glycols. These polyether alcohols can be made water soluble by addition of ethylene oxide. The resulting block polymers are nonionic surfactants, and are used as detergents and emulsifiers. Propoxylated products of polyhydric alcohols are used on a large scale in the manufacture of polyurethane foams.

3. monohydric alcohols for solvents (alcoholysis leads to two ethers)

$$CH_3CHCH_2OH + CH_3CHCH_2OR$$
$$\quad\ |\qquad\qquad\qquad\ |$$
$$\quad OR\qquad\qquad\qquad OH$$

In analogy to hydration, the alcoholysis of propylene oxide with monohydric alcohols leads to the monoalkyl ethers of propylene glycol, which have useful solvent properties. The main products are propylene glycol monomethyl ether and its ester with acetic acid.

4. NH_3 for detergents, for gas purification, and as intermediate

Isopropanolamines can be manufactured from propylene oxide and aqueous ammonia in a method analogous to the ethanolamines. The composition of the resulting mixture of mono-, di-, and triisopropanolamines depends on the NH_3/PO ratio; a larger excess of NH_3 favors the formation of the primary amine. Isopropanolamines can replace ethanolamines in many applications (*cf.* Section 7.2.3). They can be used for the production of detergents, pesticides, pharmaceuticals, and dyes. They are also used to remove acidic constituents from gases.

5. CO_2 for specialty solvents and for gas purification

Propylene carbonate can be manufactured in 95% yield from the reaction of propylene oxide with CO_2 at $150-200\,°C$ and $50-80$ bar using tetraethylammonium bromide as catalyst. It is used as a specialty solvent in the fiber and textile industry, and, like the isopropanolamines, to remove acidic components from natural gas and synthesis gas (*cf.* Section 2.1.2).

11.1.3. Acetone

Acetone, the simplest aliphatic ketone, is also the most commercially important. In 1994, the world production capacity for acetone was 3.4×10^6 tonnes per year, of which 1.2, 1.2, and 0.39×10^6 tonnes per year were in the USA, Western Europe, and Japan, respectively.

Production figures for these countries are given in the adjacent table.

acetone production (in 1000 tonnes):

	1990	1992	1995
USA	1136	1105	1252
W. Europe	958	929	1030
Japan	334	377	396

The main manufacturing processes are:

1. Wacker–Hoechst direct oxidation of propene
2. Dehydrogenation of isopropanol
3. Coproduction in Hock phenol process (*cf.* Section 13.2.1.1)

manufacturing processes for acetone:

1. Wacker–Hoechst, direct oxidation of propene
2. dehydrogenation of isopropanol
3. Hock phenol process, acetone as coproduct

Other industrial sources, including the fermentation of starch-containing products (*Bacillus macerans;* once practiced in the USA and South Africa, but now only in the CIS) and the uncatalyzed butane/propene/naphtha oxidation (Celanese, UCC, and BP processes) are less important, since acetone is only a by-product.

acetone frequently byproduct:

1. in fermentation processes, 10% acetone along with C_2H_5OH and n-C_4H_9OH
2. in paraffin oxidation *e. g.*, Celanese, UCC, or BP process
   (from isobutane with O_2, 36% acetone in oxidation product)

The most important of these manufacturing processes for acetone is the Hock process; in 1995, its share of the manufacturing capacity was 92, 79, and 92% in the USA, Western Europe, and Japan, respectively. In the USA and Western Europe, it is followed by isopropanol dehydrogenation/oxidation, with shares of 6 and 17%; the second-most-important route in Japan is the direct oxidation of propene, with an 8% share.

11.1.3.1. Direct Oxidation of Propene

The most elegant method for manufacturing acetone is the Wacker–Hoechst process, which has been practiced commercially since 1964. In this liquid-phase process, propene is oxidized to acetone with air at 110–120 °C and 10–14 bar in the presence of a catalyst system containing $PdCl_2$:

process principles of Wacker–Hoechst route:

$PdCl_2$/redox system catalyzes selective propene oxidation at nonterminal C atom

$$CH_3CH{=}CH_2 + 0.5\,O_2 \xrightarrow{\text{[cat.]}} CH_3COCH_3 \qquad (11)$$

$$\left(\Delta H = -\ \frac{61\ \text{kcal}}{255\ \text{kJ}}\ /\text{mol} \right)$$

The selectivity to acetone, the main product, is 92%; propionaldehyde is also formed with a selectivity of 2−4%. The conversion of propene is more than 99%.

As in the oxidation of ethylene to acetaldehyde (*cf.* Section 7.3.1), $PdCl_2$ is reduced to Pd in a stoichiometric reaction, then reoxidized to divalent palladium with the redox system $CuCl_2$—$CuCl$.

two processes for direct oxidation:

1. simultaneous catalyst reaction and regeneration, *i. e.*, single step
2. separate reaction and regeneration, *i. e.*, two steps

characteristics of favored two-step process:

dilute propene (approx. 85%, but free of C_2H_2 and dienes) can be used for reaction, air for regeneration

first industrial use of two-step process in Japan

future importance of propene direct oxidation:

production of acetone as co- and byproduct reduces interest in direct manufacture

two industrial designs for acetone manufacture involve two reaction principles:

1. gas phase
1.1. oxydehydrogenation in presence of O_2 or air
1.2. straight dehydrogenation with temperature strongly dependent on catalyst used

As in the acetaldehyde process, this can be done commercially in two ways:

1. The catalyst is reacted with propene and O_2 simultaneously
2. The catalyst is reacted with propene, and then regenerated with O_2 or air

The two-step process is favored for economic reasons because a propene/propane mixture (*e. g.*, resulting from cracking processes) can be used as the feed. Propane behaves like an inert gas, that is, it does not participate in the reaction. The acetone is separated from lower- and higher-boiling substances in a two-step distillation.

The first, and so far the only, plants using this technology were built in Japan. Currently, only one production unit (Kyowa Yuka) is still in operation (capacity 36000 tonnes per year, 1995).

Expansion in use of this interesting process is hindered by the numerous sources of acetone as a by- and coproduct. Moreover, the manufacture of acetone from isopropanol has been able to maintain its position.

The Wacker–Hoechst process can also be used to convert *n*-butene to methyl ethyl ketone. This has not been practiced commercially.

11.1.3.2. Acetone from Isopropanol

In 1970, approximately 50–60% of the total acetone production in the USA was still being derived from isopropanol. Today, the cumene process — with acetone as coproduct — is the primary source of acetone worldwide (*cf.* Section 11.1.3).

As in the manufacture of formaldehyde and acetaldehyde from the corresponding alcohols, isopropanol can be oxidatively dehydrogenated at 400–600 °C using Ag or Cu catalysts, or merely dehydrogenated over ZnO catalysts at 300–400 °C, or over Cu or Cu-Zn catalysts (brass) at 500 °C and 3 bar. Dehydrogenation is the favored process. In the gas phase, using for example the Standard Oil process with ZnO, acetone is obtained in 90% selectivity at an isopropanol conversion of about 98%.

$$(CH_3)_2CHOH \xrightarrow[\text{[cat.]}]{+0.5\ O_2} (CH_3)_2CO + H_2O \qquad (12)$$
$$\left(\Delta H = -\frac{43\ \text{kcal}}{180\ \text{kJ}}/\text{mol} \right)$$

$$\xrightarrow[\text{[cat.]}]{} (CH_3)_2CO + H_2 \qquad (13)$$
$$\left(\Delta H = +\frac{16\ \text{kcal}}{67\ \text{kJ}}/\text{mol} \right)$$

With catalysts such as copper chromite, which also catalyze condensation, methyl isobutyl ketone is formed as a byproduct.

In a commercially practiced process developed by Deutsche Texaco, a gas-phase dehydrogenation with an isopropanol conversion of 80% can be achieved at $250-270\,°C$ and $25-30$ bar over a special supported Cu catalyst. Because of the high heat demand, the reaction is usually run in a multiple-tube reactor.

In a process developed by IFP, isopropanol can be dehydrogenated in the liquid phase with a selectivity to acetone of over 99% using finely divided Raney nickel or copper chromite suspended in a high-boiling solvent at about $150\,°C$ and atmospheric pressure. This technology is used commercially in four plants (1991).

There is also a liquid-phase oxidation of isopropanol that was developed by Shell and Du Pont mainly for production of H_2O_2. Isopropanol is oxidized by O_2 in a radical reaction at $90-140\,°C$ and $3-4$ bar using a small amount of H_2O_2 as initiator:

$$(CH_3)_2CHOH + O_2 \rightarrow (CH_3)_2CO + H_2O_2 \tag{14}$$

At an isopropanol conversion of about 15%, the selectivities to H_2O_2 and acetone are 87% and 93%, respectively.

After dilution with water, the reaction mixture is freed from organics by distillation and used as the aqueous solution. The last plants operated by Shell were shut down in 1980. In the CIS two plants are still in operation. The anthraquinone route (*cf.* Section 12.2.4.2) is the only process used to produce H_2O_2 in the USA and Japan; it is the principal route in Western Europe.

2. liquid phase
2.1. straight dehydrogenation

2.2. autocatalytic oxidation with acetone as byproduct and H_2O_2 as main product

$(CH_3)_2CHOH + X^\bullet \rightarrow (CH_3)_2\overset{\bullet}{C}OH +$
$XH \xrightarrow{+O_2} (CH_3)_2C{=}O + HOO^\bullet$
$HOO^\bullet + (CH_3)_2CHOH \rightarrow (CH_3)_2\overset{\bullet}{C}OH + H_2O_2$

11.1.4. Secondary Products of Acetone

Acetone is the starting material for many important intermediates. A breakdown of its use in several countries is given in the following table:

Table 11–4. Acetone use (in %).

Product	World		USA		Western Europe		Japan	
	1988	1991	1984	1995	1985	1994	1985	1994
Methyl methacrylate	26	29	32	44	32	31	37	32
Methyl isobutyl ketone	10	9	10	7	10	8	23	23
Bisphenol A	8	10	10	18	10	11	10	19
Miscellaneous	56	52	48	31	48	50	30	26
Total use (in 10^6 tonnes)	2.98	3.10	0.82	1.12	0.83	0.97	0.25	0.34

acetone use:

ca. 60–70% for products from typical ketone reactions, for example:

cyanohydrin formation
aldol reactions

The most important chemicals included under 'Miscellaneous' are aldol condensation products and secondary products such as diacetone alcohol, mesityl oxide; of lesser importance are isobutyl methyl carbinol and isophorone. A large fraction of these products, as well as acetone, are used as solvents for various materials such as natural resins, varnishes, paints, cellulose acetate, nitrocellulose, fats and oils.

Consuming about 30–40% of the acetone worldwide, methyl methacrylate (*cf.* Section 11.1.4.2) is the most important product derived from acetone. Depending on the country, it is followed by methyl isobutyl ketone or bisphenol A (*cf.* Section 13.2.1.3).

11.1.4.1. Acetone Aldolization and Secondary Products

production of MIBK (in 1000 tonnes):

	1990	1992	1994
USA	82	74	77
Japan	54	54	52

Methyl isobutyl ketone (2-methyl-4-pentanone) is the largest volume aldol reaction product of acetone. Production numbers for the USA and Japan are given in the adjacent table. In 1991, the production capacity for MIBK in Western Europe, the USA, and Japan was 120000, 95000 and 54000 tonnes per year, respectively.

MIBK manufacture from acetone in three steps:

1. aldol condensation (base-catalyzed)
2. dehydration (proton-catalyzed)
3. hydrogenation (Cu- or Ni-catalyzed)

Acetone reacts in a three-step reaction sequence through the isolable intermediates diacetone alcohol and mesityl oxide to give methyl isobutyl ketone.

The first step is the base-catalyzed aldol reaction of acetone in the liquid phase to form diacetone alcohol:

$$
\begin{array}{c}
\mathrm{H_3C} \\
\!\!\!\!\!\!\!\!\!\!\!\!\!\diagdown \\
\mathrm{H_3C}
\end{array}\!\!\!\!\mathrm{C{=}O} + \mathrm{CH_3CCH_3} \xrightarrow{[\mathrm{OH}^{\ominus}]}
\begin{array}{c}
\mathrm{H_3C} \quad \mathrm{CH_2CCH_3} \\
\mathrm{C} \\
\mathrm{H_3C} \quad \mathrm{OH\ O}
\end{array}
\tag{15}
$$

$$\left(\Delta H = -\begin{smallmatrix}4\ \mathrm{kcal}\\17\ \mathrm{kJ}\end{smallmatrix}\!/\mathrm{mol} \right)$$

In the second step, water elimination occurs with catalytic amounts of H_2SO_4 or H_3PO_4 at about 100°C:

$$
\begin{array}{c}
\mathrm{H_3C} \quad \mathrm{CH_2CCH_3} \\
\mathrm{C} \\
\mathrm{H_3C} \quad \mathrm{OH\ O}
\end{array}
\xrightarrow{-H_2O}
\begin{array}{c}
\mathrm{H_3C} \\
\diagdown \\
\mathrm{H_3C}
\end{array}\!\!\!\!\mathrm{C{=}CHCCH_3} \atop \mathrm{O}
\tag{16}
$$

Mesityl oxide is then hydrogenated to methyl isobutyl ketone and further to methyl isobutyl carbinol (4-methyl-2-pentanol) at 150–200°C and 3–10 bar using Cu or Ni catalysts:

$$
\begin{array}{c}
H_3C \\
\diagdown \\
C=CHCCH_3 \\
H_3C\diagup \quad \underset{O}{\|} \\
\end{array}
\xrightarrow[\text{[cat.]}]{+H_2}
\begin{array}{c}
H_3C \\
\diagdown \\
CHCH_2CCH_3 \\
H_3C\diagup \quad \underset{O}{\|} \\
\end{array}
$$

$$
\xrightarrow[\text{[cat.]}]{+H_2}
\begin{array}{c}
H_3C \\
\diagdown \\
CHCH_2CHCH_3 \\
H_3C\diagup \quad \underset{OH}{} \\
\end{array}
$$

(17)

The hydrogenation products, which are formed concomitantly, are separated and purified by distillation. With palladium catalysts, the carbon–carbon double bond is hydrogenated selectively, and only MIBK is formed.

Recent single-step processes employ acidic cation-exchangers, zeolites, or zirconium phosphate with the addition of platinum-group metals, generally palladium. The aldol condensation and the hydrogenation occur simultaneously with high selectivity in an exothermic reaction ($\Delta H = -28$ kcal or -117 kJ/mol). In the DEA process, at an acetone conversion of 50%, MIBK is formed with about 93% selectivity. In a Deutsche Texaco (now RWE-DEA) process used in a 10000 tonne-per-year plant in Germany, an 83% selectivity to MIBK at an acetone conversion of 40% is obtained at 30 bar and 130–140 °C. In the Tokuyama Soda process, which has been tested on pilot scale, zirconium phosphate is used with finely divided palladium metal. At 20–50 bar H_2 pressure and 120–160 °C, an acetone conversion of 30–40% is obtained; the selectivity to MIBK is 95%. All products of the aldol condensation and of the subsequent reactions are excellent solvents for the paint industry. Moreover, MIBK is used in many other areas such as in the extraction of niobium and tantalum ores, in dewaxing mineral oils, in purification of organic products by liquid–liquid extraction, etc.

Producers in Western Europe are RWE-DEA, Hüls, and Shell.

11.1.4.2. Methacrylic Acid and Ester

Methacrylic acid has a limited use in the manufacture of homo- and copolymers for application as sizing and finishing agents, and as thickeners, but the methyl ester is the most frequently used derivative. In 1995, the world capacity for methyl methacrylate was about 2.2×10^6 tonnes per year, of which 0.68, 0.66, and 0.47×10^6 tonnes per year were in the USA, Western Europe, and Japan, respectively. The largest producer worldwide is ICI, with a total capacity of about 560000 tonnes per year (1995). Production figures for these countries are given in the adjacent table.

modern MIBK manufacture single step with bifunctional catalysts for:

1. condensation (cation-exchanger)
2. hydrogenation (Pt-group metals)

applications of secondary products from acetone aldol reaction:

1. solvents for *e. g.*,
   cellulose acetate
   cellulose acetobutyrate
   acrylic and alkyd resins

2. extractants for
   inorganic salts and
   organic products

3. diacetone alcohol also hydrogenated to 'hexylene glycol'

$$
\underset{OH\ \ OH}{(CH_3)_2CCH_2CHCH_3}
$$

methacrylic acid is virtually only important as methyl ester

production of methyl methacrylate (in 1000 tonnes):

	1989	1992	1995
USA	527	548	622
W. Europe	439	472	475
Japan	360	426	425

classical manufacture (since 1937) of methacrylic acid and derivatives from acetone cyanohydrin is still competitive today with about 85% of the production capacity worldwide

Acetone cyanohydrin is still the feedstock for all methacrylic acid derivatives, especially in the USA and Western Europe, although numerous new routes have been developed and several are already being used commercially. The first industrial manufacture of methyl methacrylate from acetone was carried out by Rohm & Haas and also by ICI in the 1930s. The first step is manufacture of acetone cyanohydrin by base-catalyzed addition of HCN to acetone below 40 °C:

two-step manufacturing route:

1. base-catalyzed HCN addition to acetone under mild conditions gives cyanohydrin and minimizes side reactions

$$
\underset{H_3C}{\overset{H_3C}{>}}C{=}O \;+\; HCN \;\xrightarrow{\;[OH^{\ominus}]\;}\; \underset{H_3C}{\overset{H_3C}{>}}C\underset{OH}{\overset{CN}{<}} \tag{18}
$$

The catalysts (*e. g.*, alkali metal hydroxides, carbonates, or basic ion-exchangers) are generally used in the liquid phase. These types of processes were developed by Rohm & Haas, Degussa, Du Pont, ICI, and Rheinpreussen. The selectivity to acetone cyanohydrin is 92–99% (based on HCN) and more than 90% (based on acetone).

2. acid-catalyzed hydration of the nitrile group in cyanohydrin with amide intermediate

Acetone cyanohydrin is then reacted with 98% H_2SO_4 at 80–140 °C to form methacrylic acid amide sulfate, which is converted into methyl methacrylate and NH_4HSO_4 by reacting with methanol at about 80 °C, or at 100–150 °C under pressure:

amide usually directly converted into the methyl ester (important derivative)

$$
\underset{H_3C}{\overset{H_3C}{>}}C\underset{OH}{\overset{CN}{<}} \;+\; H_2SO_4 \;+\; H_2O \;\rightarrow\; \underset{H_3C}{\overset{H_3C}{>}}C\underset{OH}{\overset{CONH_2 \cdot H_2SO_4}{<}} \tag{19}
$$

$$
\xrightarrow[-H_2O]{+CH_3OH}\; \underset{H_3C}{\overset{H_2C}{>}}C{-}COOCH_3 \;+\; NH_4HSO_4
$$

The overall selectivity to methyl methacrylate is about 77% (based on acetone).

new Mitsubishi route avoids N-elimination as NH_4HSO_4 during C_4-synthesis from C_3 + HCN with attractive net equation:

$CH_3COCH_3 + HCOOCH_3 \rightarrow$
$H_2C{=}C(CH_3)COOCH_3 + H_2O$

A new development from Mitsubishi Gas Chemical avoids the disadvantages of the NH_4HSO_4 coproduct. Here the acetone cyanohydrin is partially hydrolyzed to α-hydroxyisobutyramide, which is then reacted with methyl formate to give the methyl ester and formamide:

$$
\underset{H_3C}{\overset{H_3C}{>}}C\underset{OH}{\overset{CN}{<}} \;\xrightarrow[{[cat.]}]{+H_2O}\; \underset{H_3C}{\overset{H_3C}{>}}C\underset{OH}{\overset{CONH_2}{<}} \tag{20}
$$

$$
\xrightarrow[-HCONH_2]{+HCOOCH_3}\; \underset{H_3C}{\overset{H_3C}{>}}C\underset{OH}{\overset{COOCH_3}{<}} \;\xrightarrow[{[cat.]}]{-H_2O}\; \underset{H_3C}{\overset{H_2C}{>}}C{-}COOCH_3
$$

In a separate dehydration the methyl ester gives methyl methacrylate and formamide yields hydrogen cyanide, which can be used for the production of acetone cyanohydrin. This new process is currently being developed in a pilot plant. Process technology details have not yet been disclosed.

Recent work has concentrated on three other manufacturing routes to methacrylic acid. These start with inexpensive isobutene, isobutyraldehyde, and propionaldehyde, and avoid formation of the NH_4HSO_4 produced in the cyanohydrin route. The C_4 basis is especially used in Japan.

alternative manufacturing routes to methacrylic acid utilize three inexpensive feedstocks:

1. $(CH_3)_2C=CH_2$ from C_4 fractions (*cf.* Section 3.3.2)
2. $(CH_3)_2CHCHO$ as byproduct of propene hydroformylation (*cf.* Section 6.1)
3. CH_3CH_2CHO as main product from the hydroformylation of ethylene (*cf.* Section 6.1)

Detailed experiments have shown that it is possible to oxidize isobutene directly (single-step process) to methacrylic acid. However, the selectivity is too low to be economical. The oxidation of the intermediate methacrolein to methacrylic acid is particularly critical. This result is in contrast to the oxidation of propene to acrylic acid.

For this reason Escambia developed a two-step oxidation process with N_2O_4 in which the α-hydroxyisobutyric acid formed initially eliminates water to give methacrylic acid:

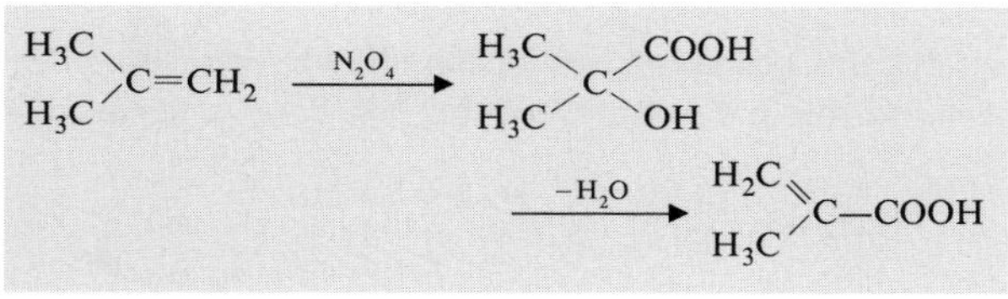

to 1:

two-step methacrylic acid manufacture by N_2O_4 oxidation of isobutene and dehydration of α-hydroxyisobutyric acid (Escambia)

$$
\begin{array}{c}
H_3C \\
\diagdown \\
C=CH_2 \\
H_3C\diagup
\end{array}
\xrightarrow{N_2O_4}
\begin{array}{c}
H_3C \quad\quad COOH \\
\diagdown\diagup \\
C \\
\diagup\diagdown \\
H_3C \quad\quad OH
\end{array}
$$

$$
\xrightarrow[-H_2O]{}
\begin{array}{c}
H_2C \\
\diagdown \\
C-COOH \\
H_3C\diagup
\end{array}
\tag{21}
$$

Since explosions occurred during industrial operation of the Escambia route this process was discontinued.

Several firms in Japan have developed a two-step oxidation of isobutene that preferentially goes through *tert*-butanol as the primary intermediate. In this process isobutene, generally in a mixture with *n*-butenes and butane (C_4 raffinate), is hydrated almost quantitatively to *tert*-butanol in the liquid phase with an acid catalyst such as an ion exchange resin.

two-step, heterogeneously catalyzed air oxidation of isobutene, directly or as *tert*-butanol, to methacrylic acid with intermediate isolation and purification of methacrolein (*e. g.*, Asahi Glass)

This is followed by the heterogeneously catalyzed oxidation to methacrolein at 420 °C and 1 − 3 bar using a promoted Mo/Fe/Ni catalyst system. The conversion of *tert*-butanol and the selectivity to methacrolein are both 94%.

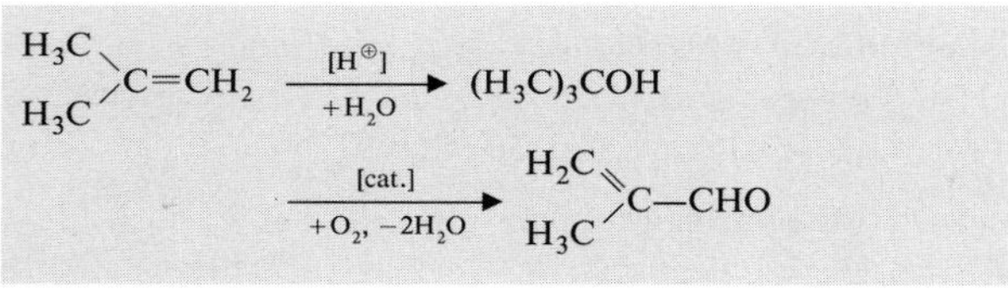

$$
\begin{array}{c}
H_3C \\
\diagdown \\
C=CH_2 \\
H_3C\diagup
\end{array}
\xrightarrow[+H_2O]{[H^{\oplus}]}
(H_3C)_3COH
$$

$$
\xrightarrow[+O_2,\, -2H_2O]{[cat.]}
\begin{array}{c}
H_2C \\
\diagdown \\
C-CHO \\
H_3C\diagup
\end{array}
\tag{22}
$$

oxidation catalyst for both steps is multi-component system (MCM = **m**ulti**c**omponent **m**etal oxide) based on metal oxides with Mo as central constituent

tandem oxidation follows without intermediate purification and therefore with high TBA yield (>99%); also, lower investment is economical

To maintain the catalyst lifetime and selectivity in the second oxidation step, the methacrolein must first be absorbed in water under pressure and freed from byproducts by distillation.

Japan Methacrylic Monomer Co. has developed a modified method which allows the oxidation to take place without purification of the material from the first step. This variant, known as tandem oxidation, is in practice in Japan.

The oxidation to methacrylic acid takes place in the presence of steam at about 300 °C and 2–3 bar over oxides of Mo, P, Sb, and W, or of Mo, P, and V, with a conversion of 89% and a selectivity of more than 96%:

$$
\begin{array}{c}
H_2C \\
\diagdown \\
H_3C \diagup
\end{array}
C-CHO
\quad \xrightarrow[+0.5\,O_2]{[cat.]} \quad
\begin{array}{c}
H_2C \\
\diagdown \\
H_3C \diagup
\end{array}
C-COOH
\qquad\qquad (23)
$$

$$
\left(\Delta H = + \begin{array}{c} 60 \text{ kcal} \\ 252 \text{ kJ} \end{array} /mol \right)
$$

The acid is extracted from the aqueous phase, purified by distillation, and esterified with methanol.

commercial use mainly in Japan due to low demand for octane-boosting gasoline additives from isobutene (TBA, MTBE)

alternative methacrylic acid and ester manufacture also by hydrolysis or alcoholysis of methacrylonitrile with intermediate methacrylamide

Plants using this technology are operated by Kyoda Monomer, Mitsubishi Rayon and Japan Catalytic/Sumitomo Chemical in Japan.

A second possible synthetic pathway to methacrylic acid and its derivatives, not based on acetone, is the ammoxidation of isobutene, *e. g.*, in the Sohio process (*cf.* Section 11.3.2). Asahi Chemical uses this technology in a plant in Japan.

to 2:

two-step methacrylic acid manufacture:

1. oxidation of isobutyraldehyde to isobutyric acid
2. two types of dehydrogenation

In the second process not based on acetone, isobutyraldehyde is first oxidized with air or O_2 to isobutyric acid. The next step, the formal dehydrogenation to methacrylic acid, can be conducted as an oxydehydrogenation in the presence of O_2:

$$
\begin{array}{c}
H_3C \\
\diagdown \\
H_3C \diagup
\end{array}
CHCHO
\quad \xrightarrow{+0.5\,O_2} \quad
\begin{array}{c}
H_3C \\
\diagdown \\
H_3C \diagup
\end{array}
CHCOOH
\qquad\qquad (24)
$$

$$
\xrightarrow[-H_2O]{+0.5\,O_2}
\begin{array}{c}
H_2C \\
\diagdown \\
H_3C \diagup
\end{array}
C-COOH
$$

2.1. homogenously or heterogeneously catalyzed oxydehydrogenation of acid, *e. g.*, process from Eastman Kodak, Cyanamid, or Mitsubishi Chemical
2.2. sulfodehydrogenation of the ester, *e. g.*, process from Asahi Kasei

This reaction can be carried out with homogeneous catalysis using HBr (Eastman Kodak) at 160–175 °C or in the presence of a Bi-Fe catalyst (Cyanamid) at 250–260 °C.

In a recent Mitsubishi Chemical process, isobutyric acid – mixed with H_2 for heat transfer – is oxydehydrogenated at

$250-350\,°C$ over a supported phosphomolybdic acid in which V is substituted for part of the Mo. At an isobutyric acid conversion of 94%, the selectivity to methacrylic acid is about 73%.

Asahi Kasei has developed another alternative for the dehydrogenation. Methyl isobutyrate is produced and is then dehydrogenated to methyl methacrylate (selectivity = 85%) in the presence of H_2S/S at about $500\,°C$.

None of the processes based on isobutyraldehyde has been used commercially.

BASF has developed another new synthesis for the production of methyl methacrylate. Propionaldehyde (produced from the hydroformylation of ethylene) is condensed with formaldehyde in the presence of a secondary amine and acetic acid at $160-210\,°C$ and $40-80$ bar to yield methacrolein. This is oxidized with oxygen to methacrylic acid which is then esterified with methanol to give methyl methacrylate:

to 3:

two-step production of methacrylic acid:

1. condensation of propionaldehyde and formaldehyde to yield methacrolein (through Mannich base with secondary amine as catalyst)
2. oxidation to methacrylic acid

$$CH_3CH_2CHO + HCHO \xrightarrow[[H^\oplus]]{[sec.\ amine]} CH_3-\underset{\underset{CH_2}{\|}}{C}-CHO + H_2O \quad (25)$$

$$\xrightarrow{+O_2} CH_3-\underset{\underset{CH_2}{\|}}{C}-COOH$$

A 40 000 tonne-per-year plant was put into production by BASF in 1990 in Germany. Similar processes have been developed in Japan but have not yet been used industrially.

commercial use by BASF in Germany

Most methyl methacrylate is used for the manufacture of Plexiglas®, a crystal-clear plastic possessing considerable hardness, resistance to fracturing, and chemical stability. It is also used in copolymers with a wide range of applications such as fabric finishes and light- and weather-resistant paints.

uses of methyl methacrylate:

1. monomer for Plexiglas®

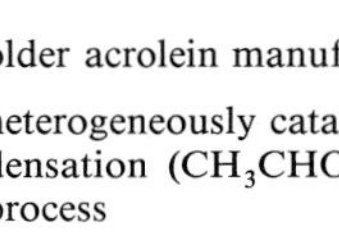

$$\left[\!\!\begin{array}{c} CH_3 \\ | \\ CH_2-C \\ | \\ \underset{O}{\overset{C}{\diagup}}\overset{}{\diagdown}OCH_3 \end{array}\!\!\right]_n$$

2. comonomer for manufacture of dispersions, *e.g.*, for paint industry and textile finishing

11.1.5. Acrolein

Degussa developed the first industrial acrolein synthesis in the early 1930s, and began production in 1942. In 1991, the capacity for acrolein manufacture worldwide was about 128 000 tonnes per year, with 54 000, 23 000, and 11 000 tonnes per year in Western Europe, the USA, and Japan, respectively. The classical process was based on the condensation of acetaldehyde with formaldehyde:

older acrolein manufacture:

heterogeneously catalyzed mixed aldol condensation (CH_3CHO, HCHO) in Degussa process

$$CH_3CHO + HCHO \xrightarrow{\text{[cat.]}} H_2C=CH-CHO + H_2O \quad (26)$$

$$\left(\Delta H = -\frac{20\ kcal}{84\ kJ}/mol\right)$$

The reaction runs at $300-320\,°C$ in the gas phase over Na silicate/SiO_2.

Today, this route has been replaced by the catalytic oxidation of propene:

$$H_2C=CH-CH_3 + O_2 \xrightarrow{\text{[cat.]}} H_2C=CH-CHO + H_2O \quad (27)$$

$$\left(\Delta H = -\frac{88\ kcal}{368\ kJ}/mol\right)$$

Shell was the first company to practice this gas-phase oxidation in a commercial unit (1958–1980). The reaction is run at $350-400\,°C$ over Cu_2O/SiC with I_2 as a promoter. At propene conversions of up to 20%, the selectivity to acrolein is $70-85\%$. The basis of a British Distillers (now BP) process is supported Cu_2O; Montecatini conducts the propene oxidation in copper plated tubes.

Since this time, other catalysts have been developed which have profited from experience gained in the manufacture of acrylic acid and acrylonitrile from propene. Components of these oxidation and ammoxidation catalysts are also suitable for the partial oxidation of propene to acrolein. Sohio's discovery of the catalytic properties of bismuth molybdate and phosphomolybdate also meant an economic breakthrough for acrolein manufacture. The Sohio acrolein process is practiced commercially in several plants. The air oxidation of propene is run at $300-360\,°C$ and $1-2$ bar over a Bi_2O_3/MoO_3 fixed-bed catalyst. Starting from this bismuth-based catalyst, various firms developed very selective multicomponent systems from the oxides of the transition, alkali, and alkaline earth metals.

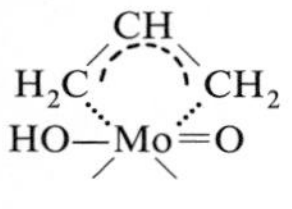

Ugine-Kuhlmann, the world's largest acrolein producer, currently operates a 50000 tonne-per-year plant in France using a Nippon Shokubai catalyst based on Mo, Bi, Fe, Co, W, and other additives. The catalyst is arranged in a fixed-bed tubular reactor, through which a heat transfer liquid flows. Propene is oxidized in high conversion (96%) with an excess of air in the presence of steam at $350-450\,°C$ and a slight pressure. A large air/propene ratio is necessary to maintain the oxidation state of the catalyst

modern acrolein manufacture:

heterogeneously catalyzed gas-phase oxidation of propene

initially with catalysts based on Cu_2O or Cu (Shell, British Distillers, Montecatini)

later with multicomponent systems based on mixed oxides of Bi-Mo, Sn-Mo, Sn-Sb (Sohio, Nippon Shokubai, etc.)

process principles of propene oxidation:

Mo-catalyzed partial gas-phase oxidation of propene with air to form acrolein through a redox mechanism with a chemisorbed π-allyl radical as intermediate

above a certain level, as otherwise the acrolein selectivity drops. Under these conditions, the selectivity can reach 90% (based on C_3H_6). The byproducts include acetaldehyde, acrylic acid, and acetic acid.

The reaction gases are washed with a limited amount of water to remove the byproducts acrylic acid and acetic acid. After intensive washing with water, the dilute aqueous acrolein solution formed in the absorption column is worked up to acrolein in a stripper. The crude product is separated from low- and high-boiling substances in a two-step distillation and is fractionated to a purity of 95–97%. Because of the tendency of acrolein to polymerize, inhibitors such as hydroquinone are added to the distillation.

process characteristics of acrolein manufacture:

air excess effects high $Mo^{6\oplus}$ share which oxidizes propene more selectively than $Mo^{5\oplus}$ and eliminates recycle operation due to high propene conversion

isolation of acrolein:

1. cooling and removal of acrylic and acetic acid from reaction gases using small amount of H_2O
2. acrolein absorption with large quantity of H_2O
3. distillation with addition of polymerization inhibitors

11.1.6. Secondary Products of Acrolein

Although the acrolein molecule has several reactive sites — the aldehyde group, the activated olefinic double bond and the conjugated double bond system — making a wide variety of reactions possible, only limited commercial use has been made of them.

For example, the reduction of acrolein to allyl alcohol and its conversion into glycerol is a commercial process (*cf.* Section 11.2.2).

The reaction of acrolein with NH_3 in the gas phase at 350–400 °C over a multicomponent catalyst based on Al_2O_3 or $Al_2O_3 \cdot SiO_2$ to form pyridine and 3-picoline is limited to Japan (*cf.* Section 7.4.5). A commercial process is operated by Daicel.

Although acrolein can be used as the starting material for acrylic acid or acrylonitrile, the direct oxidation and ammoxidation of propene are the commercially successful processes.

uses of acrolein:

1. selective hydrogenation or reduction to allyl alcohol
2. gas-phase reaction with NH_3 to form pyridine and β-picoline mixtures
3. manufacture of acrylic acid and acrylonitrile
4. manufacture of methionine
5. manufacture of 1,3-propanediol

However, the use of acrolein for the manufacture of methionine is experiencing a new and relatively rapid worldwide growth.

Commercial production of methionine began in 1950, and by 1995 world capacity had reached 492 000 tonnes per year, with 34 000 (including MHA), 160 000, and 41 000 tonnes per year in the USA, Western Europe, and Japan, respectively. Rhône-Poulenc and Degussa, the largest producers worldwide, have methionine production capacities of 140 000 and 130 000 tonnes per year, respectively. In the USA, Monsanto (now Mitsui/Nippon Soda), Rhône-Poulenc, and Du Pont manufacture **methyl hydroxythiobutyric acid** (MHA) for supplementing animal fod-

industrial production of DL-methionine (essential amino acid) with greatest predicted growth

methionine substitute

$$CH_3OCH_2CH_2CH_2C\overset{O}{\underset{SH}{\diagup}} \quad (MHA)$$

produced by three companies in the USA

der; as a methionine metabolite, MHA has the same activity as methionine.

The first step of the three-step manufacture of DL-methionine is the base-catalyzed addition of methyl mercaptan to acrolein at room temperature to form methylthiopropionaldehyde:

multistep manufacturing process for methionine from acrolein:

1. base-catalyzed CH_3SH addition to C=C double bond
2. formation of hydantoin
3. hydrolysis and separation of racemate if necessary

$$CH_3SH + H_2C{=}CHCHO \xrightarrow{[OH^\ominus]} CH_3{-}S{-}CH_2CH_2CHO \quad (28)$$

The aldehyde is then reacted with NaCN and NH_4HCO_3 in aqueous solution at 90 °C to form a hydantoin:

methionine manufacture, an example of an industrially operated Bucherer amino acid synthesis

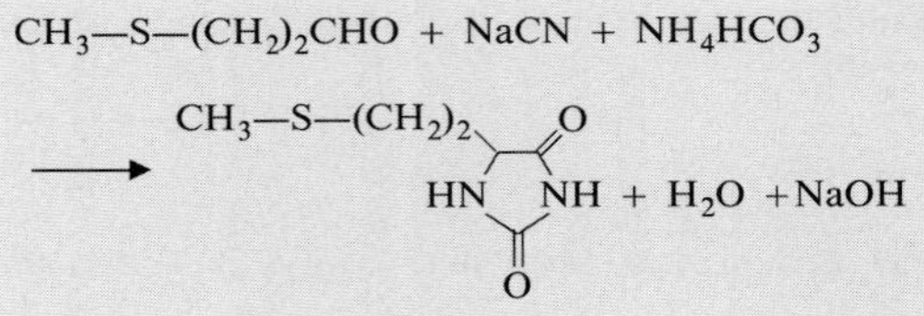

$$CH_3{-}S{-}(CH_2)_2CHO + NaCN + NH_4HCO_3$$

(29)

In the last step, the hydantoin is reacted with NaOH or K_2CO_3 under pressure at about 180 °C and, after acidification or treatment with CO_2, converted into free DL-methionine:

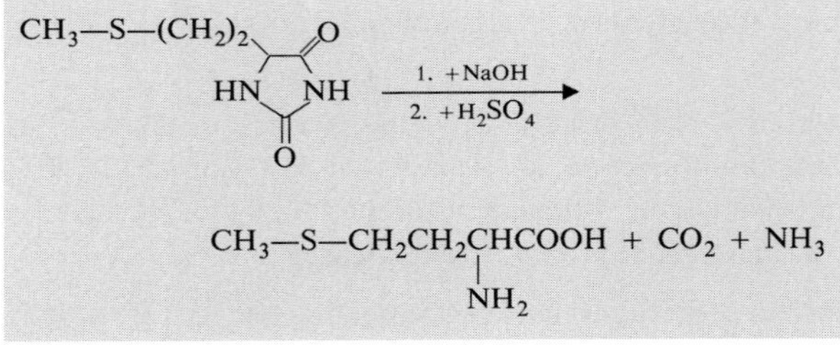

(30)

use of DL-methionine:

additive, *e. g.*, for soy flour or in the future for SCP (single cell protein) to improve biological balance (methionine deficiency)

Like lysine (*cf.* Section 10.3.1.4), synthetic methionine is an essential amino acid used for supplementing the protein concentrate in animal feed, in particular for feed optimization in poultry breeding. Since D-methionine is converted enzymatically to the L-form in the body, both isomers have the same biological activity, and the racemic mixture from the synthesis is almost always used without separation.

manufacturing process for 1,3-propanediol:

1. combined hydration and hydrogenation of ethylene oxide
2. combined hydroformylation and hydrogenation of ethylene oxide

Acrolein can also be used for the manufacture of 1,3-propanediol (trimethylene glycol). In a Degussa process, acrolein is hydrolyzed with acid catalysis to 3-hydroxypropionaldehyde, which is then hydrogenated to 1,3-propanediol over a supported Ni catalyst:

$$H_2C=CHCHO + H_2O \xrightarrow{[H^{\oplus}]} HOCH_2CH_2CHO$$

$$\xrightarrow[+H_2]{[cat]} HOCH_2CH_2CH_2OH \qquad (31)$$

Degussa has started up a 1,3-propanediol plant with a capacity of 2000 tonnes per year in Belgium. Another manufacturing route for 1,3-propanediol is based on the hydroformylation of ethylene oxide (*cf.* Section 7.2.5).

1,3-Propanediol is an intermediate for pharmaceuticals, and is also used in the manufacture of a new polyester, poly(trimethylene terephthalate), for fibers.

use of 1,3-propanediol:

pharmaceutical intermediate
component for polyester

11.1.7. Acrylic Acid and Esters

During the last few years acrylic acid and its esters have ranked first in growth among the unsaturated organic acids and esters. In 1994, the world production capacity for acrylic acid was about 2.0×10^6 tonnes per year, of which 690 000, 640 000 and 420 000 tonnes per year was in the USA, Western Europe, and Japan, respectively. Acrylic ester production numbers in these countries are given in the adjacent table. The largest producer of acrylic acid is BASF, with production facilities in a variety of locations.

together with its esters, acrylic acid is most important industrially used unsaturated carboxylic acid

acrylic ester production (in 1000 tonnes):

	1990	1992	1993
USA	470	552	587
Japan	183	231	206
W. Europe	320	330	312

Starting with the first industrial acrylic acid manufacture by Röhm & Haas (Germany) in 1901, four commercially practiced technologies were developed. However, in recent years, the oxidation of propene to acrylic acid has become so important that the traditional processes have either been discontinued or will be in the near future.

first industrial acrylic acid manufacture *via* ethylene chlorohydrin (from C_2H_4 + HOCl) and ethylene cyanohydrin (hydroxypropionitrile, from $ClCH_2CH_2OH$ + HCN)

11.1.7.1. Traditional Acrylic Acid Manufacture

In the ethylene cyanohydrin process, ethylene oxide and HCN were reacted in the presence of base to form hydroxypropionitrile, which was then reacted with an alcohol (or water) and a stoichiometric amount of H_2SO_4 to give the ester (or the free acid):

other partially outdated acrylic acid manufacturing processes:

1. cyanohydrin process, *i.e.*, addition of HCN to EO and subsequent conversion

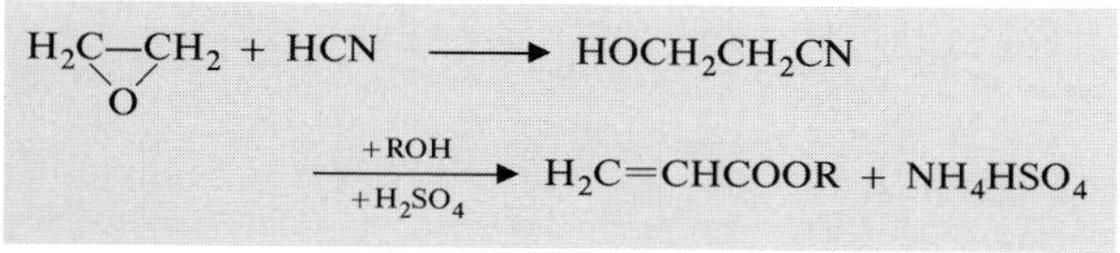

$$H_2C{-}CH_2 + HCN \longrightarrow HOCH_2CH_2CN$$
$$\underset{O}{\phantom{H_2C{-}CH_2}}$$

$$\xrightarrow[+H_2SO_4]{+ROH} H_2C=CHCOOR + NH_4HSO_4 \qquad (32)$$

This chemistry was practiced by UCC, the largest ethylene oxide manufacturer in the world, and Röhm & Haas until 1971.

2. Reppe process as hydrative carbonylation of C_2H_2 with three variants differing in the quantity of $Ni(CO)_4$ catalyst

Another method known as the Reppe process has three variants. All three are based on the carbonylation of acetylene in the presence of H_2O or alcohols to give the free acid or the ester:

$$HC \equiv CH + CO + H_2O\,(ROH) \xrightarrow{\text{[cat]}}$$
$$H_2C = CHCOO-H\,(-R)$$

(33)

2.1. traditional process with stoichiometric amount of $Ni(CO)_4$

2.2. modified process with less than stoichiometric amount of $Ni(CO)_4$

2.3. catalytic process with catalytic combination of carbonyl-forming (Ni) and non-carbonyl-forming (Cu) heavy metal, usually as iodide

Depending on the amount of catalyst (nickel tetracarbonyl) the variants are classified as stoichiometric, modified, and catalytic.

Until recently the modified Reppe process was practiced in industrial plants in the USA and Japan.

The purely catalytic Reppe reaction is operated at $220-230\,°C$ and 100 bar. $NiBr_2$ with a copper halide promoter is used as the catalyst. The reaction takes place in the liquid phase, using tetrahydrofuran which readily dissolves acetylene. The reaction mixture is worked up by distillation. Because of the small amount of catalyst introduced, recovery is usually not worthwhile. The selectivities to acrylic aicd are 90% (based on C_2H_2) and 85% (based on CO). Reppe discovered this catalytic route back in 1939, and it is still the basis of the BASF process with partial production (60000 tonnes per year) in Germany.

3. propiolactone process based on ketene and formaldehyde with subsequent β-lactone isomerization or alcoholysis

The third process, no longer used, is based on the thermolysis or alcoholysis of β-propiolactone. Ketene (*cf.* Section 7.4.2) is reacted with formaldehyde using $AlCl_3$, $ZnCl_2$, or BF_3 in a solvent or in the gas phase to give β-propiolactone with a selectivity of 90%. The lactone is then reacted with H_3PO_4 using copper powder as catalyst at $140-180\,°C$ and $25-250$ bar to form acrylic acid quantitatively. When an alcohol is present, the corresponding ester is obtained:

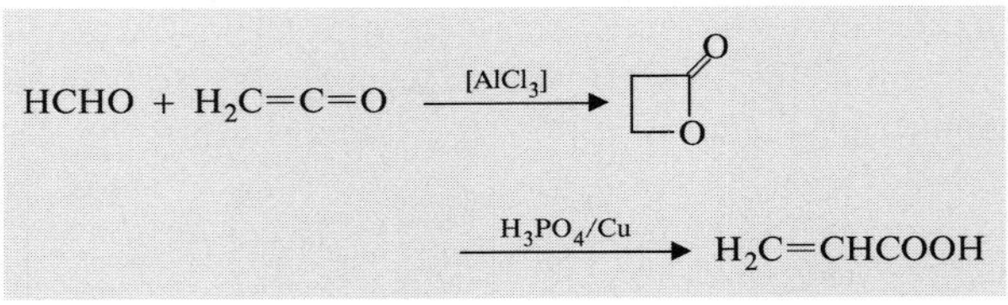

$$HCHO + H_2C = C = O \xrightarrow{\text{[AlCl}_3]}$$

$$\xrightarrow{\text{H}_3\text{PO}_4/\text{Cu}} H_2C = CHCOOH$$

(34)

Celanese operated a 35000 tonne-per-year plant for acrylic acid and its esters from 1957 to 1974.

The last of the conventional processes is based on the alcoholysis or hydrolysis of acrylonitrile. The hydrolysis is done with H_2SO_4; the acrylamide sulfate formed initially reacts with an alcohol to give the ester:

4. acrylonitrile hydrolysis or alcoholysis with acrylamide as intermediate

$$H_2C=CHCN + H_2SO_4 + H_2O \longrightarrow$$

$$H_2C=CHCONH_2 \cdot H_2SO_4 \quad (35)$$

$$\xrightarrow[-NH_4HSO_4]{+ROH} H_2C=CHCOOR$$

The selectivity to acrylic acid ester is about 90% (based on AN). This process, which produces stoichiometric amounts of NH_4HSO_4, was used by Anic, Ugilor, and Mitsubishi Petrochemical, but is currently used only by Asahi Chemical. In a more recent Mitsui Toatsu process, the saponification comprised only water over a B_2O_3-containing catalyst.

more recent variant for acrylonitrile hydrolysis:

catalytic gas-phase hydrolysis (Mitsui Toatsu)

11.1.7.2. Acrylic Acid from Propene

While most acrylic acid and its esters were still manufactured by the Reppe process at the end of the 1960s, in the newer plants propene oxidation predominates. It can be conducted catalytically either as a single-step or as a two-step process.

most modern acrylic acid manufacture:

two processes for heterogeneously-catalyzed gas-phase direct oxidation of propene:

1. single step
2. two steps

In the single-step direct oxidation, propene is reacted with air or oxygen (sometimes diluted with steam) at up to 10 bar and $200-500\,°C$, depending on the catalyst. The multicomponent catalyst consists mainly of heavy metal molybdates, and generally contains tellurium compounds as promoters. Acrolein and acrylic acid are formed together:

characteristics of single-step process:

propene oxidation with multifunctional catalyst leads to mixture of acrolein and acrylic acid

$$H_2C=CHCH_3 + O_2 \xrightarrow[-H_2O]{[cat.]} H_2C=CHCHO$$

$$\left(\Delta H = -\frac{88 \text{ kcal}}{368 \text{ kJ}}/\text{mol}\right)$$

$$\xrightarrow[+0.5\,O_2]{[cat.]} H_2C=CHCOOH \quad (36)$$

$$\left(\Delta H = -\frac{63 \text{ kcal}}{266 \text{ kJ}}/\text{mol}\right)$$

As both oxidation steps have different kinetics, uniform process conditions and a single catalyst would not lead to an optimal acrylic acid selectivity.

characteristics of two-step process:

propene oxidation in two fixed-bed reactors with different catalysts:

1st step: propene → acrolein [Bi, P, Mo, + *e. g.*, Fe, Co]

2nd step: acrolein → acrylic acid [Mo + *e. g.*, V, W, Fe, Ni, Mn, Cu]

For this reason, numerous firms developed two-step processes with an optimal coordination of catalyst and process variables, and the first industrial single-step plant (Japan Catalytic Chem. Ind.) has been converted to the two-step process for the reason given above. While the processes are essentially the same, there are marked differences in the catalyst compositions. As in the single-step process, they contain mixed oxides, the main component being molybdenum on supports of low surface area. The promoters for both stages are vastly different.

process characteristics of two-step route:

temperature in 1st oxidation step about 70 °C higher in than 2nd step, high conversions (>95%) of C_3H_6 and acrolein eliminate recycle

H_2O addition for:

1. shift of explosion limit
2. improvement in desorption from catalyst
3. facilitation of heat removal
4. lowering of acrylic acid partial pressure

isolation of acrylic acid:

extraction from 20–25% aqueous solution with *e. g.*, ethyl acetate, *n*-propyl or iso-propyl acetate, or MEK

In the first step, propene is generally oxidized to acrolein in the presence of steam and air at 330–370 °C and 1–2 bar. The exothermic reaction is conducted in a fixed-bed tubular reactor with up to 22 000 tubes per reactor. The reaction products are fed directly into a second reactor where they are further oxidized at 260–300 °C to acrylic acid. The propene and acrolein conversion are over 95%, with a selectivity to acrylic acid of 85–90% (based on C_3H_6). The large addition of water leads to an only 20–25% acrylic acid solution, from which the acid is generally isolated by extraction and separated from the byproducts acetic acid, propionic acid, maleic acid, acetaldehyde, and acetone by distillation. In a newer separation method from BASF, acrylic acid is removed from the reaction gases by absorption in hydrophobic solvents (*e. g.*, diphenyl/diphenyl ether mixtures) and purified by distillation.

The two-step processes are operated mainly in Japan, the USA, England, and France. Well-known processes have been developed by Sohio, Nippon Shokubai (Japan Catalytic Company), and Mitsubishi Petrochemical, for example.

uses of acrylic acid and its esters:

as monomer for manufacture of homo- and copolymers used as surface protectants and in surface finishing

commercially important acrylates:

$H_2C=CHCOOR$; $R=CH_3$, C_2H_5, *n*- and iso-C_4H_9, $CH_3(CH_2)_3CH(C_2H_5)CH_2$

manufacture of acrylates:

proton-catalyzed acid esterification with alcohols or transesterification of lower esters, generally methyl acrylate

Acrylic acid and its esters are important monomers for the manufacture of homo- and copolymers. They are used mainly in paints and adhesives, in paper and textile finishing, and in leather processing. One important use is as superabsorbants, that is, as polymers which can soak up extremely high amounts of liquids.

Methyl, ethyl, *n*-butyl, isobutyl, and 2-ethylhexyl esters of acrylic acid are the most common derivatives of acrylic acid. The relative importance of acrylic acid esters varies considerably between countries, as is shown for the USA and Japan in the following table:

Table 11-5. Breakdown of acrylate production (in wt%).

Product	USA		Western Europe	Japan	
	1984	1988	1987	1984	1988
Ethyl acrylate	33	34	24	16	15
n-Butyl, isobutyl acrylates	45	43	50	32	34
2-Ethylhexyl acrylate	9	5	9	29	30
Methyl acrylate	7	8	15	} 23	} 21
Remaining	6	10	2		
Total production (in 1000 tonnes)	426	360	330	110	164

Several of the classical acrylic acid processes have been modified to produce the esters directly. Modern processes lead only to acrylic acid which must then be esterified. For the esterification, the pure acid is usually reacted with the corresponding alcohol under proton catalysis (generally H_2SO_4 or cation exchange resin) in the liquid phase at $100-120°C$.

acrylic acid esterification:

generally liquid phase, lower alcohols (CH_3OH, C_2H_5OH) heterogeneously catalyzed (acidic cation exchange resins), higher alcohols homogeneously catalyzed (H_2SO_4)

In addition, acrylic esters of the higher alcohols can be easily manufactured by transesterification with the lower esters.

Acrylamides, another class of acrylic acid derivatives, are manufactured solely by the partial hydrolysis of acrylonitrile (*cf.* Section 11.3.4).

11.1.7.3. Possibilities for Development in Acrylic Acid Manufacture

At the present time, acrylic acid production is going through a changeover from an acetylene to a propene base. Modern catalyst systems, in particular those with acrylic acid selectivities of $85-90\%$ (approaching the Reppe route's 90%) have contributed to the cost advantage of the propene oxidation. Although much research has been done on the single-step oxidation of propene to acrylic acid, the low selectivities (only $50-60\%$) and short lifetime of the multicomponent catalysts favor the two-step process.

as with AcOH, VCM, AN, and vinyl acetate, acrylic acid manufacture has also been converted from acetylene to olefin base

Although nitrile saponification for acrylic acid manufacture is attractive due to the large acrylonitrile capacity available from the Sohio process (*cf.* Section 11.3.2.1), this is countered by the formation of $(NH_4)_2SO_4$. Thus, the two-step oxidation of propene could be the most economical process in the future.

inexpensive feedstocks and high acrylic acid selectivity of propene oxidation point to the superiority of this route over alternative processes

However, Union Oil has presented a novel process development in which ethylene is oxidatively carbonylated to acrylic acid in the presence of $PdCl_2$—$CuCl_2$:

oxycarbonylation of ethylene, *i.e.*, C_3 synthesis from C_2H_4 + CO (Union Oil)

$$H_2C=CH_2 + CO + 0.5\ O_2 \xrightarrow[\text{[cat.]}]{Ac_2O} H_2C=CHCOOH$$

$$+ AcOCH_2CH_2COOH$$

(37)

The reaction takes place in the liquid phase with, for example, acetic acid as solvent and acetic anhydride to remove the water formed in side reactions which could otherwise affect the main reaction. The selectivity to acrylic acid and β-acetoxypropionic acid is said to reach 85% (based on C_2H_4).

Developments in olefin pricing will be decisive in determining the economic viability of synthesis of a C_3 structure from ethylene and CO, or the direct use of propene.

11.2. Allyl Compounds and Secondary Products

starting materials for manufacture of allyl compounds are:

$H_2C=CH-CH_3$

$H_2C=CH-CH_2Cl$

of potential importance:

$H_2C=CH-CH_2OAc$

Allyl compounds are characterized by the $H_2C=CH-CH_2$-group. They are thus substituted products of propene, which is also the most important industrial feedstock for this class of compound, taking advantage of the ready exchangeability of the allyl hydrogen. Their industrial significance arises from their secondary products, which are essentially based on allyl chloride. Allyl acetate, which could be economically manufactured by propene acetoxylation, has not been used extensively as a commercial intermediate (*cf.*, Section 11.2.2), although compounds could be synthesized from it which are presently being produced from allyl chloride.

11.2.1. Allyl Chloride

Chlorine can be more readily added to propene than to ethylene. However, unlike dichloroethane, the resulting dichloropropane is not commercially important.

manufacture of allyl chloride:

1. 'hot chlorination' of propene (free radical chain reaction) leads to substitution in CH_3 group

$Cl \rightleftharpoons 2\ Cl^{\bullet}$

$CH_3CH=CH_2 \xrightarrow{+Cl^{\bullet}} {}^{\bullet}CH_2CH=CH_2$
$+ HCl$

${}^{\bullet}CH_2CH=CH_2 \xrightarrow{+Cl_2} ClCH_2CH=CH_2$
$+ Cl^{\bullet}$

In 1936, Shell discovered that addition to the double bond is replaced by free-radical substitution in the allyl position, the so-called 'hot chlorination', at temperatures above 300 °C. At 500 – 510 °C, conversion is nearly quantitative with a selectivity to allyl chloride of 85%:

$$H_2C=CHCH_3 + Cl_2 \rightarrow H_2C=CHCH_2Cl + HCl$$

$$\left(\Delta H = -\frac{27\ \text{kcal}}{113\ \text{kJ}}/\text{mol}\right)$$

(38)

An excess of propene is used so that the chlorine reacts completely. A rapid and complete mixing of the gas streams before the reaction zone is important for a selective conversion. The byproducts are 1,3-dichloro-1-propene, 2-chloro-1-propene, and 1,2-dichloropropane. The hydrogen chloride formed in the reaction is washed out with water.

process characteristics:

propene excess allows quantitative Cl_2 conversion and, with intense mixing, optimal selectivity

In the years since 1936, other firms such as Asahi, Dow, and Shell have developed their own processes.

In 1991, the production capacity for allyl chloride in the USA, Germany, and Japan was 0.30, 0.11, and 0.09×10^6 tonnes per year, respectively.

Allyl chloride can also be obtained by oxychlorination of propene. Hoechst developed a process to pilot plant scale using a Te catalyst at 240 °C and atmospheric pressure. Other companies have also worked on oxychlorination processes. DEA for example proposes using a $PdCl_2$—$CuCl_2$ catalyst, and Lummus has developed a process based on propane. These technologies have not been used commercially.

2. oxychlorination of propene with HCl/O_2, *e. g.*, according to Hoechst, DEA, etc., not used commercially

Allyl chloride is mainly used to manufacture epichlorohydrin (more than 90% of allyl chloride worldwide), allyl alcohol, and allylamine.

uses of allyl chloride:

1. hydrolysis to $H_2C=CHCH_2OH$
2. epoxidation (direct and indirect) to H_2C—$CHCH_2Cl$ with epoxide O
3. ammonolysis to $H_2C=CHCH_2NH_2$

In the two-step route to epichlorohydrin, allyl chloride is reacted with HOCl at 25 – 30 °C in the aqueous phase to give a mixture of the two isomeric dichlorohydroxypropanes:

$$H_2C=CHCH_2Cl + HOCl$$
$$\rightarrow \underset{\underset{Cl \quad OH}{|\quad|}}{CH_2-CHCH_2Cl} + \underset{\underset{OH \quad Cl}{|\quad|}}{CH_2-CHCH_2Cl} \qquad (39)$$
$$\text{ca. 70\%} \qquad\qquad \text{ca. 30\%}$$

The crude product is then converted into epichlorohydrin with $Ca(OH)_2$ at 50 – 90 °C:

epichlorohydrin manufacture:

two-step process involving HOCl addition to allyl chloride and subsequent dehydrochlorination

$$2\ \underset{\underset{(OH)\ (Cl)}{\underset{Cl \quad OH}{|\quad|}}}{CH_2-CHCH_2Cl} + Ca(OH)_2 \rightarrow 2\ \underset{O}{H_2C-CHCH_2Cl} \qquad (40)$$
$$+\ CaCl_2 + 2\ H_2O$$

Purification is achieved by fractional distillation.

epichlorohydrin production (in 1000 tonnes):

	1989	1991	1993
USA	225	216	224
W. Europe	165	186	185
Japan	94	113	98

alternative process involves epoxidation of allyl chloride with perpropionic acid

uses of epichlorohydrin:

manufacture of bisphenol A glycidyl ethers, important precursors for epoxy resins

reaction sequence for glycidyl ether formation:

1. 'chlorohydrin ether' by addition of bisphenol A to epichlorohydrin
2. bisglycidyl ether (diepoxide) by HCl elimination
3. 'epoxy resin precondensates' (molecular weight 100–5000) by continuous reaction of diepoxide with bisphenol A monoglycidyl ether

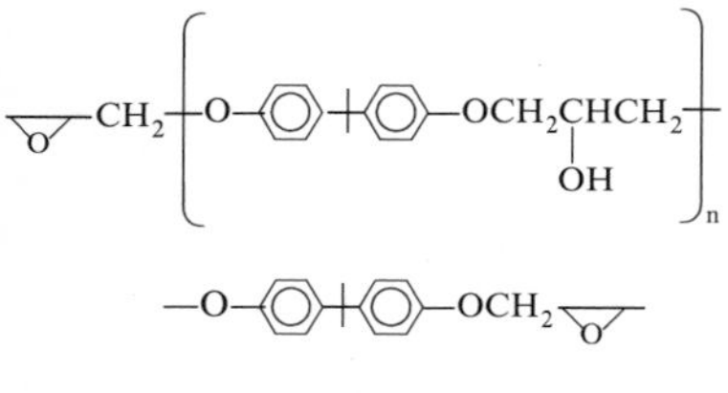

The production capacity for epichlorohydrin in 1993 in the USA, Western Europe, and Japan was 0.27, 0.24, and 0.14×10^6 tonnes per year, respectively. Production figures are listed in the adjacent table.

A newer technology for the manufacture of epichlorohydrin practiced by Laporte in a 5000 tonne-per-year plant avoids the considerable salt formation during the epoxidation of allyl chloride. This is done, like propene oxidation in the Interox process, by converting allyl chloride to epichlorohydrin with perpropionic acid from the reaction of H_2O_2 with propionic acid (cf. Section 11.1.1.2) at about $70-80°C$.

Showa Denko has been using an industrial process based on allyl alcohol in Japan since 1985. Chlorine is added to the double bond to give 2,3-dichloro-1-hydroxypropane which is then reacted with $Ca(OH)_2$ to give epichlorohydrin.

Epichlorohydrin is chiefly used to manufacture glycidyl ethers, the most significant of which are the bisphenol A glycidyl ethers. The reaction with bisphenol A is conducted in the presence of caustic soda at $100-150°C$. With an excess of epichlorohydrin the epoxide ring opens to form a 'chlorohydrin ether', which rapidly loses HCl by cleavage with NaOH:

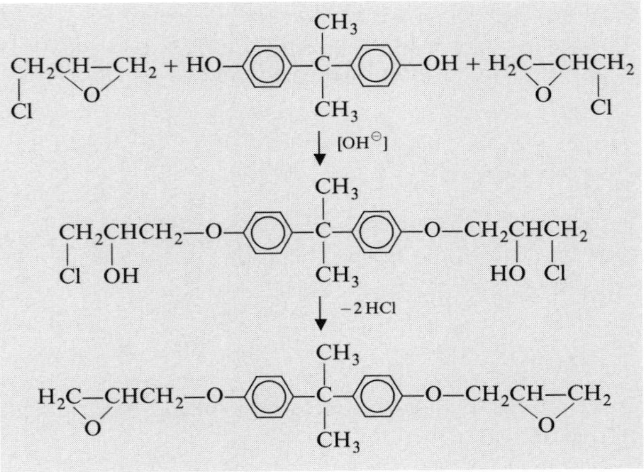

(41)

principle of epoxy resin formation:

addition of di- or polyamines or alcohols, or bisphenols, to glycidyl ethers with epoxide ring opening (curing)

The bisglycidyl ether thus obtained can continue to react with excess bisphenol A in a similar reaction to form higher molecular glycidyl ethers. The chain length depends on the mole ratio bisphenol A/epichlorohydrin. Reaction of glycidyl ethers with polyamines or polyols — usually base-catalyzed — leads, with partial crosslinking, to epoxy resins.

Many industrial materials (metal, wood, thermosetting plastics) can be glued more effectively with epoxy resins. In addition, epoxy resins are also used for surface protection, as binders, and as cast and impregnating resins in many arenas.

A third use of allyl chloride is its reaction with NH_3 to form allylamine:

$$H_2C=CHCH_2Cl + 2\,NH_3 \rightarrow H_2C=CHCH_2NH_2 + NH_4Cl \qquad (42)$$

Di- and triallylamine are formed as byproducts.

uses of epoxy resins:

adhesives
surface protectives
cast and impregnating resins

allylamine manufacture:

allyl chloride ammonolysis with two secondary products:

$(H_2C=CHCH_2)_2NH$
$(H_2C=CHCH_2)_3N$

11.2.2. Allyl Alcohol and Esters

Allyl alcohol can be manufactured from allyl chloride, propylene oxide, acrolein, or allyl acetate. The worldwide production capacity for allyl alcohol was about 50 000 tonnes per year in 1990.

The preferred processes have been those of Shell and Dow involving the alkaline hydrolysis of allyl chloride:

$$H_2C=CHCH_2Cl + NaOH \rightarrow H_2C=CHCH_2OH + NaCl \qquad (43)$$

This reaction takes place in a recycle reactor with $5-10\%$ caustic soda at $150\,°C$ and $13-14$ bar. When the conversion is quantitative, the allyl alcohol selectivity is about $85-95\%$. Byproducts include diallyl ether, propionaldehyde, and heavy ends. The complete loss of chlorine and the necessary corrosion-resistant plant equipment are obstacles to the further use of this route.

A second process for manufacturing allyl alcohol is the isomerization of propylene oxide:

$$H_3C-HC-CH_2 \xrightarrow{\text{[cat.]}} H_2C=CHCH_2OH \qquad (44)$$

Recently this reaction has been the subject of growing interest because of expanding propylene oxide capacities. The conversion can be conducted in either the liquid or gas phase. Li_3PO_4 is virtually the only catalyst used for this isomerization; BASF-Wyandotte uses Cr_2O_3.

FMC practices the Progil process in which Li_3PO_4 is finely suspended in a high boiling solvent (alkyl benzenes). Propylene oxide is then introduced at $275-280\,°C$ and up to 10 bar, and

allyl alcohol manufacture possible from four feedstocks:

1. $H_2C=CHCH_2Cl$
2. $H_2C-CHCH_3$ \; O
3. $H_2C=CHCHO$
4. $H_2C=CHCH_2OAc$

process principles of route 1:

alkaline saponification of allyl chloride under pressure with complete transfer of chlorine to NaCl

process principles of route 2:

catalytic propylene oxide isomerization in liquid phase using high-boiling solvent with suspended catalyst, or in gas phase with fixed-bed catalyst

about 60% is isomerized to allyl alcohol. The selectivity is about 92%. This process step was used by Arco for the production of 1,4-butanediol in a plant started up in 1991 (*cf.* Section 4.3).

In the gas-phase process, operated for example by Olin Mathieson or Hüls, the same reaction is carried out with a supported Li_3PO_4 catalyst at $250-350\,°C$. Higher conversions (70–75%) and slightly better selectivities (97%) are obtained. Acetone and propionaldehyde are formed as byproducts.

process principles of route 3:

selective acrolein hydrogenation with two variants:

1. heterogeneously catalyzed gas-phase hydrogenation (*e. g.*, Degussa or Celanese)
2. heterogeneously catalyzed H_2 transfer from isopropanol in gas phase (*e. g.*, Shell) or from 2-butanol

The third method is the selective gas-phase hydrogenation of the aldehyde group in acrolein. Using the Degussa process, allyl alcohol is obtained with approximately 70% selectivity over Cd—Zn catalysts. Instead of hydrogenating with H_2, hydrogen can be transferred from isopropanol:

$$H_2C{=}CHCHO + (CH_3)_2CHOH \xrightarrow{\text{[cat.]}}$$
$$H_2C{=}CHCH_2OH + (CH_3)_2C{=}O \tag{45}$$

The reaction, a type of Meerwein-Ponndorf reduction, is carried out using $MgO-ZnO$ catalysts at $400\,°C$ to give allyl alcohol with a selectivity of 80%. It is one step in a glycerol synthesis developed by Shell, which was practiced in a 23000 tonne-per-year plant in the USA for many years before being shut down for economic reasons in 1980.

Reduction of acrolein with 2-butanol is also common in industrial processes; the alcohol is simultaneously dehydrogenated to methyl ethyl ketone.

process principles of route 4:

two-step manufacture from propene, with allyl acetate intermediate and subsequent hydrolysis to allyl alcohol

In a fourth method, allyl acetate is hydrolyzed to allyl alcohol. The starting material allyl acetate can − in analogy to the vinyl acetate process − be manufactured from propene, acetic acid and O_2 over Pd catalysts:

$$H_2C{=}CHCH_3 + CH_3COOH + 0.5\,O_2 \xrightarrow{\text{[Pd cat.]}}$$
$$H_2C{=}CHCH_2OCCH_3 + H_2O \left(\Delta H = -\frac{45\,\text{kcal}}{188\,\text{kJ}}/\text{mol}\right) \tag{46}$$

manufacture of allyl acetate:

favored gas-phase acetoxylation of propene

Acetoxylation of propene, using the Hoechst or Bayer process, is done in the gas phase at $150-250\,°C$ under pressure on supported catalysts. The catalysts contain Pd metal or compounds, an alkaline metal acetate, and possibly other cocatalysts such as Fe or Bi compounds.

The selectivity to allyl acetate is over 90% for incomplete conversion of propene and acetic acid. CO_2 is virtually the only byproduct.

Allyl acetate can then be saponified to allyl alcohol in the presence of acidic catalysts such as mineral acids using known processes, or with acidic ion-exchangers using a Bayer process. Hydrolysis can also take place − using a Hoechst process − at 230 °C and 30 bar in the absence of catalyst. The recovered acetic acid can be reintroduced to propene acetoxylation resulting in the following net equation:

$$H_2C=CHCH_3 + 0.5\ O_2 \xrightarrow{\text{[cat.]}} H_2C=CHCH_2OH \qquad (47)$$

The net selectivity to allyl alcohol is 90% (based on C_3H_6). Presently, Showa Denko has a plant with a production capacity of 35000 tonnes per year.

Allyl alcohol is used primarily for the manufacture of glycerol (*cf.* Section 11.2.3). Glycidol, the intermediate in the glycerol synthesis, is also important for the synthesis of pharmaceuticals, cosmetics and emulsifying agents. In addition, esters of allyl alcohol (*e. g.*, with phthalic or maleic acid) are used in polymers.

A new use for allyl alcohol is its hydroformylation to 4-hydroxybutanal followed by hydrogenation to yield 1,4-butanediol. Daicel plans to build a plant using a further development of a Kuraray process, and Arco has begun operation of a plant using a similar process (*cf.* Section 4.3).

Allyl alcohol has also been used recently for the production of epichlorohydrin (*cf.* Section 11.2.1).

11.2.3. Glycerol from Allyl Precursors

Glycerol, the simplest trihydric alcohol, was originally merely a byproduct of fat saponification. However, when synthetic detergents began to dominate the market in the 1940s, particularly in the USA, the manufacture of soap from fatty acids was reduced, and the glycerol produced was no longer sufficient to meet the rising demand. During this period, the first synthetic processes for glycerol were developed.

---

with heterogeneous catalysts based on Pd metal or compounds

conversion of allyl acetate into allyl alcohol:
1. proton-catalyzed hydrolysis of allyl acetate in liquid phase (mineral acid or ion-exchanger as catalysts)
2. noncatalytic hydrolysis under pressure at higher temperatures

uses of allyl alcohol:
1. manufacture of glycerol *via* glycidol
2. precursor for allyl ester
3. hydroformylation and hydrogenation to 1,4-butanediol
4. chlorination and dehydrochlorination to epichlorohydrin

reasons for synthetic glycerol demand:

synthetic detergents displaced fat saponification, creating glycerol shortage

historical beginnings of synthesis of glycerol from petrochemicals:

discovery of propene 'hot chlorination'

glycerol production (% synthetic):

	1975	1980	1990	1994
W. Europe	38	29	17	13
USA	50	47	29	25
Japan	54	37	42	25

total glycerol production (in 1000 tonnes):

	1989	1992	1994
W. Europe	199	192	207
USA	133	159	159
Japan	52	43	42

present glycerol production uses two basic feedstocks:

1. $H_2C=CHCH_2Cl$ (preferred)
2. $H_2C=CHCH_2OH$

characteristics of allyl chloride route:

after manufacture of epichlorohydrin as intermediate, multistep alkaline saponification with total loss of chlorine via:

1. monochlorohydrin and
2. glycidol to
3. glycerol

basis of allyl alcohol route:

catalytic epoxidation and hydrolysis of allyl alcohol has two variations:

1. catalytic with H_2O_2 (*e. g.*, Shell or Degussa process); H_2O_2/WO_3 = Milas reagent

2. catalytic with peracetic acid (*e. g.*, FMC process, Daicel process)

However, glycerol production from petrochemicals has already passed its peak in several countries, as shown in the adjacent table. The total production in Western Europe, the USA, and Japan is given in a second table. (For hydrolysis or methanolysis of fatty acid triglycerides *cf.* Section 8.2).

The first synthetic processes, operated by IG Farben from 1943 and by Shell from 1948, were based on allyl chloride or its conversion product epichlorohydrin (*cf.* Section 11.2.1). Even today, the majority of glycerol producers (more than 80% of total glycerol capacity worldwide) use epichlorohydrin as the feedstock. It is hydrolyzed to glycerol stepwise with dilute caustic soda (10%) in a two-phase liquid reaction at elevated pressure and 100–200 °C:

$$
\begin{array}{c}
H_2C \\
\quad \diagdown O \\
HC \\
\quad | \\
H_2C-Cl
\end{array}
\xrightarrow[\text{+H}_2\text{O}]{}
\begin{array}{c}
H_2C-OH \\
| \\
HC-OH \\
| \\
H_2C-Cl
\end{array}
\xrightarrow[\text{-HCl}]{}
\begin{array}{c}
H_2C-OH \\
| \\
HC \\
\quad \diagdown O \\
H_2C \diagup
\end{array}
\xrightarrow[\text{+H}_2\text{O}]{}
\begin{array}{c}
H_2C-OH \\
| \\
HC-OH \\
| \\
H_2C-OH
\end{array}
\quad (48)
$$

The aqueous glycerol solution containing NaCl is concentrated, freed from salt, and then further concentrated to 98–99% glycerol by distillation.

This glycerin production process is in use in Western Europe, the USA, Japan, the CIS, and China.

This first commercial route involved total loss of chlorine as valueless $CaCl_2$ in the epichlorohydrin manufacture (*cf.* Section 11.2.1) and as NaCl in the subsequent saponification steps. Other glycerol processes starting with allyl alcohol were soon developed.

The allyl alcohol is hydroxylated *via* glycidol with H_2O_2 in the liquid phase at 60–70 °C in the presence of tungsten oxide or salts of tungstic acid (*e. g.*, $NaHWO_4$):

$$
H_2C=CHCH_2OH + H_2O_2 \xrightarrow[\text{[cat.]}]{-H_2O}
\begin{array}{c}
H_2C-CHCH_2OH \\
\quad \diagdown O \diagup
\end{array}
$$

$$
\xrightarrow[]{+H_2O} \begin{array}{c} HOCH_2CHCH_2OH \\ \quad\quad\quad | \\ \quad\quad\quad OH \end{array} \quad (49)
$$

The yield is about 90% based on allyl alcohol and H_2O_2. Degussa has practiced the first step, the manufacture of glycidol, in a commercial unit (capacity 3000 tonnes per year) since 1981. Instead of H_2O_2, peracetic acid in a high-boiling ketone or ester (*e. g.*, ethyl acetate) solvent can be used at 50–70 °C to epoxidize allyl alcohol. An FMC process starting with the isomerization of propylene oxide was used until 1982 to manufacture glycerol in a plant with a capacity of 18 000 tonnes per year. Daicel developed its own process which has been operating commercially since 1969. Peracetic acid serves not only in the epoxidation of propene to propylene oxide but also for the epoxidation of allyl alcohol to glycidol, which is then saponified to glycerol. Two-and-a-half times more acetic acid than glycerol is formed (*cf.* Section 7.4.1.1).

Olin Mathieson also starts with allyl alcohol produced by propylene oxide isomerization. However, HOCl is added in the next step and the resulting glycerol monochlorohydrin, $HOCH_2CH(OH)-CH_2Cl$, is then saponified to glycerol. This modification saves part of the chlorine compared to the allyl chloride route.

modified allyl alcohol route:

allyl alcohol manufacture from propylene oxide, then classical HOCl addition and saponification (*e. g.*, Olin Mathieson process)

Older processes for producing glycerol by controlled fermentation or a combination of hydrolysis and hydrogenolysis of sugar, molasses, or other carbohydrates are insignificant today (commercially operated by IG Farben from 1938, glycerogen = mixture of glycerol, propylene glycol, ethylene glycol).

The industrial uses of glycerol are determined both by its physical properties and by its chemical properties.

two aspects of glycerol usage:

1. physical properties (high boiling, highly viscous, hygroscopic) determine uses as moistening agent, antifreeze, slip additive and aid

As a high-boiling, viscous, hygroscopic substance, glycerol is used as an antifreeze, in the manufacture of pharmaceuticals and cosmetics, in tobacco moistening, and as an auxiliary agent for printing inks, inks, adhesive cements, etc.

As a triol, glycerol is used for the manufacture of alkyd resins, which are obtained by polycondensation with multibasic acids or, in the case of phthalic acid, with the anhydride (PAA). With approximately equimolar amounts of glycerine and, *e. g.*, phthalic anhydride, a linear, fusible resin is formed; with excess anhydride, this is then cross-linked by esterification of the secondary hydroxyl groups to form a nonfusible alkyd resin:

2. chemical properties (as triol) are important for reactions which require multifunctional groups, *e. g.*,
manufacture of alkyd resins, earlier known as glyptals (**gly**cerine + **phthal**ic acid) with primary OH giving linear polyesters, secondary OH crosslinking on setting or baking
manufacture of glycerol tripolyethers (by ethoxylation or propoxylation) and further reaction to polyether isocyanates

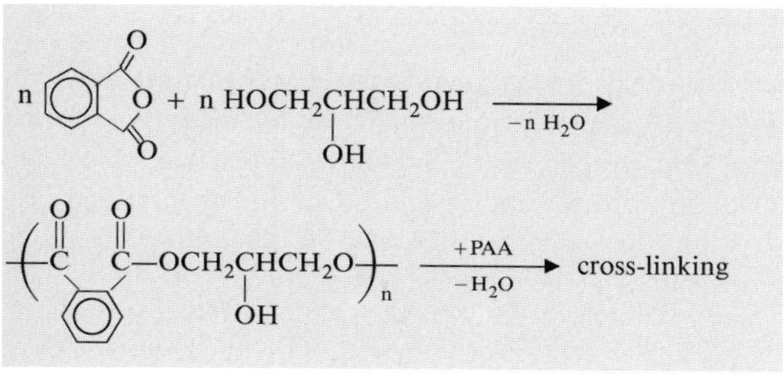

(50)

Alkyd resins are used as raw materials in the paint and varnish industry. Another product from glycerol are the glycerol tripolyethers. In their manufacture, glycerol is propoxylated with propylene oxide at 3–4 bar and 125 °C in the presence of KOH. After attaining a certain molecular weight it is generally reacted with a small amount of ethylene oxide. The resulting glycerol tripolyethers are widely used as surfactants; they are also starting materials for the reaction with diisocyanates to give polyurethanes.

Mono- and diesters of glycerol with fatty acids are used on a large scale in food production.

Today, only a small portion of the glycerol production (about 4%) is used for manufacturing glycerol trinitrate which, adsorbed on silica gel to give dynamite, was the first commercial derivative (1866) of glycerol.

11.3. Acrylonitrile

historical developments of industrial importance of acrylonitrile as monomer for homo- and copolymerization

1. buna N rubber resistant to gasoline and oil (IG Farben)
2. polyacrylonitrile fibers: Orlon (Du Pont)

Acrylic acid nitrile, usually referred to acrylonitrile (AN) today, became industrially important around 1930 in Germany and then in the USA. It was copolymerized with butadiene to form the synthetic rubber 'buna N'. Since that time, it has found numerous other applications as monomer, comonomer, and intermediate for fibers, resins, thermoplastics, and elastomers. This wide range of applications and the successful improvements in production techniques were the key reasons for the dramatic expansion in acrylonitrile production.

AN production (in 1000 tonnes):

	1990	1992	1994
USA	1374	1284	1373
W. Europe	1170	1045	1161
Japan	589	621	610

In 1996, the world production capacity for acrylonitrile was 4.3 $\times$ 10^6 tonnes per year. Of this, about 1.3, 1.4, and 0.62 $\times$ 10^6 tonnes per year were located in Western Europe, the USA, and Japan, respectively.

Production figures for acrylonitrile in these countries are summarized in the adjacent table.

AN manufacture:

1. in older processes by synthetic reactions with C_2 units + HCN
2. in more recent processes by propene ammoxidation

In the traditional process, acrylonitrile was manufactured by building the C_3 skeleton from smaller units. Today, it is chiefly produced from propene using Standard Oil of Ohio's (now BP) ammoxidation process and several other similar modified processes. With this, acrylonitrile has become second only to polypropylene as a chemical use for propene.

11.3.1. Traditional Acrylonitrile Manufacture

The older processes for manufacturing acrylonitrile used the relatively expensive C_2 building units ethylene oxide, acetylene, and acetaldehyde, which were reacted with HCN to form acrylonitrile or its precursors.

The first industrial production based on ethylene oxide was developed by IG Farben and operated by UCC from 1952 and by Cyanamid from 1970. Both plants have since been shut down. The process involved the base-catalyzed addition of HCN to ethylene oxide to form ethylene cyanohydrin, which was then dehydrated either in the liquid phase at 200 °C in the presence of alkali metal or alkaline earth salts, or in the gas phase at 250–300 °C over Al_2O_3:

traditional AN manufacture by three synthetic routes with HCN and C_2 feedstocks such as:

1. $H_2C\!-\!CH_2$
       $\diagdown\!O\!\diagup$

2. $HC\!\equiv\!CH$

3. CH_3CHO

process principles of ethylene oxide route:

two-step homogeneously catalyzed reaction to intermediate cyanohydrin followed by homogeneously or heterogeneously catalyzed dehydration

$$H_2C\!-\!CH_2 + HCN \xrightarrow[\text{[cat.]}]{} HOCH_2CH_2CN$$
$$\xrightarrow[\text{[cat.]}]{-H_2O} H_2C\!=\!CHCN \qquad (51)$$

Another industrial pathway developed by Bayer and practiced commercially by Cyanamid, Du Pont, Goodrich, Knapsack, and Monsanto involved the $CuCl\!-\!NH_4Cl$-catalyzed addition of HCN to acetylene at 80–90 °C:

process principles of acetylene route:

single-step homogeneously catalyzed hydrocyanation in liquid phase (Nieuwland system)

$$HC\!\equiv\!CH + HCN \xrightarrow[]{\text{[cat.]}} H_2C\!=\!CHCN \qquad (52)$$

In the 1950s and early 1960s, this was the preferred process for acrylonitrile in the USA and Western Europe. Since then, the route from acetylene has been abandoned.

A third process by Hoechst (Knapsack–Griesheim) has never been industrially significant. Acetaldehyde was treated with HCN in a base-catalyzed reaction to form the nitrile of lactic acid (lactonitrile), which was then dehydrated to acrylonitrile at 600–700 °C in the presence of H_3PO_4:

process principles of acetaldehyde route:

two-step reaction initially to acetaldehyde cyanohydrin (lactic acid nitrile) with subsequent catalytic dehydration

$$CH_3CHO + HCN \rightarrow \underset{\underset{OH}{|}}{CH_3CHCN} \xrightarrow[\text{[cat.]}]{-H_2O} H_2C\!=\!CHCN \qquad (53)$$

The first stage of this reaction was practiced by Sterling in the USA until 1996, and is still run by Musashino in Japan; however, the lactonitrile is used for the manufacture of lactic acid by hydrolysis at about 100 °C in the presence of H_2SO_4. Lactic acid

lactic acid nitrile presently used only as intermediate in lactic acid manufacture (DL-form)

$$\underset{\underset{OH}{|}}{CH_3CHCN} \xrightarrow[]{} \underset{\underset{OH}{|}}{CH_3CHCOOH}$$

is isolated and purified as its methyl ester. Currently, world production of lactic acid is estimated at about 50000 tonnes per year. About two-thirds of this comes from fermentation of products containing sugar or starch. Lactic acid is mainly used in the food industry as a flavoring acid (*cf.* Section 13.2.3.4), as a baking agent, and in emulsifiers as, *e. g.*, its glycerol ester. A new application for lactic acid is the production of biodegradable polymers.

AN manufacture based on propene, via nitrosation, insignificant today

Another process, no longer in use, provides the transition to the modern manufacturing routes to acrylonitrile from propene. Du Pont developed this process and operated it for a period in a pilot plant in the USA. In this process, propene was catalytically reacted with NO using either Ag_2O/SiO_2 or alkali metal oxides with thallium or lead compounds:

$$4\,H_2C{=}CHCH_3 + 6\,NO \xrightarrow{\text{[cat.]}} 4\,H_2C{=}CHCN + 6\,H_2O + N_2 \quad (54)$$

11.3.2. Ammoxidation of Propene

characteristic of ammoxidation:

activated CH_3 groups, *e. g.*, in:

$$H_2C{=}CHCH_3 \quad H_2C{=}\underset{\underset{CH_3}{|}}{C}{-}CH_3$$

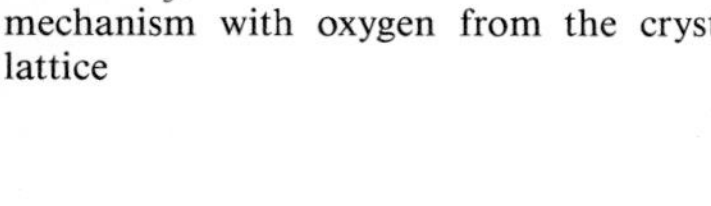

are converted into nitrile groups

Ammoxidation denotes the catalytic oxidative reaction of activated methyl groups with NH_3 to form a nitrile group. With propene, acrylonitrile is obtained:

$$H_2C{=}CHCH_3 + NH_3 + 1.5\,O_2 \xrightarrow{\text{[cat.]}}$$
$$H_2C{=}CHCN + 3\,H_2O \left(\Delta H = -\frac{120\ \text{kcal}}{502\ \text{kJ}}\big/\text{mol}\right) \quad (55)$$

mechanism of ammoxidation:

abstraction of H from chemiadsorbed C_3H_6 and NH_3, and H-oxidation in a redox mechanism with oxygen from the crystal lattice

The course of the reaction is construed as follows: Two H atoms each are eliminated stepwise from propene and NH_3 chemiadsorbed on the catalyst surface. The C_3H_4 and NH species formed react to give acrylonitrile:

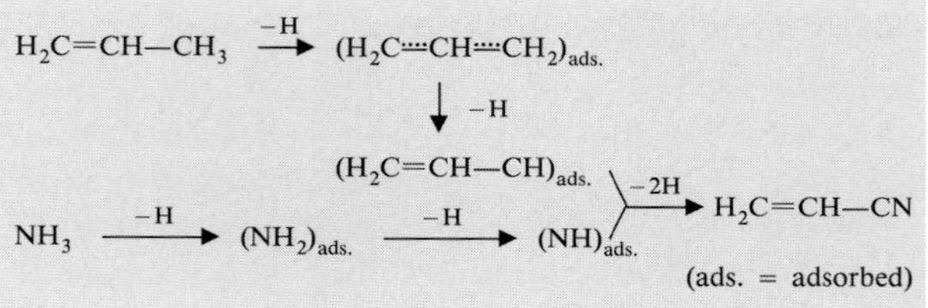

$$(56)$$

$$(\text{ads.} = \text{adsorbed})$$

The hydrogen is oxidized to water with oxygen from the crystal lattice, and the catalyst is reoxidized with oxygen from the gas phase.

Ammoxidation to give a nitrile group is only possible with those olefins having an activated methyl group that don't allow oxidative dehydrogenation as a preferred alternate reaction. For example, of the isomeric butenes only isobutene can be converted into methacrylic nitrile; the *n*-butenes are oxydehydrogenated to butadiene. Alkyl aromatics such as toluene (*e. g.*, in the Japan Catalytic process) and the isomeric xylenes can likewise be ammoxidized to the corresponding nitriles (*cf.* Section 14.2.3). Heterocyclics can also be ammoxidized; for example, β-picoline can be used to manufacture nicotinic acid nitrile, from which the economically important nicotinic acid amide (niacinamide; Vitamin B_5) is easily obtained (*cf.* Section 7.4.5).

In 1947, the principle of ammoxidation was described by Allied in a patent. However, economical selectivities were first possible in 1957 through the development of more effective catalysts by Distillers and, in particular, by Standard Oil of Ohio (Sohio). This led to start of the first commercial acrylonitrile unit in 1960.

1957: first Sohio patent on $Bi_2O_3 \cdot MoO_3$

1960: term 'ammoxidation' first used to describe oxidative amination; first commercial plant

11.3.2.1. Sohio Acrylonitrile Process

The Sohio process (now part of BP) is the most important of all the ammoxidation processes commercially; today most of the world production capacity for acrylonitrile is based on the Sohio process. Plant capacities can be as high as 180 000 tonnes per year.

process principles of Sohio AN manufacture:

heterogeneously catalyzed single-step gas-phase oxidation of propene in presence of NH_3, air, and H_2O using $Bi_2O_3 \cdot MoO_3$ catalysts in fluidized bed

The bismuth molybdate catalyst used originally was later complemented by an uranyl antimonate catalyst, making it possible to reduce the byproduct acetonitrile considerably. Further advances were made by using modified bismuth molybdate catalysts containing iron compounds (among others) to increase the selectivity. This modification with iron is based on research work conducted by Hoechst (Knapsack); these catalysts have been used commercially since 1972.

catalyst developments for Sohio process:

1. catalyst: $Bi_2O_3 \cdot MoO_3$
2. catalyst 21: $UO_2 \cdot Sb_2O_3$
3. catalyst 41: $Bi_2O_3 \cdot MoO_3$ with additives including Fe compounds
4. catalyst 49: not yet disclosed

In the industrial Sohio process, approximately stoichiometric amounts of propene and NH_3 are reacted with a slight excess of air and added H_2O in a fluidized bed at about 450 °C and 1.5 bar. Roughly 160 kcal or 760 kJ/mol are released (greater than theoretical amount) due to partial combustion of propene. This heat is removed from the fluidized-bed reactor by heat exchangers (perpendicularly arranged coiled pipes through which water is circulated) and used to generate superheated steam.

technological characteristics:

heat from exothermic main, side and secondary reactions used, through fluidized bed and heat exchanger, in steam generation

isolation of AN:

1. neutralization of small amounts of un-converted NH_3 to $(NH_4)_2SO_4$
2. separation of inerts by H_2O washes accompanied by resin formation from acrolein
3. multistep distillation of aqueous AN solution

uses of byproducts:

1. CH_3CN as solvent and intermediate; possibly as future raw material for AN
2. HCN for methacrylic acid, NaCN, methionine, oxamide, etc.

Processing of the reaction gases begins with a water wash, which removes all organic products from the inerts N_2 and propane (from the approx. $92-93\%$ propene). The residual NH_3 must be neutralized with H_2SO_4 to prevent base-catalyzed secondary reactions such as the addition of HCN (another byproduct of propene ammoxidation) to acrylonitrile. By choosing an appropriate pH value for the wash, the byproduct acrolein is converted into a resin.

Acrylonitrile is obtained in considerably greater than 99% purity (synthetic fiber grade) from the aqueous solution after a multistage distillation. The selectivity to acrylonitrile is over 70% (based on C_3H_6) with an almost complete propene conversion.

Per 1000 kg acrylonitrile, $30-40$ kg acetonitrile and $140-180$ kg HCN are formed. Acetonitrile is usually burned as a waste product; however, with additional equipment it can be isolated and purified (*e. g.*, Du Pont, BP). Acetonitrile is used as a selective solvent and intermediate. Its possible conversion into acrylonitrile by oxidative methylation has been investigated many times (*cf.* Section 11.3.3). HCN can be further converted in a variety of ways, *e. g.*, to acetone cyanohydrin and methacrylic acid (*cf.* Section 11.1.4.2), to methionine (*cf.* Section 11.1.6), to NaCN, and to oxamide (*cf.* Section 2.3.4).

11.3.2.2. Other Propene/Propane Ammoxidation Processes

AN manufacture by alternate ammoxidation processes:

1. BP (Distillers)-Ugine route:

   two-step propene reaction with interim isolation of acrolein not used; single-step BP route in use

2. Montedison route:

   single-step propene conversion in fluidized bed with Te-, Ce-, Mo-oxides used commercially following further development by UOP

Besides the Sohio process, there are a number of modified ammoxidation processes. In the BP (Distillers)-Ugine process, propene is oxidized on a Se/CuO catalyst to acrolein, which is then converted into acrylonitrile in a second stage with NH_3 and air over a MoO_3 fixed-bed catalyst. This two-step conversion leads to higher acrylonitrile selectivity of about 90% (based on $H_2C=CHCHO$). This process has, however, not been important commercially; another single-step version developed by BP is in use in several plants.

In the Montedison process, propene is ammoxidized in a fluidized-bed reactor at $420-460\,°C$ over a catalyst consisting of Te-, Ce- and Mo-oxides on SiO_2. One thousand kilograms of acrylonitrile, 50 kg HCN, 25 kg acetonitrile, and 425 kg $(NH_4)_2SO_4$ are obtained from 1200 kg propene and 560 kg NH_3; the acrylonitrile yield is thus 66% (based on propene). Further development of the Montedison process by UOP with respect to catalyst and reprocessing steps improved the acrylonitrile yield to over 80% at a propene conversion of about 95%. Moreover, the NH_3 feed could be reduced, so less $(NH_4)_2SO_4$ byproduct is formed.

Several plants practicing the improved technology are in use.

A third process was developed and used commercially by Snamprogetti/Anic. Using a fixed-bed catalyst based on Mo/V or Bi, 1260 kg propene are converted into 1000 kg acrylonitrile, 240 kg HCN, and 25 kg acetonitrile at 440–470 °C and 2 bar.

3. Snamprogetti/Anic route:

single-step propene conversion in fixed-bed reactor with Mo-, V- or Mo-, Bi-catalyst

Other commercially practiced processes come from Ugine Kuhlmann and the Austrian nitrogen producers. Both processes use fixed-bed reactors, and the heat of reaction is removed with a salt bath and used for steam generation.

Like propene, propane should also be suitable for ammoxidation. Monsanto, Power Gas, BP, and ICI have developed processes based on propane. At the required higher temperatures of 485–520 °C, propane is initially dehydrogenated to propene.

AN manufacture with other feedstocks for the ammoxidation:

propane, with developments by, *e. g.*, Monsanto, Power Gas, ICI

The selectivities are 30%, markedly lower than with direct use of propene.

Lummus has also developed an acrylonitrile manufacturing process based on propane or propene, NH_3 and O_2 in a salt melt of, for example, $KCl-CuCl-CuCl_2$. This process has not been practiced commercially but a demonstration plant using a fluidized-bed reactor for the one-step process is in development by BP.

propane or propene/propane mixtures used with salt melts in Lummus development

BP pilot plant currently being built

11.3.3. Possibilities for Development of Acrylonitrile Manufacture

The global activity in building C_1 chemistry, that is, the use of synthesis gas or methane as the simplest basis products which can be produced from all fossil raw materials or from biomass, has also pointed to new routes to acrylonitrile.

C_1 chemistry offers new, two-step route to acrylonitrile:

1. CO/H_2 reaction with NH_3 to give acetonitrile
2. oxidative methylation of acetonitrile to acrylonitrile

For example, Monsanto has developed a process for the manufacture of acetonitrile from synthesis gas and NH_3 in which an 85% selectivity to acetonitrile is reached at 350–600 °C and pressures up to 35 bar over Mo- or Fe-oxide-based catalysts with promoters such as Mn, Sr, Ba, Ca, or alkali metal compounds:

$$NH_3 + 2\,CO + 2\,H_2 \xrightarrow{\text{[cat.]}} CH_3CN + 2\,H_2O \qquad (57)$$

Acetonitrile can then be converted to acrylonitrile in an oxidative methylation with CH_4 according to the simplified equation below. A selectivity to acrylonitrile of up to 70% with a conver-

additional advantages of the C_1 route to acrylonitrile:

use of the byproduct acetonitrile from the Sohio process

sion of 45% can be obtained over catalysts based on, *e. g.*, alkali or alkaline earth halides and oxides of Bi, Mo, or Zn on supports in the presence of water at temperatures over 750 °C:

$$CH_3CN + CH_4 + O_2 \xrightarrow{\text{[cat.]}} H_2C{=}CHCN + 2\,H_2O \qquad (58)$$

The acetonitrile byproduct from the ammoxidation of propene to acrylonitrile could also be used here, directly in the aqueous solution, which would contribute to the overall selectivity of this process.

11.3.4. Uses and Secondary Products of Acrylonitrile

uses of AN:

1. as reactive monomer (CN conjugated double bond) for the manufacture of homo- and copolymers, *e. g.*, fibers, plastics, and synthetic rubber (ABS = acrylonitrile-butadiene-styrene; SAN = styrene-acrylonitrile)

The main use for acrylonitrile is as a monomer for the manufacture of homo- and copolymers in synthetic fibers, rubbers, and plastics.

Comparative percentages of acrylonitrile use in several countries, predominantly for polymers, is shown in the following table:

Table 11-6. Uses of acrylonitrile (in %).

Product	World		USA		Western Europe		Japan	
	1984	1991	1984	1991	1985	1992	1984	1990
Acrylic fibers	73	59	52	34	73	66	65	52
ABS and SAN plastics	12	20	21	26	11	15	15	26
Nitrile rubber	3	4	3	3	3	4	4	4
Adiponitrile	6	9	13	25	8	4	4	6
Remaining uses, including acrylamide	6	8	11	12	5	12	12	12
Total use (in 10^6 tonnes)	3.15	3.85	0.60	0.60	1.13	1.20	0.55	0.65

2. as intermediate for:

    electrohydrodimerization to adiponitrile
    cyanoethylation of compounds with labile H atom
    partial hydrolysis to acrylamide
    complete hydrolysis to acrylic acid and alcoholysis to esters (*cf.* Section 11.1.7.1)

example of commercial cyanoethylation:

pharmaceuticals and dyes

Under the heading 'remaining uses', the use of acrylonitrile as an intermediate is seen as having a relatively great potential for growth. For example, electrohydrodimerization leads to adiponitrile (*cf.* Section 10.2.1.1). Furthermore, with its activated double bond, acrylonitrile is a versatile reaction component. Compounds with labile hydrogen atoms such as alcohols, amines, amides, aldehydes, and ketones can be added to the double bond, usually with base catalysis. The reaction, known as cyanoethylation, is important in the pharmaceutical and dye industries.

Acrylonitrile is also used to modify starch and cellulose. The resulting cyanoethylated products are, however, of only minor importance.

On the other hand, partial hydrolysis of acrylonitrile to acrylamide is already operated on a larger industrial scale:

three processes for hydrolysis of acrylonitrile to acrylamide:

$$H_2C{=}CHCN + H_2O + H_2SO_4 \rightarrow H_2C{=}CHCONH_2 \cdot H_2SO_4$$
$$\xrightarrow{+2NH_3} H_2C{=}CHCONH_2 + (NH_4)_2SO_4 \quad (59)$$

In the classical process, the hydrolysis is conducted with stoichiometric amounts of H_2SO_4 and additionally, because of the ready polymerizability of acrylonitrile and acrylamide, with polymerization inhibitors. The first step is formation of acrylamide sulfate, which is then reacted with NH_3 to free acrylamide. Ammonium sulfate, the coproduct, and acrylamide are separated from one another in several costly crystallization steps. The selectivity to acrylamide is about 80%.

1. H_2SO_4 process through acrylamide sulfate, *e. g.*, by Mitsui-Toatsu process with costly separation of coproduct $(NH_4)_2SO_4$

Recent developments by Cyanamid, Dow, Mitsui Toatsu, and Mitsubishi Chemical focus on the heterogeneously catalyzed, acid-free partial hydrolysis of acrylonitrile to acrylamide. The advantage is that, after filtering off the catalyst and distilling over the unreacted acrylonitrile, a virtually pure aqueous acrylamide solution is obtained which can be used either directly or after concentration. The catalyst used in the Mitsui-Toatsu process is metallic copper (*e. g.*, Raney copper) at $80-120\,°C$. At an acrylonitrile conversion of $60-80\%$, a selectivity to acrylamide of 96% can be attained. A 5000 tonne-per-year plant using this technology was built in 1973, and expanded several years later. Cyanamide also started operation of a 27000 tonne-per-year plant using a copper metal catalyst at the end of the 1970s.

2. catalytic hydrolysis over Cu-containing catalysts used in numerous processes, advantage — simpler workup compared to H_2SO_4 process

Dow has developed a modified copper chromite catalyst which, after a certain operating period, must be treated oxidatively and then reactivated with H_2.

Mitsubishi Chemical discontinued its classical process and started operation of a new plant using the catalytic process in 1975. The original capacity of 25000 tonnes per year has since been expanded to 45000 tonnes per year (now Mitsubishi Kasei).

Future paths to acrylamide will also use biotechnological hydrolysis processes for acrylonitrile. In 1985 Nitto Chemical Industry began operation of a 4000 tonne-per-year plant (since

3. biocatalytic hydrolysis on supported enzymes, i.e., with nitrile hydratase

increased to 20000 tonnes per year) in which a dilute aqueous solution of acrylonitrile (2–5 wt%) is hydrolyzed with a nearly 100% yield to acrylamide using an enzyme fixed to a modified polyacrylamide at 20 °C and atmospheric pressure.

Additional enzyme-catalyzed regioselective hydrolysis processes for the conversion of acrylonitrile to acrylamide are being developed in Japanese universities.

uses of acrylamide:

monomer for manufacture of polymers for use as flocculating agents and as constituent of dispersions, resins, and paints

as methylol derivative for spontaneously cross linking polymers

Acrylamide is used as a monomer for polyacrylamide, which is increasingly important as a flocculating agent in water purification, ore flotation, and paper processing, and for the manufacture of polymers for dispersions, resins, and paints. Polyacrylamide is also being used increasingly in tertiary oil recovery. In the form of methylol acrylamide compounds, it is also a cross linking agent in polymers.

12. Aromatics – Production and Conversion

12.1. Importance of Aromatics

The key industrial products of the extensive and varied chemistry of aromatics are benzene, toluene, ethylbenzene, and the *o*-, *m*-, and *p*-isomers of xylene.

The condensed aromatics such as naphthalene and anthracene, however, are of limited significance. In the USA, Western Europe, Japan, and the CIS, the following aromatics were produced in recent years:

order of importance of high-volume aromatics:

benzene, toluene, and xylenes – the BTX fractions in aromatic petrochemistry – are easily the largest-volume products, followed by ethyl benzene

naphthalene and anthracene are of only limited importance

Table 12-1. Production of aromatics (in 10^6 tonnes).

Product	USA		Western Europe		Japan		CIS	
	1985	1995	1985	1995	1985	1994	1984	1995
Benzene	4.49	7.24	4.95	7.51	2.08	3.62	1.96	2.42
Toluene	2.26	3.05	2.43	3.05	0.80	1.22	0.40	0.53
o-/*m*-/*p*-Xylene	2.40	3.34	2.03	2.41	1.52	3.63	0.85	0.72
Ethylbenzene	3.88	6.19	3.25	4.89	1.65	2.59	0.62	1.30
Naphthalene (pure and crude; *cf.* Section 12.2.3.1)	0.14	0.09	0.20	0.24	0.13	0.19	0.13	0.09

Aromatics are among the most widespread and important chemical raw materials; they are a significant portion of all plastic, synthetic rubber, and synthetic fiber manufacture.

Although it is possible in principle to synthesize aromatics from simple aliphatic units, *e. g.*, benzene from acetylene or xylenes from diisobutene (2,4,4-trimethylpentene), so far these types of processes have not been used commercially.

principal processes for production of aromatics:

1. specific synthesis of aromatics possible from smaller units, but uneconomical

Aromatics are obtained almost exclusively from the fossil fuels, coal and oil. However, it is not economically feasible to isolate the limited quantities of aromatics present in unprocessed coal or oil.

2. direct isolation of aromatics from coal or paraffin–aromatic crude petroleum insignificant

Suitable feedstocks for economical production of aromatics result from thermal or catalytic conversion processes in coke plants and refineries.

3. today production of aromatics usually from coal or oil by thermal or catalytic processes followed by isolation

12.2. Sources of Feedstocks for Aromatics

feedstock sources for production of aromatics from coal or oil:

1. products of dry distillation of hard coal (coking)
2. reformate gasoline from gasoline conversion processes
3. pyrolysis gasoline from steam reforming of naphtha

coal tar, coupled product from coke manufacture, is receding as source of feedstock due to:

1. more efficient use of coke in iron smelting
2. change from town gas to cracked gas and natural gas
3. coke replaced by fuel oil in iron smelting in the future

production of tar from hard coal (in 10^6 tonnes):

	1975	1980	1990	1992
W. Europe	4.0	3.2	2.4	1.7
Japan	2.5	2.6	2.1	1.9
USA	2.9	2.4	1.4	1.1
world	16.0	15.0	~15	15−16

benzene production from coal tar (in %):

	1975	1980	1994
W. Europe	23	10	6
Japan	13	10	4
USA	6	6	2

BTX raw material sources (in wt %)

	Oil		Coal	
	1986	1990	1986	1990
USA	97	98	3	2
W. Europe	96	95	4	5
Japan	90	93	10	7

There are essentially three sources of feedstocks available for the commercial isolation of aromatics:

1. Products from coking of hard coal

2. Reformate gasoline from processing of crude gasoline

3. Pyrolysis gasoline from ethylene/propene production

The traditionally most important sources of aromatics − coal tar, coke water, and coke-oven gas − from the coking of hard coal are becoming less and less significant. The demand for coke, which is the main product (nearly 80%) from the coking of hard coal, has dropped due not only to advances in blast furnace smelting technology, but also to the conversion of the gas supply from municipal coke works to cracked gas from petroleum and natural gas. The use of heavy fuel oil in steel production is of minor significance today.

Along with the drop in coke production, the amount of tar produced has also decreased; in Western Europe, for example, from 6.5×10^6 tonnes per year in 1960 to 1.7×10^6 tonnes per year in 1992. The increase of hard coal coking in developing and former Eastern Bloc countries has kept global tar production at $15-16 \times 10^6$ tonnes.

This picture could change if the many projects for coal gasification and low-temperature coking are used commercially; the low-temperature tar coproduct would become an important raw material source as amounts increased.

Today, oil is used to meet the rapidly increasing demand for aromatics in most industrialized countries. In 1960, few aromatics were derived from petroleum in Western Europe; in the USA, 83% were already isolated from this source. After 1990, the largest industrialized countries obtained only a small amount of their benzene from coal tar, as shown in the adjacent table. Toluene and xylenes were obtained almost exclusively from petroleum.

The development and the current situation for BTX raw material sources are given in the adjacent table.

The production of BTX aromatics in 1995 in the USA, Western Europe, and Japan was about 13.6, 13.0, and 9.5×10^6 tonnes, respectively.

12.2.1. Aromatics from Coking of Hard Coal

Hard coal contains aromatic hydrocarbons in the form of high molecular weight compounds (average molecular weight about 3000 and according to new measurements up to 500000) which are cracked, or depolymerized, and rearranged by the coking process at 1000–1400 °C (high-temperature coking). The low molecular aromatics which are formed in the three coking products – crude gas, coke-oven water, and coke tar – can be isolated using special techniques. The residue, coke, is essentially pure carbon. Coking can therefore also be seen as a dismutation of H-rich and H-poor products.

principle of coal coking:

high temperature pyrolysis of mainly aromatic high molecular substances leads to formation of:

1. crude gas or coke-oven gas
2. coke-oven water
3. coke-oven tar or coal tar
4. coke

The aromatics are removed from the crude gas or coke-oven gas either by washing with higher boiling hydrocarbons (anthracene oil) or by adsorption on active carbon. The aromatic mixture is then distilled from the wash liquid or desorbed from the carbon with steam. Unsaturated hydrocarbons and N- and S-containing compounds are then removed either by refining with sulfuric acid or, as is preferred today, in a hydrogenative catalytic reaction. Distillation gives a product known as 'crude benzene' or 'coke-oven benzene' with a typical composition shown in the following table:

to 1:

BTX aromatic extraction from coke-oven gas with gas washes and purification by H_2SO_4 refining or hydrogenation to remove *e. g.*, thiophene or methylcyclopentadiene:

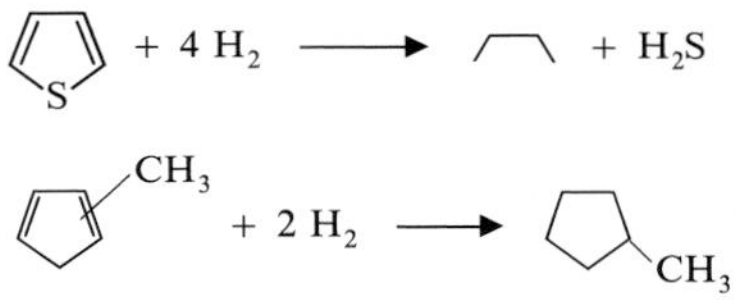

Table 12–2. Typical composition of coke-oven gas extract (crude benzene).

Product	vol%
Light ends	2
Benzene	65
Toluene	18
Xylene	6
Ethylbenzene	2
Higher aromatics	7

Most of the naphthalene is present in the residue.

The coke-oven water contains about 0.3 wt% of a mixture of phenol and its homologues which is isolated from the aqueous phase by extraction, *e. g.*, with benzene or butyl acetate (Phenosolvan process). It consists mainly of phenol (52 wt%), cresols, xylenols, and higher phenols (*cf.* Section 13.2.1).

to 2:

phenol/cresol/xylenol isolation from coke-oven water by solvent extraction

to 3:

isolation of several aromatic parent substances from coke-oven tar by distillative preseparation with subsequent pure isolation

Coal tar contains other important aromatic parent substances which can be separated by fractional distillation into the following product mixtures:

Table 12-3. Typical composition of coal tar distillate.

B. P. Limit (°C)	Fraction	Wt%	Main constituents
180	Light oil	up to 3	BTX-aromatics, pyridine bases
210	Carbolic oil	up to 3	Phenols
230	Naphthalene oil	10−12	Naphthalene
290	Wash oil	7−8	Methylnaphthalenes, acenaphthene
400	Anthracene oil	20−28	Anthracene, phenanthrene, carbazole
>400	Pitch	50−55	

The further processing of individual fractions to enrich or recover important components, *e. g.*, naphthalene and anthracene, is done by crystallization, extraction, or other separation processes (*cf.* Sections 12.2.3.1 and 12.2.3.2).

12.2.2. Aromatics from Reformate and Pyrolysis Gasoline

aromatic-rich refinery products from secondary processes:

reformate gasoline
pyrolysis gasoline

In the process of refining oil, fractions rich in aromatics are obtained from refining (reforming) of gasoline and from cracking processes in olefin manufacture. This reformate gasoline and pyrolysis or cracked gasoline represent valuable sources of aromatics.

principle of reforming process:

combination of isomerizations, aromatizations (dehydrogenations), and cyclizations with bifunctional catalysts, *e. g.*,

$Pt/Al_2O_3 \cdot SiO_2$ or $Pt-Re/Al_2O_3 \cdot SiO_2$ as well as recently introduced multimetal cluster catalysts

Platforming process:

UOP process currently used in 73 plants

Rheniforming process:

Chevron process currently used in more than 70 plants

Reformate gasoline is formed both from paraffinic crude oils, which comprise more than 50% branched and unbranched alkanes, and from naphthalenic oils, which are mainly cycloalkanes. Distillation of both types of crude oil gives low-octane fractions which must be reformed before being used as gasoline. This reforming process is a special type of catalytic modification using bifunctional catalysts containing an acidic component such as $Al_2O_3 \cdot SiO_2$ and a hydrogenation-dehydrogenation component, *e. g.*, platinum in the Platforming process or the bimetallic system platinum/rhenium in the Rheniforming process (Chevron catalyst). The addition of rhenium increases catalyst stability by preventing sintering processes and thus maintaining the metal dispersivity. In a more recent development made by Exxon, multimetal cluster catalysts of undisclosed composition exhibit substantially higher activity. They are already being used commercially.

At reforming conditions of 450−550 °C and 15−70 bar, alkyl-cyclopentanes, for example, are isomerized to substituted cyclohexanes, which are then aromatized by the dehydrogenative component of the catalyst. Ring-forming dehydrogenations of alkanes to cycloalkanes and isomerizations of *n*-alkanes to iso-alkanes also take place.

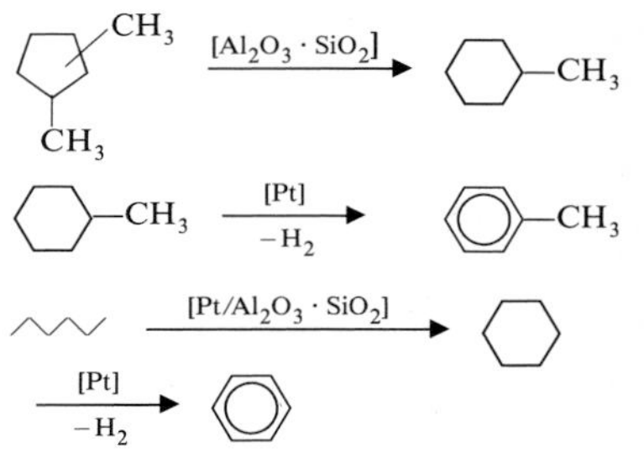

Distillation of the reformed crude product leads to a fraction rich in aromatics, which, due to the level of higher boiling aromatics, is especially suitable for the production of toluene and xylene isomers.

Pyrolysis gasoline comes from the steam cracking of naphtha for the production of ethylene, propene, and higher olefins. Its relatively high benzene content makes it one of the leading feedstocks for benzene.

While reformate gasoline can be used directly for the production of aromatics, cracked gasoline and crude benzene from coke plants must first be freed from polymerizable mono- and diolefins by hydrogenation. S, N, and O compounds are also removed by this hydrogenation, which is usually done in two stages. A typical commercial process is the IFP process, which uses Pd and Ni−W catalysts. Other processes use $CoO-MoO_3/Al_2O_3$ catalysts at 300-400 °C and 20-40 bar.

If one neglects the effect on the aromatic distribution arising from different reforming or cracking conditions (two-step) and the boiling point ranges of the naphtha cuts, a simplified picture of the distribution is obtained:

examples of reactions:

principles of steam cracking process:

combination of isomerizations, aromatizations, and cyclizations, catalyst-free in presence of H_2O, today with mainly naphtha as feedstock

crude product pretreatment (refining) to isolate BTX:

with reformate: unnecessary

with cracked gasoline: hydrogenative treatment required to remove double bonds and S-, N-, O-containing substances

aromatic content and distribution determined by various factors, *e. g.*, reforming and cracking severity, type of naphtha fraction

Table 12-4. Typical composition of reformate and pyrolysis gasolines (in wt %).

Product	Reformate gasoline	Pyrolysis gasoline
Benzene	3	40
Toluene	13	20
Xylene	18	4−5
Ethylbenzene	5	2−3
Higher aromatics	16	3
Non-aromatics	45	28−31

Products of coal coking, reformate and pyrolysis gasoline are three feedstocks for aromatics which, with their varying availability and composition from country to country, contribute differently to the total supply of individual aromatics. The origins of benzene are very different in the USA, Western Europe, and Japan, for example:

the supply of individual aromatics depends like no other major chemical complex on availability and composition of several basic products

Table 12–5. Share of various sources in benzene production (in wt%).

Product	USA		Western Europe		Japan	
	1985	1995	1986	1994	1984	1994
Pyrolysis gasoline	20	17	59	52	59	34
Reformate gasoline	47	43	15	17	23	54
Hydrodealkylation	28	26	18	24	9	–
(*cf.* Section 12.3.1)						
Coal and other sources[1]	5	14	8	7	9	12

[1] e.g., disproportionation of toluene.

principal feedstocks:

	USA	W. Europe, Japan
benzene	reformate gasoline	pyrolysis gasoline
toluene/ xylene	reformate gasoline	

other constituents of reformate and cracked gasoline, *e. g.*, oligomethylbenzenes such as:

pseudocumene   durene

as important precursors for:

'tri'   'pyro-'

mellitic anhydride

manufacture of benzenecarboxylic acids:

catalyzed oxidation of methylbenzenes, heterogeneous in gas phase or homogeneous in liquid phase

commercial oxidation processes for methylbenzenes, *e. g.*, Bergwerksverband, Bofors Nobel Kemi, Amoco

alternate three-step manufacture of trimellitic anhydride:

1. carbonylation of *m*-xylene
2. oxidation of methyls, aldehyde
3. dehydration

From this it is clear that the preferred manufacture of ethylene in Western Europe and Japan through naphtha cracking, which produces pyrolysis gasoline as a coproduct, also provides the largest source of benzene. This is in contrast to reformate gasoline in the USA. Worldwide, toluene and xylene isomers are mainly produced from reformate gasoline. Today, coking processes are an insignificant source for both of these products.

Recently, industrial use of higher alkylated benzenes, in particular the methylbenzenes such as 1,2,4-trimethylbenzene (pseudocumene) and 1,2,4,5-tetramethylbenzene (durene) has been increasing. Both products are inexpensive and can be obtained in sufficient quantities both by isolation from higher boiling fractions of reformate or pyrolysis gasoline and by chemical processing of coal.

The importance of these products lies mainly in their oxidation to trimellitic anhydride, the anhydride of 1,2,4-benzenetricarboxylic acid, and to pyromellitic dianhydride, the dianhydride of 1,2,4,5-benzenetetracarboxylic acid.

Oxidation of the methylbenzenes can be done either with oxygen or air in the gas phase at $350-550\,°C$ over modified V_2O_5 catalysts, or in the liquid phase.

Dilute nitric acid (ca. 7%) at $170-190\,°C$ (Bergwerksverband and Bofors process) or air with Mn catalysts and bromine in acetic acid (Amoco process, *cf.* Section 14.2.3) can be used as oxidation agents.

Increasing interest in trimellitic anhydride has led to the development of other manufacturing routes. In a process from Mitsubishi Gas Chemical and Amoco, this anhydride is obtained from an $HF \cdot BF_3$-catalyzed carbonylation of *m*-xylene to give 2,4-dimethylbenzaldehyde, which is oxidized in the presence of aqueous $MnBr_2/HBr$ and then dehydrated:

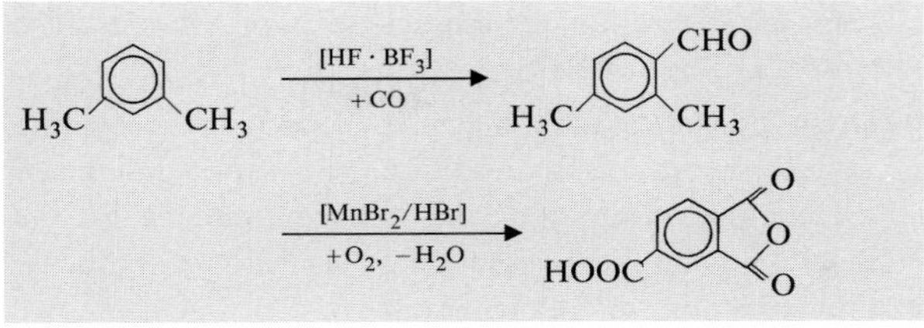

(1)

The yield of the carbonylation step is over 96% (based on *m*-xylene); of the oxidation step, about 91%. Commercial use of this process is planned.

The global capacity for production of trimellitic anhydride was about 50 000 tonnes per year in 1985; capacity for pyromellitic dianhydride in 1995 was about 8000 tonnes per year. Due to work carried out by Du Pont, both anhydrides have become important components for the manufacture of extremely heat resistant polyimides (*e. g.*, 'Kapton film' from pyromellitic dianhydride and 4,4-diaminodiphenyl ether).

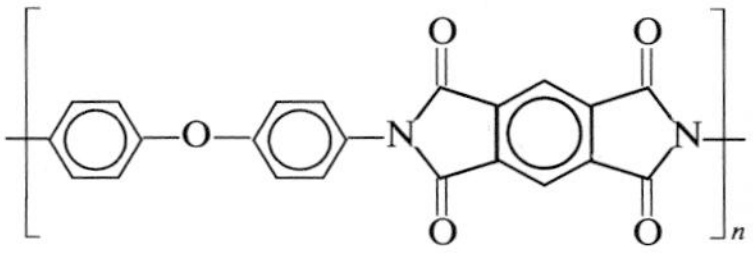

However, most trimellitic acid (60–80%) is used for the manufacture of esters for plasticizers, adhesives, and printing inks.

uses of benzenecarboxylic acids:

components for polyamides, polyimides, esters for plasticizers, adhesives, printing inks

example of commercial polyimide:

12.2.2.1. Isolation of Aromatics

The isolation of aromatics from reformate and pyrolysis gasolines consists essentially of stages for the separation of the non-aromatics followed by separation of the aromatic mixture into its individual components.

general principles for isolation of aromatics:

1. removal of non-aromatics
2. separation into individual components

For technological and economic reasons, fractional distillation is rarely appropriate for either of these separations. For example, cyclohexane, *n*-heptane, and other alkanes produce azeotropes with benzene or toluene which are not separable by distillation. In addition, the minor differences in boiling points between, for example, the xylene isomers and the attendant ethylbenzene demand an extremely involved distillation which cannot be economically justified. Thus distillations are usually limited to the production of mixed cuts.

fractional distillation to isolate aromatics limited to special cases for following reasons:

1. azeotrope formation of non-aromatics with aromatics hinders separation
2. minor differences in boiling points between C_8 components:

	b. p. °C
ethylbenzene	136.2
p-xylene	138.3
m-xylene	139.1
o-xylene	144.4

for 98% *o*-xylene, a column with 120–150 theoretical plates and high reflux ratio of 5–8 to 1 is required

The separation of aromatic/non-aromatic mixtures therefore is carried out by means of very special separation processes due to the varying separation problems of the differing feedstock mixtures and the increasingly higher demands on purity of the individual components.

development of special separation processes for aromatic fractions of varying content necessary

12.2.2.2. Special Separation Techniques for Non-Aromatic/ Aromatic and Aromatic Mixtures

Five different processes have been developed for recovering aromatics from hydrocarbon mixtures. The separation problems, the techniques employed, and the requirements for the technological or economical operation of the individual processes are summarized in the following table:

Table 12-6. Processes for aromatic recovery.

Process	Separation problem	Requirements for basic or economical operation
1. Azeotropic distillation	BTX separation from pyrolysis gasoline	High aromatic content ($>90\%$)
2. Extractive distillation	BTX separation from pyrolysis gasoline	Medium aromatic content ($65-90\%$)
3. Liquid–liquid extraction	BTX separation from reformate gasoline	Lower aromatic content ($20-65\%$)
4. Crystallization by freezing	Isolation of p-xylene from m/p-mixtures	Distillate preseparation of o-xylene and ethylbenzene from C_8 aromatic fractions
5. Adsorption on solids	Isolation of p-xylene from C_8 aromatic fractions	Continuous, reversible, and selective adsorption

principles of aromatic/non-aromatic separation by azeotropic distillation:

strongly polar auxiliary substances, such as CH_3COCH_3 or CH_3OH, increase volatility of non-aromatics (azeotrope formation) and distill with them overhead, aromatics remain in bottoms

To 1:

In *azeotropic distillation*, the addition of strongly polar auxiliary agents (amines, alcohols, ketones, H_2O) facilitates the removal of alkanes and cycloalkanes as lower boiling azeotropes.

The use of azeotropic distillation assumes that the fractions are narrow cuts, from which non-aromatics are removed with added acetone in the case of the benzene fraction, or with methanol for the toluene or xylene fraction. Acetone or methanol is then extracted from the distillate with water, recovered by distillation, and recycled to the azeotrope column.

advantage of azeotropic distillation:

when aromatic content is high, bulk of material needs not be volatilized in separation stage (energetically favorable)

Azeotropic distillation is economical when the aromatic content is greater than 90%; *i.e.*, only small amounts of non-aromatics are to be separated, such as with pyrolysis gasoline or crude benzene from coal coking (*cf.* Section 12.2.1).

principles of aromatic/non-aromatic separation by extractive distillation:

selective aromatic solvents increase boiling points of aromatics and remain with them in bottoms, non-aromatics obtained overhead

To 2:

In *extractive distillation*, an additive is also used to increase the differences in boiling points between the aromatics and the non-aromatics. In this case, however, the auxiliary substance acts on the aromatic fraction to decrease its volatility.

Thermally stable, noncorrosive substances such as di- and tri-
chlorobenzene, benzyl alcohol, polyglycols, phenols, amines,
nitriles, N-methylpyrrolidone (NMP), and sulfolane are all
suitable selective solvents for aromatics. Just as in the azeotropic
distillation, single, narrow cut fractions (benzene, toluene cuts)
are used for the extractive distillation; highly selective solvents
(e. g., N-methylpyrrolidone), however, allow the simultaneous
recovery of benzene and toluene.

Commercial extractive distillations utilize N-methylpyrrolidone
(Distapex process), N-formylmorpholine (Morphylane process,
Octenar process), dimethylformamide, or sulfolane. In the
extraction column, the non-aromatics are distilled overhead, and
the aromatics remain with the solvent in the bottoms, which are
then separated in a stripping column; steam is often added. This
is usually followed by treatment with Fuller's earth to improve
color and remove traces of unsaturated products.

characteristics of extractive distillation:

in contrast to azeotropic distillation, the
sump product must be distilled further to
give aromatic product overhead and selective
solvent as bottoms

The extractive distillation is particularly economical for the
isolation of aromatics from pyrolysis gasoline (approximately
65–90% aromatics).

To 3:

Liquid-liquid extraction is more widely applied than either of the
previously described methods. The aromatics present in a mixture can be simultaneously extracted, even at highly varying concentrations. There is one fundamental difference from the extractive distillation process: in extractive distillation there is only one
liquid phase, while in liquid-liquid extraction the extracting solvent must be so polar that at all stages a readily separable two-
liquid-phase system is retained. Many solvents and mixtures have
been used in commercial processes (cf. Table 12-7).

principles of aromatic/non-aromatic separation by liquid-liquid extraction:

selective aromatic extraction with polar
solvent mixtures, leading to formation of
two immiscible phases

contrast to extractive distillation:

aromatic/non-aromatic/solvent mixture remains one phase

The numerous modifications of the liquid-liquid extraction are
all characterized by countercurrent operation. Extraction columns are used more often than cascades of coupled mixing and
separating vessels (mixer-settler). In an extraction column, the
solvent is added at the top of the extractor, and the mixture
to be separated is introduced in the middle. The non-aromatics
leave the extractor at the upper end, while the solvent charged
with aromatics is removed in the lower part. To increase the
separation, the column is run with an 'aromatic reflux'; that is,
part of the pure aromatics are introduced to the lower part of the
column to force the non-aromatics completely from the extract.

characteristics of liquid–liquid extraction:

countercurrent extraction in mixer–settler
batteries or extraction columns with reflux
of aromatics to displace non-aromatics
from extract (solvent and aromatics)

After the extraction, the non-aromatics are obtained as raffinate,
while the aromatics and the solvent comprise the extract. The
extract can either be:

processing of extracted phase:

Table 12-7. Commercially used solvent extractions for production of aromatics.

Process	Company	Solvent	Extraction conditions
Udex	UOP-Dow	Mono-, di-, tri- or tetraethylene glycol/H_2O and mixtures	130–150 °C, 5–7 bar
Tetra	UCC	Tetraethylene glycol/H_2O	not disclosed
Sulfolane	Shell-UOP	Tetrahydrothiophene dioxide (sulfolane)	50–100 °C
Arosolvan	Lurgi	N-Methylpyrrolidone/H_2O	20–40 °C, 1 bar
DMSO	IFP	Dimethyl sulfoxide/H_2O	20–30 °C
CIS	–	Propylene carbonate	20–50 °C
Duo-Sol	Milwhite Co.	Propane/cresol or phenol	not disclosed
Formex	Snamprogetti	N-Formylmorpholine/H_2O	40 °C, 1 bar
Aromex	Koppers	N-Formylmorpholine/H_2O	80 °C, 2 bar
Morphylex	Krupp-Koppers	N-Formylmorpholine/H_2O	not disclosed
Mofex	Leuna-Werke	Monomethylformamide/H_2O	20–30 °C, 0.1–0.4 bar
Arex	Leuna-Werke	N-Methyl-ε-caprolactam	60 °C, no pressure

aromatics and selective solvent are separated by

1. direct distillation
2. extraction into a second solvent

stripping with inexpensive low-boiling solvents saves the often expensive extractant and is more energetically favorable

example of process:

IFP extraction process with DMSO as solvent and high yield, *i.e.*, practically no losses in separation

1. Directly distilled, often with a steam strip
2. Removed from the selective solvent by stripping, and separated from the extraction solvent by distillation

In stripping, the aromatics are dissolved out of the extract with a light hydrocarbon such as pentane, and freed from the paraffinic solvent by simple distillation.

The results obtained with the IFP solvent extraction process with dimethyl sulfoxide (DMSO) demonstrate the efficiency of this method. Extraction of a catalytic reformate containing, *e.g.*, 10% benzene, 27% toluene, and 18% xylenes can give the following extraction yields (in wt%):

Benzene	99.9%
Toluene	98.5%
Xylenes	95%

Liquid–liquid extraction is by far the most common method for recovering pure benzene and toluene. The Udex process (UOP–Dow), for example, is utilized in more than 50 plants. Recently the Tetra process (UCC) has frequently been preferred because of its improved economics (lower energy use, higher throughput due to more effective extraction).

To 4:

basis of separation of xylene isomers by crystallization:

Although the *crystallization method* is mainly used to separate xylene isomers, it is also used to recover durene, naphthalene, and, in special cases, benzene. The *p*-xylene crystallization is generally preceded by a fractional distillation which yields a mixture of the three xylene isomers (*o, m, p*) and ethylbenzene. As the boiling points of the components of this C$_8$ aromatic fraction are so close together, the separation of *m*- and *p*-xylene in particular is extremely difficult. Their distillative separation would require a column with roughly 800 theoretical plates and a high reflux ratio, making it completely uneconomical.

greater differences in melting points contrast with close boiling points, facilitating an economical separation process

melting points of C$_8$ components:

	m. p. (°C)
p-xylene	+13.3
o-xylene	−25.2
m-xylene	−47.9
ethylbenzene	−95.0

On the other hand, the crystallization temperatures are sufficiently different that they can be used as the basis of a commercial separation process.

other commercial uses of crystallization method:

production of

The first pilot plant for fractional crystallization was based on work by Standard Oil of California and started up by the Oronite Chem. Co. in 1950. Later, similar processes for recovering *p*-xylene were developed by Humble Oil, Sinclair, Phillips Petroleum, Krupp, Arco, Maruzen Oil, and IFP.

In modern plants, the C$_8$ aromatic fraction is first separated into the high-boiling *o*-xylene and the overhead mixture ethylbenzene/*p*-xylene/*m*-xylene, from which the low-boiling ethylbenzene is removed in an involved distillation. An intermediate fraction consisting of a roughly 2:1 ratio of *m*- and *p*-xylene remains. This mixture is carefully dried on Al$_2$O$_3$ or SiO$_2$. The water content must be reduced to about 10 ppm in order to avoid blockages arising from ice deposition. The mixture is then cooled to between −20 and −75 °C by evaporation of ethylene, propane or NH$_3$. The *p*-xylene which deposits is scraped from the cold walls by scrape chillers and removed as a crystal sludge, and separated into a crystal cake and a filtrate using either centrifuges or filters. After the first stage, the *p*-xylene concentration in the crystal cake can be as high as 70%; the *m*-xylene content in the filtrate is about 80%.

pretreatment of a C$_8$ aromatic fraction for the crystallization:

1. distillative preseparation into overhead mixtures and high-boiling *o*-xylene
2. involved distillative removal of low-boiling ethylbenzene from intermediate fraction (2/3 *m*- and 1/3 *p*-xylene)
3. predrying of intermediate fraction with Al$_2$O$_3$ or SiO$_2$ to 10 ppm H$_2$O

process operation:

multistage crystallization, separation, washing, and remelting results in *p*-xylene of 99.5% purity

p-Xylene is obtained with a purity of 99.5% by means of a series of melting and crystallization operations with a coupled system of heat exchangers enabling an optimal utilization of the expensive cooling energy. The yield is almost 100% relative to the mixture introduced. The maximum purity of the *m*-xylene obtainable from the filtrates is about 85% (eutectic mixture with *p*-xylene). This mixture is generally isomerized to *o*- and *p*-xylene (*cf.* Section 12.3.2).

process differences of industrial crystallization methods:

system of heat exchangers as well as crystallization methods determine form and size of *p*-xylene crystals

alternate method for *p/m*-xylene separation:

Amoco indirect separation process of *m, p*-xylene by oxidation to

$$
\text{COOH} \qquad \text{or} \qquad \text{COOH}
$$

principle of separation of aromatics by adsorption on solids:

solids with high surface area exhibit adsorption specificity for certain types of molecules

traditional applications of adsorption of aromatics:

1. isolation of benzene from coke-oven gas
2. Arosorb process for isolation of benzene/toluene

modern application of adsorption method:

p-xylene isolation from isomeric xylene mixtures or from C_8 reformate fractions

process examples:

1. Toray process 'Aromax' (**aro**matics **max**imum recovery), used commercially

2. UOP process 'Parex' (**par**axylene **ex**traction) operated in several plants

characteristics of *p*-xylene separation:

alternating adsorption and desorption by lower boiling hydrocarbons, using several adsorbers or a single adsorber and continuously alternating admission of xylene mixture and desorbent (UOP)

Amoco has developed a process for oxidizing *m*-xylene-rich mixtures to isophthalic and terephthalic acids, which can then be separated (*cf.* Section 14.3). Pure *m*-xylene can be obtained by extraction with $HF \cdot BF_3$ using the Japan Gas–Chemical Co. process (*cf.* Section 12.3.2).

To 5:

Selective separation of mixtures can be achieved by *adsorption on solids* where the solid surface and its pore structure have an adsorption specificity for certain molecules (*cf.* Section 3.3.3.1).

One classical example is the isolation of benzene from coke-oven gas using activated charcoal; another is the Arosorb process for the separation of aromatics (benzene and toluene) from mixtures with non-aromatics by adsorption on silica gel.

Recently, the Aromax and Parex processes have been the subject of commercial interest. They involve the isolation of *p*-xylene from xylene mixtures or from C_8 reformate fractions which still contain some non-aromatics.

In the Aromax process (Toray), *p*-xylene is selectively adsorbed (shape selectivity) from mixtures of its isomers onto modified zeolites (molecular sieves) from the liquid phase at 200 °C and 15 bar, and then desorbed using a solvent. After distillation, *p*-xylene can be obtained in 99.5% purity.

In the Parex process (UOP) the same principle of adsorption on porous solid adsorbents, usually molecular sieves, is employed (VEB-Luna also called the separation of paraffins from petroleum fractions with molecular sieves which is used in many plants the Parex process; *cf.* Section 3.3.3.1). The continuous UOP process can be used to extract *p*-xylene from C_8 aromatic cuts or xylene isomerization mixtures. Adorption is done at 120 to 175 °C in the liquid phase and the desorption is conducted by washing with toluene or *p*-diethylbenzene. The purity of *p*-xylene obtained after distillation is said to be about 99.5%, with an extraction yield of 95%.

The first large industrial plants (70–80000 tonnes per year) went on stream in 1972 and by 1993 more than 50 were in operation. The crystallization method for recovery of *p*-xylene has been replaced in large part by this adsorption technique.

12.2.3. Possibilities for Development of Aromatic Manufacture

Since about 85 – 90% of all BTX aromatics currently are derived from petroleum, the oil crisis has contributed to improving known methods for obtaining aromatics from coal, and to developing new ones.

The first processes that present themselves are those which, with more specificity than allowed by purely thermal decomposition of the macromolecular carbon structure of coal, preserve the aromatic structure, such as hydrogenation of coal or coal extraction. The second principle relates to the most extensive gasification of coal possible (*cf.* Section 2.1.1.1), followed by selective synthesis of aromatics.

Processes from Bergius/Pier and Pott/Broche from the 1920s and 1930s provide the basis for new developments in coal hydrogenation and hydrogenative extraction. New objectives are a direct liquid-phase hydrogenation of a suspension of hard coal or lignite in a heavy oil run in the presence of iron oxide. In this way, the reaction conditions can be moderated to 450 °C and 300 bar. With the incorporation of about 6% H_2, for example, a yield of up to 45% of a coal oil with a 60% aromatic fraction in the boiling range up to about 300 °C is obtained.

Many variations of this basic principle have been investigated up to the pilot plant scale, mainly to improve the economics.

The extraction of coal with solvents such as tetralin or hydrogenated anthracene oil at up to 450 °C gives – depending on the type of coal – an extract of up to 80 wt% of the raw material amount by transfer of H from the solvent to the radical formed by cracking.

This method has been extended mainly by the use of supercritical gases, such as toluene, at 350 – 450 °C and 100 – 300 bar. The coal extracts are usually rich in aromatics, but are still mixtures of high molecular weight compounds which can be converted to low molecular weight aromatics by catalytic hydrocracking.

The third principle route from coal, through synthesis gas, to aromatics has already been used commercially by Sasol in South Africa for another purpose (*cf.* Section 2.1.2). In the Fischer-Tropsch synthesis, aromatics are only byproducts of the higher aliphatic hydrocarbons, and are only isolated as raw phenol extract.

Aromatics can be obtained selectively from synthesis gas by using the intermediate methanol. If, for example, the process conditions and zeolite catalyst of the Mobil MTG process (*cf.*

expanded use of coal for obtaining aromatics by selective depolymerization (hydropyrolysis), solvolysis (extraction), or total decomposition to CO/H_2 with specific synthesis of aromatics

hydrogenation of coal as hydropyrolysis with new emphases:

1. addition of Fe oxide catalysts for moderation of reaction conditions, *e.g.*, H_2 pressure 700 → 300 bar
2. increase of reaction rate by use of a grinding oil as H_2 carrier
3. better workup by vacuum distillation

used commercially to pilot plant stage, *e.g.*, in USA, Western Europe, and Japan

extraction of coal as solvolysis, with simultaneous, though limited, thermolysis of the aromatic skeleton:

1. addition of tetralin or hydrogenated aromatics with secondary function as H-donor
2. addition of supercritical gases, *e.g.*, toluene

coal extract as thermoplastic with specific uses, *e.g.*, electrographite, or further hydrocatalytic cleavage to BTX aromatics

CO/H_2 for synthesis of aliphatic/aromatic hydrocarbons:

1. directly, by Fischer-Tropsch synthesis with low aromatic share
2. indirectly, through methanol/dimethyl ether and homologization/dehydrogenation, *e.g.*, to olefins and aromatics in Mobil process
3. directly, on bifunctional catalyst systems with intermediate formation of CH_3OH, *e.g.*, to methylbenzenes

Section 2.3.1.2) are modified, the aromatic content of the total hydrocarbon product can reach over 40 wt%. The main components are toluene and xylenes.

Other firms have also looked into this possibility for aromatics manufacture and have formulated single-step routes directly from H_2/CO in the presence of bifunctional catalysts. For example, methanol can be formed from synthesis gas and converted directly to aromatics over a zinc chromite-containing zeolite of the type ZSM5. At 425 °C and 84 bar, aromatic mixtures consisting mostly of tri- and tetramethylbenzenes can be obtained with about 60% selectivity.

LPG components for the formation of aromatics by a multistep reaction on a multi-functional zeolite catalyst with continuous regeneration:

1. dehydrogenation to olefins
2. oligomerization to unsaturated higher olefins
3. cyclization to cycloaliphatics
4. dehydrogenation to aromatics

A further production process for aromatics is based on the natural gas/LPG components propane, *n*-butane, isobutane, *n*-pentane and isopentane. In a development by BP and UOP, these low molecular weight aliphatics are converted to aromatics and the coproduct hydrogen in the Cyclar process with an aromatic yield of about 65 wt% and a BTX content of about 95 wt%. Process conditions have not yet been disclosed. The first plant went into operation in 1990 in England.

12.2.4. Condensed Aromatics

commercially important condensed aromatics and coal tar aromatics

Naphthalene and the considerably lower volume anthracene belong to the commercially most significant class of condensed aromatics. Their methyl derivatives from coal tar and oil fractions are used only – by dealkylation – to manufacture the parent substances. Although phenanthrene, another aromatic with three rings, is obtained in greater amounts (about 5%) than anthracene from the high-temperature coking of bituminous coal tar, it is of very minor importance, serving mainly as another suitable precursor for anthracene. The isomerization can be conducted after partial hydrogenation of phenanthrene to *sym*-octahydrophenanthrene.

indene/cumarone application:

polymerized in tar distillates to hydrocarbon resins with uses as components in paints, printing inks, and rubber

x = CH₂ (indene unit)
  = O (cumarone unit)

Indene (1 wt% in tar) is a fourth tar hydrocarbon with two rings whose reactive olefinic double bond is utilized in polymerizations of low molecular weight liquid and solid resins. In a commercial process, tar fractions containing indene and its homologues along with other unsaturated olefinic aromatics from tar (cumarone) are converted in the presence of Friedel-Crafts catalysts (*e.g.*, BF_3) into thermoplastic indene/cumarone resins. The world production of indene/cumarone resins originating from tar is currently about 100 000 tonnes per year. They are used in the formulation of adhesives, paints, printing inks, and rubber. The C_5 hydrocarbon resins (*cf.* Section 5.4) can be substituted for indene/cumarone resins.

12.2.4.1. Naphthalene

Because of the similarities of the boiling points of other components, isolation of naphthalene from coal tar or particular petroleum fractions leads to various purity grades; for example, the product known as 'petronaphthalene' generally has a lower S-content and therefore a higher purity. The differentiation is made between pure naphthalene with a solidification point of 79.6 °C (content $\geqslant 98.75$ wt%) and types of crude naphthalene with solidification points up to 78 °C (content $\geqslant 95.6$ wt%).

In 1994, the production capacities for pure and crude naphthalene in Western Europe, the USA, and Japan were about 0.23, 0.13, and 0.24×10^6 tonnes per year, respectively.

Until the beginning of the 1960s, coal tar was an adequate source of naphthalene in the USA. In the following years, additional naphthalene was isolated from distillation residues of catalytic reformates, and from aromatic-rich middle oil fractions from cracked gasoline. In 1965, about 43% of the naphthalene was derived from petroleum. The fraction of naphthalene from petroleum and total production have decreased significantly in the USA, as shown in the adjacent table.

Outside the USA, naphthalene production from oil fractions is not significant; almost all the naphthalene is isolated from coal tar produced in coal coking.

Naphthalene and its homologues are separated from oil raffinates by liquid–liquid extraction or adsorption.

Pure naphthalene can be prepared from its homologues by dealkylation in the presence of H_2, either thermally at over 700 °C, or catalytically at 550–650 °C and 7–70 bar over Cr_2O_3/Al_2O_3 or Co/Mo-oxide catalysts. Hydrodealkylation processes were developed by Union Oil (Unidak process) and by Sun Oil in the USA.

When isolating naphthalene from coal tar (*cf.* Section 12.2.1), the tar is first fractionally distilled, and then the naphthalene fraction is washed alternately with sulfuric acid and caustic soda to remove tar bases and tar acids. Pure naphthalene is obtained by further distillation or crystallization. A naphthalene yield of about 90%, based on the content of the coal tar, can be achieved commercially.

feedstocks for naphthalene production:

USA: bituminous coal tar,
      reformate residues, and
      cracked gasoline fractions
Europe: mainly coal tar

naphthalene production (USA):

	1965	1970	1993
total	367	326	91
(in 1000 tonnes)			
wt% petro-naphthalene	43	40	10

naphthalene production from oil fractions using two processes:

1. liquid–liquid extraction
2. adsorption

expansion of feedstock base:

hydrodealkylation of naphthalene homologues to naphthalene

isolation of naphthalene from coal tar:

combination of distillations, acid–base extractions and crystallizations uses of naphthalene:

precursor for phthalic anhydride basis for intermediates and solvents (decalin, tetralin)

products from naphthalene:

naphthalenesulfonic acids
β-naphthol
α-naphthol
naphthoquinone
chloronaphthalenes

for dyes, tanning agents, pharmaceuticals, wetting agents, insecticides, *e.g.*, in USA, 7–10% of naphthalene is used for carbaryl (tradename, *e.g.*, Sevin®)

β-naphthol production (in 1000 tonnes):

	1985	1987	1989
W. Europe	20	21	22
Japan	4	4	3

production of anthracene:

from coal tar by distillative enrichment, crystal formation, redissolving and, if required, distillation to pure product

interference in commercial production of anthracene by:

similar boiling points of:

uses of anthracene:

manufacture of anthraquinone has two variations of oxidation:

Its formerly limited supply led the largest naphthalene user – the manufacture of phthalic anhydride – to use *o*-xylene as an additional feedstock (*cf.* Section 14.2.1), so that in 1974 only 25% (and, by the beginning of the 1990s, only 11%) of the phthalic anhydride produced in the world (1991, 2.6×10^6 tonnes) was manufactured by the oxidation of naphthalene. The status varies greatly from country to country: in 1983, the percentage of naphthalene used for phthalic anhydride production in the USA, Japan, and Western Europe was about 74, 70, and 39%, respectively. In 1993, these percentages had changed to 65, 73, and 36%, respectively.

In Western Europe, most of the naphthalene is used for intermediates for, *e.g.*, the dye industry, or for the solvents decalin and tetralin. One important product is β-naphthol, which is obtained by sulfonation of naphthalene at 150–160°C and then treating the sodium naphthalene-β-sulfonate with an alkaline melt. β-Naphthol was also produced from β-isopropylnaphthalene by American Cyanamid for several years before 1982. The process went through the hydroperoxide, analogous to the Hock phenol route.

12.2.4.2. Anthracene

With respect to world demand, anthracene is present in sufficient quantity but in low concentration (1.5–1.8 wt%) in tar. It is obtained in 95% purity by various enrichment and purification processes such as distillation, crystallization, and redissolving. The yield obtained is only about 40–50% relative to the original content, as the commercial recovery is made difficult by the other constituents – phenanthrene and carbazole – which have similar boiling points.

Anthracene is not important in commercial synthetic processes, since its main product – anthraquinone – can be obtained by other routes.

World production of anthracene is currently about 30–40000 tonnes per year, of which about 10000 tonnes per year are produced in Western Europe.

Almost all anthracene is oxidized to anthraquinone, one of the most important intermediates for the manufacture of mordant, vat, dispersion, and reactive dyes.

The world production capacity of anthraquinone is about 34 000 tonnes per year, with about 17 900, 1900 and 5000 tonnes per year in Western Europe, the USA, and Japan, respectively.

Oxidation of anthracene in the 9,10-position is easily achieved. Commercially, this is usually (currently about 85% of world production) done either in the liquid phase at $50-105\,°C$ with CrO_3 or in the gas phase with air at $340-390\,°C$ over iron vanadate catalysts. The selectivity is greater than 90% for a virtually quantitative conversion:

1. liquid-phase oxidation with CrO_3
2. gas-phase oxidation with air over Fe vanadate

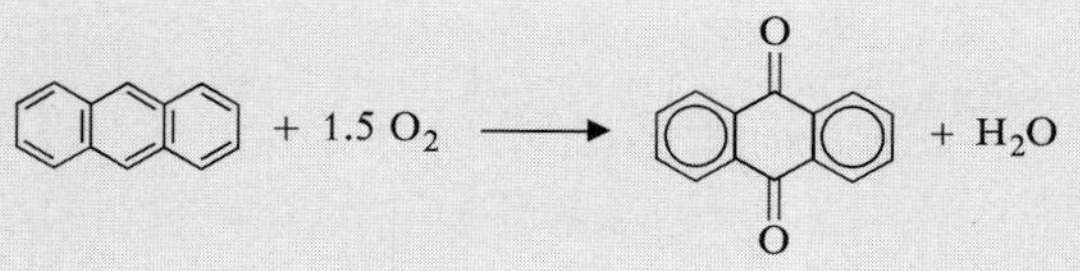

$$+ \; 1.5 \; O_2 \longrightarrow \qquad + \; H_2O \qquad (2)$$

Phthalic acid is formed as a byproduct.

Due to the growing demand for anthraquinone as the basis of textile dyes, many attempts have been made to synthesize it from other precursors.

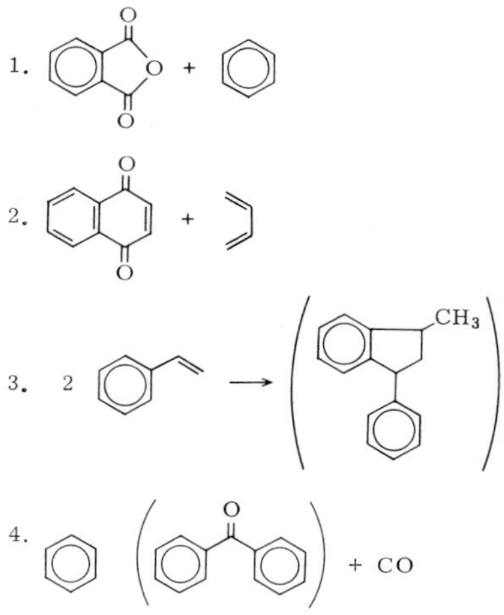

Manufacturing processes are based on, for example, the Friedel–Crafts reaction between phthalic anhydride and benzene, the Diels–Alder reaction of 1,4-naphthoquinone with butadiene (Bayer and Kawasaki processes), and the styrene dimerization with an indane derivative intermediate (BASF process). A newer route developed by American Cyanamid involves the reaction of benzene or benzophenone with CO in the presence of stoichiometric amounts of $CuCl_2$, $FeCl_3$, or platinum chlorides at $215-225\,°C$ and $25-70$ bar CO pressure. Quantitative yields of anthraquinone are reported.

Of these four synthetic processes, the Bayer process was scheduled for commercial operation, but was abandoned for technical and economic reasons. The Kawasaki process has been in operation in a 3000 tonne-per-year plant since 1980.

The main products of anthraquinone are the nitro- and dinitroanthraquinones, and the amino- and diaminoanthraquinones. These are important raw materials for a multitude of pigments.

In addition, anthraquinone can be used for the production of H_2O_2. For this, anthraquinone is catalytically hydrogenated to anthrahydroquinone using Ni or Pd catalysts at $30-35\,°C$. After catalyst separation, an extremely rapid autoxidation ensues to form the endoperoxide, which reacts quantitatively to regenerate the anthraquinone and produce H_2O_2:

alternative processes for anthraquinone:

uses of anthraquinone:

1. intermediate for pigments
2. manufacture of H_2O_2

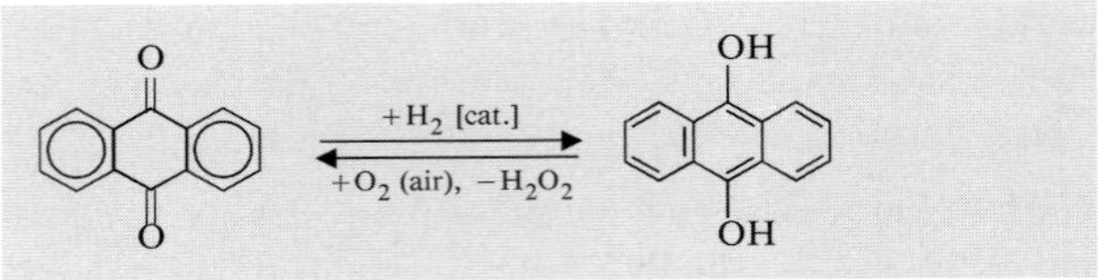

$$\text{(3)}$$

process principles for H_2O_2 manufacture:

reversible conversion of anthraquinone-anthrahydroquinone with H_2O_2 formation in autoxidation step with intermediate endoperoxide

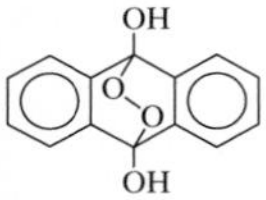

process operation:

industrial process generally with 2-ethyl, 2-*tert*-butyl- or 2-amylanthraquinone

thus, three commercial H_2O_2 processes exist:

1. anthraquinone route
2. isopropanol oxidation
3. electrolysis

production of H_2O_2 (in 1000 tonnes):

	1991	1993	1995
W. Europe	505	610	680
USA	254	301	327
Japan	155	142	138

significant production increase, *e. g.*, as ecologically friendly bleach and cleaning agent to replace Cl_2

The first H_2O_2 pilot plant, operated by IG Farben during World War II, was based on this principle. The first commercial scale unit was brought into operation by Du Pont in the USA in 1953. Due to its better solubility in solvent mixtures (for example, with benzene and higher secondary alcohols), 2-ethylanthraquinone is usually used; 2-*tert*-butylanthraquinone and, in Japan, 2-amylanthraquinone are also used.

H_2O_2 is extracted from the oxidation product with water. The crude solution is purified and concentrated in several process steps. In Europe, most H_2O_2 is produced by this process. Other manufacturing processes are isopropanol oxidation (*cf.* Section 11.1.3.2), which is no longer used in the USA, and electrolysis of peroxodisulfuric acid ($H_2S_2O_8$). Two firms in Western Europe currently practice the electrolysis route. More than 90% of the total H_2O_2 is produced by organic processes, since they have considerably more favorable space-time and energy yields than the electrochemical processes.

In 1994, the world production capacity for H_2O_2 was 1.95×10^6 tonnes per year, of which 0.97, 0.35, and 0.21×10^6 tonnes per year were in Western Europe, the USA, and Japan, respectively. World production in 1993 was estimated at 1.3×10^6 tonnes. Production figures for several countries are given in the adjacent table. Solvay (formerly Interox) is the world's largest producer with a total capacity of about 0.45×10^6 tonnes per year (1993). Other manufacturers include Degussa and Atochem.

The main use for H_2O_2 is in the manufacture of bleaching agents such as perborates and percarbonates; only a small fraction is used in the chemical industry for manufacture of products such as glycerine, peroxides, and epoxides. Amounts used in paper production, wastewater treatment and hydrometallurgical processes (*e. g.*, for extraction of uranium by oxidation to U_3O_8) are increasing rapidly.

12.3. Conversion Processes for Aromatics

The differences in amounts of pyrolysis gasoline, reformate and coal tar in different countries along with their characteristically dissimilar aromatic content make conversion processes necessary in order to counterbalance the oversupply of, *e. g.*, toluene and the demand for benzene and xylenes. The most significant processes are:

conversion of alkyl aromatics by three processes:

1. Hydrodealkylation of toluene
2. Isomerization of *m*-xylene
3. Disproportionation of toluene and transalkylation with trimethylbenzenes

1. hydrodealkylation, *e. g.*,
   toluene → benzene
2. isomerization, *e. g.*,
   m-xylene → *o*-, *p*-xylene
3. disproportionation, *e. g.*,
   toluene → benzene + xylenes and
   transalkylation, *e. g.*,
   trimethylbenzenes + toluene → xylenes

12.3.1. Hydrodealkylation

The hydrogenative degradation of side chains on aromatic rings is a generally applicable method for the manufacture of the non-substituted parent substance, *e. g.*, benzene from toluene or xylenes and naphthalene from methyl- or dimethylnaphthalene.

process principles of **hydrodealk**ylation (Hydeal process) of alkylaromatics:

hydrogenative cleavage of alkyl side chains in aromatics

The first pilot-scale dealkylation process was started up in 1961 by Ashland Oil. It was used for the manufacture of pure naphthalene from alkylnaphthalenes. Benzene was formed as a byproduct. UOP developed the Hydeal process by utilizing a license from Ashland Oil.

feedstocks for Hydeal process:

alkylnaphthalenes
toluene

Today, the most important industrial processes are those for the manufacture of benzene from toluene:

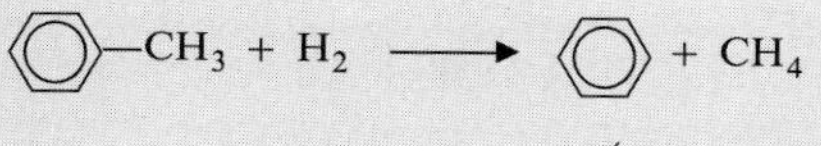

$$\left(\Delta H = -\frac{30 \text{ kcal}}{126 \text{ kJ}} \big/ \text{mol} \right) \tag{4}$$

The percentage of toluene production used for the hydrodealkylation to benzene varies between countries (*cf.* adjacent table).

use of toluene for hydrodealkylation (in %):

	1989	1991	1993
USA	67	65	57
W. Europe	52	52	53
Japan	23	30	15

The hydrodealkylation is conducted either purely thermally at 550–800 °C and 30–100 bar or catalytically at somewhat lower temperatures of 500–650 °C and 30–50 bar over Cr_2O_3, Mo_2O_3, or CoO on supports (*e. g.*, Al_2O_3) or, as in a recent development, at 400–480 °C over Rh/Al_2O_3.

process variations of toluene hydrodealkylation:

1. thermal at 580–800 °C
2. catalytic at 550–650 °C over Cr-, Mo-, or Co-oxide catalysts

characteristics of catalytic operation:

coke deposition on catalyst makes two reactor systems for regeneration after 2000–4000 operating hours expedient

In the commercial catalytic process, toluene is passed over the catalyst together with a recycle gas consisting of a mixture of H_2 and CH_4. The high working temperatures result in relatively strong coking, and two parallel reactors are generally used so the reaction can proceed in one while the catalyst is being regenerated in the other.

processing of recycle gas:

separation of CH_4 from recycle gas mixture (H_2/CH_4) by low-temperature distillation

Because of the large amount of CH_4 formed – about 250 scm/tonne toluene – special measures are necessary to remove it from the recycle gas. The CH_4/H_2 mixture can, for this purpose, be totally or partially subjected to low temperature separation. By cracking with H_2O, the CH_4 can be converted into H_2 to cover the process requirements for the hydrodealkylation.

processing of condensable phase:

benzene isolated by distillation, toluene feed and higher boiling substances recycled

recoverable as byproducts:

biphenyl

fluorene

After dealkylation, the benzene is isolated by distillation from the unreacted toluene, which is recycled. Higher boiling byproducts such as biphenyl and condensed aromatics (*e.g.*, fluorene) can be separated off or recycled to the process, since they are also subject to cleavage.

examples of commercially operated hydrodealkylations:

1. thermal processes:
   THD   – Gulf
   HDA   – Arco–Hydrocarbon Research
   MHD   – Mitsubishi Petrochemical

In thermal processes, *e.g.*, THD (Gulf), HDA (Arco–Hydrocarbon Research), and MHD (Mitsubishi Petrochemical), benzene selectivities of about 95% are attained at a 60–90% toluene conversion.

2. catalytic processes:
   Detol and Pyrotol – Houdry
                       ABB Lummus Crest
   Hydeal – UOP
   Bexton – Shell
            BASF

In the catalytic processes, *e.g.*, Detol and Pyrotol (Houdry) and Bextol (Shell and BASF), the conversions and selectivities are significantly higher. The benzene purity is greater than 99.9%. This could compensate for the cost of the catalyst. The very low sulfur content makes the benzene particularly suitable for hydrogenation to cyclohexane (*cf.* Section 13.1.5).

12.3.2. *m*-Xylene Isomerization

process principles of isomerization of alkylaromatics. usually *m*-xylene:

equilibration to mixture of C_8 aromatics starting with a nonequilibrium mixture high in *m*-xylene

The isolation of pure *p*-xylene by fractional low-temperature crystallization of *m*/*p*-xylene mixtures (*cf.* Section 12.2.2.2) is stopped at a content of approximately 85% *m*-xylene because a eutectic *m*/*p*-xylene mixture begins to crystallize out. The mother liquor can then be catalytically isomerized to an equilibrium mixture of C_8 aromatics and recycled to the processing step for the *p*-xylene isolation.

The following concentrations (mol%) are present at the thermodynamic equilibrium of the C_8 aromatics in the temperature range of the commercial processes (200–500 °C):

Ethylbenzene 8%
o-Xylene 22%
m-Xylene 48%
p-Xylene 22%

There are basically three types of catalytic isomerizations which differ according to the catalyst used and process conditions:

1. Hydrocatalytic isomerization with Pt/Al$_2$O$_3$ · SiO$_2$ catalysts in the presence of H$_2$ at 400 – 500 °C and 10 – 25 bar

2. Isomerization under cracking conditions with Al$_2$O$_3$ · SiO$_2$ at 400 – 500 °C and atmospheric pressure

3. Friedel–Crafts isomerization in the liquid phase with HF · BF$_3$ at 100 °C and atmospheric pressure, or with zeolite catalysts under various conditions

characterization of three conventional types of isomerization by type of catalyst:

1. bifunctional system:

   Al$_2$O$_3$ · SiO$_2$ for isomerization. Pt/H$_2$ for hydrogenation, *i.e.*, decrease in dehydrogenation with C deposition

2. Al$_2$O$_3$ · SiO$_2$ component from 1, only for isomerization

3. HF · BF$_3$ Friedel–Crafts system very active and selective for isomerizations, zeolite or molecular sieve bifunctional with protonic (Brönstedt) and Lewis acid activity

To 1:

So far, the widest application of this principle is to be found in the octafining process (Arco/Engelhard). The first plant was built by Mitsui in 1960. Another variation commonly used commercially is the Isomar process (UOP; more than 30 plants in 1993), which also uses a Pt-Al silicate catalyst. In the Isoforming process (Exxon), a Pt-free catalyst is employed under otherwise similar conditions.

A new process for the hydrocatalytic isomerization of industrial C$_8$ aromatic fractions (Aris process) was developed by Leuna-Werk and Petrolchemische Kombinate Schwedt and has been used in a commercial unit since 1976. The catalyst consists of a mixture of Al$_2$O$_3$ and a naturally-occuring zeolite (mordenite), together with platinum as the hydrogenation/dehydrogenation component. One special feature is the regulating of the mordenite acidity, and therefore the activity and selectivity, by controlled addition of NH$_3$; this also extends the catalyst life. Under the reaction conditions (425 °C, 12.5 bar), non-aromatic C$_8$-C$_{10}$ compounds found in the aromatic cut are degraded into low-boiling C$_3$-C$_4$ products.

The advantages of this type of isomerization are long catalyst life and low losses of C$_8$ aromatics by disproportionation or transalkylation.

process characteristics with catalyst type 1:

longer catalyst life than type 2, little disproportionation and transalkylation

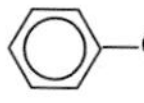

⬡—C$_2$H$_5$ is also isomerized

process characteristics with catalyst type 2:

cracking catalysts $Al_2O_2 \cdot SiO_2$ are inexpensive for isomerizations, but disadvantageous due to side reactions such as disproportionation, transalkylation (*cf.* Section 12.3.3), and carbon deposition

To 2:

This type of isomerization is used in processes from, for example, ICI, Maruzen Oil (XIS process), and Arco. The advantages of an inexpensive catalyst must be weighed against disadvantages such as catalyst deactivation by coke deposition, and disproportionation and transalkylation.

To 3:

process characteristics with catalyst type 3:

$HF \cdot BF_3$ suitable for both *m*-xylene complex formation and *m*-xylene isomerization

high activity means reduced reaction temperature

The first process of this type was introduced by Japan Gas (now Mitsubishi Gas Chemical). $HF \cdot BF_3$ is used to form a selective complex with *m*-xylene in a solvent, allowing it to be almost completely extracted from a C_8 aromatic mixture. The resulting raffinate, consisting of ethylbenzene and *o/p*-xylene, is then separated by distillation.

xylene disproportionation and transalkylation are therefore low ($<4\%$)

If pure *m*-xylene is required, then the extracted *m*-xylene–$HF \cdot BF_3$ complex can be cleaved. If there is little or no demand for *m*-xylene, then corresponding amounts of the complex extract are subjected to an isomerization which takes place under mild conditions in the presence of a $HF \cdot BF_3$ catalyst until thermodynamic equilibrium is reached.

Mitsubishi currently has three plants using this technology in Japan, and other plants using similar technology (*e. g.*, from Amoco) in the USA.

In its LTI process (**l**ow **t**emperature **i**somerization), Mobil Chemical Co. employs a zeolite fixed-bed catalyst which also isomerizes *m*-xylene at $200-260\,°C$ and 14 bar to a $95-98\%$ equilibrium mixture without loss of C_8 aromatics.

zeolites (ZSM5) in H/Pt form enable a higher *p*-xylene yield due to structure selectivity, depending on the reaction temperature

In a newer variation, the Mobil HTI process (**h**igh **t**emperature **i**somerization), a C_8 aromatic fraction is isomerized with hydrogen at $427-460\,°C$ and $14-18$ bar over an acidic ZSM5 zeolite catalyst that has been partially exchanged with platinum. *p*-Xylene is obtained up to 4% over the equilibrium concentration. This process is used in eleven plants (1993).

12.3.3. Disproportionation and Transalkylation

alkylation of alkylaromatics according to two types of reaction:

1. disproportionation
2. transalkylation

In this context, disproportionation or dismutation denotes a transfer of alkyl groups to form a mixture of higher and lower alkylated aromatics. The major process is the toluene disproportionation to benzene and xylenes:

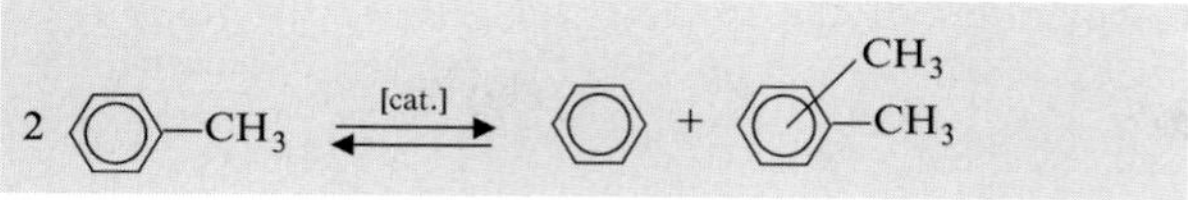

(5)

In this equilibrium reaction, the maximum theoretical conversion of toluene is only 58%; in practice, the conversion of toluene is set at about 40% to avoid side reactions.

The term transalkylation generally refers to the reaction of polyalkylated aromatics by transfer of alkyl groups and formation of lower alkylated aromatics:

reaction principle of 1:

conversion of alkylaromatics into equal parts of higher and lower alkylaromatics or dealkylated aromatics

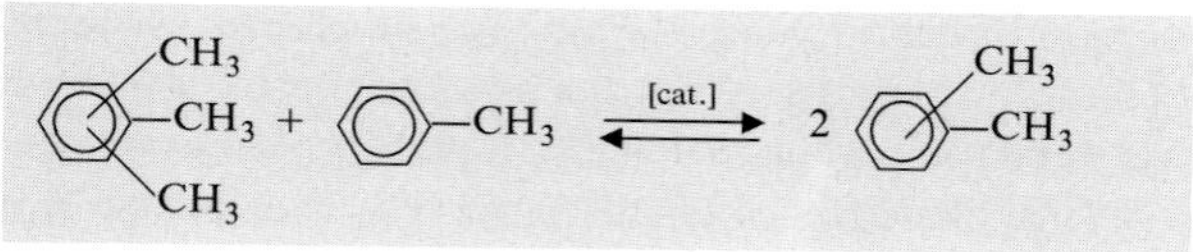

(6)

Both types of reaction are avoided as far as possible in the xylene isomerization, although they can be made the main reaction using similar catalyst systems and somewhat more severe process conditions.

reaction principle 2:

transfer of alkyl groups from higher to lower alkylated aromatics

The equilibria are usually stopped by the presence of a small amount of hydrogen, e. g., with $AlCl_3$ or BF_3 and the addition of HCl, HF or 1,2-dichloroethane at $80-125\,°C$ and $35-70$ bar. In this way Mobil's LTI process, at $260-320\,°C$ and about 45 bar with zeolite catalysts in the liquid phase, is converted into the LTD (low temperature disproportionation) process. One catalyst particularly suitable for this toluene disproportionation is the ZSM5 type, which is effective up to $480\,°C$. With a high SiO_2/Al_2O_3 content, this catalyst stands out for its thermal stability.

In a further development of Mobil (STDP = selective toluene disproportionation process), toluene is converted to a xylene mixture containing 87% p-xylene. This takes place in the presence of hydrogen over a specially pretreated ZSM5 catalyst. The reaction proceeds in the gas phase over the fixed-bed catalyst with a toluene conversion of 30%. The first of three plants now in operation was started up in 1988 by EniChem.

catalyst systems for both reaction principles:

Brönstedt acid
Lewis acid
bifunctional, specially modified systems, primarily zeolites

Disproportionation and transalkylation can occur simultaneously if a mixture of toluene and C_9 aromatics is used. Their mole ratio determines the ratio of benzene to xylenes.

combination of 1 and 2 leads, with toluene/C_9 aromatics, to controllable benzene/xylene ratio

uses of disproportionation products of toluene:

main products:
benzene for use in hydrogenation to cyclohexane

xylenes for use in isomerization for *o, p*-xylene production

byproducts:

polymethylbenzenes useful for oxidations, *e. g.*:

durene → pyromellitic anhydride
(*cf.* Section 12.2.2)
or recycled to transalkylation

examples of industrially operated disproportionation–transalkylation processes:

Tatoray – Toray/UOP
xylene–plus – Arco
LTD – Mobil

After separation of the reaction mixture by distillation, the feedstocks — toluene and possibly C_9 aromatics — are fed back to the disproportionation–transalkylation. The reaction products — benzene and the xylene mixture — can be isolated in a very pure state. Benzene is thus particularly suitable for subsequent catalytic hydrogenation to cyclohexane (*cf.* Section 13.1.5) and the xylenes for subsequent xylene isomerization (*cf.* Section 12.3.2).

With cracking catalysts, an undesired side reaction — dealkylation of ethylbenzene to benzene and ethylene — can be counteracted by higher pressures. Another side reaction leads to polymethylbenzenes, *e. g.*, to 1,2,4,5-tetramethylbenzene (durene).

A disproportionation process was, *e. g.*, developed by Toray/UOP (Tatoray process) and was first practiced in a 70000 tonne-per-year Japanese plant at $350-530\,°C$ and $10-50$ bar in 1969. Other plants have been built since. A similar process (xylene–plus process) was introduced by Arco for disproportionation and transalkylation. Worldwide, plants with a total capacity of about 0.6×10^6 tonnes per year are currently in operation using this process. The Mobil disproportionation process (LTD) has been used since 1975 in several plants.

13. Benzene Derivatives

In industrial chemistry, benzene is the most important precursor for the many various aromatic intermediates and for cyclo-aliphatic compounds. It is remarkable that the bulk of benzene production is consumed by the manufacture of only three secondary products. The breakdown of benzene use can be appreciated from the following summary:

benzene is the most important feedstock for aromatics and alicyclics

75–90% of benzene produced is for only three products, rest is for remaining benzene derivatives and secondary products

Table 13-1. Pattern of benzene use (in %).

Product	USA		Western Europe		Japan	
	1985	1994	1986	1993	1984	1994
Ethylbenzene (styrene)	57	56	46	52	54	59
Cumene (phenol)	20	22	21	21	13	16
Cyclohexane	12	11	16	12	22	18
Nitrobenzene	5	5	8	8	2	2
Maleic anhydride	1	—	3	1	2	2
Alkylbenzenes	3	3	5	4	4	2
Miscellaneous (*e. g.,* chlorobenzene, benzenesulfonic acid)	2	3	1	2	3	1
Total use (in 10^6 tonnes)	5.20	7.17	5.18	5.79	2.11	3.65

It can be seen that the alkylation of benzene to ethyl- and isopropylbenzene (the precursors of styrene and phenol, respectively), and the hydrogenation to cyclohexane and further processing to the various intermediates for polyamides, are the most economically important secondary reactions of benzene.

most important secondary reactions of benzene:

1. alkylation with ethylene and propene as preliminary steps to

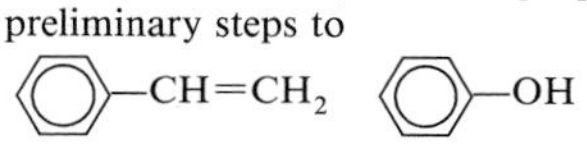

   ⬡—CH=CH$_2$ ⬡—OH

2. hydrogenation as preliminary step to

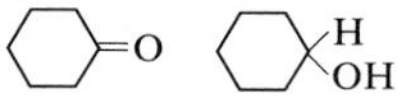

   ⬡=O   ⬡ $_{OH}^{H}$

13.1. Alkylation and Hydrogenation Products of Benzene

13.1.1. Ethylbenzene

Most ethylbenzene is manufactured by the alkylation of benzene with ethylene. Small amounts are still obtained in several countries from superfractionation of C_8 aromatics cuts; for

production of ethylbenzene:

1. mainly by ethylation of benzene
2. partially by distillation from C_8 aromatic cuts

example, 300 000 tonnes were produced in the USA in 1986. In 1989 this process was no longer practiced in the USA, but in 1991 it was used to manufacture ca. 170 000 tonnes and 130 000 tonnes in Western Europe and Japan, respectively.

In 1991, world manufacturing capacity for ethylbenzene was about 19.7×10^6 tonnes per year, of which about 5.3, 4.8, and 3.3×10^6 tonnes per year were located in the USA, Western Europe, and Japan, respectively. Since almost all ethylbenzene is used for the production of styrene, these correspond to those production figures (*cf.* Section 13.1.2).

benzene ethylation by two process routes:

Two commercial processes for the ethylation of benzene have developed alongside one another:

1. liquid-phase ethylation catalyzed with Lewis acids preferred

1. Liquid-phase ethylation with Friedel–Crafts catalysts such as $AlCl_3$, BF_3, $FeCl_3$, $ZrCl_4$, $SnCl_4$, H_3PO_4, or alkaline earth phosphates (*e. g.,* BASF, CdF Chimie (now Orkem), Dow, Monsanto, or UCC process).

2. Gas-phase ethylation catalyzed with supported acid catalysts or with Lewis acids

2. Gas-phase ethylation with H_3PO_4 carrier catalysts or Al silicates (*e. g.* Koppers, Phillips, and UOP process) or with zeolites (new Mobil–Badger or ABB Lummus Crest/UOP process), or with $BF_3/\gamma\text{-}Al_2O_3$ (UOP process).

To 1:

The alkylation of benzene with ethylene is carried out at $85 - 95\,^{\circ}C$ and slight pressure (1 – 7 bar):

$$
\begin{array}{c}
\text{\large\bigcirc} + H_2C{=}CH_2 \xrightarrow{\text{[cat.]}} \text{\large\bigcirc}{-}CH_2CH_3 \\[2mm]
\left(\Delta H = -\begin{array}{l}27\ \text{kcal} \\ 113\ \text{kJ}\end{array}\Big/\text{mol} \right)
\end{array}
\tag{1}
$$

process characteristics of liquid-phase ethylation:

countercurrent operation of a two-phase system consisting of catalyst/promoter and benzene/ethylene at atmospheric pressure in bubble reactor

A 0.6 : 1 mixture of ethylene and benzene is fed continuously into the bottom of an alkylation column, while the catalyst, generally $AlCl_3$, is introduced periodically at the head of the column. At the same time, a little ethyl chloride is added a as promoter; under the reaction conditions, it reacts with benzene to form ethylbenzene. The coproduct HCl acts as a cocatalyst for $AlCl_3$.

effective catalyst system:

$HAlCl_4 \cdot n$ $\text{\large\bigcirc}{-}C_2H_5$ from:

Lewis acid	$AlCl_3$
promoter	$HCl(C_2H_5Cl)$
ligand	$C_6H_5{-}C_2H_5$

with constant decrease in activity

The actual catalytically active system is formed from $AlCl_3$ and ethylbenzene as the particularly strong addition product $HAlCl_4 \cdot n\ C_6H_5C_2H_5$. The gradual decrease in catalytic activity can be counteracted by partially removing the catalyst and replacing it with fresh catalyst. About 1 kg $AlCl_3$ is required per 100 kg ethylbenzene.

The selectivity is enhanced by limiting the benzene conversion to approximately $52-55\%$ and having a high benzene/ethylene ratio, since only an ethylene deficiency prevents further alkylation to diethyl- and polyethylbenzenes. Higher ethylbenzenes are recycled to the reaction where they displace the alkylation equilibrium towards ethylbenzene and yield additional monoethyl product by transalkylation with benzene. Multiethylated benzenes can also be dealkylated in a special reaction with $AlCl_3$ at $200\,°C$ before being recycled to the reaction. In this way, an ethylbenzene selectivity of $94-96\%$ (based on C_6H_6) and of $96-97\%$ (based on C_2H_4) can be attained.

The reaction mixture, consisting of about 45% benzene, 37% ethylbenzene, 15% diethylbenzenes, 2% polyethylbenzenes, and 1% tarry residues, is separated into its components in a four-column distillation unit.

A more recent variation of benzene ethylation in the presence of $AlCl_3$ was developed by Monsanto–Lummus. Using a lower $AlCl_3$ concentration than the classical processes, it is possible to conduct the alkylation in a single-phase system at temperatures of $140-200\,°C$ and $3-10$ bar with a high selectivity to ethylbenzene (ca. 99%). The $AlCl_3$ consumption decreases to 0.25 kg per 1000 kg ethylbenzene. An important prerequisite is, however, the controlled introduction of ethylene in amounts such that it is never present in excess. This very economical process is operated in several plants throughout the world with a total capacity of about 3.1×10^6 tonnes per year.

The necessity, in the $AlCl_3$-catalyzed liquid-phase ethylation of benzene, of ensuring that numerous parts of the plant are corrosion resistant, of separating the suspended or dissolved catalyst by aqueous and alkaline washes, and of drying the benzene to be recycled led to the development of gas-phase alkylations on solid catalysts. However, in a new development from CdF Chimie-Technip (now Orkem) already in use in several plants, the customary washing of the reaction product to remove the catalyst and the subsequent neutralization can be replaced by anhydrous treatment of the alkylation mixture with gaseous ammonia.

optimization of ethylbenzene selectivity attainable by:

1. limitation of benzene conversion
2. less than stoichiometric ratio C_2H_4: C_6H_6
3. additional ethylbenzene from polyethylbenzenes such as

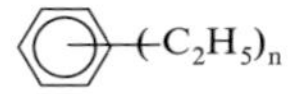

$n = 2$ to 6

by:

transalkylation after recycling or external dealkylation at about $200\,°C$

process characteristics of Monsanto liquid-phase ethylation:

smaller amount of $AlCl_3$ leads to homogeneous system $AlCl_3$/benzene/ethylene and lower $AlCl_3$ consumption when ethylene excess is avoided

disadvantages of liquid-phase ethylation:

1. corrosive catalyst
2. alternating $AlCl_3$ washes and drying of recycled benzene

process characteristics of gas-phase ethylation are dependent on catalyst used:

1. $Al_2O_3 \cdot SiO_2$ and H_3PO_4/SiO_2 are inactive for trans- and dealkylations of oligoethylbenzenes, thus optimization of selectivity only possible through low $C_2H_4 : C_6H_6$ ratio
1.1. crystalline $Al_2O_3 \cdot SiO_2$ with high SiO_2/Al_2O_3 content in the form of modified zeolites (ZSM5) with high activity for alkylation and transalkylation
1.2. additional $Mg^{2\oplus}$ exchange after H_3PO_4 treatment increases the shape selectivity; *i. e.*, *p*-ethylation of toluene based on the smallest molecular diameter is possible selectively by adjusting the zeolite pore diameter

	minimum diameter (in Å)
o-ethyltoluene	7.7
m-ethyltoluene	7.6
p-ethyltoluene	7.0

2. $BF_3/\gamma\text{-}Al_2O_3$ is also active for trans- and dealkylations, therefore high selectivity and no corrosion

To 2:

The gas-phase ethylation of benzene has been particularly successful in the USA. Benzene is reacted with ethylene at about 300 °C and 40–65 bar over acidic catalysts such as $Al_2O_3 \cdot SiO_2$ (Koppers) or H_3PO_4/SiO_2 (UOP). To avoid further alkylation at the high reaction temperature of the gas-phase process, the mole ratio of ethylene to benzene must be adjusted to 0.2 : 1. The catalysts used are not capable of dealkylating recycled oligoethylbenzenes.

In the Mobil–Badger process for the ethylation of benzene, the catalyst is a crystalline aluminium silicate in the form of a modified zeolite (ZSM5). It is used at 435–450 °C and 14–28 bar. The catalyst is also suitable for transalkylations. However, it must be regenerated after 2–4 weeks of operation, so the reactor section in the plant must be in duplicate. At an 85% conversion, ethylbenzene selectivities of 98% (based on C_6H_6) and 99% (based on C_2H_4) can be reached. The first 500 000 tonne-per-year plant was brought on-line by American Hoechst (now Huntsman) in 1980. Since then 21 plants have been licensed worldwide to produce more than 6.5×10^6 tonnes annually. Production capacity of individual plants can reach 800 000 tonnes per year. In a similar Mobil process using an $H_3PO_4/Mg^{2\oplus}$-modified ZSM5 zeolite, a shape-selective ethylation (up to 99%) of toluene to *p*-ethyltoluene can be done at, *e. g.*, 425 °C and 7 bar. In 1982, Mobil/Hoechst started up a 16 000 tonne-per-year plant in the USA. *n*-Ethyltoluene is dehydrogenated to give *p*-methylstyrene, a monomer and comonomer analogous to styrene, but with special advantages including higher thermal stability (higher glass temperature).

In a new process developed by Mobil for the ethylation of benzene or toluene on modified ZSM5 zeolites, an aqueous ethanol solution can also be used directly without the need for expensive extraction or distillation.

Another zeolite-catalyzed benzene ethylation using a fixed-bed, liquid-phase technology has been operated by ABB Lummus Crest/UOP in a 220 000 tonne-per-year plant in Japan since 1990. Since then, other plants have been started up or are being planned.

The Alkar process was developed by UOP in the 1950s. Instead of H_3PO_4/SiO_2, BF_3 — active for trans- and dealkylation, and noncorrosive — is used in combination with a $\gamma\text{-}Al_2O_3$ catalyst at 290 °C and 60–65 bar. At a total conversion of ethylene, high ethylbenzene selectivities (98–99%) are attained. A considerable advantage could arise from the possible use of dilute ethylene, or ethylene containing either O_2 or CO_2, to avoid damage to the catalyst. The Alkar process has been in use in several commercial plants since 1966.

Ethylbenzene is used almost exclusively for the manufacture of styrene; only a small fraction is used directly as a solvent, or as an intermediate for, *e. g.*, the manufacture of diethylbenzene or acetophenone.

uses of ethylbenzene:

mainly for dehydrogenation to styrene; <2% as solvent or intermediate

13.1.2. Styrene

In 1996, world production capacity for styrene was about 19.2×10^6 tonnes per year, with 5.3, 4.6, and 3.2×10^6 tonnes per year in the USA, Western Europe, and Japan, respectively. Dow Chemical is the largest producer, with a total capacity of more than 1.8×10^6 tonnes per year in the USA, Canada, and Western Europe (1996). Production figures for several countries are given in the adjacent table.

styrene production (in 10^6 tonnes):

	1990	1992	1995
USA	3.64	4.06	5.16
W. Europe	3.68	3.64	4.00
Japan	2.21	2.18	2.94

Older styrene processes based on the chlorination of the side chain of ethylbenzene followed by dehydrochlorination are no longer in use today. The oxidation of ethylbenzene to acetophenone followed by reduction to the carbinol and its dehydration has also decreased in importance; it is now only being practiced in a single plant in Spain.

manufacture of styrene from ethylbenzene:

older processes:

1. side chain chlorination and dehydrochlorination
2. oxidation to acetophenone, hydrogenation, and dehydration (UCC route)

One stage of the oxidation route, the dehydration of methylphenylcarbinol, is still used commercially in another context in a modification of the Halcon process (*cf.* Section 11.1.1.2). In this process, involving the indirect oxidation of propene to propylene oxide, ethylbenzene (in its hydroperoxide form) can be used as an auxiliary oxidation system. After being converted into methylphenylcarbinol, it is dehydrated to styrene. About 2.5 kg styrene is obtained per kilogram propylene oxide. Currently, about 15% of the styrene produced worldwide is made with this process.

modern processes:

1. indirect method:

   Halcon process with ethylbenzene as auxiliary system, oxidation to hydroperoxide, O-transfer with carbinol formation, and dehydration

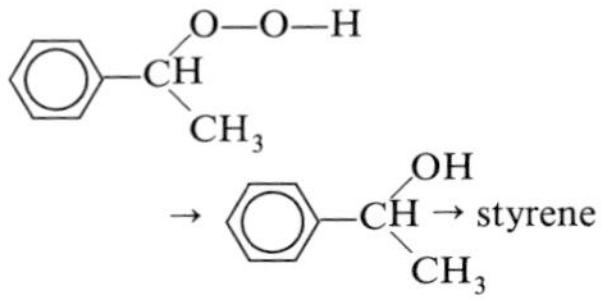

The main manufacturing route to styrene is, however, the direct catalytic dehydrogenation of ethylbenzene:

2. direct method:

   catalytic dehydrogenation

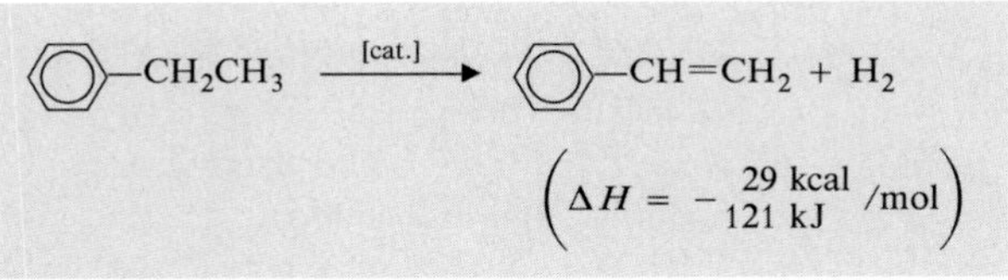

$$\left(\Delta H = -\frac{29 \text{ kcal}}{121 \text{ kJ}} /\text{mol} \right) \quad (2)$$

process characteristics of ethylbenzene dehydrogenation:

endothermic, heterogeneously catalyzed, H_2 elimination − requires high temperatures

heat supply variation:

indirect with combustible gas (BASF process) requires tube oven and thus isothermal reaction pathway

direct by means of superheated steam in reaction mixture (Dow process) allows blast furnace and thus adiabatic reaction pathway (similar in processes from Monsanto, UCC, and others)

optimization of selectivity:

1. temperature limitation and thus conversion limitation
2. H_2O addition and thus lowering of partial pressure of ethylbenzene, *i.e.*, displacement of equilibrium to styrene addition of water also lessens catalyst coking via the water gas shift:

$$C + 2 H_2O \rightarrow CO_2 + 2 H_2$$

styrene processing and purification:

fractional condensation causes deposition of tar (polymer)

costly secondary purification due to:

1. tendency to polymerize
2. similar boiling points (ethylbenzene 136 °C, styrene 145 − 146 °C)

thus addition of inhibitor, low temperature in vacuum columns, little pressure drop *i.e.*, small number of plates, high reflux

modifications of the ethylbenzene dehydrogenation:

1. catalytic gas-phase dehydrogenation in presence of O_2 (oxydehydrogenation) for the oxidation of H_2, for displacement of equilibrium and autothermic heat supply

One of the first industrial dehydrogenation processes was developed in the IG Farben plant in Ludwigshafen in 1931, and practiced in a 60 tonne-per-year plant. The 'styrene catalysts' were based on three-component systems (ZnO, Al_2O_3, and CaO). The iron oxide catalysts which are generally preferred today were introduced in 1957. They usually contain Cr_2O_3 and potassium compounds such as KOH or K_2CO_3 as promoters. The catalyst is placed in a shell-and-tube reactor, and, in the BASF process, the heat of reaction is supplied externally by a combustible gas. In the USA (*e.g.*, Dow process), the energy for the cleavage is introduced directly by means of superheated steam; here, the fixed-bed catalyst is in a vertical kiln. In this adiabatic mode for the endothermic dehydrogenation, the amount and initial temperature of the steam must be relatively high (2.5 − 3 kg steam/kg ethylbenzene at ca. 720 °C) to ensure sufficiently high temperatures for the dehydrogenation at the end of the catalyst bed.

A temperature of 550–620 °C is necessary for the dehydrogenation. In order to limit side reactions, the partial pressure of ethylbenzene is reduced by admixing an equal amount of steam; this is also done in process with indirect addition of heat. The ethylbenzene conversion is about 40% with conventional catalysts and 60—65% with modern catalysts (*e.g.*, Shell); the selectivity to styrene is greater than 90%. The byproducts are toluene, benzene, and a small amount of tarry substances. The reaction products are cooled rapidly to prevent polymerization. On reacting 100 °C, tar-like substances, then styrene and unreached ethylbenzene are fractionally condensed. The hydrogen formed during the reaction is burned to generate the dehydration temperature. The trend to larger reactors has led to capacities of 70 000 to 100 000 tonnes per year per unit.

After adding a polymerization inhibitor (formerly sulfur, now usually a phenol), the styrene is vacuum distilled. The purification for polymerization applications is very difficult due to the similar boiling points of styrene and ethylbenzene. The required purity of more than 99.8% is reached using four columns.

Processes for dehydrogenation and oxydehydrogenation of ethylbenzene, *i.e.*, with simultaneous addition of oxygen in amounts such that the reaction takes place isothermally, have been developed by many companies. The catalyst used by Distillers for the oxidative dehydrogenation contains, for example, the oxides of V, Mg, and Al. A selectivity to styrene of

95% with an increased conversion of ethylbenzene up to 90% is reported for this catalyst. Other catalysts are based on phosphates of Ce/Zr, Zr or alkaline earth/Ni. Oxydehydrogenation of ethylbenzene — for example the Styro-Plus process (now the SMART SM process) of UOP — have been piloted and will soon be used commercially.

A recent process development in the CIS involves a combination of the endothermic ethylbenzene dehydrogenation with the strongly exothermic hydrogenation of nitrobenzene to aniline in a single process step. The resulting weakly exothermic reaction gives styrene and aniline in a 3.3:1 ratio by weight. The selectivity is over 99% for both products. No commercial application has been reported.

2. catalytic H_2 transfer from ethylbenzene to nitrobenzene with exothermic formation of styrene and aniline

Single-step processes (circumventing the ethylbenzene stage) are unlikely to be introduced in the near future because of the mature technology and high selectivity of styrene manufacture from ethylbenzene. A single-step process was described in the literature by Japanese workers who conducted an oxyalkylation of benzene with ethylene and oxygen in the presence of rhodium catalysts.

alternative routes for styrene manufacture:

1. oxyalkylation

$$\bigcirc \;+\; H_2C{=}CH_2 \;+\; 0.5\,O_2$$

2. toluene methylenation

$$\bigcirc^{CH_3} + \text{C}_1\text{-cmpds} \longrightarrow \text{styrene}$$

Similarly, synthetic reactions — *e. g.*, starting with toluene and C_1 compounds such as methane, methanol or formaldehyde, or with two butadiene units in a (4+2) Diels-Alder cycloaddition to 4-vinylcyclohexene, or by the oxidative dimerization of toluene to stilbene followed by metathesis with ethylene — have not been used commercially. However, Dow has further developed the cyclic dimerization of butadiene to vinylcyclohexene, and began piloting in 1994. Here the butadiene in a C_4-fraction is dimerized in the liquid phase at about 100 °C and 19 bar over a Cu-zeolite catalyst. Selectivity to 4-vinylcyclohexene of over 99% is achieved with a conversion of about 90%. This is then oxydehydrogenated in the gas phase over a Sn/Sb-oxide catalyst at ca. 400 °C and 6 bar with a selectivity to styrene of about 92% at 90% conversion.

3. butadiene cyclodehydrogenation

$$2\,H_2C{=}CH{-}CH{=}CH_2 \rightarrow \bigcirc$$

$$\rightarrow \text{styrene}$$

with technical development by Dow

4. toluene oxydehydrogenative dimerization/C_2H_4 metathesis

$$2\,\bigcirc^{CH_3} \xrightarrow{-2H_2O} \bigcirc{=}\bigcirc$$

$$\xrightarrow{+H_2C{=}CH_2} \text{styrene}$$

Toray's Stex process — the **st**yrene **ex**traction from pyrolysis gasoline — enjoys a special position among the styrene manufacturing processes. With about 6–8% styrene in pyrolysis gasoline, styrene capacities are tied to the capacities of the cracking plants supplying the pyrolysis gasoline.

After ethylene and vinyl chloride, styrene is the most important monomer for the manufacture of thermoplastics. It is also used to a large extent in the production of elastomers, thermosetting plastics, and polymer dispersions for a variety of applications.

uses of styrene:

as monomer for thermoplastics, elastomers, dispersions, and thermosetting resins

The distribution of styrene use worldwide and in the USA, Western Europe, and Japan is shown in the following table:

Table 13-2. Styrene use (in %).

Product	World		USA		Western Europe		Japan	
	1991	1995	1986	1995	1985	1995	1985	1995
Polystyrene	66	65	65	59	66	66	63	64
Styrene-butadiene-rubber (SBR)	6	5	6	4	13	12	9	8
Acrylonitrile-butadiene-styrene (ABS)	9	9	9	8	9	8	14	12
Styrene-acrylonitrile (SAN)	2	1	1	1	2	1	4	2
Polyester resins (unsaturated)	5	5	7	6	6	5	4	6
Miscellaneous	12	15	12	22	4	8	6	8
Total use (in 10^6 tonnes)	13.9	15.6	3.04	4.37	2.95	4.01	1.63	2.08

13.1.3. Cumene

cumene production (in 10^6 tonnes):

	1990	1992	1994
USA	1.96	2.07	2.37
W. Europe	1.97	1.77	1.90
Japan	0.67	0.72[1]	0.64

[1] 1991

Cumene, or isopropylbenzene, has become an important intermediate product since the discovery of its elegant conversion into phenol and acetone (Hock process), which was widely used in industry in the 1950s.

In 1995, world production capacity was more than 8.0×10^6 tonnes per year, with 2.7, 2.4 and 0.88×10^6 tonnes per year in the USA, Western Europe, and Japan, respectively. Production figures for several countries are given in the adjacent table.

two process variations for cumene manufacture by benzene propylation:

Cumene is manufactured exclusively by alkylation of benzene with propene. In the most important industrialized countries, propene used for production of cumene accounts for 7–8% of the total propene consumption:

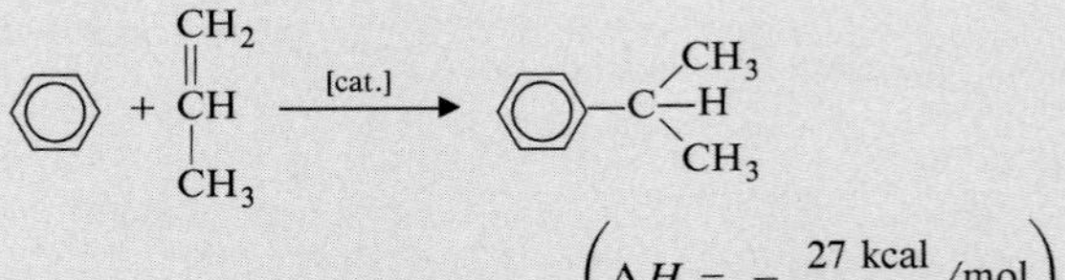

$$\left(\Delta H = -\frac{27 \text{ kcal}}{113 \text{ kJ}} / \text{mol} \right)$$ (3)

1. liquid-phase alkylation in presence of H_2SO_4, $AlCl_3$, or HF at slightly raised temperature and pressure

The reaction takes place either in the liquid or gas phase. As in the ethylation of benzene, Friedel–Crafts systems or proton donors are used as catalysts. Reaction conditions for the alkylation with propene are generally milder than with ethylene, since propene is more easily protonated.

In the liquid-phase process, H_2SO_4 or $AlCl_3$ is used at 35–40 °C (*e.g.,* Kellogg/Monsanto), or alternatively HF at 50–70 °C (Hüls process), with a low propene pressure of up to 7 bar.

The gas-phase alkylation, *e.g.,* UOP process, is done with H_3PO_4/SiO_2 catalysts promoted with BF_3 at 200–250 °C and 20–40 bar, using propene/propane mixtures with only a very small concentration of ethylene or other olefins, since these would also lead to alkylation. Propane and any other saturated constituents do not affect the reaction as they are removed unaltered. Steam is simultaneously passed over the catalyst so that:

2. gas-phase alkylation with fixed-bed H_3PO_4/SiO_2 catalyst (BF_3 promoted) under pressure at higher temperatures

1. The exothermic reaction is controlled via heat absorbed by the water

2. The phosphoric acid is better attached to the support due to hydrate formation

process characteristics of 2:

H_2O addition has two effects:

1. better temperature control
2. stabilization of catalyst activity

Moreover, an excess of benzene is normally used to minimize the facile further propylation of cumene. The selectivities are 96–97% (based on C_6H_6) and 91–92% (based on C_3H_6). Di- and triisopropyl benzene and *n*-propylbenzene are the main by-products. After distillation, cumene is obtained with a purity of more than 99.5%. Process improvements from, *e.g.,* Monsanto/Lummus include downstream transalkylation in order to use the more highly propylated products to increase the cumene yield.

benzene excess to avoid oligo-propylation results in benzene recycle

The majority of the currently operating processes use the UOP route, which in 1991 accounted for about 90% of cumene production worldwide.

More recent process developments concern the use of zeolite catalysts (*e.g.,* ABB Lummus Crest/Nova, Enichem, etc.), which have the advantages of a noncorrosive system without the release of acidic catalyst components. In addition, yields of nearly 100% can be attained, thereby increasing the capacity of the plant.

modern process variation with zeolite catalysts

Cumene is used almost exclusively for the manufacture of phenol by the Hock process (*cf.* Section 13.2.1.1). Cumene-containing alkylation products of benzene are employed to improve the octane rating of motor gasoline.

uses of cumene:

1. mainly for manufacture of phenol (Hock process)
2. in mixture with other alkylbenzenes for increased octane rating

13.1.4. Higher Alkylbenzenes

Monoalkylbenzenes with 10–14 carbon atoms in the alkyl side chain are starting materials for the manufacture of alkylbenzene-

higher alkylbenzenes are basis for manufacture of alkylarylsulfonates

$$\left[R\!\!-\!\!\bigcirc\!\!-\!\!SO_3 \right]^{\ominus} Na^{\oplus}$$

isododecylbenzenesulfonate based on tetra-propene is completely losing its importance as anion-active raw material for detergents

isoalkylbenzenesulfonates replaced by *n*-al-kylbenzenesulfonates (more biodegradable) – LAS = linear alkylbenzenesulfonates

types of cleaning agents:
anionic
nonionic
cationic
amphoteric

production of linear alkylbenzenes (in 1000 tonnes:

	1988	1990	1994
W. Europe	508	524	485
USA	166	173	180[1)]
Japan	197	212	197

[1)] 1991

feedstocks:

n-C_{10} to C_{14} olefins obtainable from *n*-paraffins by four processes:

1. cracking of wax
2. chlorination and dehydrochlorination
3. dehydrogenation
4. C_2H_4 oligomerization

alkylation in liquid phase with:

1. olefins (HBF_4, HF, $AlCl_3$)
2. alkyl chlorides ($AlCl_3$)

$$R^1\!\!-\!\!CH\!\!-\!\!CH_2R^2$$
$$\underset{Cl}{|}$$

sulfonates which have gained considerable importance as raw materials for anion-active detergents.

Until the mid 1960s, 'tetrapropene' ('iso-$C_{12}H_{24}$') was the most important olefin feedstock for the manufacture of alkylbenzenes. This olefin could be manufactured along with other oligomers from propene-containing refinery gases using H_3PO_4/support catalysts at 200–240 °C and 15–25 bar. However, its strong branching makes isododecylbenzenesulfonate very difficult to degrade biologically.

Because of heightened legal requirements in many countries for biodegradable detergents, there was a relatively rapid change-over to unbranched C_{10}-C_{14} olefins as feedstocks for benzene alkylation.

For example, in 1970, 134000 tonnes of isododecylbenzene and 269000 tonnes of linear alkylbenzenes were used for surfactants in Western Europe; in 1975, the corresponding amounts were 3000 tonnes of isododecylbenzene and 496000 tonnes of linear alkylbenzenes. Currently, only linear alkylbenzenes are used for the manufacture of alkylbenzenesulfonates for cleaning agents in Western Europe, the USA, and Japan. In 1995, the production capacity for linear alkylbenzenes in these countries was about 0.53, 0.38, and 0.15 × 10^6 tonnes per year. Production figures for linear alkylbenzenes in these countries are given in the adjacent table.

Most *n*-olefin feedstocks for alkylation are manufactured from *n*-paraffins in one of four industrial processes (*cf.* Section 3.3.3.1): by thermal cracking, chlorination and dehydrochlorination, dehydrogenation, or by Ziegler oligomerization of ethylene. To obtain high selectivity to the linear monoolefins, the paraffins are usually incompletely reacted. The resulting paraffin/olefin mixture can be directly used for benzene alkylation; the olefins react and the paraffins are recycled to the olefin manufacture. The distillation of the alkylation mixtures is simpler than separation of the olefin/paraffin mixture.

Alkylation of benzene with *n*-olefins is generally done in the liquid phase at 40–70 °C with HF, HBF_4 or $AlCl_3$, but it is also done in the gas phase. In the presence of Brönstedt or Lewis acid catalysts, and depending on the reaction conditions, a partial double bond isomerization takes place so that the benzene nucleus is distributed statistically along the chain of the linear olefin. Cracked olefins from paraffin thermolysis (wax cracking) and α-olefins give similar alkylbenzene mixtures.

In the next stage, the alkylbenzene is sulfonated with SO_3 or oleum and converted to the sodium salt:

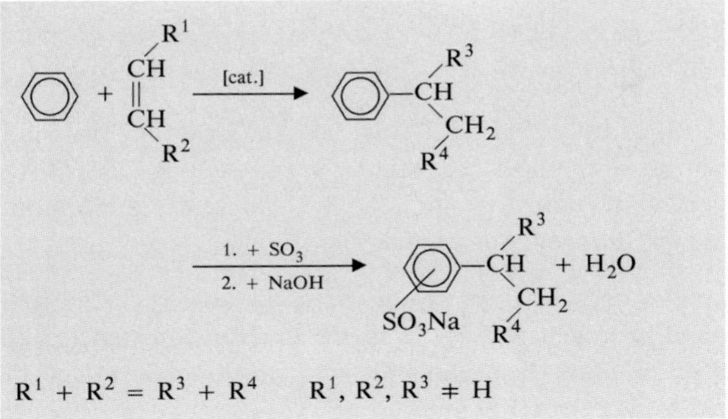

$$R^1 + R^2 = R^3 + R^4 \qquad R^1, R^2, R^3 \neq H \tag{4}$$

In other commercial processes, though still practiced by only a few companies, e. g., in Germany, Italy, and Japan, benzene is reacted directly with monochloroalkanes, or with mixtures of paraffins and monochloroalkanes, in the presence of $AlCl_3$. The reaction takes place with the elimination of HCl.

13.1.5. Cyclohexane

Cyclohexane was first obtained directly by fractional distillation of suitable crude gasoline cuts; the purity, however, was only 85%. The product quality was improved to almost 98% by the simultaneous isomerization of methylcyclopentane to cyclohexane, as first practiced in plants in the USA. Because of the markedly increasing demand for cyclohexane as a feedstock for nylon 6 and nylon 6,6, this covers only a small fraction of the demand for cyclohexane. The greater portion (ca. 80–85%) is obtained from the hydrogenation of benzene.

In 1995, the production capacity for cyclohexane worldwide was 5.1×10^6 tonnes per year, with 2.0, 1.5, and 0.74×10^6 tonnes per year in the USA, Western Europe, and Japan, respectively. Production figures for these countries are summarized in the adjacent table.

Of the numerous methods for the hydrogenation of benzene, the older high pressure processes using sulfur-resistant catalysts such as $NiS-WS_2$ and a benzene feedstock which generally contained sulfur are insignificant today.

Newer processes over nickel or platinum metal catalysts require extremely pure benzene with less than 1 ppm sulfur in order that

production of cyclohexane:

older process:

isolation from crude gasoline by fractional distillation

modern processes:

1. isolation from crude gasoline with simultaneous isomerization

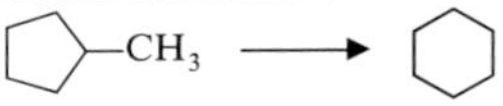

2. hydrogenation of benzene

cyclohexane production (in 1000 tonnes):

	1990	1992	1994
USA	1119	1002	890
W. Europe	861	923	945
Japan	629	632	640

to 2:

benzene hydrogenation requires two hydrogenation methods (depending on S content):

2.1. high-pressure hydrogenation with sulfur-resistant catalysts (insignificant today) the catalysts are effective in the liquid phase under mild conditions (20 – 40 bar, 170 – 230 °C):

2.2. medium-pressure hydrogenation with sulfur-sensitive Ni- or *e. g.*, Pt–Li/Al$_2$O$_3$ catalysts (commonly used today)

$$\bigcirc + 3\,H_2 \longrightarrow \bigcirc \quad \left(\Delta H = -\begin{array}{c} 51\ \mathrm{kcal} \\ 214\ \mathrm{kJ} \end{array} /\mathrm{mol} \right) \qquad (5)$$

two process variations for medium-pressure hydrogenation of benzene with low S content:

In the exothermic hydrogenation, careful heat removal and observation of an upper temperature limit of about 230 °C with exact residence times are imperative to prevent equilibration between cyclohexane and methylcyclopentane.

1. liquid-phase hydrogenation, stepwise with initial and finishing reactor

In several processes, there is only 95% conversion in the first stage, and hydrogenation is completed in a finishing reactor. In this way, residual benzene and methylcyclopentane can be reduced to less than 100 ppm.

Liquid-phase hydrogenations are used commercially by, *e. g.*, Mitsubishi Chemical, Hydrocarbon-Sinclair, and IFP.

process characteristics:

exothermic hydrogenation requires careful temperature and residence time control to attain complete conversion with high selectivity

In the IFP process, a Raney nickel catalyst is used in a bubble column reactor at 200 – 225 °C and 50 bar. The nickel suspension is circulated to improve heat removal. Unreacted benzene (> 5%) is completely hydrogenated in the gas phase in a coupled fixed-bed reactor. For example, in 1993 the process was used worldwide in 19 plants with a total annual production capacity of 1.5×10^6 tonnes.

2. gas-phase hydrogenation, *e. g.*, in coupled reactors

Recently, numerous gas-phase hydrogenations have been developed *e. g.*, by Arco, DSM, Toray (Hytoray), Houdry, and UOP. The Hydrar process (UOP) uses a series of three reactors with a supported Ni packing and a stepwise temperature increase (from 400 to 600 °C) at 30 bar.

process characteristics:

gas phase means simple separation of substrate and catalyst, combined with short residence time, which, despite high temperature, prevents isomerization

Despite higher reaction temperatures, the isomerization between cyclohexane and methylcyclopentane does not equilibrate due to the short residence times.

The Arco and DSM processes use a noble metal catalyst with which a complete conversion of benzene and almost 100% selectivity to cyclohexane can be achieved in a single hydrogenation step at about 400 °C and 25 – 30 bar.

If necessary, benzene homologues can be hydrogenated to the corresponding cyclohexane homologues by hydrogenation of the aromatic nucleus.

benzene hydrogenation, alone or in combination with isomerization, also used to improve quality of motor fuels

To reduce the benzene in gasoline, many companies use hydrogenation processes such as the UOP Ben-Sat process to completely hydrogenate benzene to cyclohexane. Others use a process such as the Tenex-Plus process, which combines hydrogenation and isomerization to increase the octane number.

The importance of cyclohexane lies mainly in its conversion to the intermediate cyclohexanone, a feedstock for nylon precursors such as adipic acid, ε-caprolactam, and hexamethylenediamine (*cf.* Section 10). Currently, adipic acid consumes about 58, 13, and 42% of the cyclohexane produced in the USA, Japan, and Western Europe, respectively; about 38, 84, and 43%, respectively, is used in the production of ε-caprolactam.

uses of cyclohexane:

starting material for cyclohexanone/cyclohexanol precursor for:

$HOOC(CH_2)_4COOH$

$H_2N(CH_2)_6NH_2$

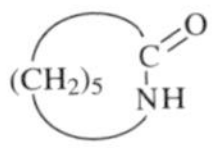

13.2. Oxidation and Secondary Products of Benzene

13.2.1. Phenol

Phenol, the second largest volume chemical derived from benzene in the USA and Western Europe, currently consumes about 20% of the total benzene production. Besides synthesis, other sources include tar and coke-oven water from coal coking and low temperature carbonization of brown coal, and the so-called 'cresylic acids' present in waste water from cracking plants. Phenols, cresols, and xylenols are recovered by washing cracked gasolines with alkaline solutions and treating the wash solution with CO_2. A newer method for obtaining phenolic products, especially from tar distillation fractions, is the Lurgi Phenoraffin process, which uses aqueous sodium phenolate solution and isopropyl ether extractant. Other processes, such as Phenosolvan, use butyl acetate or other esters for the extraction.

phenol production:

1. synthesis based on benzene
2. isolation from bituminous and brown coal tar and wash liquor from cracked gasoline

The amount of phenol produced from tars, coke-oven water and wastewater from cracking units has risen slightly in the past decade in the USA and Western Europe, but is still insignificant when compared to synthetic phenol production. The following table indicates the development of phenol production in the USA, Western Europe, and Japan:

Table 13-3. Phenol production (in 1000 tonnes).

Country	Phenol source	1972	1973	1985	1991	1995
USA:	Synthetic phenol	865		1260	1553	1873
	From tar and wastewater	20		24	27	27
	Total	885		1284	1580	1900
Western Europe:	Synthetic phenol		891	1157	1460	1493
	From tar and wastewater		14	14	28	14
	Total		905	1171	1488	1507
Japan:	Synthetic phenol			255	568	
	From tar and wastewater			2	2	
	Total			257	570	

synthetic phenol production (in 1000 tonnes):

	1988	1990	1992
USA	1594	1592	1664
W. Europe	1205	1358	1495
CIS	520	515	435[1)]
Japan	358	403	538

[1)] 1991

Additional production figures for synthetic phenol in several countries are summarized in the adjacent table. In 1996, world production capacity for synthetic phenol was about 4.9×10^6 tonnes per year, with about 1.9, 1.8, and 0.93×10^6 tonnes per year in the USA, Western Europe, and Japan, respectively. Phenolchemie is the largest producer worldwide with a capacity of 920000 tonnes per year (1997).

13.2.1.1. Manufacturing Processes for Phenol

principles of phenol syntheses from benzene or its derivatives:

indirect oxidation processes with numerous precursors and intermediates, as direct oxidation of benzene → phenol is unselective *i. e.*, total oxidation of phenol is preferred to partial oxidation of benzene

Although a direct oxidation of benzene to phenol could well be the most economical route, only the indirect manufacturing processes have been implemented, since the total oxidation of phenol is preferred to the partial oxidation of benzene. Moreover, larger amounts of diphenyl are formed as a byproduct.

Thus numerous indirect processes have been developed for the manufacture of phenol; of these, five have attained commercial significance:

Table 13-4. Routes for manufacturing phenol.

Precursor	Intermediate	Process
1. Benzene	Benzenesulfonic acid	Classical
2. Benzene	Chlorobenzene	Dow and Bayer (alkaline hydrolysis) Raschig–Hooker (acidic hydrolysis)
3. Toluene	Benzoic acid	Dow and California Research
4. Cyclohexane	Cyclohexanone–cyclohexanol	Scientific Design
5. Benzene/propene	Isopropylbenzene (cumene)	Hock

To 1:

principles of benzenesulfonic acid route:

1. introduction of —SO_3H
2. exchange for —OH

additional steps for neutralization and formation of free phenol

The classical sulfonation process has the advantage over other phenol manufacturing methods that small plants with capacities below 4000 tonnes per year can also operate economically if there is a market for the byproducts, in particular for Na_2SO_3. The plant equipment required is simple and suitable for other chemical processes. Phenol is no longer produced by this process; the last phenol plant in the USA to use benzene sulfonylation was shut down in 1978. The process proceeds in four steps:

industrial operation in four steps:

1. sulfonation, discontinuous or continuous
2. neutralization, *e. g.*, with $CaCO_3$ in older processes, with Na_2SO_3 or NaOH in more recent processes

1. Sulfonation of benzene at 110–150 °C with an excess of H_2SO_4 or oleum ($H_2SO_4 + SO_3$)
2. Neutralization of benzenesulfonic acid and excess H_2SO_4 with Na_2SO_3:

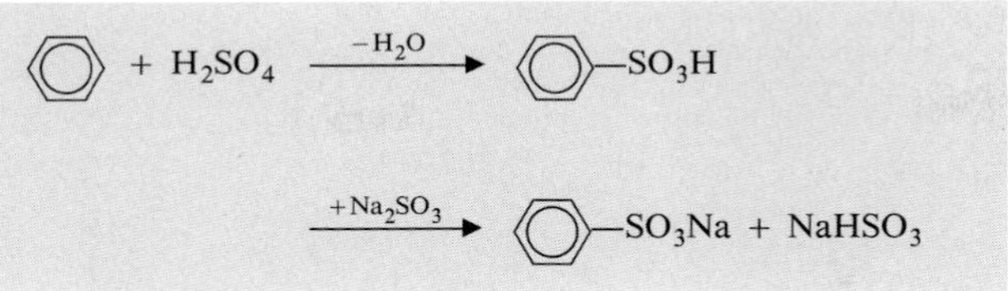

$$\text{C}_6\text{H}_6 + \text{H}_2\text{SO}_4 \xrightarrow{-\text{H}_2\text{O}} \text{C}_6\text{H}_5\text{--SO}_3\text{H} \tag{6}$$

$$\xrightarrow{+\text{Na}_2\text{SO}_3} \text{C}_6\text{H}_5\text{--SO}_3\text{Na} + \text{NaHSO}_3$$

3. NaOH melt with Na salt or with sulfonic acid directly (then step 2 eliminated)
4. acidification with SO_2 from process as well as from other sources

3. Reaction of sodium salt in a NaOH melt a $320-340\,°C$ to form sodium phenolate and sodium sulfite
4. Release of phenol by acidification with SO_2, including SO_2 from the second stage:

benzenesulfonic acid is used for a multitude of other intermediates, independent of phenol manufacture

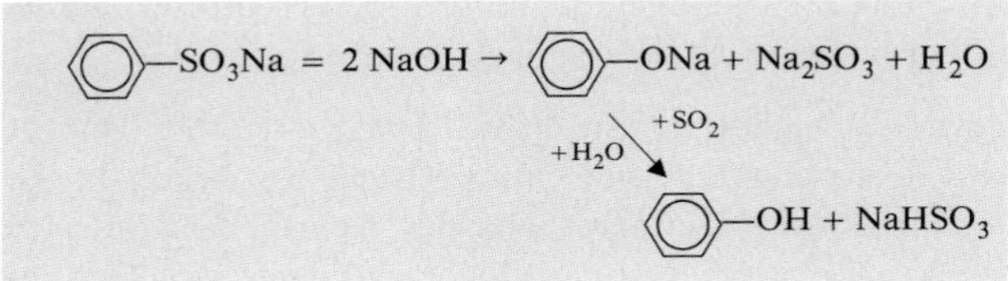

$$\text{C}_6\text{H}_5\text{--SO}_3\text{Na} = 2\,\text{NaOH} \rightarrow \text{C}_6\text{H}_5\text{--ONa} + \text{Na}_2\text{SO}_3 + \text{H}_2\text{O}$$

$$\xrightarrow[+\text{H}_2\text{O}]{+\text{SO}_2} \tag{7}$$

$$\text{C}_6\text{H}_5\text{--OH} + \text{NaHSO}_3$$

One tonne of phenol requires the following feedstocks and produces the following byproducts:

disadvantage of benzenesulfonic acid route:

very high (unavoidable) formation of salts, high energy use in individual process steps

Feedstocks:

Byproducts:

Feedstocks:		Byproducts:	
1 tonne	benzene	1.35 tonnes	Na_2SO_3
1.75 tonnes	H_2SO_4 (100%)	2.1 tonnes	Na_2SO_4 (from excess H_2SO_4 in
1.7 tonnes	NaOH (100%)		stage 1)

The phenol is then purified by three-stage distillation. The yield is about 88%.

To 2:

The second route to phenol involves the intermediate monochlorobenzene, which can be manufactured from benzene in two ways.

principles of two-step chlorobenzene route:

1. chlorobenzene manufacture from benzene by chlorination or oxychlorination
2. chlorobenzene hydrolysis in liquid or gas phase

The classical chlorination of benzene, as was still run, for example, by Dow and Bayer until 1975 and 1977, respectively, was done at $25-50\,°C$ in the liquid phase with $FeCl_3$ catalyst. About 5% higher chlorinated benzenes are obtained as byproducts.

process characteristics of benzene chlorination (*e. g.*, Dow and Bayer):

In the more economical Raschig–Hooker process — first operated in 1932 by Rhône–Poulenc in a 3000 tonne-per-year plant — chlorobenzene is obtained by oxychlorination of benzene with HCl/air mixtures at about $240\,°C$ and atmospheric pressure:

$FeCl_3$-catalyzed liquid-phase reaction in bubble column reactor

amount of dichlorobenzene dependent on:

1. temperature
2. mole ratio C_6H_6/Cl_2
3. residence time
4. catalyst
5. conversion of C_6H_6

process characteristics of benzene oxychlorination (Raschig–Hooker):

catalyzed gas-phase reaction with Deacon system, conversion restricted due to:

1. limited heat removal in fixed bed
2. desired high selectivity to monochlorobenzene

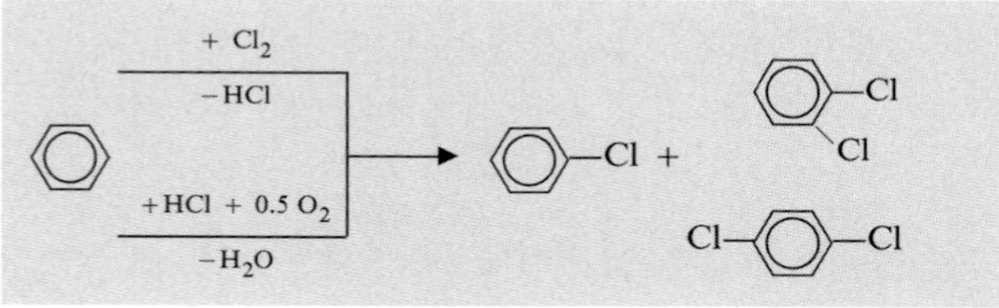

process characteristics of classical chlorobenzene hydrolysis:

liquid-phase hydrolysis with aqueous NaOH or Na_2CO_3 (*e. g.*, Dow and Bayer); disadvantage – total loss of chlorine

byproducts:

gas-phase hydrolysis with H_2O using fixed-bed catalyst (Raschig–Hooker),

advantage:

HCl assumes catalyst function

disadvantages:

highly corrosive oxychlorination and hydrolysis, requires catalyst regeneration, high energy consumption

$$(8)$$

$CuCl_2$–$FeCl_3/Al_2O_3$ is used as the catalyst. The benzene conversion is limited to 10–15% to control the evolution of heat at the fixed-bed catalyst and to suppress the formation of dichlorobenzenes. HCl is converted as completely as possible. In oxychlorination, the fraction of dichlorobenzenes is 6–10%, somewhat higher than with the classical chlorination.

The classical processes also differ from the Raschig–Hooker process in the conversion of chlorobenzene to phenol. In the classical processes, hydrolysis of chlorobenzene is done with 10–15% caustic soda or sodium carbonate solution at 360–390 °C and a pressure of 280–300 bar. Diphenyl ether, and *o*- and *p*-hydroxydiphenyl, are formed as byproducts.

HCl from the benzene chlorination is used to release the phenol from its sodium salt.

In the Raschig–Hooker process, hydrolysis takes place catalytically over $Ca_3(PO_4)_2/SiO_2$ with H_2O at 400–450 °C. Due to carbon deposition, the catalyst must be regenerated frequently. The conversion is about 10–15%, with an overall selectivity of 70–85% for both steps. In the USA, all plants using this technology have been shut down; however, this classical chlorobenzene/phenol route is still used to some extent in several countries, including Argentina, India, Italy, and Poland.

Based on the net equation for Dow and Bayer process:

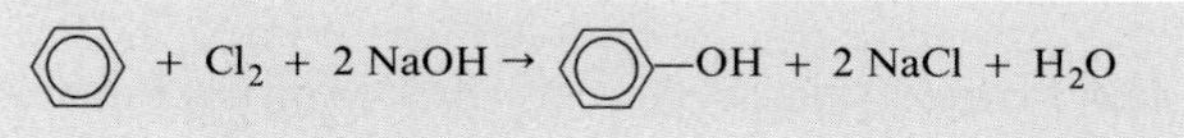

$$(9)$$

and for the Raschig–Hooker process, in which HCl from the hydrolysis of chlorobenzene can be used for the oxychlorination of benzene, one would expect a large economic advantage for the latter:

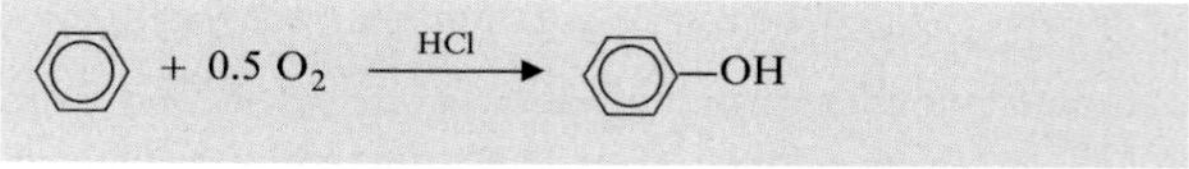

$$(10)$$

Only about 0.02 tonnes of HCl are consumed per tonne of phenol. However, hydrochloric acid is highly corrosive under the hydrolysis and oxychlorination conditions, calling for large investments for corrosion-resistant plants. Furthermore, due to the required vaporization of dilute hydrochloric acid, the Raschig–Hooker process uses large amounts of energy.

Chlorobenzene is not only a suitable starting material for phenol, but is also used in the manufacture of aniline (*cf.* Section 13.3.2) and as a raw material for numerous other aromatic intermediates. In addition, chlorobenzene is an important solvent.

use of chlorobenzene:

hydrolysis to phenol
ammonolysis to aniline
nitration to chloronitrobenzenes
condensation with chloral to DDT

In 1996, the production capacities for chlorobenzenes in the USA, Western Europe, and Japan were about 0.26, 0.20, and 0.12 tonnes per year, respectively.

Production figures for chlorobenzene in these countries are summarized in the adjacent table.

production of chlorobenzenes (in 1000 tonnes):

	1989	1991	1994
W. Europe	149	128	115
USA	134	146	151
Japan	69	79	65

Dichlorobenzenes are solvents, disinfectants, and intermediates in the manufacture of dyes and insecticides.

A commercially important secondary reaction of chlorobenzene is its nitration to a mixture of the three possible *o-*, *m-* and *p-*chloronitrobenzenes. The isomer ratio is essentially independent of conditions, with about 65% *p-*, 34% *o-* and only about 1% *m-*chloronitrobenzene. The mixture is separated by a combination of fractional distillation and crystallization. The chloronitrobenzenes are mainly processed to dyes.

One well-known product of chlorobenzene is DDT [1,1-bis(*p-*chlorophenyl)-2,2,2-trichloroethane], which is manufactured by acid-catalyzed condensation of chlorobenzene with chloral:

principle of DDT synthesis:

proton-catalyzed condensation of chlorobenzene with chloral to **d**ichloro**d**iphenyl-**t**richloroethane (*cf.* Section 13.2.1.3)

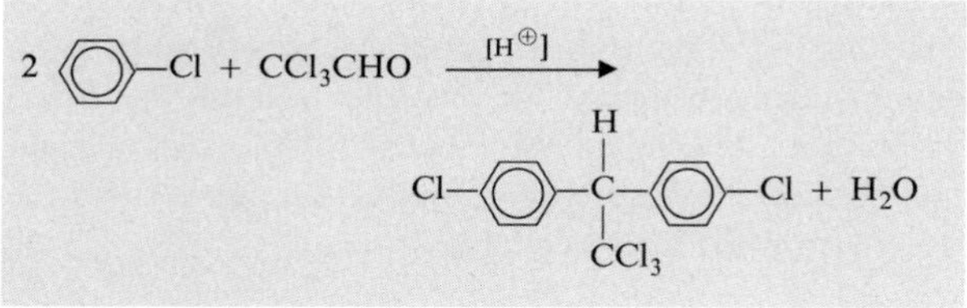

$$2 \; \text{C}_6\text{H}_5{-}\text{Cl} + \text{CCl}_3\text{CHO} \xrightarrow{[\text{H}^{\oplus}]}$$

$$\text{Cl}{-}\text{C}_6\text{H}_4{-}\overset{\overset{\displaystyle H}{|}}{\underset{\underset{\displaystyle CCl_3}{|}}{C}}{-}\text{C}_6\text{H}_4{-}\text{Cl} + \text{H}_2\text{O} \qquad (11)$$

DDT is an effective insecticide which was used very successfully several decades. Due to its extremely low rate of decomposition and its accumulation in human and animal fat tissues its use was severely restricted or forbidden in many countries, and its importance dropped sharply. However, since there was no inexpensive, effective substitute – especially for combating the anopheles mosquito, which carries malaria – DDT is produced (ca. 30000 tonnes in 1984 and ca. 12000 tonnes in 1989) and used in third-world countries; forbidding its use in the tropics could have catastrophic consequences. However, the anopheles mosquito has become resistant to DDT in several countries.

To 3:

Dow and California Research operate a two-step phenol manufacturing process. Toluene is first oxidized to benzoic acid, which is then oxidatively decarboxylated to phenol:

principle of toluene–benzoic acid route:

1. toluene oxidation to benzoic acid intermediate by radical chain mechanism
2. oxidative decarboxylation of benzoic acid to phenol

$$\text{[C}_6\text{H}_5\text{]}-CH_3 \xrightarrow[\text{[cat.]}]{+1.5\ O_2,\ -H_2O} \text{[C}_6\text{H}_5\text{]}-COOH \tag{12}$$

$$\xrightarrow[\text{[cat.]}]{+0.5\ O_2} \text{[C}_6\text{H}_5\text{]}-OH + CO_2$$

examples of commercial toluene oxidation processes:

liquid-phase oxidations (Dow, Amoco, or Snia Viscosa)

Benzoic acid can be produced from toluene in various ways. Dow, for example, oxidizes toluene in the liquid phase with air at $110-120\,°C$ and $2-3$ bar in the presence of Co salts. A similar process is used by Snia Viscosa with a Co-acetate catalyst at $165\,°C$ and 9 bar in several plants in Italy and the CIS . The selectivity to benzoic acid is about 90%. Another process is analogous to the Amoco terephthalic acid manufacture (*cf.* Section 14.2.2).

benzoic acid use other than phenol manufacture:

e. g., Snia Viscosa
→ hexahydrobenzoic acid →
ε-caprolactam

benzoic acid production (in 1000 tonnes):

	1989	1991	1994
USA	47	46	54

Benzoic acid is not only a precursor to phenol; it also has general importance as an intermediate, *e. g.*, in the Snia Viscosa route to ε-caprolactam (*cf.* Section 10.3.1.2), and in the Henkel process to terephthalic acid (*cf.* Section 14.2.3). Benzoic acid is also an intermediate for the manufacture of dyes and perfume, an auxiliary agent for the rubber industry, and a preservative. In 1995, the manufacturing capacity for benzoic acid in the USA, Western Europe, and Japan was about 107000, 162000, and 11000 tonnes per year, respectively. Production numbers for the USA are given in the adjacent table.

process characteristics of oxidative benzoic acid decarboxylation:

homogeneously catalyzed liquid-phase reaction (melt or solvent) with Cu salts as main catalyst and promoters

In the second step of the phenol manufacture, the purified benzoic acid is oxydecarboxylated to phenol either as a melt or in a high boiling solvent at $220-250\,°C$ in the presence of a steam/air mixture and copper salts and other promoters, *e. g.*, Mg salts. The total selectivity is about $70-80\%$ (based on toluene).

most important mechanistic step:

o-attack of the benzoate anion and decarboxylation of the salicylic acid ester

The catalytic effect of the copper (II) ion involves the intermediate formation of copper benzoate, which is thermally cleaved to CO_2, copper (I) ion, and phenyl benzoate. The ester is hydrolyzed to phenol and benzoic acid, and the copper (I) ion is reoxidized with air:

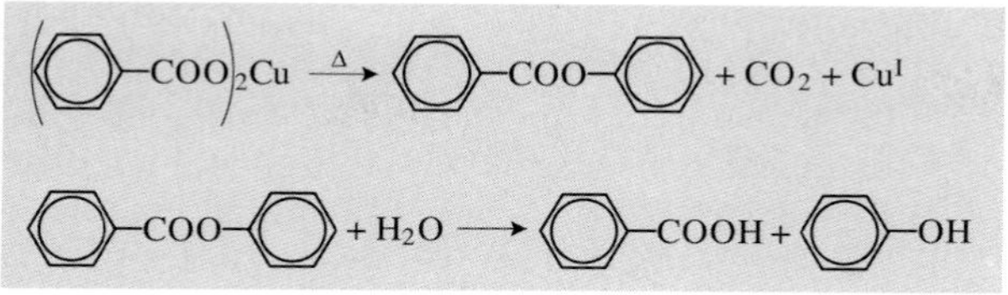

$$\text{C}_6\text{H}_5\text{-COO}\big]_2\text{Cu} \xrightarrow{\Delta} \text{C}_6\text{H}_5\text{-COO-C}_6\text{H}_5 + \text{CO}_2 + \text{Cu}^{\text{I}}$$

$$ \tag{13}$$

$$\text{C}_6\text{H}_5\text{-COO-C}_6\text{H}_5 + \text{H}_2\text{O} \longrightarrow \text{C}_6\text{H}_5\text{-COOH} + \text{C}_6\text{H}_5\text{-OH}$$

Alkali and alkaline earth metal additives, particularly magnesium and cobalt, have proven to be of value in increasing activity and selectivity.

examples of toluene–benzoic acid–phenol processes used commercially

Plants in the USA (Kalama), Canada, and the Netherlands (DSM) using the Dow process have a total capacity of 175 000 tonnes per year (1991). A new plant based on toluene with a capacity of 120 000 tonnes per year was brought on line in Japan by Nippon Phenol in 1991.

Dow (USA, Canada)

DSM/Dow (Netherlands)

Nippon Phenol (Japan)

To 4:

In the Scientific Design process, the mixture of cyclohexanone and cyclohexanol resulting from the cyclohexane oxidation is dehydrogenated to phenol at 400 °C (*cf.* Section 10.1.1):

principle of cyclohexanone/cyclohexanol route:

heterogeneously catalyzed gas-phase dehydrogenation of ketone/alcohol mixtures

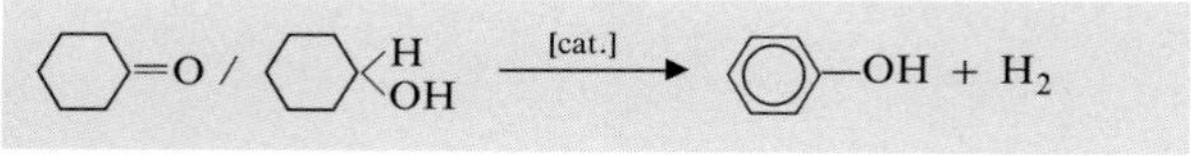

$$\text{C}_6\text{H}_{10}{=}\text{O} \,/\, \text{C}_6\text{H}_{10}\langle\substack{\text{H}\\\text{OH}} \xrightarrow{\text{[cat.]}} \text{C}_6\text{H}_5{-}\text{OH} + \text{H}_2 \tag{14}$$

Noble metal catalysts such as Pt/charcoal or Ni–Co can be used for the dehydrogenation. Since phenol forms an azeotropic mixture with cyclohexanone, an extractive workup of the reaction product is necessary, requiring a high rate of conversion (about 80%).

extractive workup of mixture from dehydrogenation to separate azeotrope of ketone and phenol

Monsanto practiced this technology for three years in a commercial plant in Australia; in 1968 it was converted to the cumene process for economic reasons.

To 5:

The cumene process, *i.e.*, the proton-catalyzed cleavage of cumene hydroperoxide to phenol and acetone, was discovered by Hock and Lang in 1944. Although the earliest commercialization was in an 8000 tonne-per-year Gulf plant in Canada in 1953, about 97% of the total synthetic phenol production in the USA until 1987, and 100% in Japan until 1990, was manufactured by this process. At that time, a new process based on toluene was introduced, and this route is now used for about

principles of Hock process (most important phenol manufacturing method throughout the world):

three-step process:

1. propylation of benzene to cumene
2. cumene oxidation to hydroperoxide
3. hydroperoxide cleavage

91% of phenol production in Western Europe. The world production capacity for phenol using the Hock process is currently about 5×10^6 tonnes per year. Despite the 0.62 tonnes of acetone produced per tonne phenol, this process has dominated because of its economics.

The principle of the combined phenol/acetone manufacture by the Hock process is based on the oxidation of cumene to cumene hydroperoxide and its subsequent cleavage in acidic media to phenol and acetone:

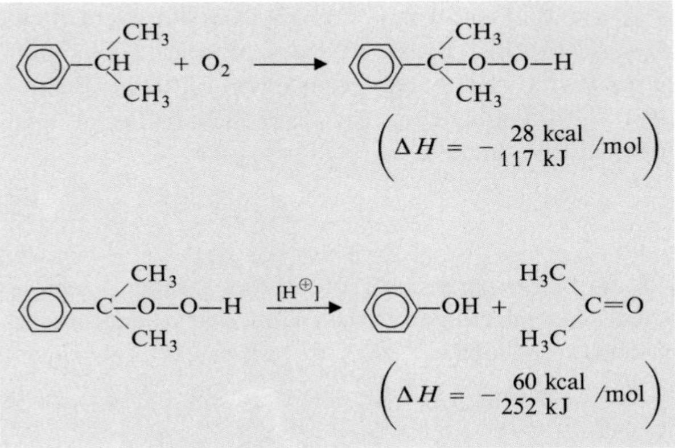

cumene oxidation to hydroperoxide, two process variations:

1. air oxidation in aqueous emulsion (Na_2CO_3)
2. air oxidation, undiluted, with redox catalysts

$$C_6H_5-CH(CH_3)_2 + O_2 \longrightarrow C_6H_5-C(CH_3)_2-O-O-H \qquad (15)$$

$$\left(\Delta H = -\frac{28 \text{ kcal}}{117 \text{ kJ}} \middle/ \text{mol} \right)$$

$$C_6H_5-C(CH_3)_2-O-O-H \xrightarrow{[H^{\oplus}]} C_6H_5-OH + (H_3C)_2C=O \qquad (16)$$

$$\left(\Delta H = -\frac{60 \text{ kcal}}{252 \text{ kJ}} \middle/ \text{mol} \right)$$

process characteristics:

limitation of cumene conversion (35–40%) to prevent side and secondary reactions, followed by concentration of dilute hydroperoxide solution, stabilization of acid-sensitive hydroperoxide by presence of Na_2CO_3 (otherwise premature cleavage to phenol, an effective inhibitor of radical oxidation)

In the industrial process, cumene from benzene propylation (*cf.* Section 13.1.3) can be oxidized to the hydroperoxide in a bubble column either by the BP (formerly Distillers), Hercules and Kellogg process with air in an aqueous emulsion containing Na_2CO_3 at pH 8.5–10.5, 90–130 °C and 5–10 bar, or undiluted in a homogeneous phase at 120 °C by the Hercules process. Cu, Mn, or Co salts can be used as catalysts. Although these catalysts can decrease the induction period at the start of the reaction, a lower hydroperoxide concentration may result due to its further oxidation to CO_2; thus catalysts have not been used commercially. The oxidation is discontinued when the hydroperoxide content is 35–40% in order to ensure low formation of byproducts such as dimethylphenylcarbinol and acetophenone. The oxidized product is concentrated to about 65–90% in vacuum columns and cleaved to phenol and acetone.

two process variations of proton-catalyzed hydroperoxide cleavage:

1. homogeneous in acetone or phenol with dilute H_2SO_4
2. heterogeneous with 40% H_2SO_4

The cleavage can be conducted either in homogeneous solution with 0.1–2% H_2SO_4 at 60–65 °C in acetone according to the BP and Hercules processes or in phenol, according to the Rhône-Poulenc approach. Phenolchemie operates a two-phase process with 40% H_2SO_4 at about 50 °C.

Numerous byproducts such as α-methylstyrene (isopropenyl-benzene), mesityl oxide, and higher boiling components such as α-cumylphenol are also formed in the strongly exothermic cleavage reaction. The phenol and acetone selectivity is about 91% (based on cumene). α-Methylstyrene is generally hydrogenated to cumene and recycled. Older processes used Ni catalysts with low selectivity. In a new process from Engelhardt, high selectivities to cumene are obtained with a quantitative conversion of α-methylstyrene over noble metal catalysts. The first commercial unit was brought on-line in 1979 by Shell, USA.

byproducts from:

1. hydroperoxide decomposition:

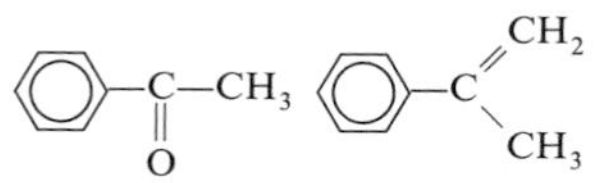

$C_6H_5-\overset{O}{\underset{\|}{C}}-CH_3$ $C_6H_5-C\overset{CH_2}{\underset{CH_3}{\diagdown}}$

2. secondary reactions:

$(CH_3)_2C=CHCOCH_3$

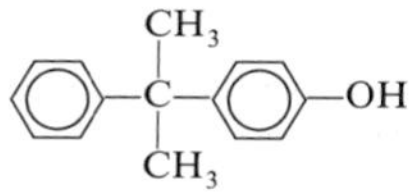

$C_6H_5-\overset{CH_3}{\underset{CH_3}{\overset{|}{\underset{|}{C}}}}-C_6H_4-OH$

α-Methylstyrene is also purified for use in styrene copolymers with increased thermal stability. α-Cumylphenol can be thermally cleaved to α-methylstyrene and phenol.

In 1993, the world capacities of phenol plants using the BP–Hercules–Kellogg process totalled more than 2.4×10^6 tonnes per year. These plants account for more than fifty percent of world phenol production. Phenol manufacture from cumene has also been developed to commercial processes by other firms, including UOP (Cumox process) and Allied/Lummus Crest.

In analogy to the Hock process, *m*- and *p*-isopropyltoluene (cymene) can, after oxidation and cleavage, yield *m*- and *p*-cresol along with acetone. Commercial cymene/cresol processes have been in operation by Sumitomo Chemical and Mitsui since 1969. Hydroquinone and resorcinol have also been made commercially from the corresponding diisopropylbenzenes by oxidation and cleavage (*cf.* Section 13.2.2).

analogues to Hock process for:

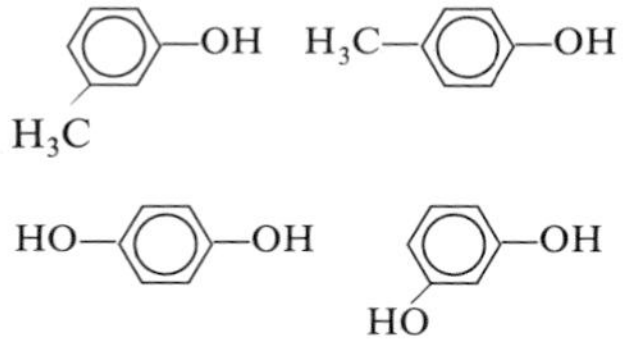

$H_3C-C_6H_4-OH$ $H_3C-C_6H_4-OH$

$HO-C_6H_4-OH$ $HO-C_6H_4-OH$

13.2.1.2. Potential Developments in Phenol Manufacture

As with all large-volume chemicals, considerations for less expensive or more environmentally sound manufacture come from two directions:

1. Improvements in existing processes

2. Establishing new processes using the cheapest raw materials and latest process technology

With phenol, both advancements point to the optimization of known indirect routes through intermediates, since a direct benzene-oxidation route has remained economically unsatisfactory.

phenol manufacture by indirect processes characterized by promotion of thrifty manufacture of the intermediate and economical use of byproducts and coproducts

The situation for the existing processes is as follows:

The sulfonation route is no longer used because of the unavoidable salt formation and the high energy requirements of individual process steps.

sulfonic acid route at most useable in small units with use for Na_2SO_3/Na_2SO_4

chlorobenzene route only acceptable if NaCl reconverted to Cl₂ and NaOH

One disadvantage of the Dow–Bayer chlorobenzene route is the expensive chlorine feedstock. This phenol process would require inexpensive electrical energy in order to recycle the NaCl product to the necessary chemicals Cl₂ and NaOH in a chlor-alkali electrolysis.

Raschig–Hooker route almost salt-free therefore attractive for increasing profitability. Three possibilities:

1. Hooker
2. Hoechst
3. Gulf

The Raschig–Hooker modification of the chlorobenzene route has high capital costs, but, ignoring the higher chlorinated by-products, produces fairly nonpolluting salts. For this reason, suggestions for improvement have centered on this process.

to 1:

Hooker catalyst allows utilization of di-chlorobenzenes for additional phenol

The Hooker modification uses a special hydrolysis catalyst which also dechlorinates dichlorobenzene to monochlorobenzene so that it can also be converted into phenol.

However, Hooker has shut down two of its three phenol plants using this process (total capacity 90 – 95 000 tonnes per year), and now uses the cumene process.

to 2:

Hoechst development allows utilization of higher chlorinated benzenes for CCl₄ manufacture by chlorolysis

Another possibility for utilizing dichlorobenzenes and higher chlorinated benzenes was developed by Hoechst. In this process, the oxychlorination of benzene can be carried out with higher conversions than the usual 10 – 15%, thus requiring less benzene recycle. The inevitably larger fraction of polychlorinated benzenes can be efficiently and noncatalytically converted into the important intermediate carbon tetrachloride by chlorolysis with an excess of chlorine at 400 – 800 °C and 20 – 200 bar (*cf.* Section 2.3.6.1).

to 3:

Gulf development allows technologically attractive mild, selective oxychlorination and hydrolysis route; presupposes a noncorrosive plant

A new liquid-phase route described by Gulf for the oxychlorination of benzene, alkylbenzenes, and other aromatics with aqueous HCl and air under mild conditions, followed by hydrolysis to phenols, presents another possibility for a closed chlorine cycle.

This catalytic oxychlorination is carried out either heterogeneously in the aqueous phase with dilute nitric acid, or homogeneously in acetic acid with Pd salts and nitrate ions. Benzene can, for example, be oxychlorinated with 80% conversion to monochlorobenzene (95% selectivity).

In the second step, the chlorinated aromatics are hydrolyzed to phenols over rare earth phosphates, preferably La- and CePO₄ with Cu salts as cocatalysts. The HCl released is used for the oxychlorination in the usual manner. 65% of the chlorobenzene is hydrolyzed to phenol in a single pass. A commercial plant has not yet been built.

acetoxylation of benzene to

as alternative to oxychlorination hardly corrosive, either valuable ketene as coproduct or AcOH recycle

benzoic acid decarboxylation route also worth efforts to increase profitability, since toluene is cheapest aromatic

developments relating to increased selectivity involve the catalyst ($Mo^{3\oplus}$ instead of $Cu^{2\oplus}$) and cocatalyst (*e. g.*, $Mg^{2\oplus}$) and the gas-phase process

ketone/alcohol dehydrogenation route appears uneconomical even with large scale production of feedstocks

Hock process remains optimal as long as acetone supplies a credit

Hock process with other feedstocks and thus coproducts not used commercially

Many firms have studied the acetoxylation of benzene, *i. e.*, the oxidative reaction of benzene with acetic acid over Pd catalyst, as an alternative to oxychlorination. Phenol could be obtained from the intermediate acetoxybenzene (phenylacetate) by hydrolysis to regenerate the acetic acid, or by thermolysis to produce the valuable intermediate ketene as a coproduct.

This route has not been used commercially.

In the Dow–California process, the benzoic acid feedstock is obtained from toluene, the least expensive aromatic compound. However, tar formation and the catalyst regeneration it necessitates were a source of technical difficulties for a long time.

Systematic catalyst research, in particular the work of an Italian team, showed that molybdenum benzoate, analogous to copper benzoate, can be thermolyzed to phenol with an increased selectivity of about 96%. With cocatalysts such as magnesium benzoate, selectivities of almost 99% are reported. The toluene–phenol route has been developed further by other firms such as DSM, which began operation of an expanded plant in the Netherlands in 1976, and Nippon Phenol, which started up a new unit in 1991.

Lummus has designed an oxidative decarboxylation of benzoic acid in the gas phase in the presence of a Cu-containing catalyst. The short residence time of the phenol leads to a sharp decrease in tar formation and allows for a selectivity to phenol of about 90%.

Phenol manufacture from cyclohexanol/cyclohexanone appears to be uneconomical, now and in the future, even if it could be operated in combination with large plants for caprolactam production based on cyclohexane oxidation. This is emphasized by the changeover of the lone process to a different operation.

The Hock process will maintain its dominant position as long as its coproduct acetone has a market. This situation is currently optimal. In previous years, the demand for phenol was greater than for acetone, but today the need for both products is similar since the markets for products made from acetone, especially methyl methacrylate, and acetone itself as a nontoxic solvent have increased. Of the many attempts to use feedstocks other than isopropylbenzene for the hydroperoxide in order to produce more attractive coproducts after cleavage to phenol, none have been used commercially.

These include ethylbenzene to give the coproduct acetaldehyde, and cyclohexylbenzene to give cyclohexanone, which can also be converted to phenol.

use of phenol:

1. phenolic resins (Bakelite, novolacs) by polycondensation with HCHO, with hydroxymethylphenols as precursors

13.2.1.3. Uses and Secondary Products of Phenol

The use of phenol in the world, the USA, Western Europe, and Japan is distributed roughly as follows:

Table 13-5. Use of phenol (in %).

Product	World		USA		Japan		Western Europe	
	1989	1995	1985	1995	1986	1994	1985	1994
Phenol resins	36	37	40	30	36	33	41	29
ε-Caprolactam	7	15	18	17	–	–	17	16
Bisphenol A	20	32	22	35	29	39	22	27
Adipic acid	1	2	1	1	–	–	1	2
Alkylphenols	5	2	4	6	4	4	4	6
Miscellaneous[1]	21	12	15	11	26	24	24	20
Total use (in 10^6 tonnes)	4.70	5.23	1.07	1.79	0.25	0.50	1.05	1.25

[1] e.g., aniline, chlorophenols, plasticizers, antioxidants.

production of phenoplasts (in 1000 tonnes):

	1988	1990	1993
USA	760	749	710
W. Europe	567	574[1]	569
Japan	356	385	328

[1] 1991

According to the above, phenolic resins or phenoplasts – the polycondensation products of phenol with formaldehyde – are the most important products from phenol in these countries. They are used, for example, in the manufacture of paints, adhesives, molding materials, and foam plastics. The production figures for phenolic plastics in several countries are given the adjacent table.

2. bisphenol A, A = acetone (bisphenol B, B = butanone, bisphenol F, F = formaldehyde, less important commercially)

2,2-Bis-(4-hydroxyphenyl)propane, the second largest consumer of phenol, is known as bisphenol A because of its production from phenol and acetone. Production figures for bisphenol A in several countries are listed in the adjacent table.

production of bisphenol A (in 1000 tonnes):

	1988	1990	1993
USA	489	518	583
W. Europe	247	365	345
Japan	124	148	211

In 1994, production capacities for bisphenol A in the USA, Western Europe, and Japan were 720000, 530000, and 320000 tonnes per year, respectively.

manufacture of bisphenol A:

acetone–phenol condensation:
catalyst: H_2SO_4
        HCl (Hooker)
        ion exchange resin (UCC)

The condensation of acetone with excess phenol can be conducted either continuously or discontinuously in the presence of H_2SO_4 or, in the Hooker process, with dry HCl as catalyst and methylmercaptan as promoter. The reaction at 50 °C is almost quantitative. According to the UCC process, a heterogeneous catalyst system (e.g., ion exchange resin) can also be used:

$$2\ HO-\!\!\!\bigcirc\!\!\!- + \ O\!=\!C\!\!\begin{array}{c}CH_3\\CH_3\end{array}\quad\xrightarrow{[H^{\oplus}]}$$

$$HO-\!\!\!\bigcirc\!\!\!-\overset{\overset{\displaystyle CH_3}{|}}{\underset{\underset{\displaystyle CH_3}{|}}{C}}-\!\!\!\bigcirc\!\!\!-OH\ +\ H_2O \qquad (17)$$

Bisphenol A is widely used in the manufacture of synthetic resins and thermoplastics. Bisphenol A glycidyl ethers are made from the reaction between bisphenol A and epichlorohydrin. These ethers are the basis of epoxy resins (*cf.* Section 11.2.1). Thermoplastic polycarbonates are obtained by polydehydrochlorination of bisphenol A with phosgene or by transesterification with dimethyl carbonate. These are the most commercially important type of polyesters of carbon dioxide, for which other aliphatic or aromatic dihydroxy compounds can also be used. Polycarbonates are mainly used in electrical engineering, in the building trade, and − for the past few years − in the rapidly growing market for compact disks.

The world production capacity for polycarbonates in 1995 was about 0.96×10^6 tonnes per year, of which approximately 0.40, 0.31, and 0.19×10^6 tonnes per year were in the USA, Western Europe, and Japan, respectively.

Transparent, thermally stable polysulfones are obtained by reaction of bisphenol A with 4,43-dichlorodiphenylsulfone. Transparent aromatic polyesters with high resistance to UV and temperatures to 150 °C are manufactured from phthalic acid and bisphenol A. The largest use for bisphenol A – currently 90–93% in the USA, Western Europe, and Japan – is in the manufacture of epoxides and polycarbonates.

Production of ε-caprolactam ranks third in the USA and Western Europe. In Japan, however, all ε-caprolactam is derived from cyclohexane. For this reason, manufacture of adipic acid from phenol is not found in Japan, since both ε-caprolactam and adipic acid are produced from phenol via the same intermediate, cyclohexanol.

Higher alkylphenols are obtained from the addition of olefins containing six or more carbon atoms to phenol. The reaction is generally run with equimolar amounts of the reactants in the liquid phase at 50 °C, and using HF, H_2SO_4, $AlCl_3$, BF_3 or acid ion exchange resin as a catalyst. The alkyl group is found

use of bisphenol A in reactions:

2.1. with epichlorohydrin to form epoxy resin precursors

2.2. with $COCl_2$ or $(CH_3O)_2CO$ to polycarbonates

2.3. with

$$Cl-\!\!\!\bigcirc\!\!\!-SO_2-\!\!\!\bigcirc\!\!\!-Cl$$

to polysulfones

2.4. with phthalic acid to polyarylates

3. ε-caprolactam with ketone/alcohol precursor

$$\bigcirc\!\!=\!O \quad\xrightarrow{\text{multistep}}$$

$$\underset{(CH_2)_5}{\bigcirc}\begin{array}{c}C\!=\!O\\ \\NH\end{array}$$

4. alkylphenols

4.1. higher alkylphenols:

manufacture by catalytic phenol alkylation with olefins in liquid phase

HO—⟨◯⟩ ⟶ HO—⟨◯⟩—R

R generally iso-C$_8$H$_{17}$ ⎫ for
        iso-C$_9$H$_{19}$ ⎬ commercial
        iso-C$_{12}$H$_{25}$ ⎭ use
or linear C$_6$–C$_{20}$ olefins

uses of alkylphenols:

after EO addition (ethoxylation):

surfactants
emulsifying agents ⎱ (nonionic)
detergents ⎰

sulfonation increases surfactant properties

4.2. methyl- and dimethylphenols:

manufacture by:

heterogeneously catalyzed gas- and liquid-phase methylation of phenol with
methanol

process characteristics:

dimethylation favored over mono, when

1. temperature
2. pressure
3. ratio CH$_3$OH/C$_6$H$_5$OH

are increased

primarily para, but also ortho, to the hydroxyl group. Since the alkylation catalyst is also active for isomerization, the proportion of the more stable para isomer in the reaction mixture depends on the reaction conditions. Higher temperatures and catalyst concentration favor the desired *p*-isomer.

Isooctyl, isononyl, and isododecylphenols are the most important alkylphenols, accounting for ca. 60% of the alkylphenol production worldwide, since they are derived from the inexpensive olefins diisobutene, and tri- and tetrapropene. They are used mainly in industry as drilling oil additives, antioxidants, and other aids in rubber and plastic processing.

Linear olefins (C$_6$–C$_{20}$) with terminal or inner double bonds are generally used for phenol alkylation to produce raw materials for nonionic detergents, surfactants, and emulsifiers, since these are more biodegradable. The hydrophobic long-chain phenols are ethoxylated with ethylene oxide for these purposes (*cf.* Section 7.2.2). The surface-active characteristics can be strengthened further by sulfonylation.

The most significant of the commercially important lower alkylphenols are the methyl derivatives, *i. e.*, the cresols and xylenols.

Of these, the demand for *o*-cresol and 2,6-xylenol has recently increased, so that demand for *o*-cresol could no longer be met solely from petroleum and tar distillate sources (*cf.* Section 13.2.1.1). And since it is not found in natural products, manufacturing processes had to be developed for 2,6-xylenol.

Cresols and xylenols can be obtained by methylation of phenol with methanol in the gas or liquid phase. Several firms in the USA, England, and Japan use a gas-phase process over metal oxide catalysts (*e. g.*, Al$_2$O$_3$) at 300–450 °C and atmospheric or slightly elevated pressure. The product composition (mainly *o*-cresol and 2,6-xylenol) is very dependent on the catalyst and the reaction conditions.

Liquid-phase methylation of phenol is used in Germany. Here also the ratio of *o*-cresol to 2,6-xylenol can be very widely varied by altering the process conditions. *o*-Cresol is favored at 40–70 bar and 300–360 °C over an Al$_2$O$_3$ catalyst. At higher temperature and pressure, 2,6-xylenol is formed preferentially, along with a little 2,4-xylenol. *p*-Cresol is a byproduct of *o*-cresol:

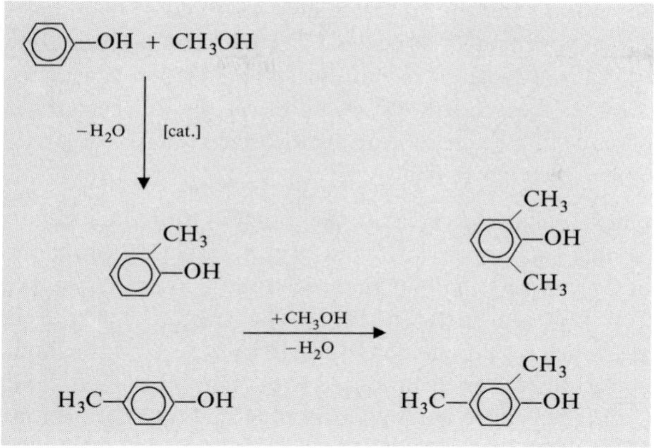

(18)

o-Cresol is used to a large extent in the manufacture of herbicides and insecticides such as − after nitration − 4,6-dinitro-*o*-cresol (DNOC). However, it is primarily used for the chlorination to *p*-chloro-*o*-cresol and its further reaction to selective herbicides. *p*-Cresol is also alkylated with isobutene in the manufacture of 2,6-di-*tert*-butyl-4-methylphenol (BHT, Jonol), which is used as an antioxidant and preservative for plastics, motor oil, and foodstuffs.

use of methylphenols:

o-cresol as starting material for herbicides and insecticides

2,6-Xylenol is the starting material for polyphenylene oxide, a thermoplastic with high heat and chemical resistance and excellent electrical properties developed by General Electric.

2,6-xylenol for polyphenylene oxide

The manufacture of the linear polyether takes place in the presence of a basic copper complex catalyst and air or oxygen in accordance with the oxidation coupling principle:

manufacture of polyphenylene oxide by oxidative polycondensation ('polyoxidation')

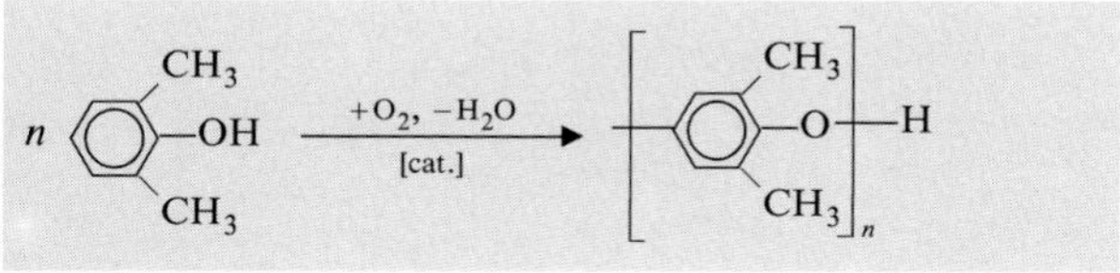

(19)

Other products of phenol include salicylic acid from the direct carboxylation of sodium phenolate with CO_2, and more recently phenoxycarboxylic acids, which are used as selective herbicides.

13.2.2. Dihydroxybenzenes

Resorcinol and hydroquinone are more important industrially than pyrocatechol, the third isomeric dihydroxybenzene.

resorcinol and hydroquinone are most important dihydroxybenzenes commercially

resorcinol production (in 1000 tonnes):

	1988	1990	1993
USA	15.4	14.5	19.4
W. Europe	8.0	8.0	–
Japan	6.0	7.5	10.7

In 1993, world capacity for resorcinol was about 45 000 tonnes per year, of which about 20 000 and 21 000 tonnes per year were in the USA and Japan, respectively. The production figures for resorcinol in these countries are listed in the adjacent table. Capacities and production of hydroquinone are similar or slightly less than resorcinol.

resorcinol manufacture in five steps in classical method:

1. monosulfonation
2. disulfonation
3. Na salt formation
4. NaOH melt reaction
5. release of resorcinol from salt

improved manufacture by sulfonation in melt of disulfonic acid avoids step 2 (Hoechst process)

Resorcinol is manufactured in the classical sulfonate fusion process only by Koppers (now Indspec), the sole US producer. Hoechst, previously the only manufacturer in Western Europe, ceased production early in 1992. The starting material is benzene, which is sulfonated in two steps using an older method. The monosulfonic acid is produced with 100% H_2SO_4 at about 100 °C and then converted to m-disulfonic acid with 65% oleum at 80–85 °C. In a separate continuous Hoechst process, benzene and SO_3 are reacted in a m-benzenedisulfonic acid melt at 140–160 °C. The melt is neutralized directly. This shortens the older resorcinol route by one step; in addition, the disodium salt of the resorcinol contains very little Na_2SO_4.

The reaction of m-benzenedisulfonate to resorcinol takes place in an alkaline melt at about 300 °C:

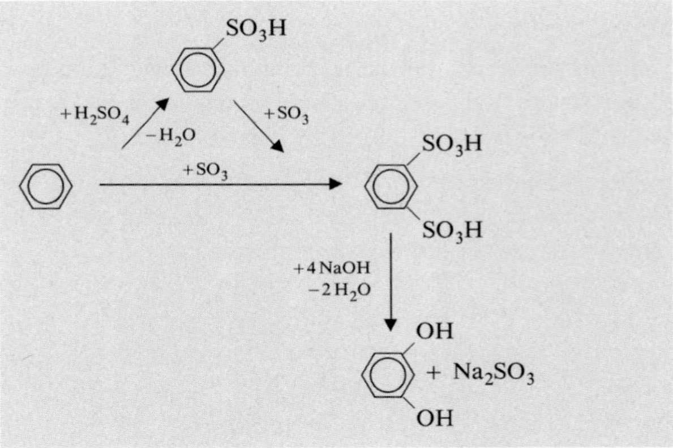

(20)

disadvantages of benzenedisulfonic acid–resorcinol route:

Na_2SO_3 and Na_2SO_4 are coproducts contaminated with organic substances

Resorcinol is obtained from its disodium salt after neutralizing the excess NaOH in the aqueous solution of the melt. In addition to the Na_2SO_3 from the actual reaction, Na_2SO_4 is formed by acidification with H_2SO_4. The crude resorcinol is extracted from the acidified aqueous solution with diisopropyl ether and purified by vacuum distillation. The selectivity can be as high as 82% (based on C_6H_6).

attempts to manufacture resorcinol more economically in analogy to Hock phenol process have shown:

The unfavorable economic and ecological aspects of this classical resorcinol manufacture have prompted several firms to attempt to transfer the Hock phenol synthesis to resorcinol. In this analogy, *m*-diisopropylbenzene would be oxidized to dihydroperoxide and then converted, by acidic cleavage, into acetone and resorcinol:

dihydroxylation of benzene is possible in principle, however several disadvantages arise:

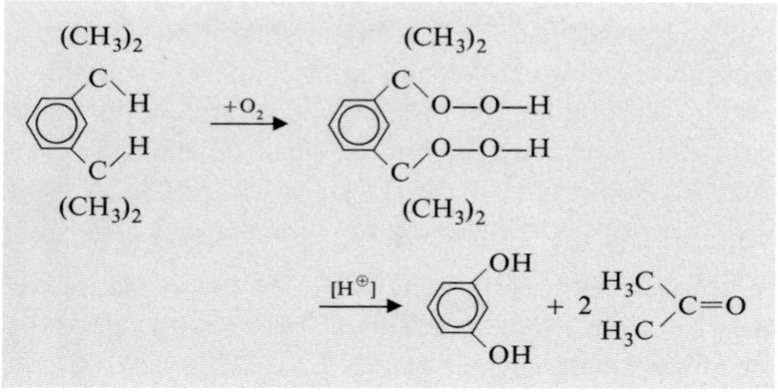

(21)

It was found that the two hydroperoxide groups in the molecule give rise to a much greater number of byproducts from parallel and secondary reactions than with cumene hydroperoxide (*cf.* Section 13.2.1.1). In addition, the rate of reaction is considerably lower, so that only low space–time yields can be attained.

1. formation of numerous byproducts from stepwise or simultaneous side and secondary reactions of both hydroperoxide groups
2. low reaction rate for dihydroperoxide formation reduces space – time yield

In Japan, several of these disadvantages were overcome, and in 1981 Sumitomo Chem. Co. began operation of a plant for the production of resorcinol (original capacity 5000, later expanded to 17 000, tonnes per year) using a modified Hock process.

first commercial resorcinol manufacture analogous to Hock process for phenol in Japan

The manufacture of hydroquinone from the oxidation of *p*-diisopropylbenzene is similar. However, in contrast to resorcinol, hydroquinone has been produced successfully by Signal Chemical in the USA in a 2700 tonne-per-year plant since 1971 using a modified Hock process. Goodyear took over this hydroquinone production and currently produces a technical grade material (95%). Mitsui Petrochemical also started production of hydroquinone using a similar process in a 5000 tonne-per-year plant in Japan in 1975.

hydroquinone manufacture by Hock process possesses, compared to resorcinol process, advantages such as:

1. *p*-diisopropylbenzene can be more rapidly and selectively oxidized than *m*-isomer

2. dihydroperoxide can be cleaved with higher selectivity

As in the USA and Europe, in Japan most hydroquinone is still produced by reducing *p*-quinone with iron at $50-80\,^\circ$C. The *p*-quinone feed is obtained from aniline through a complex reaction involving oxidation with MnO_2 or CrO_3 in a solution acidified with H_2SO_4:

classical hydroquinone manufacture from aniline involves two steps:

1. oxidation to quinone
2. reduction to hydroquinone

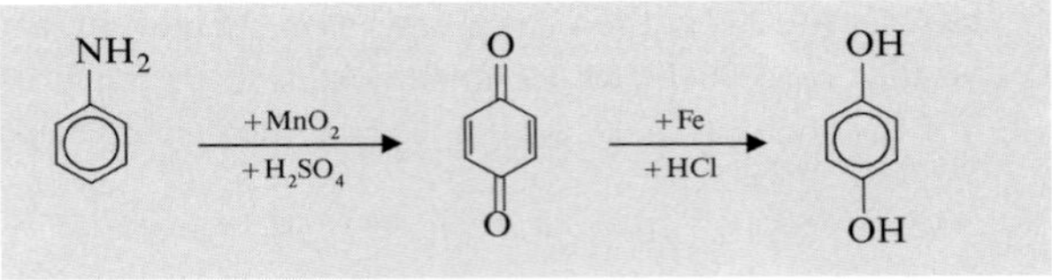

$$\text{(22)}$$

The process is burdened by considerable formation of manganese, chromium and iron salts, and ammonium sulfate. For this reason, all hydroquinone processes based on aniline in Japan have been shut down, and the last US plant (Eastman Kodak) was converted from aniline to the Hock process in 1986. A 4000 tonne-per-year plant is still in operation in China.

Rhône-Poulenc hydroquinone process:

phenol oxidation with percarboxylic acids or H_2O_2 and mineral acids as electrophilic hydroxylation

process characteristics:

characterized by limited control of hydroquinone/pyrocatechol ratio by reaction conditions

Ube hydroquinone process:

phenol oxidation with H_2O_2

Another process (Rhône-Poulenc) involves the oxidation of phenol with perfomic acid, or mixtures of H_2O_2 with carboxylic acids or mineral acids such as H_3PO_4 or $HClO_4$, at 90 °C and with a low phenol conversion ($<10\%$). Pyrocatechol, the coupled product formed along with hydroquinone, has only a small market. Depending on reaction conditions, the hydroquinone/pyrocatechol ratio varies from about 60:40 to 40:60. The selectivity to both hydroxyphenols is 85–90% (based on H_2O_2). Rhône-Poulenc operates a plant with a capacity of 18 000 tonnes per year (12 000 tonnes per year hydroquinone). A similar process, the oxidation of phenol with H_2O_2 developed by Ube Industries, is used commercially in Japan.

Brichima pyrocatechol/hydroquinone process:

phenol oxidation with H_2O_2 as radical hydroxylation (Fe/H_2O_2 = Fenton's reagent)

In Italy, Brichima runs a process for the oxidation of phenol with H_2O_2 in a 5000 tonne-per-year plant. In the presence of small amounts of iron- and cobalt-salt initiators, phenol is oxidized with 60% H_2O_2 at 40 °C in a radical reaction. The ratio of hydroquinone to pyrocatechol is between 40:60 and 20:80, with a total selectivity of over 90% and a phenol conversion of about 20%.

Reppe hydroquinone process:

metal carbonyl-catalyzed ring-forming carbonylation of acetylene

An older Reppe hydroquinone synthesis involving the ring-forming carbonylation of acetylene with Fe or Co complexes has been revised in various ways, with an emphasis on catalyst modification:

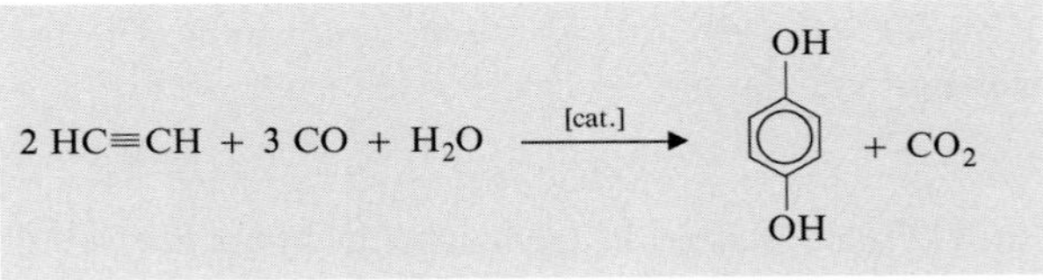

$$2\ HC{\equiv}CH + 3\ CO + H_2O \xrightarrow{\text{[cat.]}} \text{(hydroquinone)} + CO_2 \quad \text{(23)}$$

advance in industrial development due mainly to Lonza pilot plant

main problem:

Du Pont uses Ru and Rh catalysts at 600–900 bar in the presence of H_2 instead of H_2O. Ajinomoto and Lonza use Rh and $Ru(CO)_4$ catalysts, respectively, at 100–300 °C and 100–350 bar. Lonza (now Alusuisse-Konzern) operated a small pilot plant for several years, but no industrial plant has been constructed. The quantitative recovery of expensive noble metal catalysts was — as is usual in homogeneously catalyzed reactions — a very important but obviously unsolved problem.

quantitative recovery of expensive Ru catalyst

It has been known from the literature since about 1900 that benzene can be oxidized electrochemically to hydroquinone. This manufacturing route has been piloted several times. In principle, this is a two-step process. First, benzene dispersed in $2N$ H_2SO_4 is oxidized to p-quinone using PbO_2 as the anode; this is then fed continuously to the cathode region, where it is reduced to the hydroquinone with Pb. The selectivity to hydroquinone is about 80% (based on benzene), with a purity of 99%. The main byproducts are CO_2 and a small amount of pyrocatechol. With an optimal space–time yield, the electric current yield is 40%. A production plant has not yet been constructed.

electrochemical hydroquinone process by UK–Wesseling route:

process principles:

two-step electrochemical conversion of benzene:

1. anodic oxidation to p-quinone
2. cathodic reduction to hydroquinone

process characteristics:

electrolysis cell equipped with cation-exchange membrane, Pb/PbO_2 electrodes, and a C_6H_6 dispersion in $2N$ H_2SO_4 produced by turbulent circulation

Hydroquinone's main uses are as a photographic developer, polymerization inhibitor, and antioxidant. It is also an important intermediate for numerous dyes. Commercially important simpler derivatives include the mono- and di-ethers of hydroquinone in particular, and the alkylhydroquinones and their ethers.

uses of hydroquinone:

developer
polymerization and oxidation inhibitor
chemical intermediate

Resorcinol is also used as an intermediate for dyes, and for UV stabilizers for polyolefins and pharmaceutical products. By far the largest amount (currently about 70% in the USA, more than 55% in Western Europe, and almost 40% in Japan) is used in the form of resorcinol/formaldehyde co-condensates, which may also contain other comonomers such as butadiene, styrene, or vinylpyridine. These are used to promote adhesion between steel or cord and rubber. In addition, polycondensation products with formaldehyde, or with formaldehyde and phenol, are used as stable, water-resistant adhesives for wood.

uses of resorcinol:

adhesive for steel-belted and textile radial-ply tires
special adhesive
UV light absorber
chemical intermediate

Pyrocatechol is a starting product for a series of important fine chemicals for pest control, pharmaceuticals, and flavors and aromas.

use of pyrocatechol:

chemical intermediate

13.2.3. Maleic Anhydride

Until the beginning of the 1960s, benzene was the only raw material used for the manufacture of maleic anhydride (MA). With increasing demand due to its use in polyester resins, paint raw materials, and as an intermediate (e.g., for γ-butyrolactone, 1,4-butanediol, and tetrahydrofuran), more economical manufacturing routes based on C_4 compounds were developed.

maleic anhydride (MA) production/recovery:

1. oxidative degradation of benzene (oldest commercial process)
2. oxidation of C_4-feedstocks (increasing)
3. isolation as byproduct from the oxidation of:

naphthalene (commercial)
o-xylene (commercial)
toluene (possible in principle)

example of commercially tested MA isolation from product of naphthalene/*o*-xylene oxidation:

UCB process
BASF process

In 1991 roughly 36% of the worldwide maleic acid capacity was still based on benzene and the rest was from C_4 compounds, chiefly *n*-butane.

Maleic anhydride is also a byproduct in the oxidation of aromatic feedstocks such as naphthalene or *o*-xylene (to phthalic anhydride) or toluene (to benzoic acid). These sources of maleic anhydride have been exploited recently; UCB, for example, improved the processing of the oxidized naphthalene product with the result that in a conventional 80000 tonne-per-year phthalic acid plant, 4000–5000 tonnes per year maleic anhydride can be isolated with a high degree of purity.

Another continuous process for recovery of pure maleic anhydride from wastewater produced in the oxidation of naphthalene or *o*-xylene to phthalic anhydride was developed by BASF. About 5% by weight of pure MA (99.5%) based on the phthalic anhydride capacity can be recovered by evaporation of the wash water, dehydration of maleic acid to the anhydride, and distillation. In 1984, the total capacity of plants using this process was already 20000 tonnes per year. This byproduct isolation has since been used in other countries such as Japan. The aqueous maleic acid can also be isomerized to fumaric acid in a process from Chemie Linz AG.

maleic anhydride production (in 1000 tonnes):

	1990	1992	1995
USA	193	193	251
W. Europe	200	175	188
Japan	101	104	113

Production figures for maleic anhydride in USA, Western Europe, and Japan are summarized in the adjacent table.

World production capacity of maleic anhydride in 1995 was about 870000 tonnes per year, with about 280000, 290000, and 150000 tonnes per year in the USA, Western Europe, and Japan, respectively.

13.2.3.1. Maleic Anhydride from Oxidation of Benzene

process principles of benzene oxidation to MA:

strongly exothermic V_2O_5-catalyzed oxidative degradation of benzene in gas phase

characteristics of benzene oxidation:

fixed-bed catalyst in tube reactor with circulating fused salt for heat removal and production of high pressure steam

The many processes for the oxidation of benzene to maleic anhydride all use a similar catalyst based on V_2O_5, which may be modified with, *e. g.*, MoO_3 or H_3PO_4. Due to the strongly exothermic reaction, tube bundle reactors with, *e. g.*, 13000 externally cooled tubes in a reactor diameter of about 5 m are used.

Using fused salts as the circulating heat-exchange liquid, the heat of reaction is removed and used to generate high pressure steam.

A benzene/air mixture is oxidized to maleic anhydride over the catalyst at 2–5 bar, 400–450 °C, and with a residence time of about 0.1 s:

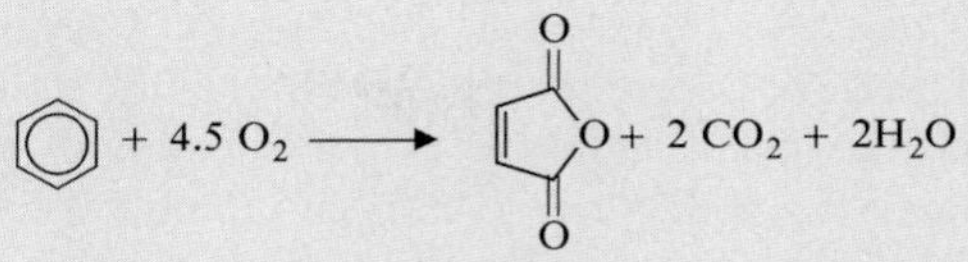

$$\text{benzene} + 4.5\,O_2 \longrightarrow \text{maleic anhydride} + 2\,CO_2 + 2\,H_2O \qquad (24)$$

$$\left(\Delta H = -\frac{447 \text{ kcal}}{1875 \text{ kJ}} \Big/ \text{mol} \right)$$

mechanism of MA formation from benzene: stepwise incorporation of oxygen

through the intermediate $O{=}\langle\text{ring}\rangle{=}O$

follows a redox mechanism, *i.e.*, catalyst supplies lattice oxygen for selective oxidation, and is regenerated with atmospheric oxygen

The benzene conversion reaches $85-95\%$. The selectivity to maleic anhydride is only about $60-65\%$ (up to 75% using newer developments). About a quarter of the benzene is completely oxidized so that the total evolution of heat is about $6500-7000$ kcal ($27\,200-29\,400$ kJ) per kilogram of converted benzene.

In 1982, about two-thirds of world capacity for maleic anhydride was based on a process developed by Scientific Design. Currently, however, many of these plants have been shut down. Thus, for example, all of the benzene-based plants in the USA were either idled or converted to butene or butane oxidation in the beginning of the 1980s.

In the Alusuisse–UCB process, in which the dehydration and purification is a UCB development, selectivities of up to 95% (based on maleic anhydride) are obtained with an improved catalyst at $355-375\,°C$. In a $70\,000$ tonne-per-year plant (1994) operated in Italy by SAVA Ftalital, a subsidiary of Alusuisse, about $12\,000$ tonnes per year can be manufactured per reactor.

example of a modern commercial benzene oxidation:

Alusuisse–UCB with multiple-train plant in Italy

As described in the oxidation of butene/butane to maleic anhydride, fixed-bed technology for the oxidation of benzene has also been replaced by more economical fluidized-bed or moving-bed processes.

technological development from fixed bed to fluidized bed and moving bed

The reaction gas is cooled in several heat exchangers. Since the temperature in the last cooler is below the condensation temperature of the anhydride, about $50-60\%$ is obtained directly as an anhydride melt. The remainder is washed out with water in the form of maleic acid and converted into maleic anhydride in a dehydration column or a thin-film evaporator.

isolation of maleic anhydride:

1. $50-60\%$ as MA directly
2. remainder as aqueous maleic acid which is dehydrated to MA by distillation with or without entraining agent (*o*-xylene)

This last stage can be run continuously using *o*-xylene as a water entraining agent. Both crude products are then fractionated to a purity of 99.7%. The distillation residue contains fumaric acid and higher boiling products.

13.2.3.2. Maleic Anhydride from Oxidation of Butene

feedstocks for butene oxidation to maleic anhydride:

1. butenes as byproducts from butane dehydrogenation to butadiene

2. C_4 fractions from steam cracking of light gasoline with characteristic properties of its components

In 1962, Petro–Tex (now Denka) in the USA was the first company to manufacture maleic anhydride by the oxidation of *n*-butenes on a commercial scale. After several years, this plant was converted to benzene, and then to butane, because of pricing. The butenes result as byproducts from the dehydrogenation of butane to butadiene. Today, along with propene and ethylene, the steam cracking of light gasoline also yields various amounts of unsaturated C_4 hydrocarbons, depending on reaction conditions. After separating the butadiene, which is used elsewhere, the C_4 fraction is oxidized to maleic anhydride without isolating isobutene, which is totally oxidized. *n*-Butane remains practically unchanged during the reaction, just like an inert gas. It is combusted at 800 °C along with other off-gas and used for heat generation.

characteristics of oxidation of a C_4 fraction:

isobutene is totally oxidized
n-butane is inert
n-butenes are oxidized to maleic anhydride
butadiene is oxidized to maleic anhydride

mechanism of MA formation from *n*-butenes:

lattice oxygen forms
crotonaldehyde $CH_3CH{=}CHCHO$
intermediate with ring closure to furan and oxidation to MA

In principle, it is also possible to oxidize butadiene to maleic anhydride. A particular advantage over all previously mentioned feedstocks is that butadiene has the lowest reaction enthalpy, *i. e.*, -237 kcal (-995 kJ)/mol compared to *e. g.*, benzene with -447 kcal (-1875 kJ)/mol. However, butadiene is a valuable feedstock for other secondary products (*cf.* Section 5.1.4) and is used preferentially for these.

operation of butene oxidation:

1. fixed-bed tube reactor (up to 21 000 vertical tubes) with catalyst based on V_2O_5 (similar to benzene oxidation) and circulating salt melt for heat removal

The catalytic oxidation of a C_4 fraction without butadiene is also carried out using air and a fixed-bed catalyst in a tube reactor. The principle of the process is therefore very similar to that of the benzene process as far as the reaction section is concerned. The basic components of most industrial catalysts are vanadium or phosphoric oxides on carriers with a low surface area, modified with other oxides, *e. g.*, Ti, Mo or Sb. The oxidation takes place at 350–450 °C and 2–3 bar with the formation of maleic acid and its anhydride:

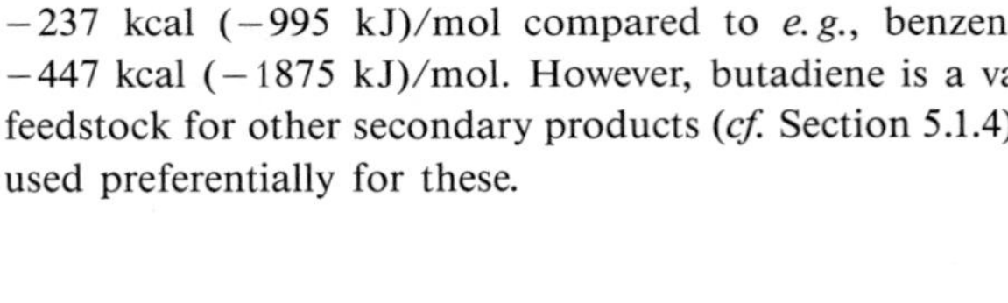

$$CH_3{-}CH{=}CH{-}CH_3$$

$$+\ 3\ O_2 \longrightarrow \qquad \qquad \qquad O + 3\ H_2O \qquad (25)$$

$$H_2C{=}CH{-}CH_2{-}CH_3$$

$$\left(\Delta H = -\frac{314\ \text{kcal}}{1315\ \text{kJ}}\Big/\text{mol} \right)$$

The selectivity, relative to the butene content which can be oxidized to maleic anhydride, is usually only 45−60%. The byproducts are CO_2, CO, formaldehyde, and acetic, acrylic, fumaric, crotonic, and glyoxylic acids.

byproducts:

CO_2, CO, CH_3COOH, $H_2C=CHCOOH$

$$HOOC \diagdown C=C \diagup COOH$$
(with H on the HOOC carbon and H on the other carbon)

$CH_3CH=CHCOOH$ $OHC—COOH$

Mitsubishi Chemical uses a V_2O_5–H_3PO_4 catalyst whirled in a fluidized bed for the same oxidation. This technology − used for the first time with maleic anhydride − has the great advantage of more facile heat removal at a uniform reaction temperature. Mitsubishi has used this process in an 18 000 tonne-per-year plant since 1970. This plant has since been expanded to 21 000 tonnes per year (1995).

2. fluidized-bed reactor (Mitsubishi process) with V_2O_5 catalyst similar to fixed-bed process

In contrast to the benzene oxidation, the workup consists merely of washing the reaction gas with dilute aqueous maleic acid solution; *i.e.*, there is no partial condensation of maleic anhydride. The dilute (∼40%) maleic acid solution is concentrated either under vacuum or with the help of a water entraining agent (*o*- or *p*-xylene). The acid is then dehydrated to the anhydride in a rotary evaporator or a column, either batch or continuous. The maleic anhydride is then separated from lower and higher boiling substances in a two-stage process. The final anhydride is about 99% pure.

isolation of maleic anhydride:

no maleic anhydride directly from process, only aqueous acid, which is concentrated and dehydrated to anhydride

13.2.3.3. Maleic Anhydride from Oxidation of Butane

The transition from benzene to *n*-butene as the raw material for oxidative manufacture of maleic anhydride meant a fundamental improvement, since the total oxidation of two extra carbon atoms was no longer necessary. However, the lower selectivity of the *n*-butene oxidation and, especially, the increasing demand and resulting pricing situation for benzene and *n*-butene led to the development of the oxidation process for *n*-butane as the most economical raw material for maleic anhydride.

increase in demand and price of benzene, *e. g.*, as octane booster, led to *n*-butene, and then *n*-butane, as attractive raw material for MA production

The first commercial plant (Monsanto) based on *n*-butane began operation in the USA in 1974. Other firms including Amoco, BP, Denka, Halcon/Scientific Design, and Sohio/UCB either already use *n*-butane as a feedstock, or plan to convert from benzene to *n*-butane. The process conditions are similar to those for benzene oxidation. The catalysts used − also for the fluidized bed − are still based on vanadium oxides; they differ in promoters such as phosphorous and the oxides of Fe, Cr, Ti, Co, Ni, Mo, and other elements. The selectivities are 50−60% at a butane conversion of 10−15%.

commercial plants initially only fixed-bed, tube reactors; now also fluidized-bed reactors with the advantages: lower capital costs, isothermal reaction, higher MA concentration with larger fraction condensed directly

fundamental catalyst functions such as oxidative dehydrogenation of butane and oxidation of terminal CH_3 groups to MA provided by V−P−O complex

example of a commercial fluidized-bed MA process:

Alma process (**Alusuisse Italia Lummus Crest maleic anhydride**)

In Europe in 1975, Alusuisse (now Lonza, Italy) began operation of a 3000 tonne-per-year maleic anhydride plant using their own butane-based process. This was later converted to use an organic solvent, *e. g.*, *o*-xylene, to absorb the maleic anhydride without water. This plant has since been converted, together with Lummus, to a fluidized-bed process (Alma process). In 1994, the world's largest plant (capacity, 50000 tonnes per year) was brought on line by Lonza in Italy using this process.

A variation of the fluidized-bed process has been developed by Du Pont. In this, butane is oxidized with a vanadium phosphate catalyst which is continously transferred to a regenerator by a moving-bed reactor. This oxidation, performed without free oxygen, increases the average selectivities from 50–60% to 70–75%. Maleic acid is isolated as an aqueous solution. This process is now being developed in pilot plants.

Even though the use of C_5 cuts containing cyclopentene, 1,3-pentadiene, 1-pentene, and isoprene has been explored for the oxidation to maleic anhydride by Nippon Zeon in Japan, *n*-butane will certainly be the most important basis in the future.

13.2.3.4. Uses and Secondary Products of Maleic Anhydride

uses of maleic anhydride:

30–60% for unsaturated polyester resins (thermosetting plastics) and for modifying alkyd resins

Maleic anhydride consumption in several countries is distributed roughly as follows:

Table 13-6. Use of maleic anhydride (in %).

Product	USA		Western Europe		Japan	
	1986	1995	1986	1995	1986	1995
Unsaturated polyesters	57	64	56	50	41	32
Fumaric/malic acid	10	3	4	10	14	11
Pesticides	10	3	2	1	3	3
Lubricating oil additives	9	11	5	6	3	3
Miscellaneous[1]	14	19	33	33	39	51
Total use (in 1000 tonnes)	164	206	157	201	64	104

[1] *e. g.*, 1,4-butanediol

3–14% for fumaric acid/malic acid

up to 10% as intermediate for fungicides and insecticides, *e. g.*,:

malathion (insecticide)

$(CH_3O)_2-P-S-CHCOOC_2H_5$
$\quad\quad\quad \| \quad\quad\quad |$
$\quad\quad\quad S \quad\quad CH_2COOC_2H_5$

Maleic anhydride is used mainly (40–60%) in the manufacture of unsaturated polyester resins, *e. g.*, thermosetting plastics, especially with glass fiber reinforcement (*cf.* Section 14.1.3). About 10–15% is converted into the *trans* isomer of maleic acid, *i. e.*, fumaric acid, and its secondary product malic acid. Other uses of maleic anhydride include the manufacture of pesticides, *e.g.*, malathion for rice plantations or in fruit growing, and maleic hydrazide as an herbicide for specific grasses and to regulate growth, *e.g.*, in tobacco growing; of reactive plasticizers such as dibutyl maleate; and of lubricating oil additives.

Another commercially important series of maleic anhydride derivatives has been made available by Japanese firms and by ICI and UCC/Davy McKee, generally through multistep hydrogenation of maleic anhydride via succinic anhydride, dimethyl maleate, or diethyl maleate. The products are those typical of Reppe reactions, *e.g.*, γ-butyrolactone, 1,4-butanediol, and tetrahydrofuran (*cf.* Section 4.3). With a Ni catalyst, the mole ratio of γ-butyrolactone to tetrahydrofuran can be varied from 10:1 to 1:3 with 100% maleic anhydride conversion by changing reaction temperature and pressure.

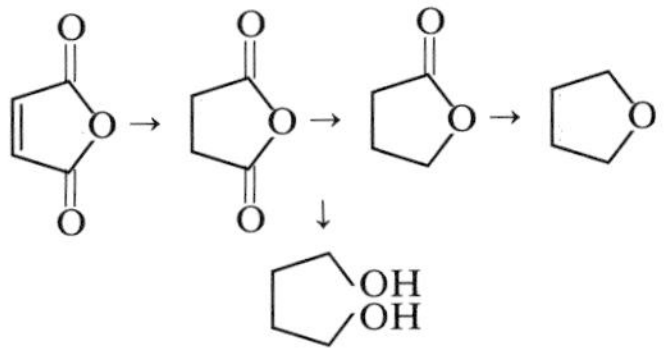

A Ni-Co-ThO$_2$/SiO$_2$ catalyst at 250 °C and 100 bar is used for the hydrogenation of γ-butyrolactone to 1,4-butanediol. At 100% conversion, the selectivity is about 98%. The main by-product is tetrahydrofuran.

This maleic anhydride hydrogenation has not been used for the production of 1,4-butanediol and tetrahydrofuran for a long time.

Due to the increasing demand for 1,4-butanediol several plants are planned or in use chiefly in Japan and South Korea but also in Western Europe. γ-Butyrolactone is produced by Mitsubishi Kasei in a plant with a 4000 tonne-per-year capacity (*cf.* Section 4.3).

A new process for the production of tetrahydrofuran has been developed by Du Pont. An aqueous maleic acid solution obtained from oxidation of *n*-butane (*cf.* Section 13.2.3.3) is hydrogenated directly to tetrahydrofuran on a rhenium-modified palladium catalyst with approximately 90% selectivity. This second step is also being developed in pilot plants. A 45000 tonne-per-year plant was planned for 1995 in Spain.

In a Degussa process, DL-tartaric acid can be obtained with 97% selectivity from the reaction of maleic anhydride with H$_2$O$_2$ in the presence of a molybdenum- or tungsten-containing catalyst. The conversion takes place via an epoxytartaric acid intermediate, which is hydrolyzed. No commercial unit has been built.

maleic hydrazide (herbicide)

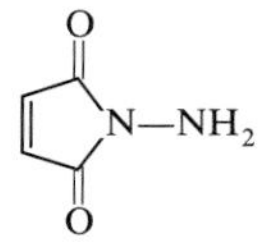

maleic acid ester for plasticizers and additives

maleic anhydride stepwise hydrogenation to:

succinic anhydride
γ-butyrolactone
1,4-butanediol
tetrahydrofuran

reaction of maleic anhydride with H$_2$O$_2$ to give DL-tartaric acid

An analogous production of DL-tartaric acid based on maleic anhydride has been operated by Butakem in South Africa in a 2000 tonne-per-year plant since 1974:

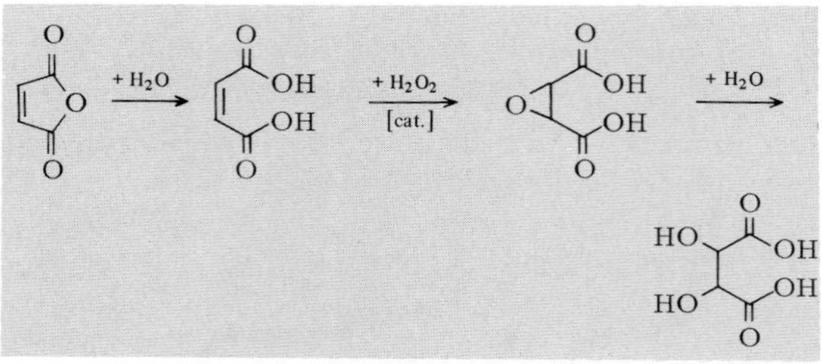

(26)

maleic acid dimethyl ester ozonolysis and hydrogenation/hydrolysis to glyoxylic acid

One new use for maleic acid is oxidative cleavage with ozone (*cf.* Section 7.2.1.3).

two routes for isomerization of maleic acid to fumaric acid:

1. thermal, higher temperature and longer residence times
2. catalytic (in practice often thiourea), at lower temperature

The isomerization of maleic to fumaric acid, which is almost quantitative, is done either in aqueous solution without catalyst by heating for a long period at 150 °C, or with H_2O_2, thiourea, ammonium persulfate, etc., at 100 °C. Since fumaric acid has such a low solubility in water, it is almost completely precipitated from the aqueous reaction solution as a crystal:

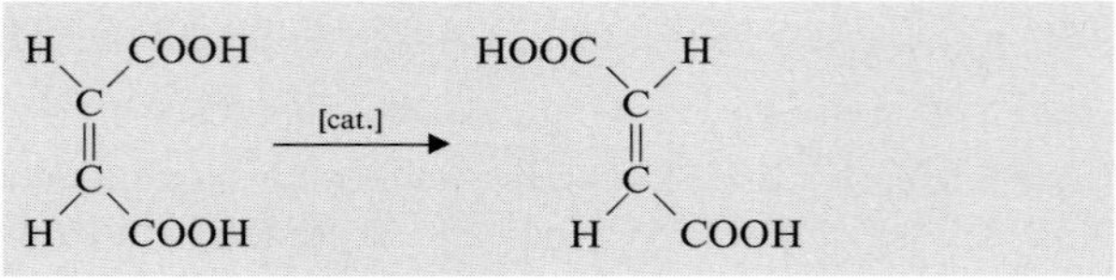

(27)

An industrial process developed by Alusuisse is currently used in several plants. It is based on the catalytic isomerization of a 25% aqueous maleic acid solution at about 100 °C.

Fumaric acid can also be manufactured, as in the USA for example, by the fermentation of sugars or starches.

uses of fumaric acid:

component for polyesters, precursors for DL-maleic acid

More than 40% of the fumaric acid production is used for polyesters for use in the paper industry. 10−20% is used to manufacture DL-maleic acid (hydroxysuccinic acid) by proton-catalyzed hydration:

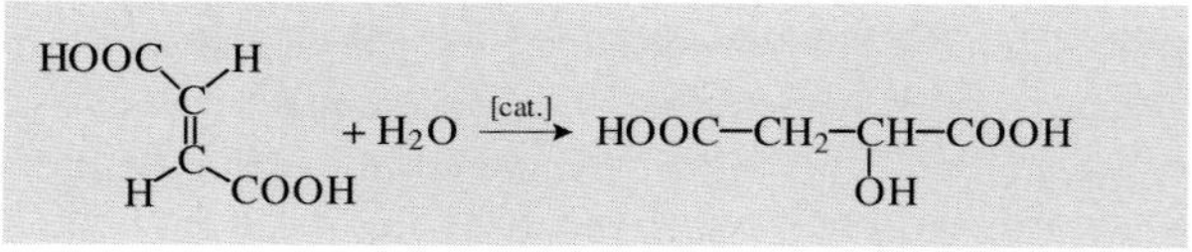

(28)

The Allied process — that of the greatest malic acid producer in the world until production was stopped in 1980 — is based on the addition of H_2O to maleic anhydride. The only producer in the USA is Miles, with a capacity of 5000 tonnes per year malic acid and 3000 tonnes per year fumaric acid. In 1991, the total production capacity for fumaric acid in the USA was about 18 000 tonnes per year. Other producers of malic acid are Fuso in Japan and Bartek in Canada; Lonzo plans to start up a 10 000 tonne-per-year plant in Italy in 1997.

Depending on the country, malic acid is used either as its racemate or as the naturally occuring D(−)-malic acid[1] in the food industry as an acidulant to adjust tartness.

malic acid manufacture:

hydration of fumaric acid or maleic anhydride (Allied process)

world consumption of flavor acids (in 1000 tonnes):

	1976	1983	1990
citric acid	200	300	500
phosphoric acid	80	200	200
acetic acid	70	100	n. a.
malic acid	20	20	n. a.
tartaric acid	10	25 – 30	n. a.
lactic acid	10	18	25
fumaric acid	3	3	n. a.

n. a. = not available

13.3. Other Benzene Derivatives

13.3.1. Nitrobenzene

The classical pathway to nitrobenzene has remained basically unaltered since the first nitration of benzene in 1834 by E. Mitscherlich.

The batch processes have, however, been complemented by continuous processes in order to meet the greater demand for its secondary product, aniline, economically. In 1994, manufacturing capacity for nitrobenzene in Western Europe, the USA, and Japan was roughly 970 000, 860 000, and 150 000 tonnes per year, respectively. Production figures are summarized in the adjacent table.

The nitration process is usually conducted using nitrating acid, a mixture of nitric acid and concentrated sulfuric acid. The sulfuric acid has several functions. It promotes the formation of the nitrating agent (nitronium ion, *cf.* eq 29) and prevents the dissociation of nitric acid into an oxidizing $NO_3^{\ominus}$ ion by binding water as a hydrate (*cf.* eq 30). It also enhances the solubility between the aqueous and organic phases:

nitrobenzene manufacture:

nitration of C_6H_6 with HNO_3 in H_2SO_4 (nitrating acid or nitrosulfuric acid)

nitrobenzene production (in 1000 tonnes):

	1989	1991	1993
W. Europe	749	656	765
USA	554	523	565
Japan	147	145	140

H_2SO_4 has several functions:

1. promotes $NO_2^{\oplus}$ formation for electrophilic nitration
2. blocks the HNO_3 dissociation to oxidizing $NO_3^{\ominus}$ ion
3. enhances solubility

$$HNO_3 + 2\ H_2SO_4 \rightleftarrows NO_2^{\oplus} + H_3O^{\oplus} + 2\ HSO_4^{\ominus} \qquad (29)$$

$$HNO_3 + H_2O \rightleftarrows NO_3^{\ominus} + H_3O^{\oplus} \qquad (30)$$

---

[1] formerly known as L(−)-malic acid

two manufacturing variations:

1. batchwise with strong stirring because of two-phase exothermic reaction
2. continuous with three-stage cascade with increasing reaction temperatures

In the batch process, benzene and nitrating acid, *i. e.*, a mixture of $32-39$ wt% HNO_3, $60-53$ wt% H_2SO_4, and 8 wt% H_2O, are reacted in cast-iron nitrating vessels at $50-55\,°C$ with strong stirring due to the formation of two phases and the required mass and heat exchange. In this process, the nitrating acid is allowed to flow into the benzene:

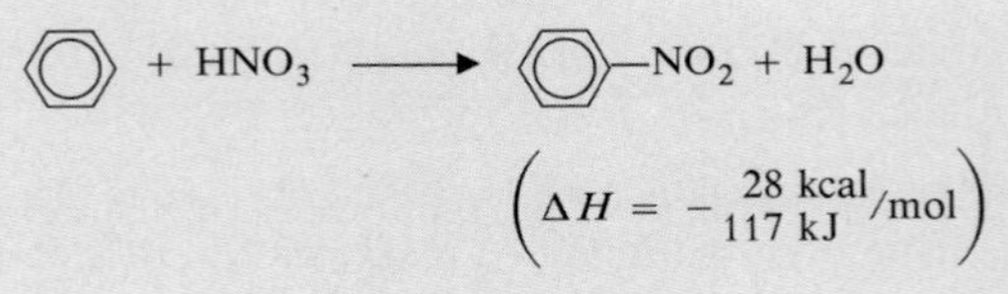

$$\bigcirc + HNO_3 \longrightarrow \bigcirc\!-\!NO_2 + H_2O \qquad (31)$$

$$\left(\Delta H = -\,{28\ kcal \atop 117\ kJ}/mol\right)$$

byproduct formation:

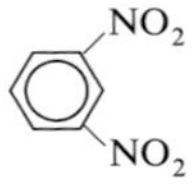

is minor, as further nitration is a factor of 10^4 slower due to electron withdrawing effect of the nitro group

After a residence time of several hours, the nitrating acid is largely exhausted and the more dense nitrobenzene is separated, washed and distilled. The selectivity reaches $98-99\%$. Some *m*-dinitrobenzene is obtained as a byproduct.

uses of nitrobenzene:

1. aniline (90–95% of production worldwide)
2. intermediate, *e. g.,* for dyes, pesticides, pharmaceuticals
3. solvent, *e. g.,* for $AlCl_3$ in Friedel-Crafts reaction (limited due to decomposition danger)
4. mild oxidizing agent, *e. g.,* for manufacture of triarylmethane dyes

The increasingly used continuous nitration plant consists of a cascade of stirred vessels (usually three nitrating vessels) with a stepwise, slowly increasing temperature (35–40 °C in the first vessel, 50 °C in the second, and 55–60 °C for the final reaction). The largest units are the final settling and washing vessels.

The workup and purification of nitrobenzene take place in a manner analogous to the batch process. The waste sulfuric acid is concentrated for reuse; in adiabatic processes, the heat of reaction from the nitration is utilized. Nitrobenzene is further substituted to give intermediates such as chloronitrobenzene and nitrobenzenesulfonic acid, and used as a solvent and oxidizing agent. However, it is used mainly for aniline manufacture.

13.3.2. Aniline

aniline production (in 1000 tonnes):

	1990	1992	1993
W. Europe	460	500	537
USA	430	457	508
Japan	148	168	184

In 1994, the production capacity for aniline in the USA, Western Europe, and Japan was roughly 630000, 750000, and 260000 tonnes per year, respectively. The largest producer is Bayer, with a capacity of approximately 385000 tonnes per year worldwide (1992). Production figures for several countries are summarized in the adjacent table.

feedstocks for aniline manufacture:

1. nitrobenzene
2. chlorobenzene
3. phenol

aniline manufacture by reduction of nitrobenzene with two different reducing agents:

Nitrobenzene is the classical feedstock for aniline manufacture. Recently less chlorobenzene and phenol are being used in aniline manufacturing processes in several countries.

The reduction of nitrobenzene with iron turnings and water in the presence of small amounts of hydrochloric acid is the oldest form of industrial aniline manufacture. It would certainly have been replaced much earlier by more economical reduction methods if it had not been possible to obtain valuable iron oxide pigments from the resulting iron oxide sludge. However, the increasing demand for aniline has far surpassed the market for the pigments, so that not only catalytic hydrogenation processes (both liquid- and gas-phase) but also other feedstocks have been used for aniline production.

The modern catalytic gas-phase hydrogenation processes for nitrobenzene can be carried out using a fixed-bed or a fluidized-bed reactor:

$$\text{\Large ⬡}-NO_2 + 3\,H_2 \xrightarrow{\text{[cat.]}} \text{\Large ⬡}-NH_2 + 2\,H_2O \qquad (32)$$

$$\left(\Delta H = -\frac{117 \text{ kcal}}{443 \text{ kJ}} \big/ \text{mol} \right)$$

Bayer and Allied work with nickel sulfide catalysts at 300–475 °C in a fixed bed. The activation of the hydrogenation catalysts with Cu or Cr, and the use of different supports and catalyst sulfidization methods with sulfate, H_2S or CS_2 all belong to the expertise of the corresponding firms. The selectivity to aniline is more than 99%. The catalytic activity slowly decreases due to carbon deposition. However, the catalyst can be regenerated with air at 250–350 °C and subsequent H_2 treatment. Similar processes are operated by Lonza with Cu on pumice, by ICI with Cu, Mn, or Fe catalysts with various modifications involving other metals, and by Sumitomo with a Cu—Cr system.

The gas-phase hydrogenation of nitrobenzene with a fluidized-bed catalyst is used in processes from BASF, Cyanamid and Lonza. The BASF catalyst consists of Cu, Cr, Ba, and Zn oxides on a SiO_2 support; the Cyanamid catalyst consists of Cu/SiO_2. The hydrogenation is conducted at 270–290 °C and 1–5 bar in the presence of a large excess of hydrogen (H_2: nitrobenzene = ca. 9:1). The high heat of reaction is removed by a cooling system which is built into the fluidized bed. The selectivity to aniline is 99.5%; the nitrobenzene conversion is quantitative. The catalyst must be regenerated with air periodically.

An alternate manufacturing route for aniline is the ammonolysis of chlorobenzene or of phenol. For example, in the Kanto Electrochemical Co. process, chlorobenzene is ammonolyzed to

1. classical (stoichiometric with Fe/H_2O (+ traces of HCl) and formation of Fe oxide

2. catalytic with H_2 with two process variations in the gas phase

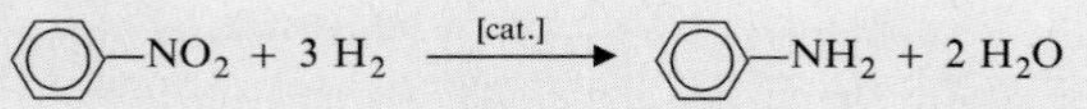

2.1. fixed-bed hydrogenation with sulfur-containing catalysts generally based on Ni or Cu

process examples:

Bayer, Allied, Lonza, ICI, Sumitomo

2.2. fluidized-bed hydrogenation with modified Cu catalysts

process examples:

BASF, Cyanamid, Lonza

aniline manufacture by ammonolysis of:

1. chlorobenzene
2. phenol

aniline with aqueous NH_3 at $180-220\,°C$ and $60-75$ bar in the presence of CuCl and NH_4Cl ("Niewland catalyst", *cf.* Section 5.3):

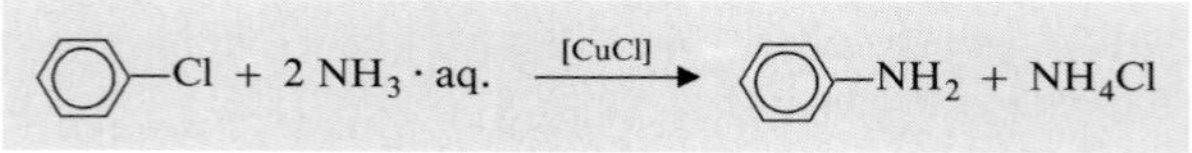

$$\langle\!\!\bigcirc\!\!\rangle\text{—Cl} + 2\,NH_3\cdot aq. \xrightarrow{\text{[CuCl]}} \langle\!\!\bigcirc\!\!\rangle\text{—}NH_2 + NH_4Cl \qquad (33)$$

to 1:

process characteristics:

two-phase, catalytic substitution of Cl by NH_2 with loss of Cl as NH_4Cl

Aniline can be isolated with 91% selectivity from the organic phase of the two-phase reaction product.

Dow stopped operation of a similar process for aniline in 1966.

Phenol can also be subjected to gas-phase ammonolysis with the Halcon/Scientific Design process at 200 bar and 425 °C:

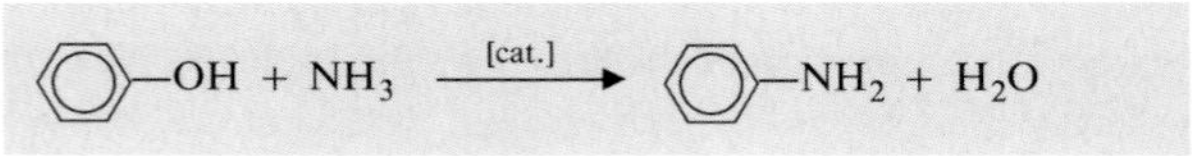

$$\langle\!\!\bigcirc\!\!\rangle\text{—OH} + NH_3 \xrightarrow{\text{[cat.]}} \langle\!\!\bigcirc\!\!\rangle\text{—}NH_2 + H_2O \qquad (34)$$

to 2:

process characteristics:

heterogeneously catalyzed gas-phase ammonolysis of phenol over special Lewis acids with metal promotors in fixed-bed reactor

byproducts:

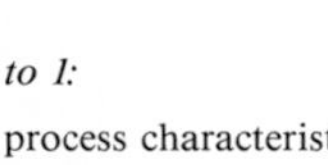

$Al_2O_3 \cdot SiO_2$ (possible as zeolites) and oxide mixtures of Mg, B, Al, and Ti are used as catalysts; these can be combined with additional cocatalysts such as Ce, V, or W. The catalyst regeneration required previously is not necessary with the newly developed catalyst. With a large excess of NH_3, the selectivity to aniline is 87–90% at a phenol conversion of 98%. The byproducts are diphenylamine and carbazole. This process has been operated since 1970 by Mitsui Petrochemical in a plant which has since been expanded to 45000 tonnes per year. A second plant with a capacity of 90000 tonnes per year was started up by US Steel Corp. (now Aristech) in 1982.

In 1977, Mitsui Petrochemical started production of *m*-toluidine by the reaction of *m*-cresol with ammonia in a 2000 tonne-per-year plant, analogous to the phenol ammonolysis. Thus, there is another manufacturing path besides the conventional route (nitration of toluene and hydrogenation of *m*-nitrotoluene).

potential aniline manufacture:

NiO/Ni-catalyzed ammonodehydrogenation of benzene with simultaneous reduction of NiO and subsequent reoxidation of Ni

Du Pont has developed an interesting new manufacturing process for aniline. Benzene and ammonia can be reacted over a NiO/Ni catalyst containing promoters including zirconium oxide at 350 °C and 300 bar to give a 97% selectivity to aniline with a benzene conversion of 13%:

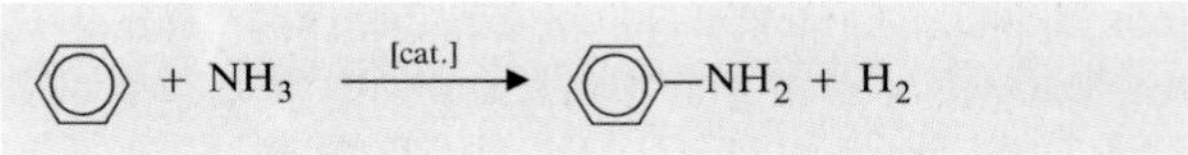

$$\langle\!\!\bigcirc\!\!\rangle + NH_3 \xrightarrow{\text{[cat.]}} \langle\!\!\bigcirc\!\!\rangle\text{—}NH_2 + H_2 \qquad (35)$$

Since the hydrogen formed in the reaction reduces the NiO part of the catalyst, a catalyst regeneration (partial oxidation) is necessary. Despite inexpensive feedstocks, industrial implementation is still thwarted by the low benzene conversion and the necessary catalyst reoxidation.

Aniline is one of the most significant key compounds in aromatic chemistry. Many commodity chemicals, including cyclohexylamine, benzoquinone, alkylanilines, acetanilide, diphenylamine, and 4,4′-diaminodiphenylmethane, are manufactured from aniline.

uses of aniline:

precursor for numerous secondary products

In the USA in the early 1970s, aniline and its secondary products were used primarily for the production of rubber additives such as vulcanization accelerators and antioxidants. Since that time, isocyanates based on aniline have assumed first place, with 4,4′-diphenylmethane diisocyanate (MDI) the primary component (*cf.* Section 13.3.3). The distribution of aniline usage in the USA, Western Europe, and Japan is found in the following table:

Table 13–7. Aniline use (in %).

Product	USA		Western Europe		Japan	
	1980	1993	1979	1993	1980	1993
Isocyanates (MDI)	62	78	65	78	52	80
Rubber chemicals	22	12	22	10	31	9
Dyes, pigments	4	2	⎫ 13	3	9	4
Hydroquinone	3	–	⎬	2	⎫ 8	–
Miscellaneous (*e. g.*, pharmaceuticals, pesticides)	9	8	⎭	7	⎬	7
Total consumption (in 10^6 tonnes)	0.29	0.49	0.30	0.59	0.07	0.16

13.3.3. Diisocyanates

Organic isocyanate compounds have been known for a long time, but first became commercially interesting in the last decades based on development work by O. Bayer (1937). It was shown that the reaction of di- and polyisocyanates with di- and polyols formed polyurethanes with many uses. The preferred use of polyurethanes in the automobile industry, in construction, and in refrigeration technology led to a considerable increase in the production capacity for feedstock diisocyanates.

organic monoisocyanates for preparative uses insignificant compared to commercial importance of diisocyanates for polyaddition

Toluene diiocyanate (TDI), in the form of its 2,4- and 2,6-isomers, has been the most significant diisocyanate.

industrially important diisocyanates:

1. HDI: $O=C=N-(CH_2)_6-N=C=O$

2. TDI: $O=C=N-$⟨⟩$-CH_3$ +
$N=C=O$

$N=C=O$
⟨⟩$-CH_3$
$N=C=O$

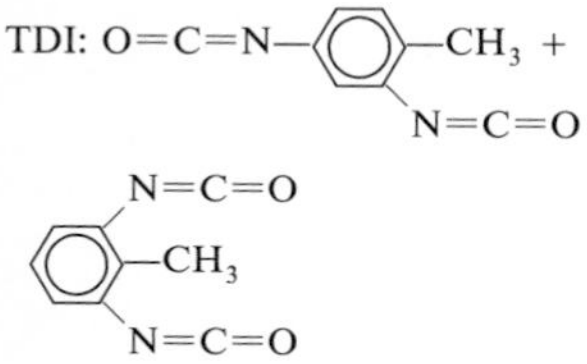

3. MDI:

$O=C=N-$⟨⟩$-CH_2-$⟨⟩$-N=C=O$

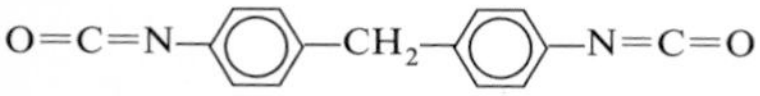

TDI production (in 1000 tonnes):

	1989	1991	1993
USA	333	336	359
W. Europe	331	336	350
Japan	101	136	137

MDI production (in 1000 tonnes):

	1989	1991	1993
USA	510	570	540
W. Europe	436	455	514
Japan	165	170	205

industrial TDI manufacture in three-step continuous reaction of toluene:

1. nitration of toluene to *o*-, *p*-, *m*-nitrotoluene, separation of *m*-isomer and further nitration of *o*-, *p*-mixture to 2,4- and 2,6-dinitrotoluene

characteristics of toluene nitration:

electron pressure of the CH_3 group (+I effect) means lower $NO_2^\oplus$ concentration is required for nitration than with benzene; *i.e.*, H_2SO_4 can contain more H_2O

In the last few years, however, 4,4'-diphenylmethane diisocyanate (**m**ethane **di**phenyl**di**isocyanate, MDI), whose precursor 4,4'-diaminodiphenylmethane is obtained from the condensation of aniline with formaldehyde, has overtaken TDI. In 1994, capacities worldwide, in Western Europe, the USA, and Japan for MDI were 1.9, 0.81, 0.53, and 0.22 $\times$ 10^6 tonnes per year, respectively; for TDI, they were 1.2, 0.45, 0.43, and 0.15 $\times$ 10^6 tonnes per year.

Another important component of polyurethanes is **h**examethylene-1,6-**di**isocyanate (HDI, formerly HMDI), whose precursor hexamethylenediamine and its manufacture were described in Section 10.2.1.

Toluene diisocyanate is generally manufactured in a continuous process involving three steps:

1. Nitration of toluene to dinitrotoluene
2. Hydrogenation of dinitrotoluene to toluenediamine
3. Phosgenation to toluene diisocyanate

To 1:

The continuous nitration of toluene can be done under milder conditions than are necessary for benzene due to the activating effect of the methyl group. For example, the H_2O content of the nitrating acid can be 23%, compared to 10% for the nitration of benzene. The mixture of mononitrated products of toluene consists of the three isomers *o*-, *p*- and *m*-nitrotoluene, whose distribution is influenced only slightly by reaction conditions. A typical composition is 63% *o*-, 33 – 34% *p*- and 4% *m*-nitrotoluene. The mixture can be separated by distillation or crystallization.

Nitrotoluenes are intermediates for dyes, pharmaceuticals, and perfumes, and precursors for the explosive 2,4,6-trinitrotoluene (TNT). A mixture of *o*- and *p*-nitrotoluene can be nitrated to dinitrotoluenes, the feedstocks for the manufacture of diisocyanates. The isomeric 2,4- and 2,6-dinitrotoluenes are obtained in a ratio of roughly 80:20:

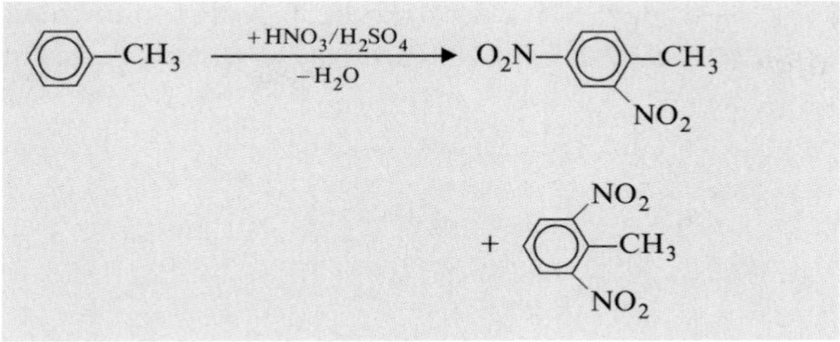

(36)

Nitration and workup are done in the usual manner, *e. g.* analogous to manufacture of nitrobenzene (*cf.* Section 13.3.1).

To 2:

The hydrogenation of the dinitrotoluene mixture to the two toluenediamines is once again a standard process in aromatic synthesis. This reaction can be carried out with iron and aqueous hydrochloric acid like the reduction of nitrobenzene, but catalytic hydrogenation — for example in methanol with a Raney nickel catalyst at about 100 °C and over 50 bar, or with palladium catalysts — is preferred.

2. dinitrotoluene reduction to toluene diamine:

  2.1. with Fe/HCl
  2.2. catalytic with H_2 in the liquid phase using Raney nickel or
  2.3. with Pd/C catalyst

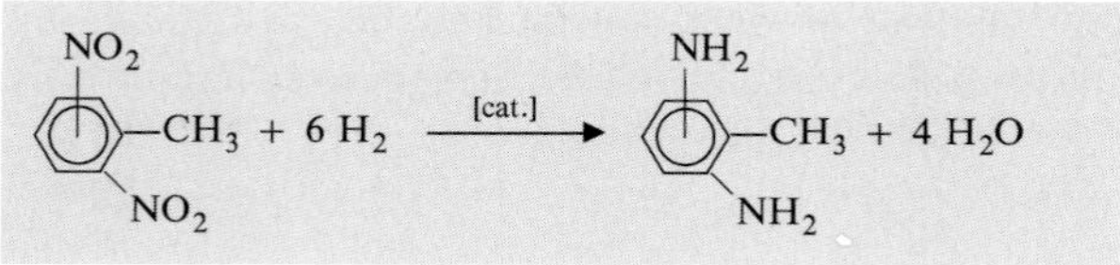

(37)

The dinitrotoluenes are reduced quantitatively in a succession of pressure hydrogenations. The selectivity to the toluenediamines is 98 – 99%. In contrast to the manufacture of aniline from nitrobenzene, gas-phase hydrogenations are not used commercially due to the ready explosive decomposition of the dinitrotoluenes at the required reaction temperatures. Purification is done in a series of distillation columns.

To 3:

Phosgenation of the toluenediamines can be carried out in several ways. 'Base phosgenation', *i. e.*, the reaction of the free primary amine with phosgene, is the most important industrially. In the first step, the amine and phosgene are reacted at 0 – 50 °C in a solvent such as *o*-dichlorobenzene to give a mixture of carbamyl chlorides and amine hydrochlorides.

3. toluenediamine phosgenation by two routes:

  3.1. with free base, subdivided into:

    cold phosgenation at 0 – 50 °C and hot phosgenation at 170 – 185 °C

The reaction product is fed into the hot phosgenation tower where, at 170 – 185 °C, it is reacted further with phosgene to form the diisocyanates:

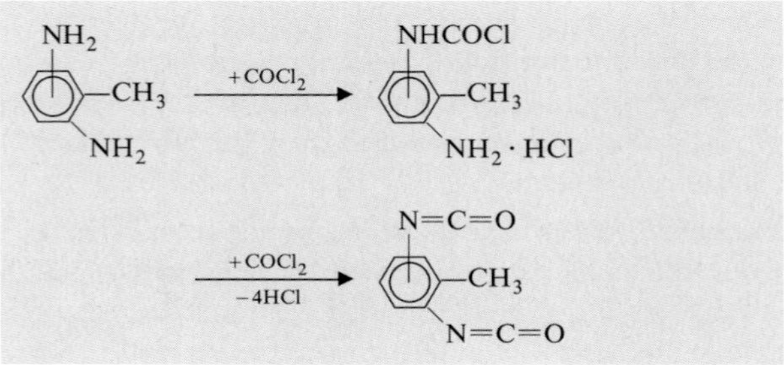

(38)

The excess phosgene can be separated from HCl in a deep freezing unit and recycled to the process.

3.2. with HCl salt, subdivided into:

salt formation (HCl) to decrease activity of free amine and hot phosgenation

The phosgenation of the toluenediamine hydrochlorides is the second possible manufacturing process for the diisocyanates. In the Mitsubishi Chemical process, for example, the toluenediamines are dissolved in *o*-dichlorobenzene and converted into a salt suspension by injecting dry HCl. Phosgene is reacted with the hydrochlorides at elevated temperatures and with strong agitation to give the diisocyanates. The HCl which evolves is removed with an inert gas stream:

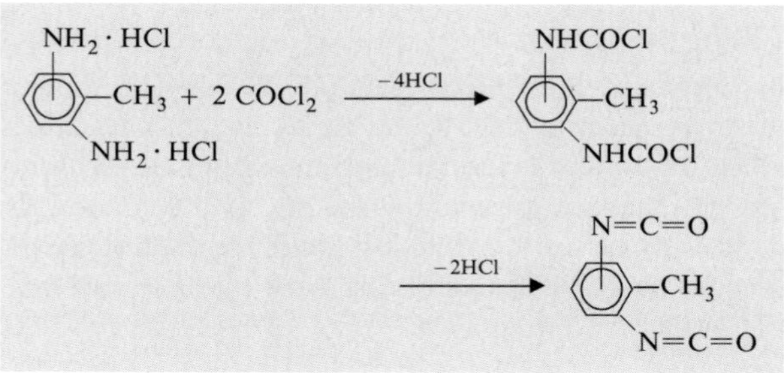

(39)

The workup and purification are done by fractional distillation. The selectivity to toluene diisocyanates is 97% (based on diamine). In the Mitsubishi process, the overall selectivity to diisocyanate is 81% (based on toluene). In addition to pure 2,4-toluene diisocyanate, two isomeric mixtures are available commercially, with ratios of 2,4- to 2,6-isomer of 80:20 and 65:35.

Because of the increased commercial interest in diisocyanates, new manufacturing routes without the costly phosgenation step − that is, without total loss of chlorine as HCl − have recently been developed. Processes for the catalytic carbonylation of aromatic nitro compounds or amines are the most likely to become commercially important.

In a development from Atlantic Richfield Company (Arco), the nitrobenzene feedstock for the manufacture of 4,4′-diphenylmethyl diisocyanate (MDI) is first reacted catalytically (*e. g.*, SeO$_2$, KOAc) with CO in the presence of ethanol to give N-phenylethyl urethane. After condensation with formaldehyde, thermolysis at 250−285 °C is used to cleave ethanol and form MDI:

newer manufacturing routes for diisocyanates without COCl$_2$ and thus without loss of Cl as HCl:

1. MDI in Arco three-step process from nitrobenzene through urethane derivative (ethyl phenyl carbamate), HCHO condensation and thermal elimination of EtOH

characteristics of nitroisocyanate transformation:

CO has two functions: reduction of the nitro group and carbonylation in the presence of ethanol (ethoxycarbonylation)

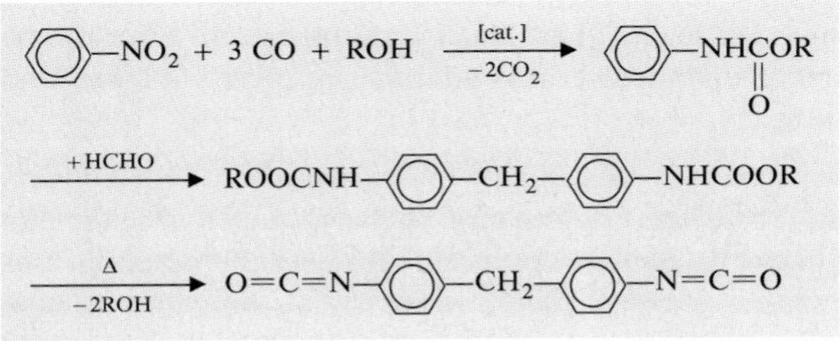

Mitsui Toatsu and Mitsubishi Chemical have formulated a similar carbonylation process for the conversion of dinitrotoluene to toluene diisocyanate (TDI).

2. TDI from Mitsui Toatsu and Mitsubishi analogous to 1.

Another manufacturing process for 4,4′-diphenylmethyl diisocyanate (MDI) has been introduced by Asahi Chemical. In contrast to the Arco route, aniline is used for the carbonylation to N-phenylethyl urethane; otherwise, the same steps are followed (*cf.* eq. 40). The oxidative carbonylation of aniline is done in the presence of metallic palladium and an alkali iodide promoter at 150−180 °C and 50−80 bar. The selectivity is more than 95 % with a 95 % aniline conversion:

3. MDI from three-step Asahi process from aniline, otherwise analogous to 1, but with higher total selectivity and without chlorinated byproducts with unfavorable effects on polyadditon of MDI to polyurethanes

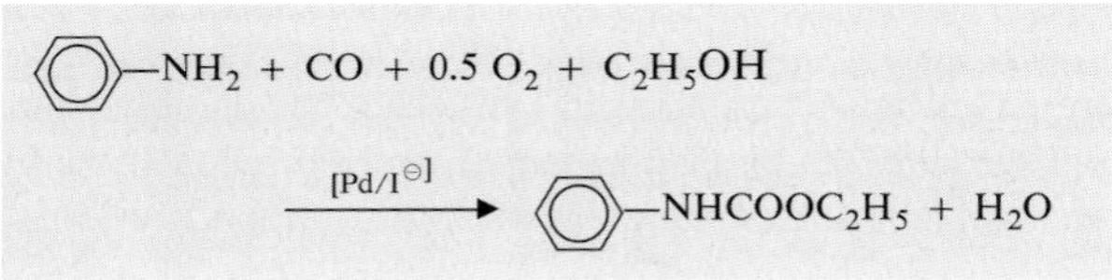

The final condensation with formaldehyde at 60−90 °C and atmospheric pressure first takes place in the presence of H$_2$SO$_4$

Pd function in aniline/CO/O$_2$ reaction:

$$C_6H_5NH\text{—}Pd\text{—}H \xrightarrow{+CO}$$
$$C_6H_5NH\text{—}CO\text{—}Pd\text{—}H \xrightarrow{+ROH}$$
$$C_6H_5NHCOOR + H\text{—}Pd\text{—}H$$
$$\xrightarrow{+O_2} Pd + H_2O$$

urethane/HCHO condensation with limited urethane conversion to avoid trimers and tetramers

two-step condensation for

conversion of $>\!N\!-\!CH_2\!-\!\bigcirc\!-$ and

$>\!N\!-\!CH_2\!-\!N\!<$ byproducts into

the desired diurethane

$-\!\bigcirc\!-CH_2\!-\!\bigcirc\!-$

4. isocyanate, diisocyanate, polycarbonate production with $(CH_3O)_2C\!=\!O$ instead of $COCl_2$ as carbonylating agent

uses of diisocyanates:

polyaddition to polyhydric alcohols, to polyether alcohols, *e. g.,*

$HO\!-\!\!\!\left[CH_2CH_2O\right]_{\overline{n}}\!H$

to polyester alcohols, *e. g.,*

$HOC_2H_4O\!-\!\!\!\left[\begin{smallmatrix}\\ CCH=CHC\\ \| \quad\quad \|\\ O \quad\quad O\end{smallmatrix}\!-\!OC_2H_4O\right]_{\overline{n}}$

addition of triols, tetrols
→ cross linking

addition of H_2O → foam formation

in two phases and then, after removal of the water phase and additional treatment with, *e. g.,* trifluoroacetic acid, in a homogeneous phase with over 95% selectivity to the diurethane at a urethane conversion of about 40%.

The last step, the thermal elimination of ethanol, is done at $230 - 280\,°C$ and $10 - 30$ bar in a solvent; the selectivity to MDI is over 93%.

The Asahi process has not been used commercially.

Another way of avoiding uneconomical and toxic phosgene is to use dimethyl carbonate for the production of isocyanates from aromatic amines, and for polycarbonates, such as with bisphenol A (*cf.* Section 13.2.1.3). To meet the growing demand for dimethyl carbonate, Ube has begun operation of the first pilot plant for the selective gas-phase carbonylation of methanol in Japan. Other similar liquid-phase processes for the production of dimethyl carbonate are in operation at Daicel and Mitsui Sekka.

Diisocyanates are used mainly in the manufacture of polyurethanes. These are produced by polyaddition of diisocyanates and dihydric alcohols, in particular the polyether alcohols, *i. e.,* polyethylene glycols, polypropylene glycols, and the reaction products of propylene oxide with polyhydric alcohols. Oligomeric esters from dicarboxylic acids and diols (polyester alcohols) are also used:

$$n\ O\!=\!C\!=\!N\!-\!R^1\!-\!N\!=\!C\!=\!O + (n + 1)\ HO\!-\!R^2\!-\!OH$$

$$\longrightarrow H\!-\!\!\!\left[OR^2\!-\!O\!-\!\underset{\underset{O}{\|}}{C}\!-\!NH\!-\!R^1\!-\!NH\!-\!\underset{\underset{O}{\|}}{C}\right]\!-\!OR^2\!-\!OH\Big]_n$$

$$e.\,g.,\ R^1 = \bigcirc\!-CH_3 \quad R^2 = \left(CH\!-\!CH_2\!-\!O\right)_{\!n}\!CH\!-\!CH_2\!-\!\\ \qquad\qquad\qquad\qquad\qquad |\qquad\qquad\qquad |\\ \qquad\qquad\qquad\qquad\qquad CH_3 \qquad\qquad CH_3$$

(42)

Polyurethanes can be crosslinked by adding tri- or polyhydric alcohols (glycerol, trimethylolpropane), and caused to foam by adding small amounts of water, which causes saponification of the isocyanate group to the amino group and CO_2.

Polyurethanes are processed to flexible and rigid foams; they are also used in textile coatings and as elastomers (*e. g.,* spandex fibers).

Other products using polyurethanes are, for example, artificial leather, synthetic rubber, dyes, paints, and adhesives. However, the greatest use of polyurethanes is for foams (*cf.* Table 13−8).

uses of polyurethanes:

flexible foams
rigid foams
coated fabrics
spandex fibers
synthetic leather, rubber
dyes, paints
adhesives

Table 13-8. Polyurethane use (in %).

Product	USA 1990	Western Europe 1987	Western Europe 1990	Japan 1990
Flexible foams	52	41	43	32
Rigid foams	27	25	24	17
Integral-skin and filling foams	21	8	33	51
Paint raw materials		7		
Elastomers		6		
Thermoplastic polyurethanes				
Artificial leathers		6		
Others		8		
Total use (in 10^6 tonnes)	1.54	1.35	1.40	0.60

14. Oxidation Products of Xylene and Naphthalene

14.1. Phthalic Anhydride

Phthalic anhydride (PA) is an important intermediate in the manufacture of phthalate plasticizers, alkyd and polyester resins, phthalocyanine dyes, and numerous fine chemicals. In 1995 the world capacity for phthalic anhydride was about 2.9×10^6 tonnes per year, with roughly 0.93, 0.47, and 0.39×10^6 tonnes per year in Western Europe, the USA, and Japan, respectively.

Production figures for phthalic anhydride in these countries are summarized in the adjacent table.

phthalic anhydride production (in 1000 tonnes):

	1990	1992	1995
W. Europe	791	753	930
USA	426	407	470
Japan	301	309	390

Until 1960, PA was manufactured almost exclusively from naphthalene from coal tar. A reduction in coal coking, leading to a shortage and therefore higher price for naphthalene, coupled with the constantly growing demand for phthalic anhydride led to o-xylene as an inexpensive, readily available, and − with regard to the stoichiometry − more economical feedstock. Differences between countries can be seen in the adjacent table. In 1975, approximately 75% of all phthalic anhydride was produced from o-xylene. This figure had risen to about 85% by the early 1980s, and remained fairly constant through the beginning of the 1990s. Differences between several countries can be seen in the adjacent table. Other raw materials for the manufacture of phthalic anhydride (e. g., acenaphthene) are no longer important.

phthalic anhydride feedstocks:

		naphthalene	o-xylene
world	1960	100	−
	1975	25	75
	1980	15	85
	1991	15	85
Japan	1991	44	56
	1994	41	59
USA	1991	13	87
	1994	17	83
W. Europe	1991	6	94
	1994	6	94

14.1.1. Oxidation of Naphthalene to Phthalic Anhydride

In analogy to the oxidative degradation of benzene to maleic anhydride, naphthalene yields phthalic anhydride:

phthalic anhydride manufacture based on naphthalene:

oxidative degradation with switch of oxidation system:

earlier: MnO_2/HCl or CrO_3 or oleum

today: air/catalyst

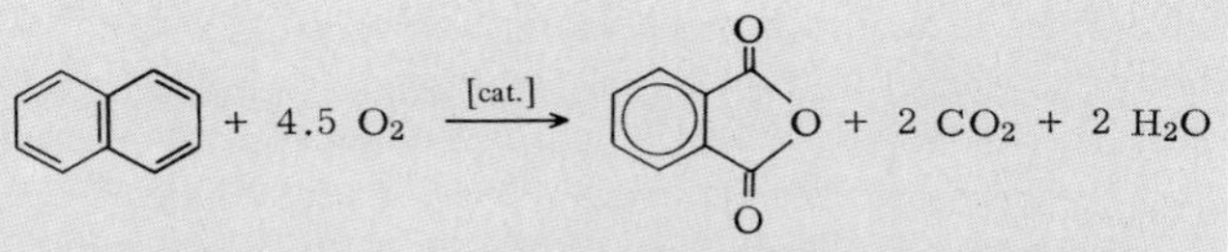

$$\left(\Delta H = - \frac{428 \text{ kcal}}{1792 \text{ kJ}} / \text{mol} \right) \qquad (1)$$

The original reaction conducted with stoichiometric amounts of oxidizing agents was replaced by BASF in 1916 by an air oxidation with V_2O_5 catalysts. IG Farben produced 12000 tonnes phthalic anhydride as early as 1941.

Two types of processes have been developed for the gas-phase oxidation:

naphthalene gas-phase oxidation with two process variations for catalyst bed:

1. fixed bed
2. fluidized bed

1. Processes with fixed-bed catalyst, which can be subdivided into a low temperature version (350–400 °C) using a pure naphthalene feed, and a high-temperature version (400–550 °C) with a lower grade of naphthalene.

2. Fluidized-bed processes at temperatures above 370 °C.

    Well-known naphthalene oxidation processes have been developed by BASF, Von Heyden, Koppers, and Sherwin Williams/Badger. They differ mainly in the type of reactor, catalyst, and phthalic anhydride isolation. Therefore the following description portrays only the reaction principles.

To 1:

principles of phthalic anhydride manufacture in fixed bed:

V_2O_5/SiO_2 (+ promoter)-catalyzed oxidation of naphthalene with air in multitube reactor using molten salt for heat removal

isolation of phthalic anhydride:

phthalic anhydride formed as crystal, dehydrated once again and purified by distillation

byproducts:

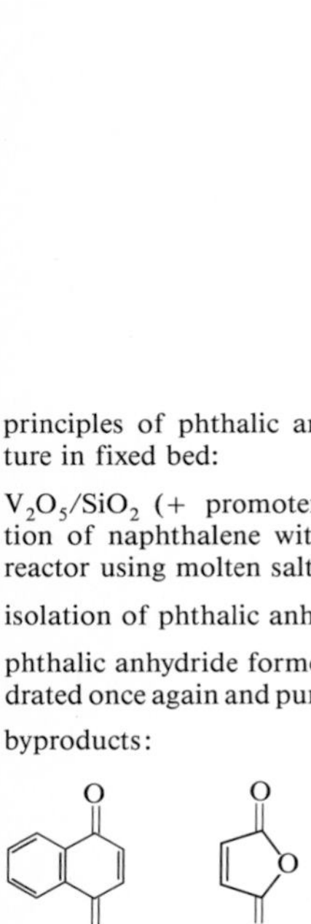

manufacturing variation for phthalic anhydride using lower grade naphthalene:

less selective oxidation at higher temperature

The catalyst – usually V_2O_5/SiO_2 with, *e. g.*, K_2SO_4 promoter – is situated in a multitube reactor cooled with a salt melt to remove the substantial heat of reaction. Naphthalene and air are introduced through an evaporator. With fresh catalyst, the reaction occurs at 360 °C; the temperature must be slowly increased as the catalyst activity decreases. The reaction gases are cooled rapidly to below 125 °C – the dew point of the anhydride. The crude product, which crystallizes as needles, is completely dehydrated in melting vessels and then distilled. The selectivity is 86–91 % at a naphthalene conversion of about 90 %.

The byproducts include 1,4-naphthoquinone, maleic anhydride, and higher molecular weight condensation products.

The high-temperature process was developed in the USA specifically for lower grades of naphthalene. V_2O_5 on a support is also used as a catalyst. Although at the higher temperature of 400–550 °C the catalyst does not lose its activity very rapidly, its selectivity is noticeably lower (60–74 %). Maleic anhydride (6–10 %) is obtained as a byproduct along with other compounds.

To 2:

principles of phthalic anhydride manufacture in fluidized bed:

The fluidized-bed manufacture of phthalic anhydride using V_2O_5 catalysts at 350–380 °C was first practiced in 1944 by

Badger/Sherwin Williams in the USA. In the commerical process, liquid naphthalene is injected into the fluidized bed where preheated air serves as the vortex gas. The advantages of this process are those characteristic of all fluidized-bed processes: uniform temperature distribution in the entire catalyst bed, possibility of rapidly exchanging catalyst, and catalyst circulation to remove heat with a secondary current. A high conversion with a large throughput − *i. e.*, a high space–time yield − is obtained. The selectivity to phthalic anhydride is as high as 74%. Part of the anhydride can be separated in liquid form.

naphthalene oxidation, analogous to the fixed-bed operation, with catalyst based on V_2O_5, however due to fluidized-bed method characterized by minor temperature gradients and high space–time yield

14.1.2. Oxidation of *o*-Xylene to Phthalic Anhydride

New plants for the manufacture of phthalic anhydride are usually based on an *o*-xylene feedstock:

phthalic anhydride manufacture based on *o*-xylene with following technological advantages compared to naphthalene basis:

1. lower O_2 requirement
2. lower reaction enthalpy
3. liquid more easily metered

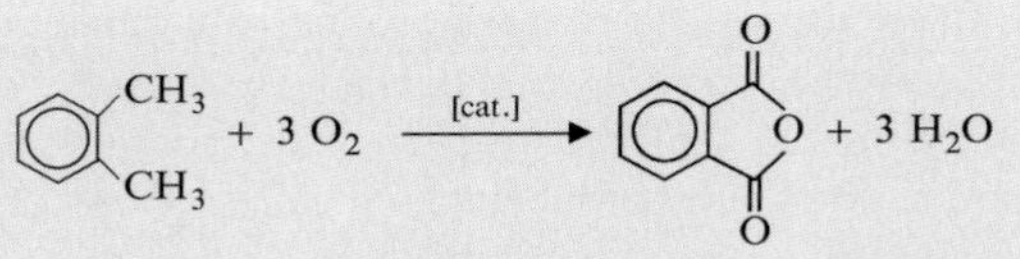

$$\left(\Delta H = -\; \frac{265 \; \text{kcal}}{1110 \; \text{kJ}} /\text{mol} \right)$$

(2)

There are several factors in favor of its use: the number of carbon atoms remains the same in the product and feedstock − unlike naphthalene, there is no oxidative degradation − and the evolution of heat is reduced due to the lower oxygen requirement. Despite this fact, many plants have been so constructed that either *o*-xylene or naphthalene can be used.

The current *o*-xylene oxidation processes can be divided into two main groups:

1. Gas-phase oxidation with a fixed- or fluidized-bed catalyst based on V_2O_5

2. Liquid-phase oxidation with dissolved metal salt catalysts

two methods for phthalic anhydride manufacture by air oxidation of *o*-xylene:

1. gas phase with V_2O_5 based catalyst in fixed or fluidized bed
2. liquid phase with homogeneous metal salt catalysts

To 1:

In commercial processes, *o*-xylene is generally oxidized in the gas phase. Two widely used processes were developed by BASF and Chemische Fabrik von Heyden (now Wacker Chemie). In 1989, the world capacity for phthalic anhydride made by the BASF and Von Heyden processes was about 1.0×10^6 tonnes per year and more than 1.5×10^6 tonnes per year, respectively.

characeristics of o-xylene oxidation using fixed-bed catalysis:

strongly exothermic reaction requires multi-tube reactors; despite narrow tubes, catalyst spheres (punctiform contact), allow high throughput with low pressure drop

byproducts of o-xylene oxidation:

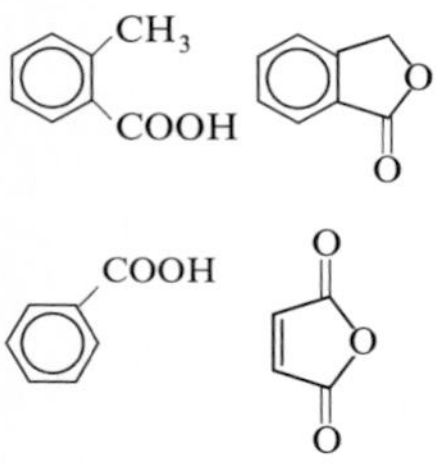

In both processes, o-xylene (95% pure) is oxidized at 375–410 °C with an excess of air over V_2O_5 catalysts arranged in multitube reactors with about 10 000 tubes.

The BASF catalyst consists of a mixture of V_2O_5 and TiO_2 with promoters such as Al and Zr phosphates which are distributed on spheres of, for example, porcelain, quartz, or silicium carbide, which have a smooth surface and are largely pore-free (shell catalysts). Phthalic anhydride is obtained with a selectivity of 78% (based on o-xylene) and, after a two-stage distillation, with a purity of at least 99.8%. The byproducts include o-toluic acid, phthalide, benzoic acid, and maleic anhydride, as well as CO_2 from the total oxidation.

Improved enlarged reactors now have an output of 40–50 000 tonnes per year phthalic anhydride per unit. In a new development in the Von Heyden process, specially constructed tube reactors are used which, using a salt melt for cooling, allow an exact control of the temperature profile, and thus a higher loading of the air with o-xylene (60 g/m³ vs. 44 g/m³). The more intense heat generation leads to substantial energy savings.

Other firms such as Nippon Shokubai and Alusuisse have also realized new technologies for a higher o-xylene/air ratio.

process development:

higher o-xylene loading in air leads to increase of catalyst productivity (STY) and decrease in energy use

Other fixed-bed processes were developed by Ftalital (now Alusuisse), Japan Gas (now Mitsubishi Gas), Pechiney–Saint Gobain, Rhône–Progil, Ruhröl (now Hüls), and Scientific Design.

characteristics of o-xylene oxidation with fluidized-bed catalysis:

compared to fixed bed: lesser excess of air, higher throughput, phthalic anhydride removed as liquid

The catalyst selectivities in the fluidized- and fixed-bed processes are almost equal. However, since the danger of explosion in the fluidized bed is considerably less, a lower excess of air can be used. As a result, part of the phthalic anhydride can be removed above its melting point, i.e., as a liquid. This type of isolation offers marked technological advantages over crystal deposition.

To 2:

characteristics of o-xylene liquid-phase oxidation (Rhône–Progil process):

high radical concentration allows low oxidation temperatures of 150 °C, AcOH as diluent increases selectivity to ca. 90%

o-Xylene can also be oxidized in the liquid phase with air in processes developed by, for example, Rhône–Progil. Soluble acetates or naphthenates of Co, Mn, or Mo are generally used with cocatalysts containing bromine. Carboxylic acids, mainly acetic acid, are added as solvents. The oxidation is conducted at about 150 °C. The phthalic acid is removed as a solution in acetic acid, separated in crystalline form by cooling, dehydrated to the anhydride and distilled. The selectivity is reported to be 90%.

Although other firms such as Hüls and Standard Oil of Indiana have also developed liquid-phase processes for phthalic anhydride manufacture, this technology has not been used commercially.

Similarly, *m*-xylene can be oxidized to isophthalic acid with the *o*-xylene oxidation process (*cf.* Section 14.2.2). This acid and its esters are becoming increasingly important as precursors for high melting plastics, *e.g.*, polybenzimidazoles.

transfer of oxidation principle to *m*-xylene → isophthalic acid possible:

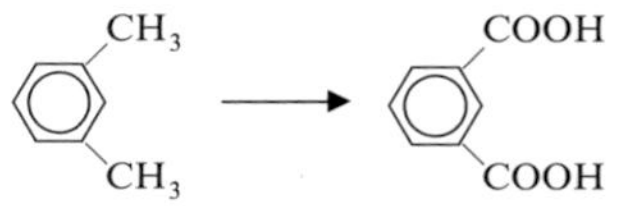

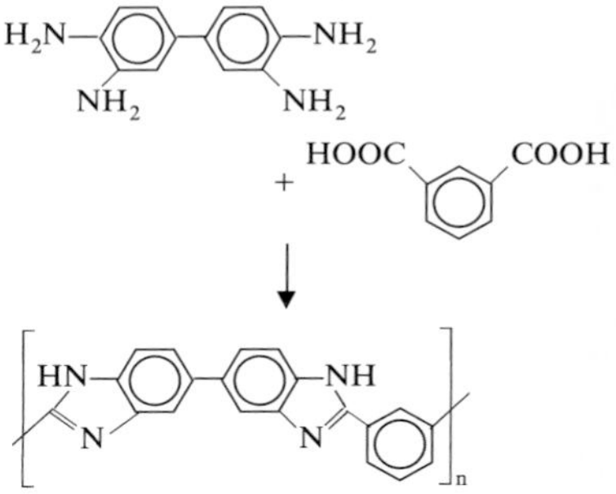

14.1.3. Esters of Phthalic Acid

The use of phthalic anhydride for a broad spectrum of phthalic acid esters is outlined in the following table:

Table 14–1. Use of phthalic anhydride (in %).

Product	USA 1991	USA 1995	W. Europe 1991	W. Europe 1995	Japan 1991	Japan 1995
Softeners	53	53	60	61	67	68
Unsat. polyesters	24	21	20	20	12	11
Alkyd resins	16	15	16	14	17	16
Others	7	11	4	5	4	5

use of PA:

1. phthalic acid esters with $C_4 - C_{10}$ alcohols
2. unsaturated polyesters with, *e.g.*, maleic anhydride, diols and cross-linking copolymerization with styrene
3. polyesters (alkyd resin) with glycerin
4. intermediates

Most phthalic acid anhydride is converted to one of three different types of esters. The largest use is for the monomeric phthalic acid esters with $C_4 - C_{10}$ alcohols, used as plasticizers. Manufacturing processes for these will be described in detail.

The second group comprises the unsaturated polyesters. In their manufacture, phthalic anhydride together with maleic anhydride (*cf.* Section 13.2.3.4) or fumaric acid is polycondensed with a diol (1,2-propylene glycol or diethylene glycol), generally in a melt. A subsequent hardening is usually accomplished through a radically initiated cross-linking with styrene. The duroplasts formed are characterized by their thermal stability and their good mechanical properties for fiber applications.

manufacturing process for unsaturated polyesters:

copolycondensation of PA and unsaturated dicarboxylic acids or their anhydrides with diols to control the degree of unsaturation in the macromolecule for the subsequent graft polymerization with styrene

The third use for phthalic anhydride − sometimes mixed with, *e.g.*, adipic acid − is for alkyd resins and glyptals from its reaction with glycerin (*cf.* Section 11.2.3). A small amount of phthalic anhydride is also used for intermediates for, *e.g.*, dye manufacture.

manufacturing principle for alkyd resins:

partial polycondensation of PA with oligofunctional alcohols (glycerin) followed by cross linkage with higher degree of condensation

The main use of phthalic anhydride is for the manufacture of plasticizers.

production of phthalate plasticizers (in 1000 tonnes):

	1990	1992	1994
W. Europe	1168	1202	1253
USA	572	573	638
Japan	453	444	444

Depending on the country, the amount of anhydride used for this application varies between 50 and 70%, and parallels the development of flexible plastics and dispersions. Production figures for Western Europe, Japan, and USA are summarized in the adjacent table.

favored alcohol components for phthalate plasticizers:

$$CH_3(CH_2)_3\underset{\underset{C_2H_5}{|}}{CH}CH_2OH$$

n, iso-C_9H_{19} — CH_2OH

Of the various phthalic acid esters used as plasticizers, bis(2-ethylhexyl) phthalate (also known as dioctyl phthalate or DOP), diisodecyl phthalate (DIDP) and diisononyl phthalate (DINP) are the highest volume products (cf. Section 6.1.4.3).

DOP share of softeners (in wt%)

	1986	1991	1994
USA	24	20	18
W. Europe	53	48	55
Japan	n.a.	66	66

n.a. = not available

The fraction of the total production of phthalate softeners that is diisooctyl phthalate is decreasing somewhat, depending on the country, as shown in the adjacent table.

Through the formation of esters with these and many other alcohols, the phthalates give rise to a broad spectrum of softeners.

BASF started up the production of bis(2-ethylhexyl)phthalate as early as 1940.

The esterification of phthalic acid with 2-ethylhexanol is a two-step process starting from phthalic anhydride which passes through the semi-ester to the diester:

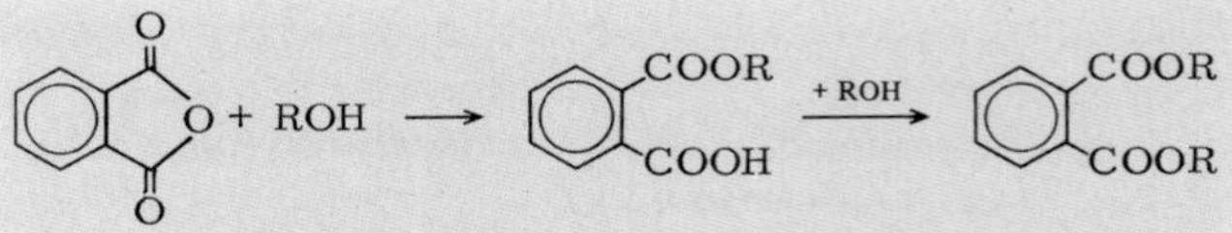

$$+ H_2O \left(\Delta H = ca. -\frac{20\ kcal}{84\ kJ}/mol \right) \quad (3)$$

two-step manufacture of phthalic acid esters from phthalic anhydride:

first step — rapid nucleophilic alcoholysis of phthalic anhydride without catalyst

second step to diester slow and thus accelerated by Brönsted or Lewis acid catalysis or higher reaction temperature

characteristics of proton catalysis:

side reactions such as olefin, ether, and isomer formation from alcohol reduce the selectivity

The first step of the adduct formation takes place very rapidly; the monoester is formed as the anhydride dissolves in the alcohol.

In the rate-determining second step, activation by either an esterification catalyst or higher reaction temperature is necessary.

In the industrial processes acidic catalysts such as H_2SO_4, p-toluenesulfonic acid, or α-naphthalenesulfonic acid have been favored until now. However, if the reaction temperature is allowed to exceed 160 °C, the selectivity drops due to proton-catalyzed side reactions of the alcohol, e. g., dehydration to the olefin or ether, and isomerizations. The process must therefore be run at reduced pressure to remove the water of reaction azeotropically with excess 2-ethylhexanol at a low temperature. The catalyst also has to be separated and an involved purification is necessary for the crude ester.

A newer development from M & T Chemicals (USA) led to the addition of aprotic catalysts based on tin, such as tin oxalate, which results in a substantial increase in the esterification rate at 200–220 °C without the byproduct formation caused by proton catalysis. The tin oxalate separates from the reaction mixture on cooling, allowing it to be easily removed.

characteristics of aprotic catalysis:

specific catalyzed esterification with temperature-dependent solubility of SnC_2O_4 leads to higher selectivity and easier catalyst separation

BASF has developed a catalyst-free esterification of phthalic anhydride at a higher temperature. Under these conditions, the proton of the phthalic acid monoester from the first esterification step can function autocatalytically in the total esterification. The esterification of phthalic anhydride with 2-ethylhexanol is conducted continuously at 185–205 °C and atmospheric pressure. The water of reaction is removed continuously by azeotropic distillation with an excess of 2-ethylhexanol.

characteristics of purely thermal esterification:

autocatalysis by phthalic acid semiester leads to higher ester quality without expensive secondary treatment

A selectivity of 98% (based on phthalic anhydride and 2-ethylhexanol) is obtained with a 97% conversion of the monophthalate to the diester in a multistep cascade of boilers. As the resulting phthalate is very pure, neither rectification nor chemical (H_2O_2) or physical treatment (U. V. irradiation, activated carbon) is necessary.

Other 'oxo' alcohols can also be esterified with phthalic anhydride using the same process.

The function of phthalic acid ester plasticizers is to transform hard, brittle thermoplastics into a soft ductile and elastic state necessary for processing and use.

function of phthalic acid esters as plasticizers:

additives for thermoplastics to reduce the Van der Waals forces between polymer chains

The properties of PVC, in particular, can be extensively modified by specific plasticizing. For example, when PVC is processed for tubing and films, then a plasticizer mixture is added in amounts of 30–70% of the weight of the untreated plastic. Phthalates account for the largest proportion of the plasticizers.

increasing production of PVC, greatest plasticizer consumer, has also caused rise in demand for plasticizers (mainly phthalates)

One result of the considerable expansion in production of PVC has been an increase in significance and growth of phthalate plasticizers.

14.2. Terephthalic Acid

greatest use of terephthalic acid (TPA) is manufacture of polyethylene terephthalate (PET):

the simplest manufacturing route to PET by polycondensation of glycol and TPA was impossible for many years due to insufficient purity of TPA

Since terephthalic acid (TPA) is used chiefly for polyethylene terephthalate (PET), it also experienced the same dramatic growth as this polyester fiber and film material during the last three decades. This growth will continue if the use of PET in the bottling industry expands as expected. World production capacity for PET in 1995 was about 3.2×10^6 tonnes per year, with about 1.6, 0.81, and 0.15×10^6 tonnes per year in the USA, Western Europe, and Japan, respectively. Eastman Chemical is the largest producer with a total capacity of about 0.7×10^6 tonnes per year (1993).

For many years, the unusual physical properties of terephthalic acid prevented the chemically simplest method − polycondensation by direct esterification of the dicarboxylic acid and diol − from being used (*cf.* Section 14.2.4). Terephthalic acid is extremely insoluble in hot water (100 °C) and the usual organic solvents, and doesn't melt, so that it could not be prepared on a large scale with a purity sufficient for polycondensation. The crude terephthalic acid was therefore converted into its dimethylester which was then brought to fiber grade by crystallization and distillation.

TPA purification difficult as TPA cannot be simply crystallized, doesn't melt, and only sublimes above 300 °C, thus crude TPA purified as dimethyl ester (DMT) up to now

In 1963, Teijin and Toray were successful in using a classical process with improved technology to manufacture a pure terephthalic acid that was polycondensable with ethylene glycol. Since then, other processes have been developed for producing fiber grade terephthalic acid which, in comparison to the production of dimethyl terephthalate (DMT), have become more important in all industrialized countries and especially in Japan where it was initially used commercially as is shown in the adjacent table.

fiber grade TPA recently made available by improved oxidation technology of *p*-xylene

TPA share of TPA/DMT capacity (in %)

	USA	W. Europe	Japan	world
1976	33	26	39	29
1980	39	34	58	41
1987	41	41	77	53
1993	60	58	80	68

In 1995, world production capacity for TPA was about 10.8×10^6 tonnes, with roughly 2.6, 1.4, and 1.7×10^6 tonnes in the USA, Western Europe, and Japan, respectively. The world capacity for DMT in 1994 was about 5.0×10^6 tonnes with 1.9, 1.2, and 0.38×10^6 tonnes in the USA, Western Europe, and Japan, respectively. Production figures for these countries can be found in the adjacent tables.

TPA production (in 1000 tonnes):

	1990	1992	1995
USA	1780	1872[1)	2570
Japan	1338	1425	1681
W. Europe	961	1010	1345

[1) 1991

The world's largest producer of TPA is Amoco with a capacity of about 2×10^6 tonnes per year (1995).

DMT production (in 1000 tonnes):

	1988	1990	1992
USA	1709	1625	1466[1)
Japan	327	366	368
W. Europe	1038	975	1000

[1) 1991

14.2.1. Manufacture of Dimethyl Terephthalate and Terephthalic Acid

The most important commercial route to ester and acid manufacture is the liquid-phase oxidation of *p*-xylene. However, if no special precautions are taken, the oxidation stops at *p*-toluic acid:

$$\text{(4)}$$

Several methods are available to convert the second methyl group into a carboxyl group. A differentiation can be made between three basic possibilities:

1. The carboxyl group of *p*-toluic acid is esterified with methanol in a separate step (Witten, Hercules, California Research) or, if methanol is used as solvent for the oxidation, esterification occurs simultaneously (BASF, Montecatini, DuPont); the second methyl group is then oxidized.

2. In addition to the metal salt catalyst (Mn or Co), a cocatalyst or promoter such as a bromine compound is used (Amoco/ Mid-Century and IFP).

3. In a co-oxidation process, an auxiliary substance which is capable of supplying hydroperoxides is simultaneously oxidized. Acetaldehyde (Eastman Kodak), paraldehyde (Toray Industries), and methyl ethyl ketone (Mobil Oil and Olin Mathieson) are all used as co-oxidizable substances.

The first method yields, of course, dimethyl terephthalate, while recently the two others have been used mainly to manufacture the pure acid (*cf.* Section 14.2.2).

The Witten process for the manufacture of dimethylterephthalate, also known as the Imhausen or Katzschmann process, was developed at about the same time by California Research in 1950/51. Hercules contributed to the industrial development of the process with a California license and Imhausen expertise. After taking over Chemischen Werke Witten, Dynamit Nobel (now Hüls) has become the leading European producer of DMT. In 1993, world production capacity for this process was about 3×10^6 tonnes per year.

Marginal notes:

manufacturing principles of terephthalic acid and ester:

catalytic liquid-phase oxidation of *p*-xylene with *p*-toluic acid as intermediate product

the inhibition towards further oxidation of *p*-toluic acid can be overcome using one of three methods:

1. masking the inhibiting carboxyl groups by catalyst-free esterification with methanol

2. increasing the activity of oxidation catalyst by addition of bromine compounds as cocatalyst

3. increase in activity of oxidation catalyst through synthesis of hydroperoxides by co-oxidation

the 1st method of further oxidation leads to DMT, 2nd and 3rd yield TPA

commercial use of esterification method (DMT manufacturing process): Imhausen (later Witten, Dynamit Nobel, now Hüls) and California Research/Hercules

principles of two-step DMT manufacture:

first step:

homogeneously catalyzed, combined liquid-phase oxidation of:

1. *p*-xylene and
2. *p*-toluic acid ester

second step:

combined esterification either catalyst-free and under pressure, or with a proton acid, of:

1. *p*-toluic acid
2. TPA monoester with methanol

Both processes are two-step liquid-phase oxidations which produce *p*-toluic acid in the first step with air at $140-170\,°C$ and $4-8$ bar in the presence of Co/Mn salts of organic acids. A small amount of terephthalic acid is also formed. After esterification with methanol, e. g., catalyst-free at $250-280\,°C$ and almost 100 bar, or at $140-240\,°C$ and up to 40 bar in the presence of a proton catalyst such as *p*-toluenesulfonic acid, the second methyl group can be oxidized to the monomethyl ester of terephthalic acid. Industrially, the oxidation and esterification steps are combined. A mixture of *p*-xylene and *p*-toluic acid ester is oxidized and the products, toluic acid plus terephthalic acid monomethyl ester, are esterified together:

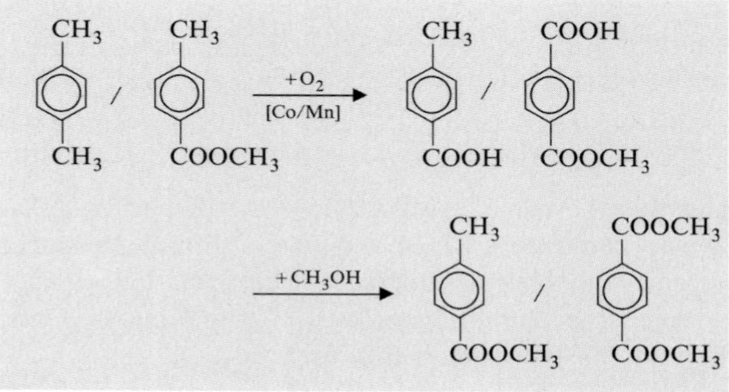

(5)

purification of DMT for use in fiber production:

1. distillation
2. crystallization
   (fiber grade DMT mp 141 °C)

The crude esters are separated into their components in a system of columns under vacuum. The *p*-toluic acid ester is recycled to the oxidation.

The dimethyl terephthalate is further purified by recrystallizing twice from methanol or xylene, melted, and converted into readily manageable flakes with a drum flaker. The dimethyl terephthalate selectivity is about 85% (based on *p*-xylene) and roughly 80% (based on CH_3OH).

Since then, Dynamit Nobel has developed their own process for the manufacture of fiber grade terephthalic acid so that the existing DMT plant could be converted to the production of TPA.

DMT manufacturing variations:

p-xylene oxidation with simultaneous esterification in liquid phase in accordance with counter-flow principle

BASF, Du Pont, and Montecatini combine the air oxidation of *p*-xylene in the liquid phase and the esterification with methanol in a single-step process:

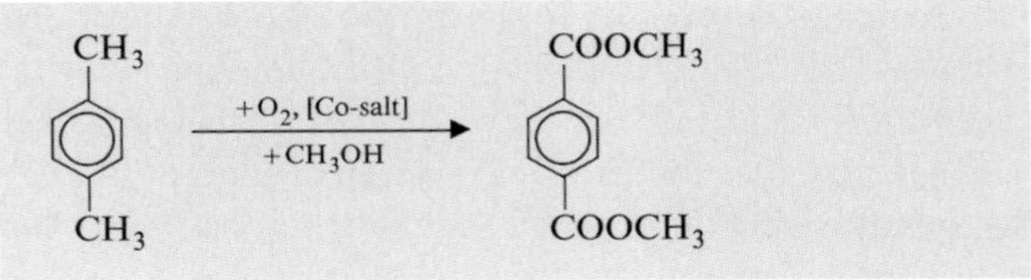

(6)

In a countercurrent reactor *p*-xylene and recycled, partially oxidized product are introduced at the top, and methanol and air at the bottom. The oxidation is done with Co salts at 100–200 °C and 5–20 bar, and a residence time of 22 hours. The crude dimethyl terephthalate selectivity is said to be greater than 90% (based on *p*-xylene) and 60–70% (based on methanol).

Older processes for oxidizing *p*-xylene with HNO_3 (BASF, Bergwerksverband, Du Pont, Hoechst, ICI, Richfield) are no longer used commercially, since nitrogen-containing impurities are difficult to remove from terephthalic acid.

older DMT production by oxidation of *p*-xylene with HNO_3 is insignificant today as N-containing byproducts are difficult to separate

14.2.2. Fiber Grade Terephthalic Acid

Dimethyl terephthalate manufactured by the first process principle (*cf.* Section 14.2.1) can be converted into pure terephthalic acid, *i.e.*, fiber grade, by pressurized hydrolysis in a simple process step.

manufacturing possibilities for fiber grade TPA:

1. pressurized hydrolysis of purified DMT
2. *p*-xylene oxidation with $Br^\ominus$ promoter
3. *p*-xylene co-oxidation with auxiliary substances

According to the second principle (cocatalysis), in direct manufacture the inhibition of the oxidation can be overcome by the addition of promoters to the intermediate (*p*-toluic acid).

industrial development of 'promoter' oxidation (route 2):

The most widely used process of this type is the Amoco process, based on an original development of Mid-Century Corp. Today, both firms are owned by Standard Oil of Indiana.

Amoco (Mid-Century)

cocatalysis principle in TPA manufacture (Amoco process):

In commercial operation, a catalyst combination of Co and Mn acetate in 95% acetic acid is used. The cocatalyst is a mixture of NH_4Br and tetrabromoethane. In variations of the Amoco process, the catalyst is $CoBr_2$, $MoBr_2$, or HBr. The bromine functions as a regenerative source of free radicals. *p*-Xylene is oxidized with air in stirred autoclaves at 190–205 °C and 15–30 bar. The equipment must be lined with titanium or Hastelloy C due to the corrosive catalyst solution. The oxidation product is cooled and the terephthalic acid, which crystallizes out, is separated.

radical-catalyzed liquid-phase oxidation of *p*-xylene in AcOH through activation with bromine

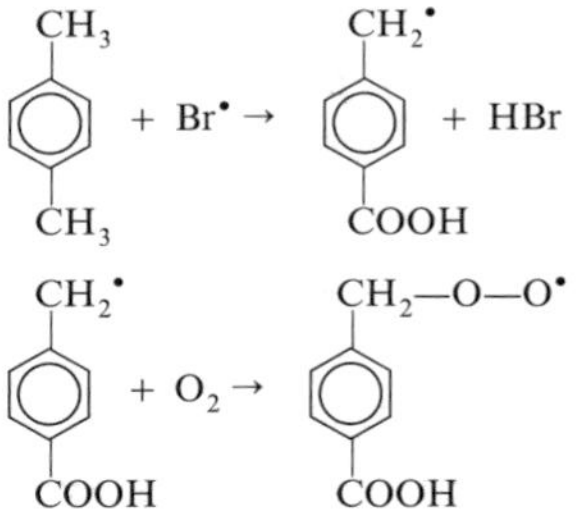

→ hydroperoxide → terephthalic acid

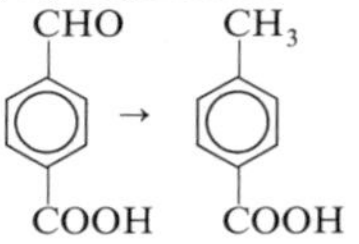

In the purification section of the plant, the crude acid is dissolved under pressure in water at 225–275 °C and hydrogenated in the presence of, for example, a Pd/C catalyst. By this means, 4-carboxybenzaldehyde (which would interfere in the polycondensation) is hydrogenated to *p*-toluic acid in the liquid phase. The aqueous solution is cooled, causing the pure terephthalic acid to crystallize out.

purification of crystalline TPA:

1. dissolving in H_2O at 225–275 °C under pressure
2. conversion of troublesome intermediate by hydrogenation

3. crystallization by concentration

range of application of Amoco process:

1. used industrially:

2. applicable in principle:

(I)   (II)

(III)

(IV)

production of isophthalic acid (in 1000 tonnes):

	1990	1992	1994
W. Europe	58	75	100
USA	79	77	86
Japan	18	23	35

variation of Amoco process:

IFP process for catalytic liquid-phase oxidation of p-xylene and other processes

principles of co-oxidation in TPA manufacture (route 3):

combined catalyzed liquid-phase oxidation of p-xylene together with a hydroperoxide-supplying, more readily oxidizable compound

The conversion of p-xylene is more than 95%, and the selectivity to the acid is over 90%, with a purity of 99.99%.

The process can be operated batch or continuously. The first commercial plant went on stream in 1958. Other plants are operated in numerous countries under license from Amoco; they manufacture about 80% of the world production of fiber grade terephthalic acid. A newer development concerns the production of a middle-grade terephthalic acid which has lower manufacturing costs due to a simplified purification, but which is satisfactory for several areas of use.

The Amoco process is also suitable for oxidizing other methylbenzenes and methylnaphthalenes to aromatic carboxylic acids. For example, benzoic acid is produced from toluene with 99% conversion and 96% selectivity in a commercial plant in England. Similarly, m-xylene can be oxidized to isophthalic acid, pseudocumene to trimellitic anhydride (I), mesitylene to trimesic acid (II), 2,6-dimethylnaphthalene to 2,6-naphthalenedicarboxylic acid (III), and 1,4-dimethylnaphthalene to naphthalene-1,4-dicarboxylic acid (IV). Like terephthalic acid, isophthalic acid is used for the manufacture of polyesters and polyester resins or, together with terephthalic acid, for the manufacture of copolyesters, e.g., with 1,4-dimethylolcyclohexane (cf. Section 14.2.4). Another use is its polycondensation with diamines to polyamides (cf. Chapter 10).

In 1994, world production capacity for isophthalic acid was about 350000 tonnes per year with 91000, 172000, and 35000 tonnes in the USA, Western Europe, and Japan, respectively. Production figures for these countries are given in the adjacent table.

IFP developed a liquid-phase oxidation of p-xylene with air in acetic acid similar to the Amoco process using a modified bromine–cobalt catalyst. The process conditions – 180 °C and 10 bar – are somewhat milder than in the Amoco process.

Other firms (e.g., Hüls, ICI, IFP, Maruzen Oil, Mitsui Petrochemical) have also developed their own direct oxidation processes.

A pure fiber-grade terephthalic acid can also be obtained using the third reaction principle – the co-oxidation of p-xylene with aldehydes such as acetaldehyde or its trimer paraldehyde, formaldehyde, or ketones such as methyl ethyl ketone. The Toray process with paraldehyde as the promoter is an example of this method. This development led, in 1971, to the first industrial plant in Japan.

In this process, *p*-xylene, paraldehyde and a cobalt acetate solution are introduced at the head of a bubble column, while air is introduced at the bottom. The oxidation is conducted at 100–140 °C and 30 bar with acetic as solvent, and the intermediate peracetic acid. The terephthalic acid formed is removed as a suspension in acetic acid, separated, and purified or esterified with methanol to dimethyl terephthalate. The selectivity is said to be greater than 97% (based on *p*-xylene). However, a use must be found for the byproduct acetic acid from the co-oxidation of paraldehyde. Each tonne of dimethyl terephthalate results in 0.21 tonnes of acetic acid.

In a similar manner, Eastman Kodak uses acetaldehyde for the co-oxidation of *p*-xylene in acetic acid solution in the presence of a cobalt salt. The coproduction of acetic acid can be varied between 0.55 and 1.1 tonne per tonne terephthalic acid. The TPA yield is 96.7%. For economic reasons, the Mobil process — using a methyl ethyl ketone auxiliary system — was abandoned in 1975.

industrial operation of co-oxidation:

Toray, Japan (paraldehyde)
Eastman Kodak, USA (acetaldehyde)

process operation:

p-xylene and paraldehyde are reacted in a bubble column in counter-flow with air to TPA and AcOH (co-oxidized product and also solvent)

comparison with other TPA process:

disadvantage: AcOH as coproduct

advantages:

1. no corrosion from $Br^{\ominus}$
2. high selectivity due to milder conditions

14.2.3. Other Manufacturing Routes to Terephthalic Acid and Derivatives

There are two well-known rearrangement processes, designated the Henkel I and the Henkel II process after the firm responsible for their discovery. The classical processes were recently developed further by various firms — Henkel I by Teijin and Kawasaki, Henkel II by Mitsubishi Chemical and Phillips Petroleum/ Rhône–Poulenc. Teijin, as already mentioned, was the first to manufacture fiber grade terephthalic acid by this route.

older manufacturing processes for terephthalic acid:

1. Henkel I isomerization
2. Henkel II disproportionation

The first step in the Henkel I process is the manufacture of dipotassium phthalate from phthalic anhydride, which is then rearranged in an isomerization step to dipotassium terephthalate at 430–440 °C and 5–20 bar CO_2 in the presence of a Zn–Cd catalyst:

principle of Henkel I route:

catalyzed isomerization of phthalic acid to TPA in form of its K salts

(7)

The potassium recycle is conducted expediently with a potassium exchange between K terephthalate and phthalic anhydride. All plants using the process have been closed down since they could not compete with other processes.

principle of Henkel II route:

catalyzed disproportionation of benzoic acid to TPA + benzene

In the Henkel II process, potassium benzoate is disproportionated at 430–440 °C in the presence of CO_2 (50 bar) and Cd or Zn benzoate to dipotassium terephthalate and benzene with 95 % selectivity:

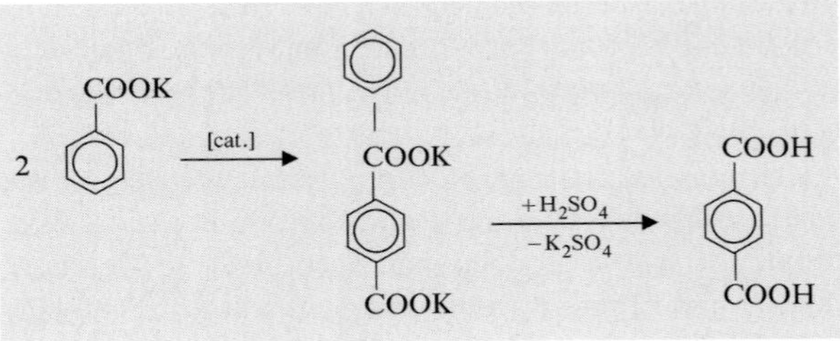

$$ \tag{8} $$

industrial use by Mitsubishi Chemical

The purification in both processes is conducted at the salt stage by decolorizing the aqueous solution with adsorbents and by recrystallization. In 1971, Mitsubishi Chemical was still using the Henkel II process in a 23 000 tonne-per-year plant in Japan; it was shut down in 1975.

more recent pilot plant application by Phillips Petroleum/Rhône–Poulenc

Recently, an improved Henkel II process has aroused interest. Phillips Petroleum and Rhône–Poulenc jointly developed a continuous process for the manufacture of fiber grade terephthalic acid using the Henkel II principle. It is characterized by a complete potassium recycle (exchange of potassium between potassium terephthalate and benzoic acid) which prevents the formation of potassium salts. The disproportionation of potassium benzoate is carried out in suspension in a terphenyl mixture in the presence of zinc oxide.

manufacturing process for terephthalic acid based on terephthalonitrile:

process principles:

1. *p*-xylene ammoxidation
2. dinitrile saponification

The chain of terephthalic acid process has been extended with a newly revised route from Lummus. It is based, like several older processes (*e. g.*, Allied, Distillers, and Showa Denko), on the ammoxidation of *p*-xylene to terephthalonitrile:

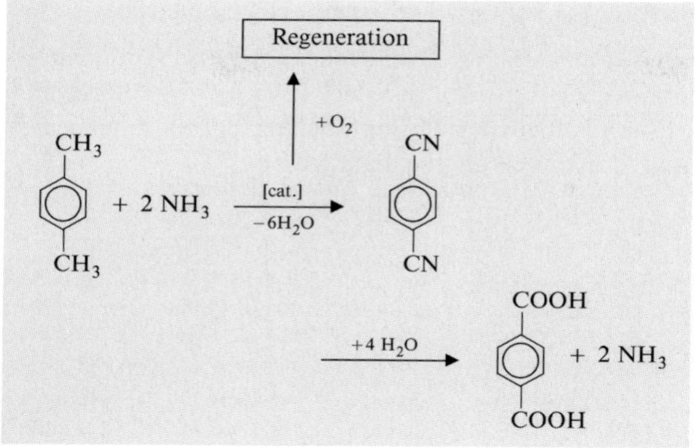

No free oxygen is used in the ammoxidation step of the Lummus process. *p*-Xylene and NH_3 react more readily at $400-450\,°C$ with a fluidized-bed metal catalyst at a higher oxidation state, generally V_2O_5/Al_2O_3. The catalyst, which becomes reduced during the process, is then reoxidized in a separate reaction at $500\,°C$ with O_2 or air. The selectivity to terephthalonitrile and its precursor *p*-tolylnitrile, which is recycled to the ammoxidation, is reported to be over 90%. The *p*-xylene conversion is adjusted to about 50%.

characteristics of Lummus *p*-xylene ammoxidation:

two-step gas-phase process with

1. *p*-xylene /NH_3 reaction with reduction of oxidation catalyst (metal oxide base)
2. oxidative regeneration of catalyst in separate process step

Terephthalonitrile is converted into pure terephthalic acid in three steps. The dinitrile is first hydrolyzed with steam to mono-ammonium terephthalate. In the second step, this ammonium salt is thermally cleaved into terephthalic acid and NH_3.

characteristics of dinitrile saponification:

three-step, uncatalyzed sequence: hydrolysis, thermolysis, and 'rehydrolysis'

In the third step, hydrolysis is used to convert traces of the semi-amide into acid. Pure terephthalic acid must have less than 10 ppm nitrogen-containing substances, since they cause a yellow coloration in the polyester:

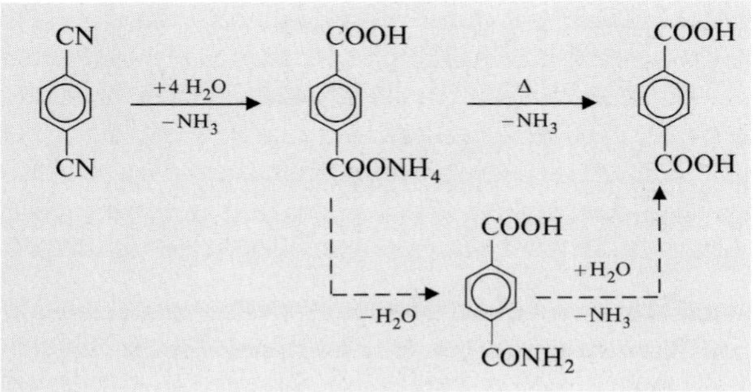

This process has not, however, been used commercially.

p-/m-xylene ammoxidation by Showa Denko with subsequent hydrogenation of terephthalonitrile and isophthalonitrile to corresponding xylylene diamines

On the other hand, Showa Denko (Japan) practices the commercial ammoxidation of *p*- or *m*-xylene to the corresponding dinitriles, terephthalonitrile and isophthalonitrile, which are generally used to manufacture the diamines:

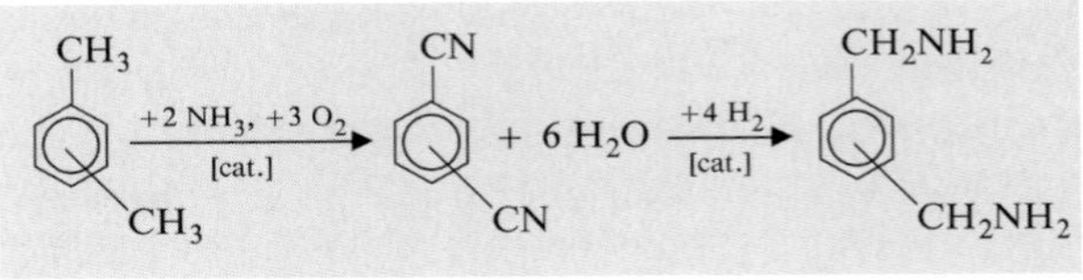

$$
\begin{array}{ccc}
\underset{\text{CH}_3}{\overset{\text{CH}_3}{\bigcirc}} & \xrightarrow[\text{[cat.]}]{+2\text{ NH}_3,\,+3\text{ O}_2} & \underset{\text{CN}}{\overset{\text{CN}}{\bigcirc}} + 6\text{ H}_2\text{O} \xrightarrow[\text{[cat.]}]{+4\text{ H}_2} \underset{\text{CH}_2\text{NH}_2}{\overset{\text{CH}_2\text{NH}_2}{\bigcirc}}
\end{array}
\tag{11}
$$

m-xylene ammoxidation by Mitsubishi Gas Chemical with commercial use in USA and Japan

Several other firms have developed processes for ammoxidation of alkylated aromatics; for example, Mitsubishi Gas Chemical has developed a process for the manufacture of isophthalonitrile from *m*-xylene which is used in two commercial plants (USA, Japan). Hydrogenation of isophthalonitrile yields *m*-xylylene diamine, which is converted to diisocyanate used in polyurethanes.

o-xylene ammoxidation by Japan Catalytic and BASF to phthalonitrile as precursor for phthalocyanine dyes

p-xylene ammoxidation by Mitsubishi Gas Chemical

Japan Catalytic Chem. Ind. and BASF also produce phthalonitrile by the ammoxidation of *o*-xylene in their own processes. Phthalonitrile is an important precursor for the manufacture of phthalocyanine dyes.

manufacturing process for terephthalic acid based on toluene according to Mitsubishi Gas Chemical via *p*-tolylaldehyde

Mitsubishi Gas Chemical has recently developed a terephthalic acid process to the pilot plant stage. The first step is the reaction of toluene with CO at 30–40 °C in the presence of HF/BF$_3$ to give *p*-tolylaldehyde. Aldehyde yields of 96% (based on toluene) and 98% (based on CO) are obtained. After catalyst separation, the *p*-tolylaldehyde is purified and then oxidized to terephthalic acid:

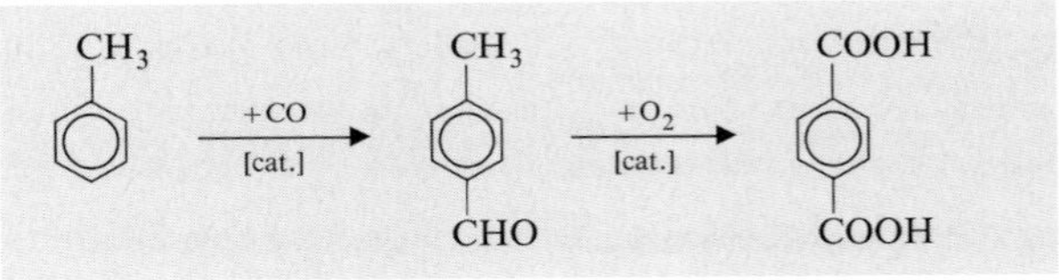

$$
\underset{\text{CH}_3}{\overset{\text{CH}_3}{\bigcirc}} \xrightarrow[\text{[cat.]}]{+\text{CO}} \underset{\text{CHO}}{\overset{\text{CH}_3}{\bigcirc}} \xrightarrow[\text{[cat.]}]{+\text{O}_2} \underset{\text{COOH}}{\overset{\text{COOH}}{\bigcirc}}
\tag{12}
$$

14.2.4. Uses of Terephthalic Acid and Dimethyl Terephthalate

uses of TPA and DMT:

dicarboxylic acid components for polycondensation with diols:

Terephthalic acid and dimethyl terephthalate are starting materials for the manufacture of polyesters used mainly in fibers (currently more than 91% worldwide). A smaller amount is used in polyester resins for the manufacture of films, paints, and adhesives. Recently, a number of companies have also used polyesters (polyethylene terephthalate) for beverage containers, a use which is increasing rapidly.

By far the most important product is polyethylene terephthalate, with ethylene glycol as the diol component. In addition, 1,4-dimethylolcyclohexane and, to an increasing extent, 1,4-butanediol (*cf.* Section 4.3) are being used in polycondensations with terephthalic acid.

In 1993, world production capacity for polyethylene terephthalate was about 11.1×10^6 tonnes per year, with about 3.0, 1.9, and 1.1×10^6 tonnes per year in the USA, Western Europe, and Japan, respectively. Production figures for these countries are summarized in the adjacent table.

In 1991, world production of polyester for fibers was 8.96×10^6 tonnes. Polyesters thus accounted for about 55% of all synthetic fibers (*cf.* Chapter 10). An increase to 60% is expected in the next few years.

In the manufacture of **p**olyethylene **t**erephthalate (PET), dimethyl terephthalate is transesterified, or terephthalic acid is esterified, with an excess of ethylene glycol at 100–150 °C and 10–70 bar in the presence of Cu, Co, or Zn acetate catalysts. The intermediate − bis(2-hydroxyethyl) terephthalate − is then polycondensed at a higher temperature (150–270 °C) under vacuum and usually with a catalyst, *e. g.*, Sb_2O_3; ethylene glycol is removed by distillation during the reaction:

diol components:

$HOCH_2CH_2OH$ (favored)

$HOCH_2$—⟨ ⟩—CH_2OH

$HOCH_2$—$(CH_2)_2$—CH_2OH

PET production (in 10^6 tonnes):

	1987	1989	1991
USA [1]	2.46	2.59	2.67
W. Europe	1.06	1.12	1.39
Japan	1.01	1.09	1.21

[1] films and fibers

principles of PET manufacture:

1st step:
catalytic transesterification of DMT with glycol and loss of CH_3OH or esterification of TPA with glycol and loss of H_2O

2nd step:
catalytic polycondensation of bis(2-hydroxyethyl)terephthalate to PET at ca. 10–20 °C above the melting point of PET (246 °C) with loss of glycol

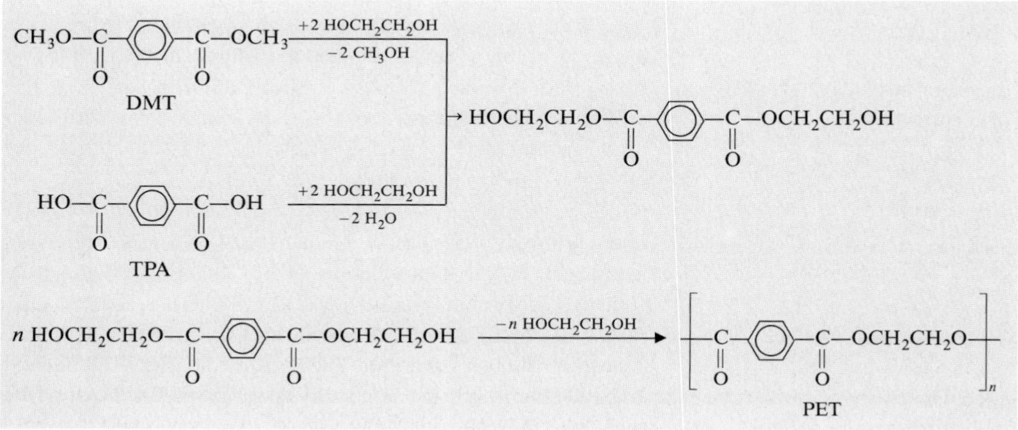

(13)

The resulting polyester melts are either spun directly or – as with the majority of manufacturers – cooled, granulated, and spun separately.

development tendencies in PET manufacture through DMT or TPA:

economic advantages lead to higher growth rate for TPA compared to DMT

As for the choice between terephthalic acid and dimethyltere-phthalate in the future, it is already recognized that an increasing fraction of polyethylene terephthalate will be manufactured by the direct route through esterification – as opposed to trans-esterification – followed by polycondensation. The advantages of direct esterification of terephthalic acid result from lack of methanol ballast, from a higher yield (15% higher than with DMT), and from a faster rate of reaction.

manufacturing variation for PET:

primary esterification of TPA with ethylene oxide and subsequent polycondensation

Another variation of polyethylene terephthalate production is based on the directly manufacturable intermediate, bis(2-hy-droxyethyl) terephthalate, and its polycondensation. Several processes have been developed for the direct esterification of terephthalic acid with ethylene oxide, leading to this interme-diate:

$$
\begin{array}{c}
COOH \\
\text{(ring)} \\
COOH
\end{array}
+ 2\ H_2C\!-\!CH_2 \xrightarrow{\ [cat.]\ }
\begin{array}{c}
COOCH_2CH_2OH \\
\text{(ring)} \\
COOCH_2CH_2OH
\end{array}
\qquad (14)
$$

$$
\left(\Delta H = -\frac{24\ \text{kcal}}{100\ \text{kJ}}\bigg/\text{mol}\right)
$$

process principles of TPA addition to ethyl-ene oxide:

solvent-free, base-catalyzed liquid-phase ethoxylation to a recrystallizable TPA ester

The ethoxylation of terephthalic acid is done in the liquid phase without a solvent at $90-130\,°C$ and $20-30$ bar in the presence of basic catalysts such as amines or quaternary alkylammonium salts. Since the resulting ester is easily recrystallized from water, the purification of the crude acid is postponed to this stage.

industrial practice of TPA/EO addition:

Teijin (100 000 tonne-per-year plant)

Along with Toyobo and Nippon Soda, Teijin in particular has worked on the development of this route and, in 1971, started up production of the intermediate (capacity ca. 100 000 tonnes per year).

other uses for DMT:

hydrogenation to 1,4-dimethylolcyclohexane

Besides its main use as a dicarboxylic acid component for poly-esters, dimethyl terephthalate is also used to a lesser extent as feedstock for the manufacture of 1,4-dimethylolcyclohexane (1,4-bis(hydroxymethyl) cyclohexane).

manufacturing principles of 1,4-dimethylol-cyclohexane:

two-step DMT hydrogenation:

1. ring hydrogenation with Pd catalyst

In the Eastman Kodak process, DMT is hydrogenated with hydrogen over a supported Pd catalyst at $160-180\,°C$ and $300-400$ bar to give *cis*- and *trans*-cyclohexane-1,4-dicarboxylic acid dimethyl ester quantitatively.

The high heat of reaction (47 kcal or 197 kJ per mol) is removed by circulating the hydrogenated ester. The melted dimethyl terephthalate is fed to the circulating ester at a controlled rate in order to maintain a low concentration. The selectivity increases to 96–98% when a fiber-grade ester is used.

The second step of the ester hydrogenation and hydrogenolysis with a copper chromite catalyst (Adkins type) is done without previous purification:

2. ester hydrogenolysis with Adkins catalyst characteristics of ring hydrogenation:

hydrogenation of aromatic to cyclohexane derivative leads to *cis/trans* mixture, exothermic reaction only selective on dilution (*e. g.*, with hydrogenated product, cyclohexanedicarboxylic acid dimethyl ester)

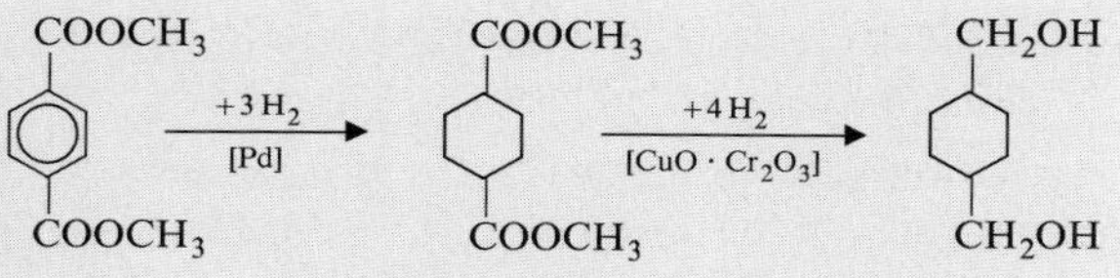

$$+ 2\ CH_3OH\ \left(\Delta H = -\frac{57\ \text{kcal}}{239\ \text{kJ}}\big/\text{mol}\right) \qquad (15)$$

The new isomeric mixture is independent of the *cis/trans* ratio of the dimethyl esters. The choice of reaction conditions and the catalyst components (*e. g.*, basic constituents) can affect the *cis/trans* ratio.

1,4-Dimethylolcyclohexane is used as a diol component for polyesters, polyurethanes and polycarbonates. For example, it is polycondensed with terephthalic acid in the manufacture of polyester fibers (*e. g.*, Kodel®, Vestan®). Since the crystal structure, and therefore the properties, of polyesters manufactured from the individual *cis/trans* isomers of 1,4-dimethylolcyclohexane are also different, the isomer ratio in the commercial process must be kept as constant as possible. In practice, a *cis/trans* ratio of 1:3 to 1:4 is chosen.

characteristics of ester hydrogenolysis:

in second hydrogenation, renewed equilibration (*cis/trans*) can be influenced, for example, by catalyst basicity

uses of 1,4-dimethylolcyclohexane:

diol component for polyesters, polyurethanes and polycarbonates

15. Appendix

15.1. Process and Product Schemes

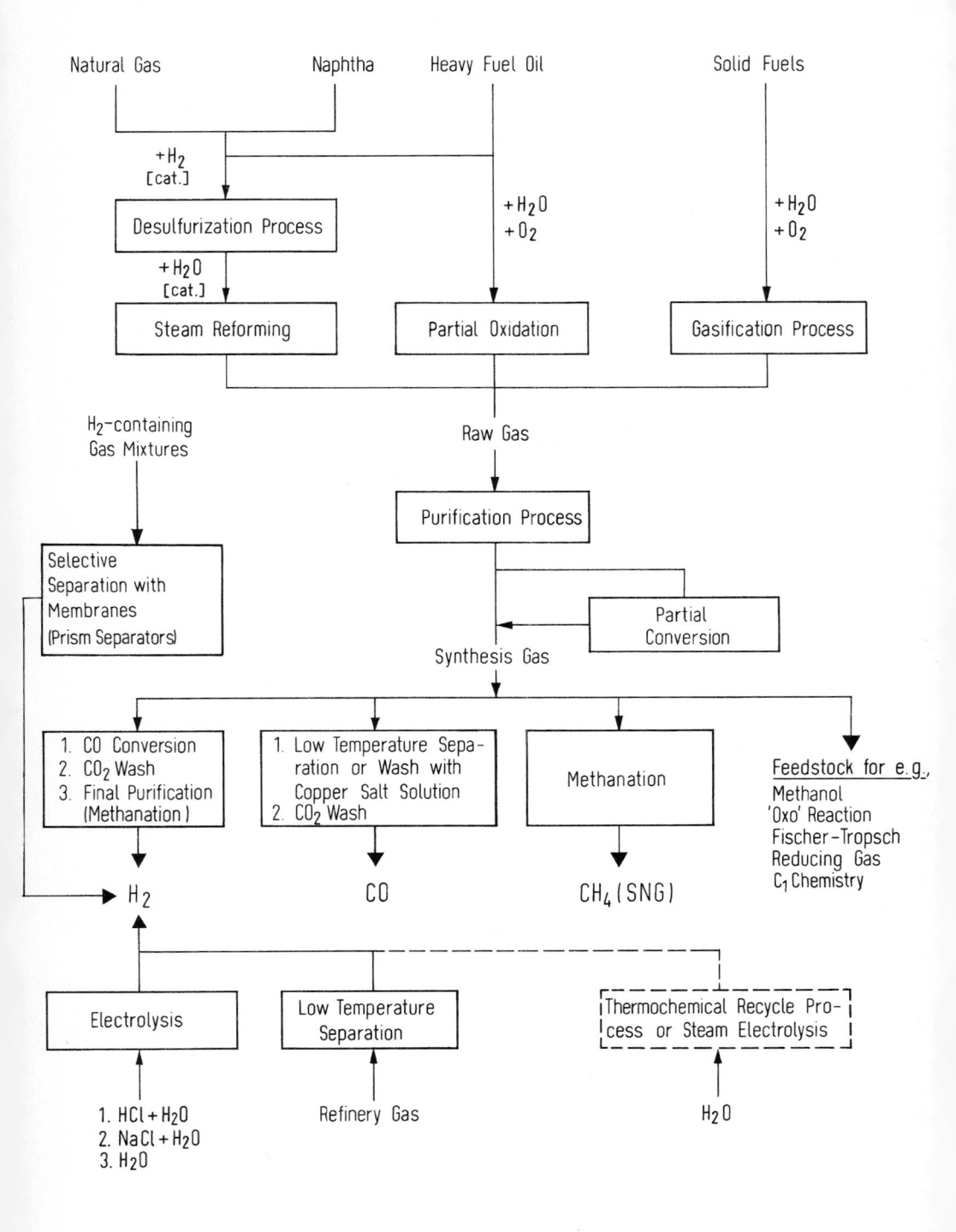

Process Scheme for Sections 2−2.2.2

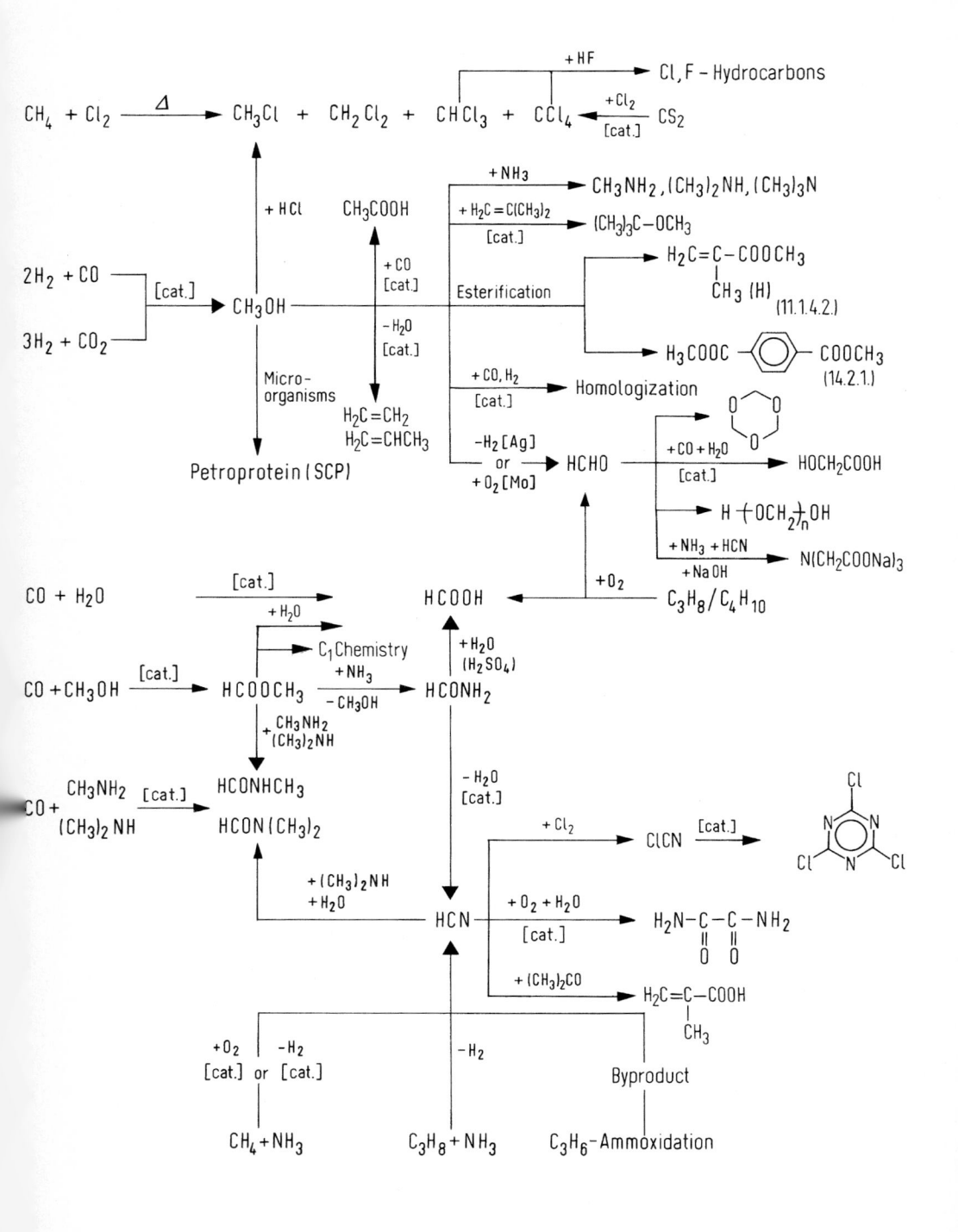

Process Scheme for Sections 3 – 3.3.2

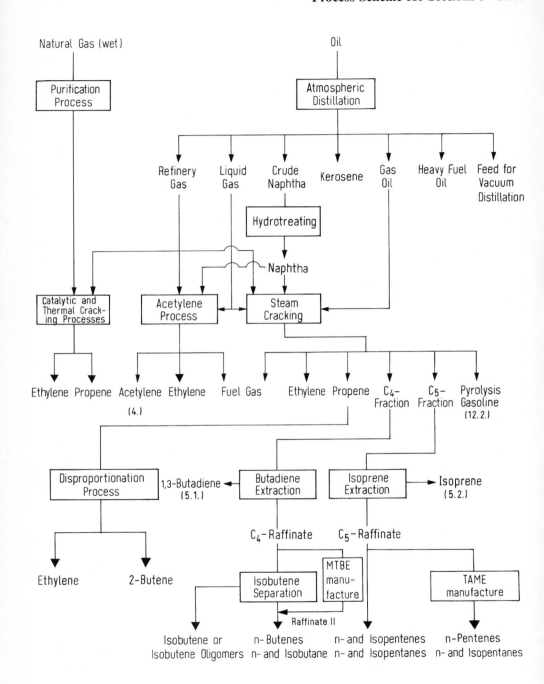

Process Scheme for Sections 3.3.3 – 3.4

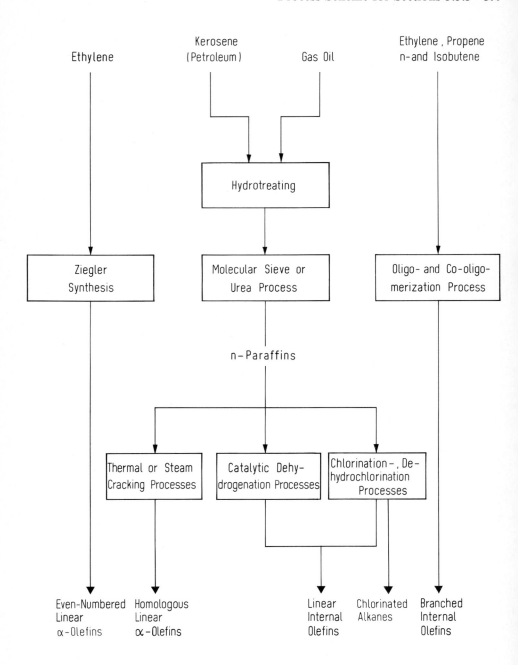

Process Scheme for Sections 3.3.3 – 3.4

Product Scheme for Chapter 4

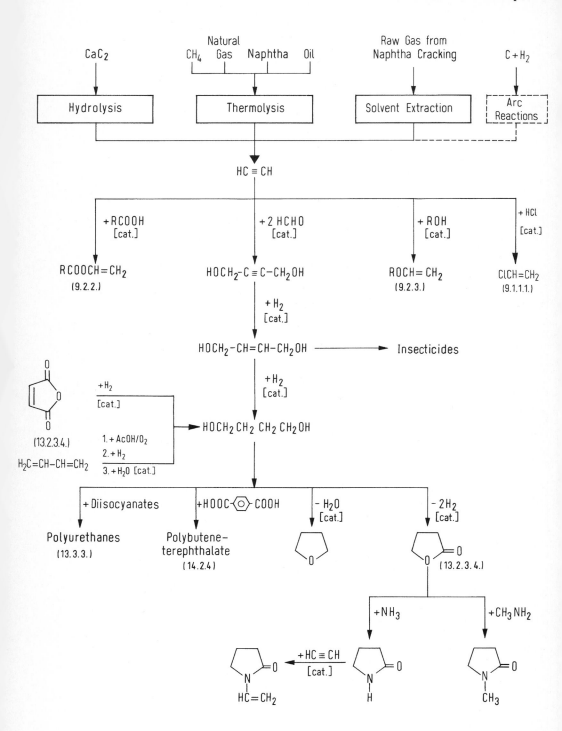

Process and Product Scheme for Chapter 5

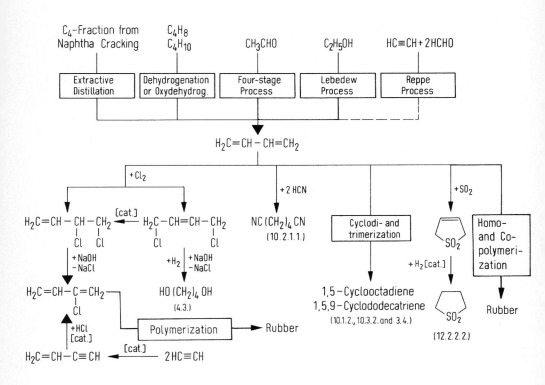

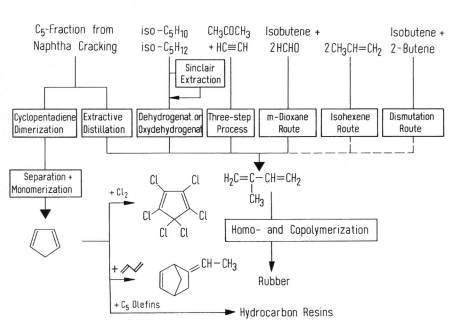

Product Scheme for Chapter 6 (basic examples)

Product Scheme for Chapter 6 (basic examples)

$$CH_3CH=CH_2 \xrightarrow[\text{[cat.]}]{+CO \ +H_2} \boxed{\begin{array}{c}\text{Hydroformy-}\\\text{lation}\end{array}}$$

$$\begin{array}{c}H_3C\\H_3C\end{array}\!\!>\!\!CHCHO$$

$$\xrightarrow{+O_2} \begin{array}{c}H_3C\\H_3C\end{array}\!\!>\!\!CHCOOH \quad (11.1.4.2.) \xrightarrow[\text{[}H^{\oplus}\text{]}]{+ROH} \begin{array}{c}H_3C\\H_3C\end{array}\!\!>\!\!CHCOOR$$

$$\xrightarrow[\text{[cat.]}]{+H_2} \begin{array}{c}H_3C\\H_3C\end{array}\!\!>\!\!CHCH_2OH \xrightarrow[\text{[}H^{\oplus}\text{]}]{+RCOOH} \begin{array}{c}H_3C\\H_3C\end{array}\!\!>\!\!CHCH_2OCR\!\!\underset{O}{\overset{\parallel}{}}$$

$$CH_3(CH_2)_2CHO \xrightarrow[\text{[cat.]}]{+H_2} CH_3(CH_2)_2CH_2OH \xrightarrow[\text{[}H^{\oplus}\text{]}]{+RCOOH} CH_3(CH_2)_2CH_2O-\underset{O}{\overset{\parallel}{C}}R$$

$$CH_3(CH_2)_2COOH \xleftarrow[\text{[cat.] [}OH^{\ominus}\text{]}]{+O_2}$$

$$CH_3(CH_2)_2CH=C-CHO \ (\underset{C_2H_5}{|}) \xrightarrow[\text{[cat.]}]{+H_2} CH_3(CH_2)_3CH-CH_2OH \ (\underset{C_2H_5}{|})$$

$$\begin{array}{c}1.+H_2\text{[cat.]}\\2.+O_2\end{array}$$

$$CH_3(CH_2)_3-CH-COOH \ (\underset{C_2H_5}{|}) \xleftarrow[\text{[cat.]}]{+O_2} \cdots \longrightarrow \text{'DOP'}$$

$$H_2C=CH_2 \xrightarrow[\text{[cat.]}]{+CO \ +H_2} \boxed{\begin{array}{c}\text{Hydro-}\\\text{formylation}\end{array}} \longrightarrow CH_3CH_2CHO$$

$$\xrightarrow[\text{[cat.]}]{+HCHO+O_2} CH_3-\underset{\overset{\parallel}{CH_2}}{C}-COOH$$

$$\xrightarrow[\text{[cat.]}]{+H_2} CH_3CH_2CH_2OH$$

$$\xrightarrow[\text{[cat.]}]{+O_2} CH_3CH_2COOH$$

$$H_2C=CH_2 \xrightarrow[\text{[cat.]}]{+CO \ +H_2O} \boxed{\begin{array}{c}\text{Reppe}\\\text{Carbonylation}\end{array}} \longrightarrow CH_3CH_2COOH$$

$$\xrightarrow[\text{[}H^{\oplus}\text{]}]{+ROH} CH_3CH_2COOR$$

$$\xrightarrow[\text{[cat.]}]{+HC\equiv CH} CH_3CH_2COO \ CH=CH_2$$

$$\begin{array}{c}H_3C\\H_3C\end{array}\!\!>\!\!C=CH_2 \xrightarrow[\text{[cat.]}]{+CO \ +H_2O} \boxed{\begin{array}{c}\text{Koch}\\\text{Carbonylation}\end{array}} \longrightarrow H_3C-\underset{\underset{CH_3}{|}}{\overset{\overset{CH_3}{|}}{C}}-COOH$$

$$\xrightarrow[\text{[}H^{\oplus}\text{]}]{+ ROH} (CH_3)_3C-COOR$$

$$\xrightarrow[\text{[cat.]}]{+ HC\equiv CH} (CH_3)_3C-COOCH=CH_2$$

Product Scheme for Sections 7 – 7.2.5

$CH_3OH + CO + O_2$ $\xrightarrow{\text{[cat.]}}$ $CH_3OCCOCH_3$ (with $\underset{O\ O}{||\ ||}$)

$+H_2$ [cat.]

$HOCH_2CH_2OH$

$+O_2$ [cat.] $\longrightarrow$ $HC-CH$ ($\underset{O\ O}{||\ ||}$) $\xrightarrow[\text{[cat.]}]{+O_2}$ $HC-COOH$ (with $\overset{O}{||}$)

1. $+H_2$
2. $+H_2O - CH_3OH$

$+HCHO$ $[H^\oplus]$

$H_3COOC \overset{O}{\diagdown} \diagup COOCH_3$ (dioxolane ring, O–O)

$+H_2O$

$HO(CH_2CH_2O)_2H$ $\xrightarrow{-H_2O}$ (dioxane ring)

$HO(CH_2CH_2O)_nH$

$+O_3$

$H_3COOCH = CHCOOCH_3$

$+HOAc$ $\longrightarrow$ $ROCH_2CH_2OAc$

$ROCH_2CH_2OH$

$\begin{array}{c}1.+NaOH\\2.+CH_3Cl\end{array}$ $ROCH_2CH_2OCH_3$

$+ROH$ [cat.]

$R = CH_3$
$\quad\ \ C_2H_5$
$=\ \ n-C_4H_9$

$RO(CH_2CH_2O)_nH$ $\xrightarrow[2.+CH_3Cl]{1.+NaOH}$ $RO(CH_2CH_2O)_nCH_3$

$H_2C = CH_2$

$+O_2$ [cat.]

H_2C-CH_2 (with O epoxide)

$+NH_3$

$+NH_3$ [cat.] $\longrightarrow$ $H_2NCH_2CH_2NH_2$

$HOCH_2CH_2NH_2$

$+H_2SO_4$ $H_2NCH_2CH_2OSO_3H$ $\xrightarrow{-H_2SO_4}$ H_2C-CH_2 (with N–H aziridine ring)

$(HOCH_2CH_2)_2NH$ $\xrightarrow[{[H^\oplus]}]{-H_2O}$ (morpholine ring O NH)

$+NH_3$

$ClCH_2-CH_2Cl$

$(HOCH_2CH_2)_3N$

$+NH_3$ [cat.]

$HO(CH_2CH_2O)_2H$

$-HCl$

$HO-CH_2CH_2Cl$

$+Cl_2$
$+H_2O$

$H_2C = CH_2$

$+CO_2$ [cat.]

(cyclic carbonate: $O \overset{}{\diagdown} C \diagup O$, with $\overset{}{\underset{||}{C}}$ and O below)

$\xrightarrow[+CH_3OH, -(CH_3O)_2CO]{+H_2O, -CO_2}$ $HOCH_2CH_2OH$

$+R-H$

$R(CH_2CH_2O)_nH$ $\longrightarrow$ $R(CH_2CH_2O)_nSO_3H$

$R = R^1$⟨◯⟩$-O-,\ R^2COO-,\ R^3-NH-$

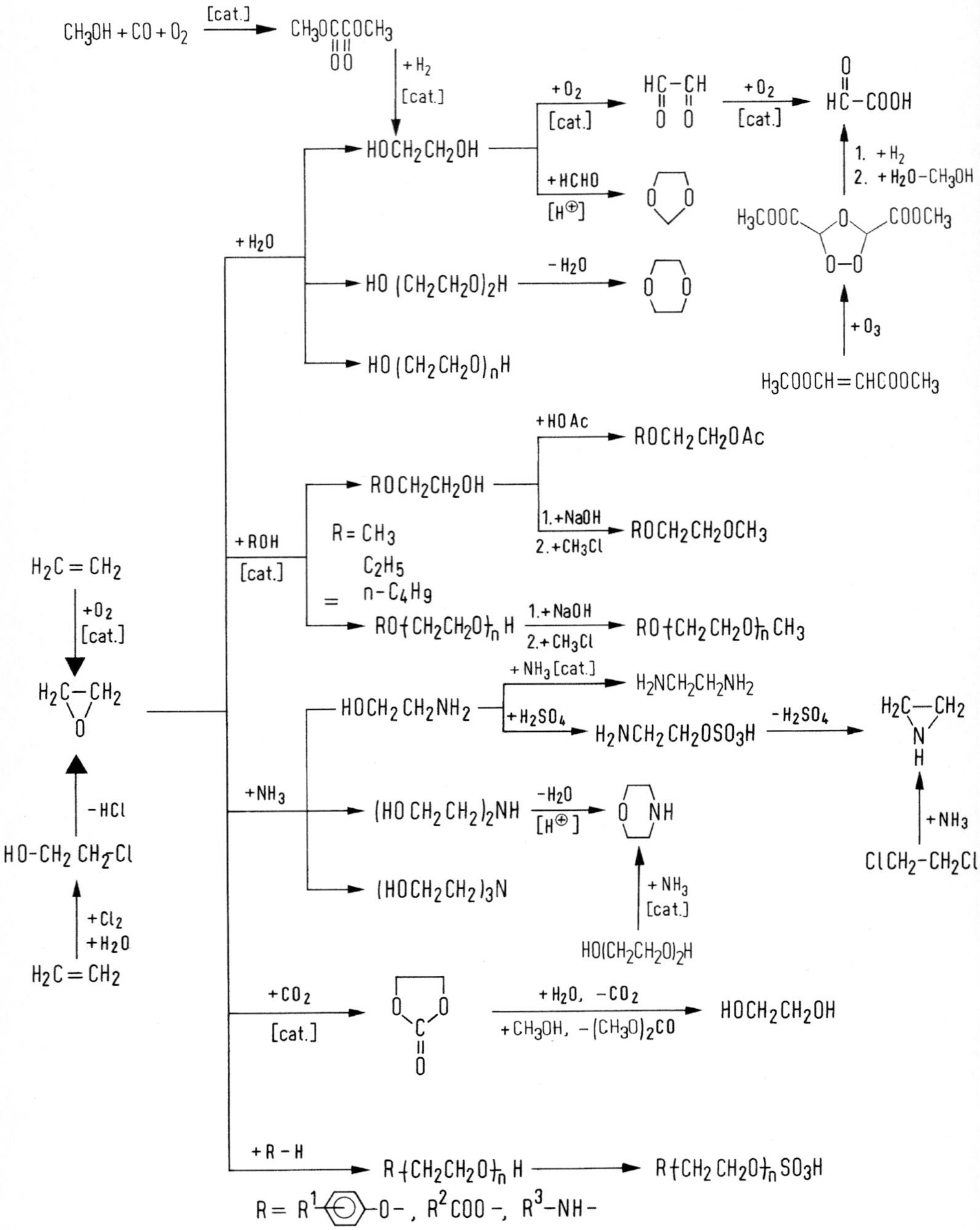

Product Scheme for Sections 7.3 − 7.4.5

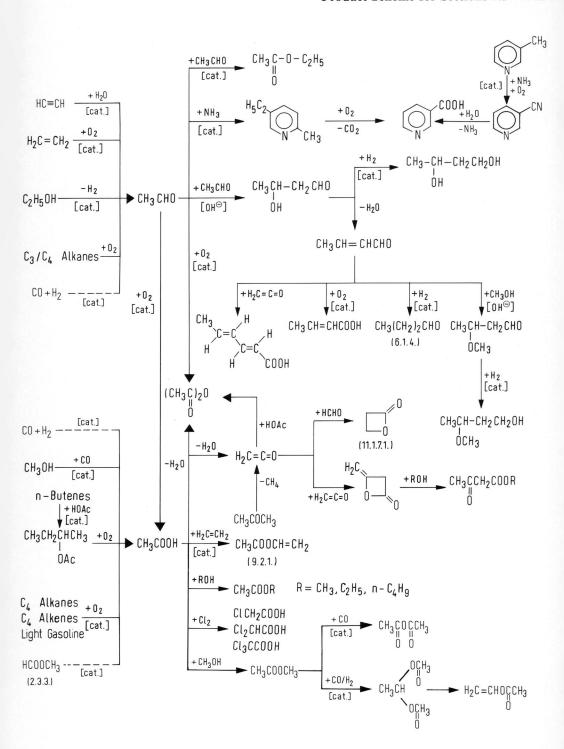

Process and Product Scheme for Sections 8−8.1.4

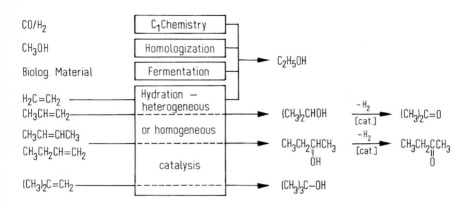

CO/H$_2$ → | C$_1$Chemistry |
CH$_3$OH → | Homologization | → C$_2$H$_5$OH
Biolog. Material → | Fermentation |

H$_2$C=CH$_2$
CH$_3$CH=CH$_2$ → | Hydration – heterogeneous | → (CH$_3$)$_2$CHOH $\xrightarrow[\text{[cat.]}]{-H_2}$ (CH$_3$)$_2$C=O

CH$_3$CH=CHCH$_3$
CH$_3$CH$_2$CH=CH$_2$ → | or homogeneous | → CH$_3$CH$_2$CHCH$_3$ $\overset{OH}{}$ $\xrightarrow[\text{[cat.]}]{-H_2}$ CH$_3$CH$_2$CCH$_3$ $\overset{O}{}$

(CH$_3$)$_2$C=CH$_2$ → | catalysis | → (CH$_3$)$_3$C–OH

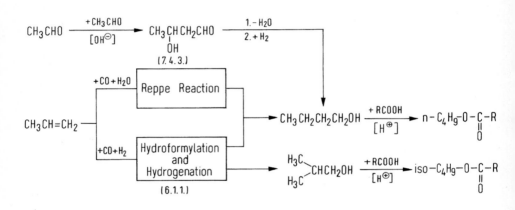

CH$_3$CHO $\xrightarrow[\text{[OH}^{\ominus}\text{]}]{+CH_3CHO}$ CH$_3$CHCH$_2$CHO $\overset{OH}{}$ $\xrightarrow[\text{2.+H}_2]{\text{1.–H}_2\text{O}}$
(7.4.3.)

CH$_3$CH=CH$_2$ → $\xrightarrow{+CO+H_2O}$ | Reppe Reaction |

→ CH$_3$CH$_2$CH$_2$CH$_2$OH $\xrightarrow[\text{[H}^{\oplus}\text{]}]{+RCOOH}$ n–C$_4$H$_9$–O–C–R $\overset{O}{}$

$\xrightarrow{+CO+H_2}$ | Hydroformylation and Hydrogenation | (6.1.1.)

→ H$_3$C⟩CHCH$_2$OH / H$_3$C $\xrightarrow[\text{[H}^{\oplus}\text{]}]{+RCOOH}$ iso–C$_4$H$_9$–O–C–R $\overset{O}{}$

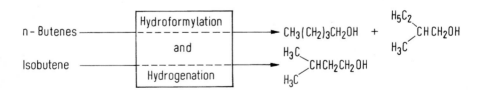

n–Butenes → | Hydroformylation and Hydrogenation | → CH$_3$(CH$_2$)$_3$CH$_2$OH + H$_5$C$_2$⟩CHCH$_2$OH / H$_3$C

Isobutene → → H$_3$C⟩CHCH$_2$CH$_2$OH / H$_3$C

Process and Product Scheme for Sections 8.2 − 8.3.3

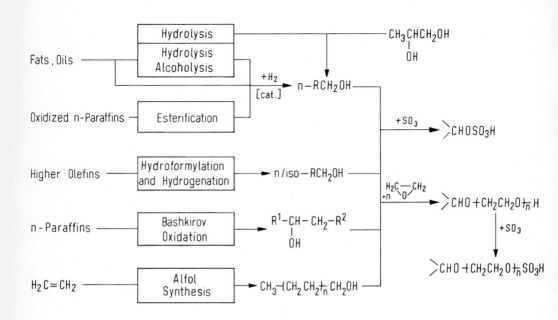

$$CH_3CHO \xrightarrow[\text{[OH}^\ominus\text{]}]{+HCHO} \begin{array}{c} HOH_2C \\ HOH_2C \end{array} \!\!>\!\! C <\!\! \begin{array}{c} CH_2OH \\ CH_2OH \end{array}$$

$$CH_3CH_2CH_2CHO \xrightarrow[\text{[OH}^\ominus\text{]}]{+HCHO} CH_3CH_2 - \underset{\underset{CH_2OH}{|}}{\overset{\overset{CH_2OH}{|}}{C}} - CH_2OH$$

$$\begin{array}{c} H_3C \\ H_3C \end{array} \!\!>\!\! CHCHO \xrightarrow[\text{[OH}^\ominus\text{]}]{+HCHO} HOCH_2 - \underset{\underset{CH_3}{|}}{\overset{\overset{CH_3}{|}}{C}} - CH_2OH$$

Product Scheme for Chapter 9

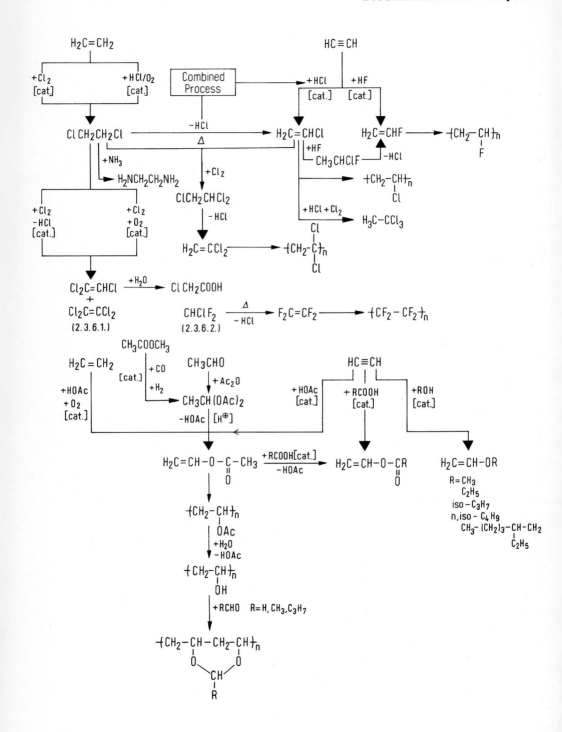

Product Scheme for Chapter 10

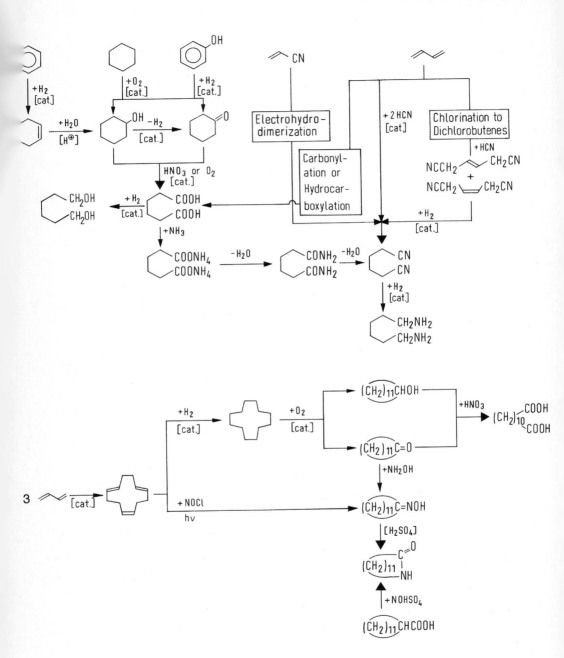

Product Scheme for Chapter 10 (cont.)

CH₃ (toluene, benzene ring)

$+O_2$ [cat.]

COOH (benzoic acid)

$+H_2$ [cat.]

COOH (cyclohexanecarboxylic acid)

$+NOHSO_4$ Oleum

cyclohexane

$+HNO_3$

NO₂ (nitrocyclohexane)

$+NOCl$

$+H_2O_2/NH_3$ [cat.]

$=O$ (cyclohexanone)

$+H_2C=C=O$

OCCH₃ ‖ O (enol acetate)

$+HNO_3$

$=O$ NO_2 (2-nitrocyclohexanone)

$+H_2O$

COOH CH₂NO₂

$+H_2$ [cat.]

COOH CH₂NH₂

$-H_2O$

$+NH_2OH \cdot H_2SO_4$ $+H_2O_2/AcOH$

$(CH_2)_5 \; C=O \; | \; O$

$+NH_3$

NOH (cyclohexanone oxime)

$-O-(CH_2)_5C \overset{\|}{\underset{O}{}} \overset{}{]_n}$

Beckmann Rearrangement

$(CH_2)_5 \overset{-C=0}{\underset{-NH}{|}}$ caprolactam

$+CH_3OH$ [cat.]

$(CH_2)_5 \overset{-C=0}{\underset{-NCH_3}{|}}$

$+H_2$ [cat.]

$(CH_2)_6 \; NH$

$+[(CH_2)_5-\overset{}{\underset{O}{C}}-NH]_n$

$CH_3(CH_2)_5CH \overset{|}{\underset{OH}{}} CH_2 \; CH=CH(CH_2)_7COOCH_3$

Δ

$CH_3(CH_2)_5 \overset{}{\underset{O}{C}}H \quad H_2C=CHCH_2(CH_2)_7COOCH_3$

$+HBr$ [Peroxide]

$Br-CH_2CH_2CH_2(CH_2)_7COOCH_3 \xrightarrow[2.+H_2O]{1.+NH_3} H_2N-(CH_2)_{10}COOH$

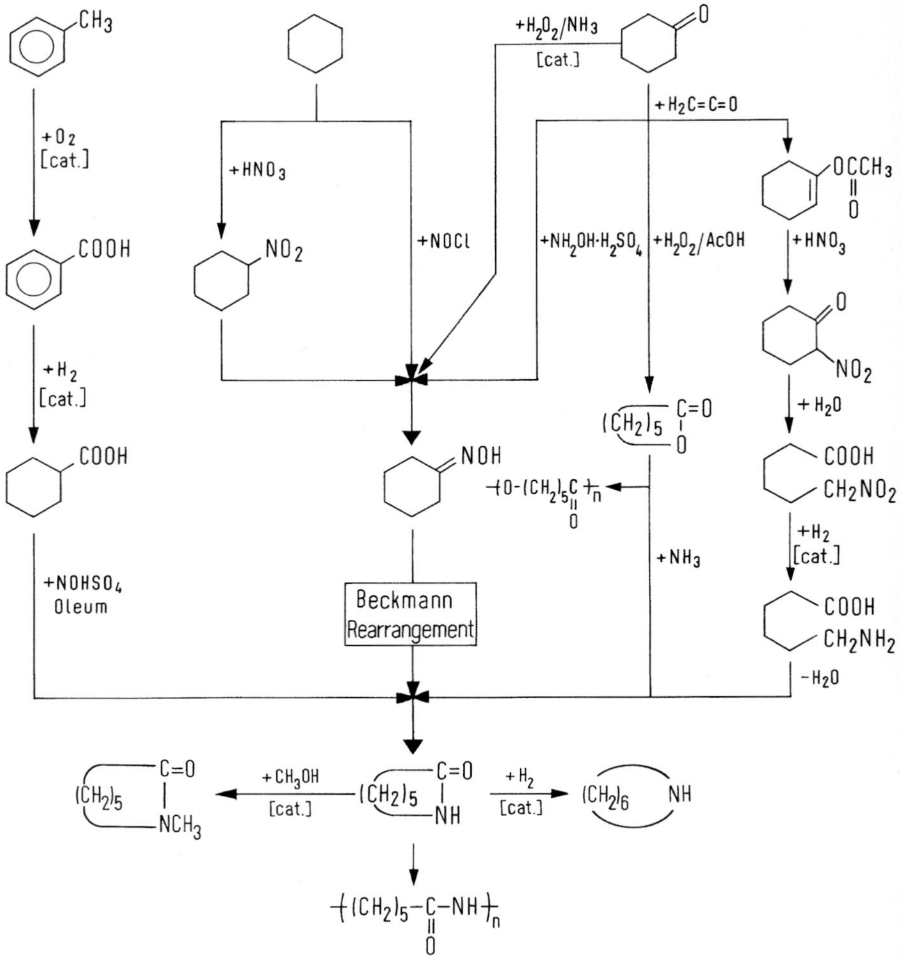

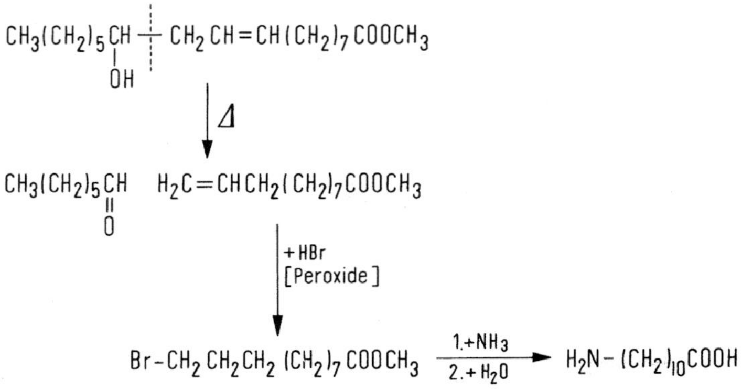

Product Scheme for Sections 11 – 11.1.7.3

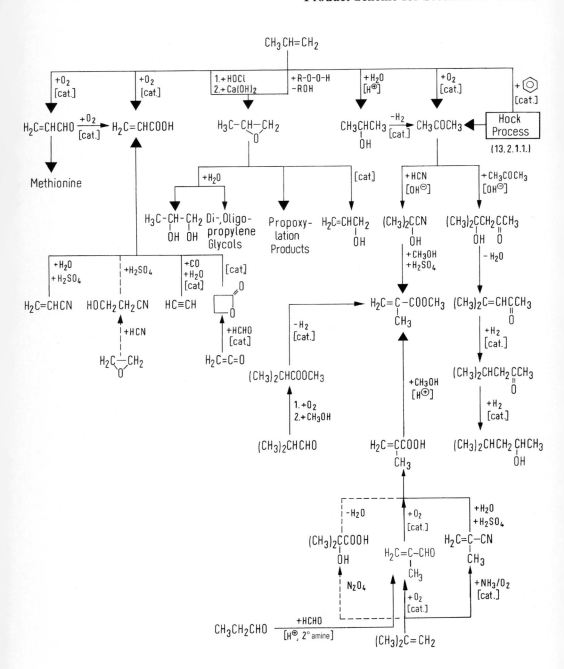

Product Scheme for Sections 11.2 – 11.3.3

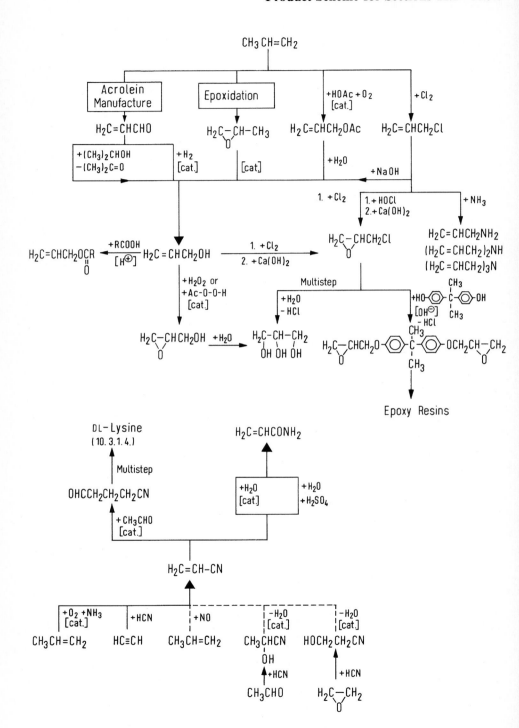

Process and Product Scheme for Chapter 12

Process and Product Scheme for Chapter 12

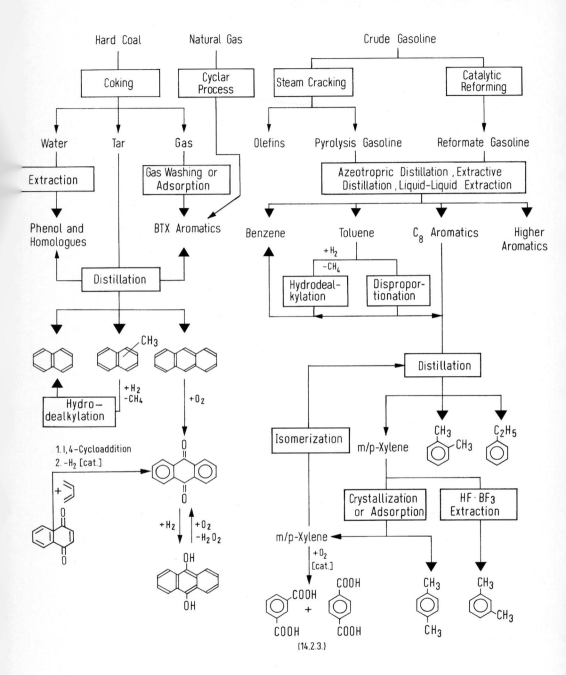

Product Scheme for Sections 13−13.2.2

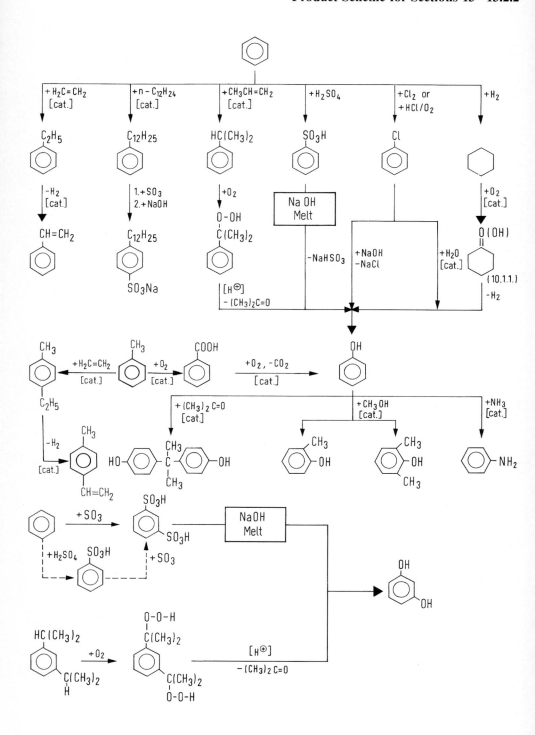

Product Scheme for Sections 13.2.3 – 13.3.3

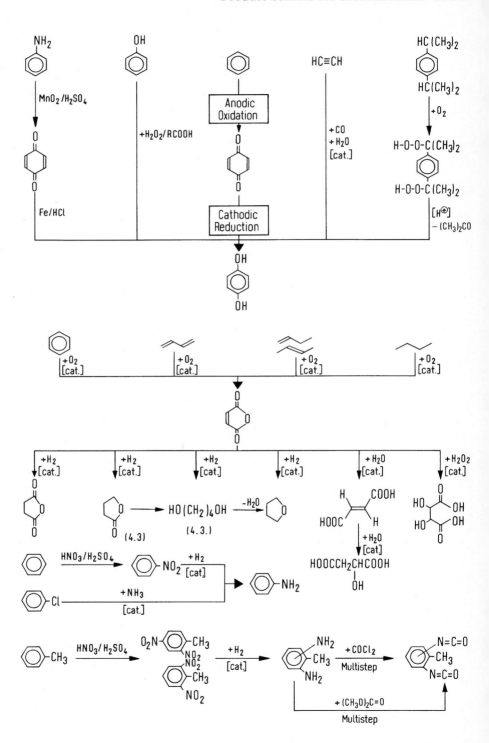

Process and Product Scheme for Chapter 14

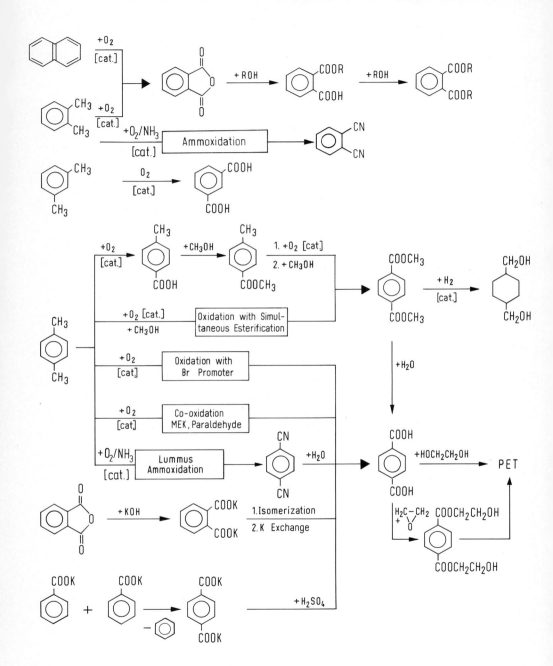

15.2. Definitions of Terms used in Characterizing Chemical Reactions

Conversion, selectivity, yield, and space-time yield are important mathematical parameters for characterizing a chemical reaction.

parameters which characterize a reaction:

1. conversion
2. selectivity
3. yield
4. space-time yield

However, as they are not uniformly employed in scientific publications, they will be defined in order to prevent any misunderstanding by the reader.

Conversion, selectivity, and yield are generally quoted in mole, volume, or weight percents. In this book, mole percent (mol%) is generally used and, for simplification, no additional units are employed. In all other cases, the percentages are clearly defined as weight percent (wt%) or volume percent (vol%).

favored unit, mole percent abbreviated to mol% in this publication i.e., only wt% and vol% are specified as such

example of reaction (ethylene acetoxylation):

$$H_2C=CH_2 + CH_3COOH + 0.5\ O_2$$

$$\xrightarrow{\text{[cat.]}} H_2C=CHOCCH_3 + H_2O$$
$$\qquad\qquad\qquad\underset{O}{\|}$$

(cf. Section 9.2.1.2)

The basis unit for amounts in all equations below, unless stated otherwise, is the mole.

The *conversion* of a reaction component is the quotient of the amount reacted Q_C and the initial quantity of this reactant (Q_F):

to conversion:

reaction components ($H_2C=CH_2$, CH_3COOH, O_2) exhibit differing conversions due to:

$$\text{Conversion (in \%)} = \frac{Q_C}{Q_F} \times 100$$

1. nonstoichiometric feedstock mixture: compared to CH_3COOH, $H_2C=CH_2$ in excess, O_2 deficiency
2. side reactions: CO_2 formation mainly from $H_2C=CH_2$ small amount from CH_3COOH

If two or more components react together then feed mixture would not necessarily be stoichiometric, should this be required by, for example, economic, technical, or safety considerations. Furthermore, the individual reaction components can participate in side reactions to a varying extent so that each component has its own conversion.

conversions in ethylene acetoxylation:

$H_2C=CH_2$	10%
CH_3COOH	20–30%
O_2	50–80%

The *selectivity* to a reaction product makes an important statement concerning competitive reactions which lead to reaction products other than those desired from the same reactants. Where possible, a differentiation has been made between the two types of competitive reactions, i.e., side or parallel reactions, and secondary reactions. The products from these reactions can be characterized as follows: A byproduct results directly from the starting materials, while a secondary product is formed by a subsequent conversion of a reaction product.

to selectivity:
desired reaction (F → P) can be accompanied by two competitive reactions:

1. side or parallel reactions

2. secondary reactions

F → P → P''

The coproduct — the third type of product — will also be dealt with in this connection although it does not affect the selectivity.

both competitive reactions reduce the selectivity to P

It is formed simultaneously with the desired reaction product, e.g., in all decomposition or elimination reactions.

formation of a coproduct:

F → P + P'''

does not affect the selectivity to P

selectivities during ethylene acetoxylation
for:

$H_2C=CHOCCH_3$:

94% relative to converted $H_2C=CH_2$
98–99% relative to converted CH_3COOH

abbreviated form:

94% ($H_2C=CH_2$) or 94% (based on $H_2C=CH_2$)
98–99% (CH_3COOH) or 98–99% (based on CH_3COOH)

yields in ethylene acetoxylation for

$H_2C=CHOCCH_3$:

9.4% relative to $H_2C=CH_2$ feedstock
19–28.5% relative to CH_3COOH feedstock

abbreviated form:

   9.4% ($H_2C=CH_2$)
19–28.5% (CH_3COOH)

yield of $H_2CHOCCH_3$, relative to ethylene:

$$\frac{10 \times 94}{100} = 9.4\% \ (H_2C=CH_2)$$

space–time yield:

frequently used dimension in industry:

kg product per liter catalyst (or per liter reactor) per hour ($kg \cdot L^{-1} \cdot h^{-1}$)

space–time yield in ethylene acetoxylation:
0.3–0.6 $kg \cdot L^{-1} \cdot h^{-1}$

The *selectivity* to a reaction product is the quotient of the amount of reaction product Q_R and the amount of a converted feedstock component Q_C:

$$\text{Selectivity (in \%)} = \frac{Q_R}{Q_C} \times 100$$

Since in a multicomponent reaction the individual starting materials can participate to varying extents in side or secondary reactions, i.e., they exhibit differing conversions, several selectivities are possible for a reaction product. In this book they are denoted according to the reference component which, for simplicity, is placed in parentheses behind the selectivity data (usually with the words 'based on').

The *yield* of a reaction product is the quotient of the amount of reaction product Q_R and the amount of a starting component Q_F:

$$\text{Yield (in \%)} = \frac{Q_R}{Q_F} \times 100$$

When there are several reaction partners, the reference component must be given with the yield, just as in the case of the selectivity.

The yield is always less than the selectivity if the conversion of the reaction component is less than 100%. The values for yield and selectivity are only identical if there is 100% conversion. Conversion, selectivity and yield are – based on the above equations – connected by the following relationship:

$$\frac{\text{Conversion} \times \text{Selectivity}}{100} = \text{Yield}$$

The space–time yield – a frequently used quantity in industry – gives the amount of reaction product formed per unit volume of the catalyst per unit time. It is also termed catalyst efficiency or catalyst productivity.

The expression space–time efficiency should be avoided as only the total quantity represents the efficiency.

15.3. Abbreviations for Firms

In general, company names have been quoted in order that they can be unambiguously identified. Frequently used but less well known abbreviations are summarized below:

Akzo	Algemene Koninklijke Zout Organon
Arco	Atlantic Richfield Company
ATO	Aquitaine Total Organico
BP	British Petroleum Company
CDF Chimie	Société Chimique Des Charbonnages, now ORKEM
CFR	Compagnie Française de Raffinage
CFP	Compagnie Française des Pétroles
DEA	Deutsche Erdöl AG, now RWE-DEA
DSM	Dutch Staats Mijnen
EC-Dormagen	Erdölchemie Dormagen
FMC	Food Machinery & Chemical Corp.
GAF	General Aniline & Film Corp., now ISP
ICI	Imperial Chemical Industries
IFP	Institut Français Du Pétrole
ISP	International Specialty Products
KFA-Jülich	Kernforschungsanlage Jülich
PCUK	Produits Chimiques Ugine Kuhlmann
PPG	Pittsburgh Plate Glass Co.
RCH	Ruhrchemie AG
Rheinbraun	Rheinische Braunkohlenwerke AG
ROW	Rheinische Olefinwerke GmbH
RWE	Rheinisch-Westfälisches Elektrizitätswerk AG
Sasol	Suid-Afrokaanse Steenkool-, Olie en Gaskorporasie Beperk
SBA	Soc. Belge De L' Azote
Sisas	Societa Italiana Serie Acetica Syntetica
SKW	Süddeutsche Kalkstickstoff-Werke AG
Sohio	Standard Oil of Ohio
UCB	Union Chimique Belge
UCC	Union Carbide Chemicals Co.
UK-Wesseling	Union Rheinische Braunkohlen Kraftstoff AG, Wesseling
UOP	Universal Oil Products Co.
USI	US-Industrial Chemicals Co. (National Distillers)
VEBA	Vereinigte Elektrizitäts- und Bergwerks-AG

15.4. Sources of Information

The evident demand for a description of modern industrial organic chemistry indicates the problem of the limited flow of information from the chemical industry to the scientific literature. Information about processes, the variations used industrially, and the complex interplay between feedstocks, intermediates, and products had to be gleaned from many sources.

The necessary background data were initially supplied by encyclopedias, reference books and supplemented mainly by monographs. However, in the main, publications and review articles in technical or industrially orientated journals provided a basic as well as an up-to-date data source. Other important sources were company publications and newsletters, and reports of lectures and congresses. Capacity and production data could generally be obtained from the above-mentioned publications as well as from reports of the US International Trade Commission, the Statistical Federal Office (Statistisches Bundesamt) in Germany, and the Japan Chemical Annual. Production data for the former Eastern bloc are available only to a very limited extent. Production and capacity numbers for Germany after 1991 include the former East Germany.

The following literature review gives, after a summary of the major reference books and encyclopedias, the most important reports and monographs relating to the individual chapters of the book.

15.4.1. General Literature

(books, encylopedias, reference works)

F. Asinger: Die Petrochemische Industrie, Akademie Verlag, Berlin 1971.

Autorenkollektiv der Technischen Hochschule für Chemie 'Carl Schorlemmer' Leuna-Merseburg: Lehrbuch der Technischen Chemie, VEB Verlag, Leipzig 1974.

A. M. Brownstein: Trends in Petrochemical Technology, Petroleum Publishing Co., Tulsa, Oklahoma 1976.

W. L. Faith, D. B. Keyes, R. L. Clark: Industrial Chemicals, John Wiley, New York, London, Sidney 1975.

L. F. Hatch, S. Matar: From Hydrocarbons to Petrochemicals, Hydrocarbon Processing, May 1977–1979.

Hydrocarbon Processing, Petrochemical Processes Mar. 1995, Gas Processes 1994, Refining Handbook Nov. 1994.

R. E. Kirk, D. F. Othmer, Encyclopedia of Chemical Technology, The Interscience Encyclopedia, Inc. New York 1978–1984, 4th ed. since 1991.

P. Leprince, J. P. Catry, A Chauvel: Les Produits Intermédiaires De La Chimie Des Dérivés Du Pétrole, Société Des Editions Technip, Paris 1967.

R. Pearce, W. R. Patterson: Catalysis and Chemical Processes, Leonard Hill, 1981.

M. Sittig: Organic Chemical Process Encyclopedia, Noyes Development Corp., Park Ridge 1969.

M. Sittig: Combining Oxygen and Hydrocarbons for Profit, Gulf Publishing, Houston 1968.

J. M. Tedder, A. Nechvatal, A. H. Jubb: Basic Organic Chemistry Part 5: Industrial Products, John Wiley & Sons, London, New York 1975.

C. L. Thomas: Catalytic Processes and Proven Catalysts, Academic Press, New York, London 1970.

Ullmanns Encyclopädie der technischen Chemie, Verlag Chemie, Weinheim 1972–1984.

Ullmann's Encyclopedia of Industrial Chemistry, VCH Weinheim 1985–1996.

A. L. Waddams: Chemicals from Petroleum, John Murray, London 1978.

K. Winnacker, H. Biener: Grundzüge der Chemischen Technik, C. Hanser Verlag, München and Wien 1974.

K. Winnacker, L. Küchler: Chemische Technologie, C. Hanser Verlag, München, 4. Auflage 1981–1986.

P. Wiseman: An Introduction to Industrial Organic Chemistry, Applied Science Publishers Ltd., London 1976.

15.4.2. More Specific Literature (publications, monographs)

Chapter 1:

H. C. Runge et al., Weltbedarf und Weltbedarfsdeckung bei Öl und Gas, Erdöl und Kohle *33*, 11 (1980).

H. C. Runge, W. Häfele, Zur Verfügbarkeit von Erdöl und Erdgas, Erdöl und Kohle *37*, 57 (1984).

Shell Briefing Service, Energie im Profil, Dec. 1984.
Raffinerien im Blickpunkt, May 1985.
Mineralöl, Erdgas und Kohle, June 1985.
Erdöl- und Erdgasförderung im Offshore-Bereich, 1/1994.
Der internationale Handel mit Rohöl und Mineralölprodukten, 2/1994.
Weltenergie, Daten und Fakten, to 1/1995.
Energie im 21. Jahrhundert, 5/1995.

G. Zürn, K. Kohlhase, K. Hedden, J. Weitkamp, Entwicklung der Raffinerietechnik, Erdöl und Kohle *37*, 62 (1984).

W. Häfele, W. Terhorst, Bereitstellung und Nutzung von Mineralöl, Gas, Kohle, Kernenergie, Chem. Industrie *XXXVII*, 9 (1985).

H. E. Hanky, Die Zukunftsaussichten der Petrochemie, Erdöl, Erdgas, Kohle *102*, 38 (1986).

BP Statistical Review of World Energy, to 1996.

BP, Das Buch vom Erdöl, Reuter und Klöckner, Hamburg 1989.

G. Ondrey, P. Hoffmann, S. Moore, Hydrogen Technologies, Chem. Engng. *May*, 30 (1992).

Main-Kraftwerke AG, Energienachrichten to 1996.

Info-Zentrale Elektrizitätswirtschaft e.V. to 1996.

Chapter 2:

G. Kaske, Petrochemische Wasserstoff-Herstellung, Transport und Verwendung, Chemie-Ing.-Techn. *48*, 95 (1976).

O. Neuwirth, Technische Entwicklung der Methanol-Synthese in der Nachkriegszeit, Erdöl und Kohle *29*, 57 (1976).

K. W. Foo, J. Shortland, Compare CO Production Methods, Hydrocarbon Processing, May, 149 (1976).

S. L. Meisel, J. P. McCullough, C. H. Lechthaler, P. B. Weisz, Gasoline from Methanol in One Step, Chemtech., Feb. 86 (1976).

H.-J. Derdulla, I. Hacker, M. Henke, R. Rebbe, Tendenzen und Fortschritte bei der Synthese von Methylaminen, Chem. Techn. *29*, 145 (1977).

H. Diem, Formaldehyde Routes bring Cost, Production Benefits, Chem. Engng. Feb., 83 (1978).

F. Obernaus, W. Droste, The New Hüls Process for MTBE, Vortrag Philadelphia, June 1978.

E. Hancock, Chemical Prospects in South Africa, Chem. and Ind. April, 276 (1979).

H. Jüntgen, K. H. van Heek, Grundlagen, Anwendung und Weiterentwicklung der Kohlevergasung II, Gwf-Gas Erdgas *121* (1980).

A. Aguilo, J. S. Alder, D. N. Freeman, R. J. H. Voorhoeve, Focus on C_1-Chemistry, Hydrocarbon Processing, March, 57 (1983).

B. Cornils, Synthesegas durch Kohlevergasung, Ber. Bunsengesellschaft Phys. Chem. *87*, 1080 (1983).

C. Brecht, G. Hoffmann, Vergasung von Kohle, Gaswärme int. *32*, 7 (1983).

F. Asinger, Methanol auf Basis von Kohlen, Erdöl und Kohle *36*, 28 (1983), *36*, 130 (1983).

K. Griesbaum, W. Swodenk, Forschungs- und Entwicklungstendenzen in der Petrochemie, Erdöl und Kohle *37*, 103 (1984).

M. Röper, Oxygenated Base Chemicals from Synthesis Gas, Erdöl und Kohle *37*, 506 (1984).

K. Kobayashi, International Trends in Methanol, Chem. Econ. & Eng. Rev. *16*, 32 (1984).

H. Teggers, H. Jüntgen, Stand der Kohlegasversorgung zur Erzeugung von Brenngas und Synthesegas, Erdöl und Kohle *37*, 163 (1984).

G. Kaske, Trends bei der Herstellung von Wasserstoff, Chem. Industrie *XXXVII*, 314 (1985).

D. L. King, J. H. Grate, Look What You Can Make From Methanol, Chemtech, April 244 (1985).

H. Hiller, E. Sup, Octamix-Verfahren, Erdöl und Kohle *38*, 19 (1985).

T. Hiratani, S. Noziri, C_1-Chemistry Based on Methyl Formate, Chem. Econ. & Eng. Rev. *17*, 21 (1985).

G. A. Mills, E. E. Ecklund, Alternative fuels: Progress and Prospects, Chemtech, Sept., 549 (1989), Oct., 626 (1989).

R. S. Hug, H. Vennen, Wohin mit gebrauchtem FCKW? Chem. Techn. *23*, 38 (1994).

Chapter 3:

K. L. Anderson, T. D. Brown, Olefin Disproportionation − New Routes to Petrochemicals, Hydrocarbon Processing, Aug., 119 (1976).

V. J. Guercio, Opportunities in Butylenes, Chem. Economy & Eng. Review *9*, 14 (1977).

T. C. Ponder, U. S. Ethylene Supply Demand: 1977−1980. Hydrocarbon Processing, June, 155 (1977).

H. Isa, α-Olefins and Its Derivatives, Chem. Economy & Eng. Review *9*, 26 (1977).

M. F. Farona, Olefin Metathesis – a Technology Begets a Science, Chem. Techn. Jan., 41 (1978).

R. A. Persak, E. L. Politzer, D. J. Ward, P. R. Pujado, Petrochemical Intermediates from C_3/C_4 Olefins, Chem. Economy & Eng. Review *10*, 25 (1978).

E. R. Freitas, C. R. Gum, Shell's Higher Olefins Process, Chem. Engng. Prog. Jan., 73 (1979).

J. Weitkamp, E. Eyde, Stand und Aussichten der Verarbeitungstechnik für Mineralöl und Erdgas, Erdöl und Kohle *33*, 16 (1980).

J. Leonard, J. F. Gaillard, Make Octenes with Dimersol X, Hydrocarbon Processing, March, 99 (1981).

R. L. Baldwin, G. R. Kamm, Make Ethylene by ACR-process, Hydrocarbon Processing, Nov., 127 (1982).

R. Streck, Olefin Metathesis and Polymer Synthesis, Chemtech., Dec., 758 (1983).

D. Commereuc et al., Dimerize Ethylene to Butene-1, Hydrocarbon Processing, Nov. 118 (1984).

L. S. Bitar, E. H. Hazbun, W. J. Diel, MTBE-Production and Economics, Hydrocarbon Processing, Oct., 63 (1984).

W. Keim, Homogene Übergangsmetallkatalyse, dargestellt am SHOP-Prozeß, Chem.-Ing. Tech. *56*, 850 (1984).

P. M. Lange, F. Martinola, S. Oeckl, Use Bifunctional Catalysts for MTBE, TAME and MIBK, Hydrocarbon Processing, Dec., 51 (1985).

B. Vora, P. Pujado, T. Imai, T. Fritsch, Production of Detergent Olefins and Linear Alkylbenzenes, Chem. and Ind. March, 187 (1990).

P. R. Sarathy, G. S. Suffridge, Etherify field butanes, Hydrocarbon Processing Jan., 89 (1993), Feb., 43 (1993).

H. Steinhauer, Pervaporation, Chemie-Technik *23*, 50 (1994).

Chapter 4:

K. Stork, J. Hanisian, I. Bac, Recover Acetylene in Olefins Plants. Hydrocarbon Processing, Nov., 151 (1976).

Y. Tsutsumi, Technological Trends in 1,4-Butanediol, Chem. Economy & Eng. Review *8*, 45 (1976).

H. Bockhorn, R. Coy, F. Fetting, W. Prätorius, Chemische Synthesen in Flammen, Chem.-, Ing.-Techn. *49*, 883 (1977).

E. Bartholomé, Die Entwicklung der Verfahren zur Herstellung von Acetylen durch partielle Oxidation von Kohlenwasserstoffen, Chem.-Ing.-Tech. *49*, 459 (1977).

A. M. Brownstein, H. L. List, Which Route to 1,4-Butanediol?, Hydrocarbon Processing, Sept., 159 (1977).

Y. Tanabe, New Route to 14BG and THF, Hydrocarbon Processing, Sept., 187 (1981).

K. Eisenäcker, Acetylen aus Carbid, Chem. Ind. Aug., 435 (1983).

J. Schulze, M. Homann, Acetylen in der zukünftigen Kohlechemie, Erdöl und Kohle *36*, 224 (1983).

G. E. Beekhuis, J. G. M. Nieuwkamp, New Process for 2-Pyrrolidone, Hydrocarbon Processing, April, 109 (1983).

R. Müller, G. Kaske, The Use of Plasma Chemical Processes for Chemical Reactions, Erdöl und Kohle *37*, 149 (1984).

Mitsubishi Kasei Corp., 1,4-Butandiol/Tetrahydrofuran Production Technology, Chemtech., Dec., 759 (1988).

A. M. Brownstein, 1,4-Butandiol and Tetrahydrofuran: Exemplary Small-volume Commodities, Chemtech., Aug., 506 (1991)
A. Budzinski, Reife Früchte, Chem. Ind. March, 34 (1993).
D. Rohe, Ein Lösemittel macht Monomer-Karriere, Chem. Ind. *3*, 12 (1995).

Chapter 5:

J. H. Prescott, Butadiene's Question Mark, Chem. Engng., Aug., 46 (1976).
A. M. Brownstein, Butadiene: A Viable Raw Material?, Hydrocarbon Processing, Feb., 95 (1976).
W. Meyer, DCPD: Abundant Resin Raw Material, Hydrocarbon Processing, Sept., 235 (1976).
T. C. Ponder, U. S. Butadiene − Coproduct or dehydro?, Hydrocarbon Processing, Oct., 119 (1976).
L. M. Welsh, L. I. Croce, H. F. Christmann, Butadiene via Oxidative Dehydrogenation, Hydrocarbon Processing, Nov., 131 (1978).
W. Günther, Welt-Kautschukmarkt, Chem. Ind. *XXXII*, 574 (1980).
B. V. Vora, T. Imai, C_2/C_5-Dehydrogenation Updated, Hydrocarbon Processing, April, 171 (1972).
A. Yoshioka, H. Hakari, R. Sato, H. Yamamoto, K. Okumura, Make Butadiene from Butene/butane Feed, Hydrocarbon Processing, Nov., 97 (1984).
D. Commereuc, Y. Chauvin, J. Gaillard, J. Léonard, Dimerize Ethylene to Butene-1, Hydrocarbon Processing, Nov., 118 (1984).
H. Gröne, G. Kuth, Moderne Produktentwicklungen festigen traditionelle Synthesekautschukmärkte, Chem. Ind. *XXXVI*, 672 (1984).
Chemical Week, Product Report (Rubber), May 10, 26 (1993).
M. Morgan, C_5 hydrocarbons and derivatives: new opportunities, Chem. and Ind. Sept., 645 (1996).

Chapter 6:

H. Weber, W. Dimmling, A. M. Desai, Make Plasticizer Alcohols this Way, Hydrocarbon Processing, April, 127 (1976).
R. Fowler, H. Connor, R. A. Baehl, Hydroformylate Propylene at Low Pressure, Hydrocarbon Processing, Sept., 274 (1976), Carbonylate with Rhodium, CHEMTECH Dec., 772 (1976).
E. A. V. Brewester, Low-Pressure Oxo Process Features Rhodium Catalyst, Chem. Engng. Nov., 90 (1976).
J. Falbe, Homogeneous Catalysis, Vortrag CHEMRAWN I, Toronto 1978.
B. Cornils, A. Mullen, 2-EH: What You Should Know, Hydrocarbon Processing, Nov., 93 (1980).
J. Falbe, New Syntheses with Carbon Monoxide, Springer Verlag 1980.
K.-H. Schmidt, Neuentwicklungen in der homogenen Katalyse. Chem. Ind. *XXXVII*, 762 (1985).
E. Wiebus, B. Cornils, Water-soluble catalysts improve hydroformylation of olefins, Hydrocarbon Processing, March, 63 (1996).

Chapter 7:

S. C. Johnson, U. S. EO/EG-Past, Present and Future, Hydrocarbon Processing, June, 109 (1976).

J. Kiguchi, T. Kumazawa, T. Nakai, For EO: Air and Oxygen Equal, Hydrocarbon Processing, March, 69 (1976).

M. Gans, B. J. Ozero, For EO: Air or Oxygen, Hydrocarbon Processing, March, 73 (1976).

B. DeMaglie, Oxygen Best for EO, Hydrocarbon Processing, March, 78 (1976).

R. Jira, W. Blau, D. Grimm, Acetaldehyde via Air or Oxygen, Hydrocarbon Processing, March, 97 (1976).

J. B. Saunby, B. W. Kiff, Liquid-phase Oxidation, Hydrocarbons to Petrochemicals, Hydrocarbon Processing, Nov., 247 (1976).

D. Forster, On the Mechanism of a Rhodium-Complex-Catalyzed Carbonylation of Methanol to Acetic Acid, J. Amer. Chem. Soc. *98*, 846 (1976).

H. Beschke, H. Friedrich, Acrolein in der Gasphasensynthese von Pyridinderivaten, Chemiker Ztg. *101*, 377 (1977).

H. Beschke, H. Friedrich, H. Schaefer, G. Schreyer, Nicotinsäureamid aus β-Picolin, Chemiker Ztg. *101*, 384 (1977).

P. Kripylo, L. Gerber, P. Münch, D. Klose, L. Beck, Beitrag zur quantitativen Beschreibung der selektivitätsverbessernden Wirkung von 1.2-Dichloräthan bei der Oxydation von Äthylen zu Äthylenoxid, Chem. Techn. *30*, 630 (1978).

W. Swodenk, H. Waldmann, Moderne Verfahren der Großchemie: Ethylenoxid und Propylenoxid, Chemie in unserer Zeit *12*, 65 (1978).

G. E. Weismantel, New Technology Sparks Ethylene Glycol Debate, Chem. Engng. Jan., 67 (1979).

G. Sioli, P. M. Spaziante, L. Guiffre, Make MCA in Two Stages, Hydrocarbon Processing, Feb., 111 (1979).

J. C. Zomerdijk. M. W. Hall, Technology for the Manufacture of Ethylene Oxide, Catal. Rev.-Sci. Eng. *23*, 163 (1981).

S. Rebsdat, S. Mayer, J. Alfranseder, Der Ethylenoxid-Prozeß und die Regenerierung des dabei verwendeten Silber-Katalysators, Chem.-Ing. Tech. *53*, 850 (1981).

A. Budzinski, Pyridin hat noch Wachstumschancen, Chem. Ind. *XXXIII*, 529 (1981).

B. I. Ozero, J. V. Procelli, Can Developments Keep Ethylene Oxide Viable, Hydrocarbon Processing, March, 55 (1984).

S. Nowack, J. Eichhorn, J. Ohme, B. Lücke, Sauerstoffhaltige organische Zwischenprodukte aus Synthesegas, Chem. Techn. *36*, 55 (1984), *36*, 144 (1984).

M. Schrod, G. Luft, Carbonylierung von Essigsäuremethylester zu Essigsäureanhydrid, Erdöl und Kohle *37*, 15 (1984).

U. Dettmeier, E. J. Leupold, H. Pöll, H.-J. Schmidt, J. Schütz, Direktsynthese von Essigsäure aus Synthesegas, Erdöl und Kohle *38*, 59 (1985).

B. Juran, R. V. Porcelli, Convert Methanol to Ethanol, Hydrocarbon Processing, Oct., 85 (1985).

R. A. Sheldon, Fine chemicals by catalytic oxidation. CHEMTECH Sept., 566 (1991).

Chemical Week, Product Focus (Ethylene Glycol/Oxide), Febr. 8, 44 (1995).

(Acetic Acid), Apr. 24, 30 (1996).

(EO-EG), Oct. 9, 41 (1996).

Chapter 8:

O. Winter, M.-T. Eng, Make Ethylene from Ethanol, Hydrocarbon Processing, Nov., 125 (1976).

N. Kurata, K. Koshida, Oxidize *n*-Paraffins for sec-Alcohols, Hydrocarbon Processing, Jan., 145 (1978).

U. Tsao, J. W. Reilly, Dehydrate Ethanol to Ethylene, Hydrocarbon Processing, Feb., 133 (1978).

Y. Onoue, Y. Mizutani, S. Akiyama, Y. Izumi, Hydration with Water, Chemtech., July 432 (1978).

M. K. Schwitzer, Fats as Source for Cationic Sufactants, Chem. and Ind. Jan. 11 (1979).

D. Osteroth, Die Bedeutung der natürlichen Fettsäuren, Chemie für Labor und Betrieb *32*, 571 (1981).

H. Bahrmann, W. Lipps, B. Cornils, Fortschritte der Homologisierungsreaktion, Chemiker Ztg. *106*, 249 (1982).

H.-P. Klein, Hochleistungsozonanlagen und ihr industrieller Einsatz, Chem.-Ing.-Tech. *55*, 555 (1983).

C. A. Houston, Marketing and Economics of Fatty Alcohols, JAOCS *61*, 179 (1984).

P. Lappe, H. Springer, J. Weber, Neopentylglykol als aktuelle Schlüsselsubstanz, Chemiker Ztg. *113*, 293 (1989).

D. H. Ambros, Fettchemie wächst, blüht und gedeiht, Chem. Ind. *31* (1991).

C. Breucker, V. Jordan, M. Nitsche, B. Gutsche, Oleochemie-Chemieprodukte auf der Basis nachwachsender Rohstoffe, Chem.-Ing.-Tech. *67*, 430 (1995).

Chapter 9:

W. E. Wimer, R. E. Feathers, Oxygen Gives Low Cost VCM, Hydrocarbon Processing, March, 81 (1976).

P. Reich, Air or Oxygen for VCM?, Hydrocarbon Processing, March, 85 (1976).

R. W. McPherson, C. M. Starks, G. I. Fryar, Vinylchloride monomer . . . What You Should Know, Hydrocarbon Processing, March, 75 (1979).

C. M. Schillmoller, Alloy Selection for VCM Plants, Hydrocarbon Processing, March, 89 (1979).

W. M. Burks, Y. W. Rao, A. Eldring, Wirtschaftlichere Vinylchloridproduktion, Chem. Ind. *XXXII*, 250 (1980).

J. L. Ehrler, B. Juran, VAM and Ac$_2$O by Carbonylation, Hydrocarbon Processing, Feb., 109 (1982).

J. Schulze, M. Weiser, Rückstandsprobleme chlororganischer Produkte, Chem. Ind. *XXXVI*, 468 (1984), 747 (1984).

P. K. Eichhorn, Fluorkunststoffe, Kunststoffe *79*, 927 (1989).

F. Nader, Perspektiven der Chlorchemie, Chemie-Technik *25*, 66 (1996).

Chemical Week, Product Focus (Vinyl Chloride), Sept. 25, 46 (1996).
(Chlorine), March 13, 38 (1996).

Chapter 10:

A. H. de Rooij, C. Dijkhuis, J. T. J. von Goolen, A Scale-Up Experience; The DSM Phosphate Caprolactam Process, Chemtech., May 309 (1977).

H. J. Naumann et al., Entwicklung und Anwendung des Verfahrens der selektiven Phenolhydrierung für die Herstellung von Cyclohexanon, Chem. Techn. *29*, 38 (1977).

W. Rösler, H. Lunkwitz, Entwicklungstendenzen und technologische Fortschritte bei der Produktion von ε-Caprolactam, Chem. Techn. *30*, 67 (1978).

M. Fischer, Photochemische Synthesen im technischen Maßstab, Angew. Chem. *90*, 17 (1978).

Y. Izumi, I. Chibata, T. Itoh, Herstellung und Verwendung von Aminosäuren, Angew. Chem. *90*, 187 (1978).

D. E. Danly, Adiponitrile via Improved EHD, Hydrocarbon Processing, April, 161 (1981).

K. Wehner et al., Entwicklung eines technischen Verfahrens zur Herstellung von N-Methyl-ε-caprolactam, Chem. Techn. *33*, 193 (1981).

A. S. Chan, New route to Adipic Acid Developed at Monsanto, Chem. Eng. News April, 28 (1984).

O. Immel, H. H. Schwarz, K. Starke, W. Swodenk, Die katalytische Umlagerung von Cyclohexanonoxim zu Caprolactam, Chem. Ing.-Techn. *56*, 612 (1984).

K. Ito, I. Dogane, K. Tanaka, New Process for Sebacic Acid, Hydrocarbon Processing, Oct., 83 (1985).

B. v. Schlotheim, Aufwärtstrend bei Chemiefasern setzte sich fort, Chem. Ind. *XXXVIII*, 127 (1986).

J. Dodgson et al., A Low Cost Phenol to Cyclohexanone Process, Chem. and Ind. Dec., 830 (1989).

Chemical Week, Product Focus (Caprolactam), Nov. 16, 48 (1994).

Chapter 11:

K. H. Simmrock, Die Herstellverfahren für Propyleneoxid und ihre elektrochemische Alternative, Chem.-Ing.-Tech. *48*, 1085 (1976).

K. Yamagishi, O. Kageyama, H. Haruki, Y. Numa, Make Propylene Oxide Direct, Hydrocarbon Processing, Nov., 102 (1976).

K. Yamagishi, O. Kageyama, Make Glycerine via Peracetic Acid, Hydrocarbon Processing, Nov., 139 (1976).

F. Matsuda, Acrylamide Production Simplified, Chemtech., May, 306 (1971).

Y. Onoue, Y. Mizutani, S. Akiyama, Y. Izumi, Y. Watanabe, Why Not Do It In One Step? The case of MIBK, Chemtech., Jan., 36 (1977).

P. R. Pujado, B. V. Vora, A. P. Krueding, Newest Acrylonitrile Process, Hydrocarbon Processing, May, 169 (1977).

J. C. Zimmer, Cut Polyester Costs – Use PO, Hydrocarbon Processing, Dec., 115 (1977).

H. Schaefer, Katalytische Ammonoxidation und Ammondehydrierung, Chemie-Technik *7*, 231 (1978).

T. C. Ponder, U. S. Propylene: Demand vs. Supply, Hydrocarbon Processing, July, 187 (1978).

K. H. Simmrock, Compare Propylene Oxide Routes, Hydrocarbon Processing, Nov., 105 (1978).

T. Hasuike, H. Matsuzawa, Make MMA from Spent-BB, Hydrocarbon Processing, Feb., 105 (1979).

K.-H. Schmidt, Aktive Sauerstoffverbindungen: Chancen auf neuen Märkten – Überkapazitäten in der Gegenwart, Chem. Ind. XXXI, 135 (1979).

S. L. Neidleman, Use of Enzymes as Catalysts for Alkene Oxide Production, Hydrocarbon Processing, Nov, 135 (1980).

J. Itakura, Present State and Prospects for Acrylic Ester Industry, Chem. Eonomy & Eng. Review July, 19 (1981).

T. Nakamura, T. Kito, A new Feedstock for the Manufacture of Methyl Methacrylate Emerges. Chem. Econ. & Eng. Rev. Oct., 23 (1983).

H. Itatani, International Technological Trends in C_1-Chemistry, Chem. Econ. & Eng. Rev. *16*, 21 (1984).

P. Kripylo, K. Hagen, D. Klose, K. M. Tu, Mechanismus und Selektivität der Oxidation von Propen zu Acrolein, Chem. Techn. *36*, 58 (1984).

K. Drauz, A. Kleemann, M. Samson, Acrolein-Baustein für neue Synthesen von Aminosäuren und Naturstoffen, Chemiker-Ztg. *108*, 391 (1984).

R. V. Porcelli, B. Juran, Selecting the Process for Your Next MMA Plant, Hydrocarbon Processing, March, 37 (1986).

E. Johnson, J. Chowdhury, New Menu for MMA Plants, Chem. Engng. March, 35 (1990).

I. Young, Weakness of Propylene in Europe, Chem. Week, March 24, 37 (1993).

D. Rohe, Masse und Klasse (Acrylsäure), Chem. Ind. *1/2*, 12 (1995).

D. Rohe, Vor dem Quantitätssprung (1.3-Propandiol), Chem. Ind. *3*, 13 (1996).

Chemical Week, Product Focus (Acetone), Dec. 14, 40 (1994).
                    (Acrylonitrile), March 27, 36 (1996).

Chapter 12:

G. Collin, Technische und wirtschaftliche Aspekte der Steinkohlenteer-chemie, Erdöl und Kohle *29*, 159 (1976).

J. E. Fick, To 1985: U. S. Benzene Supply Demand, Hydrocarbon Processing, July, 127 (1976).

T. C. Ponder, Benzene Outlook Through 1980, Hydrocarbon Processing, Nov., 217 (1976).

P. J. Bailes, Solvent Extraction in the Petroleum and Petrochemical Industries, Chem. and Ind. Jan., 69 (1977).

T. C. Ponder, Benzene Supply Demand Stable in Europe?, Hydrocarbon Processing, June, 158 (1977).

S. McQueen, Abundant Feedstock is Key to New Anthraquinone Route, Chem. Engng. Aug., 74 (1978).

H. Franke et al., Die hydrokatalytische Isomerisierung technischer C_8-Aromatenfraktionen, Chem. Techn. *31*, 402 (1979).

U. Langer et al., Gewinnung von BTX-Aromaten nach dem Arex-Verfahren, Chem. Techn. *33*, 449 (1981).

J. Lempin et al., Das Aris-Verfahren – ein hocheffektiver petrol-chemischer Prozeß zur Gewinnung von Xylenisomeren, Chem. Techn. *33*, 356 (1981).

G. Collin, Aromatische Chemiegrundstoffe aus Kohle, Erdöl und Kohle *35*, 294 (1982).

G. Preußer, G. Emmrich, Gewinnung von Reinaromaten, Erdöl und Kohle *36*, 207 (1983).

G. Kölling, J. Langhoff, G. Collin, Kohlenwertstoffe und Verflüssigung von Kohle, Erdöl und Kohle *37*, 394 (1984).

J. W. Stadelhofer, Present Status of the Coal Tar Industry in Western Europe, Chem. and Ind. March, 173 (1984).

G. Collin, Steinkohlenteerchemie: Bedeutung, Produkte und Verfahren, Erdöl und Kohle *38*, 489 (1985).

P. C. Doolan, P. R. Pujado, Make Aromatics from LPG, Hydrocarbon Processing, Sept., 72 (1989).

C. D. Gosling et al., Process LPG to BTX Products, Hydrocarbon Processing, Dec., 69 (1991).

M. Coeyman, N. Alperowicz, Future stays bright for H_2O_2, Chem. Week, Febr. 17, 42 (1993).

Chemical Week, Product Focus (Toluene), Jan. 25, 60 (1995).
                                  (Xylenes), Oct. 11, 58 (1995).
                                  (Benzene), Febr. 14, 40 (1996).
                                  (*p*-Xylene), Apr. 10, 45 (1996).
                                  (Toluene), Oct. 23, 55 (1996).

Chapter 13:

P. R. Pujado, J. R. Salazar, C. V. Berger, Cheapest Route to Phenol, Hydrocarbon Processing, March, 91 (1976).

G. Lenz, Herstellung von Maleinsäureanhydrid aus Butenen, Chemieanlagen, Verfahren July 27 (1976).

M. Gans, Which Route to Aniline?, Hydrocarbon Processing, Nov., 145 (1976).

J. A. Bewsey, Synthetic Tartaric Acid and the Economics of Food Acidulants, Chem. and Ind. Feb., 119 (1977).

T. C. Ponder, U. S. styrene: More growth with less plants, Hydrocarbon Processing, July, 137 (1977).

P. Maggioni, F. Minisci, Catechol and Hydrochinone from Catalytic Hydroxylation of Phenol by Hydrogen Peroxide, La Chimica E L'Industria *59*, 239 (1977).

A. Portes, J. Escourrou, Optimize Styrene Production, Hydrocarbon Processing, Sept., 154 (1977).

P. Fontana, Butan als Ausgangsprodukt für Maleinsäureanhydrid, Chimica *31*, 274 (1977).

G. E. Tong, Fermentation Routes to C_3 and C_4 Chemicals, Chem. Eng. Prog. April, 70 (1978).

R. Hirtz, K. Uhlig, Polyurethan-Baukasten – noch Platz für neue Bausteine, Chem. Ind. *XXX*, 617 (1978).

R. L. Varma, D. N. Saraf, Selective Oxidation of C_4-Hydrocarbons to Maleic Anhydride, Ind. Eng. Chem. Prod. Res. Dev. *18*, 7 (1979).

J. C. Bonacci, R. M. Heck, R. K. Mahendroo, G. R. Patel, Hydrogenate AMS to Cumene, Hydrocarbon Processing, Nov., 179 (1980).

T. Wett, Monsanto/Lummus Styrene Process is Efficient, Oil & Gas Journal July, 76 (1981).

R. A. Innes, H. E. Swift, Toluene to Styrene – a Difficult Goal, CHEMTECH April, 244 (1981).

F. Budi, A. Neri, G. Stefani, Future MA Keys to Butane, Hydrocarbon Processing, Jan., 159 (1982).

W. W. Kaeding, L. B. Young, A. G. Prapas, Para-Methylstyrene CHEMTECH Sept., 556 (1982).

G. S. Schaffel, S. S. Chem, J. J. Graham, Maleic Anhydride from Butane, Erdöl und Kohle *36*, 85 (1983).

S. Fukuoka, M. Chono, M. Kohuo, Isocyanate without Phosgene, CHEMTECH Nov., 670 (1984).

S. C. Arnold, G. D. Sucin, L. Verde, A. Neri, Use Fluid Bed Reactor for Maleic Anhydride from Butane, Hydrocarbon Processing, Sept., 123 (1985).

J. Schulze, M. Weiser, Rückstandsprobleme chlororganischer Produkte, Chem. Ind. *XXXVII*, 105 (1985).

D. Rohe, Maleinsäureanhydrid, ein Langweiler macht Karriere, Chem. Ind. Aug. 32 (1989).

H. Harris, M. W. Tuck, Butanediol via maleic anhydride, Hydrocarbon Processing May, 79 (1990).

D. Jackson, L. Tattum, Tougher times ahead for polyurethane, Chem. Week, Oct. 9, 32 (1991).

I. Young, Overbuilding plagues phenol, Chem. Week, June 16, 68 (1993).

M. S. Reisch, Thermoplastic Elastomers Target, Chem. Eng. News, Aug. 10 (1996).

A. Budzinski, Im sicheren Hafen (Phenol), Chem. Ind. *5*, 13 (1996).

Chemical Week, Product Focus (Toluene Diisocyanate), Sept. 13, 39 (1995).

(Maleic Anhydride), Oct. 25, 38 (1995).
(Styrene), Aug. 21, 53 (1996).
(Phenol), Sept. 11, 65 (1996).

Chapter 14:

M. C. Sze, A. P. Gelbein, Make Aromatic Nitriles this Way, Hydrocarbon Processing, Feb., 103 (1976).

K. Matsuzawa, Technogical Development of Purified Terephthalic Acid, Chem. Economy & Eng. Review *8*, 25 (1976).

H. Kawabata, Synthetic Fibers and Synthetic Fiber Materials in Japan, Chem. Economy & Eng. Review *9*, 27 (1977).

F. Obenaus, M. O. Reitemeyer, Eine einfache DMT-Reinigung durch chemische Umsetzung, Erdöl und Kohle *31*, 469 (1978).

O. Wiedemann, W. Gierer, Phthalic Anhydride Made with Less Energy, Chem. Engng. Jan., 62 (1979).

A. P. Aneja, V. P. Aneja, Process Options for Polyester, Chem. and Ind. April, 252 (1979).

A. P. Aneja, V. P. Aneja, Process Options, Feedstock Selections, and Polyesters, CHEMTECH April, 260 (1979).

L. Verde, A. Neri, Make Phthalic Anhydride with Low Air Ratio Process. Hydrocarbon Processing, Nov., 83 (1984).

A. Nitschke, Bottle-grade PET, Chem. Week, Apr. 21, 38 (1993).

Chemical Week, Markets and Economics (PET), June 7, 50 (1995)

Chemical Week, Product Focus (Phthalic Anhydride), Dec. 6, 64 (1995).

(PTA), Jan. 31, 48 (1996).
(PET), June 19, 38 (1996).

Index